SEDIMENTARY STRUCTURES AND EARLY DIAGENETIC FEATURES OF SHALLOW MARINE CARBONATE DEPOSITS

By

Robert V. Demicco, Department of Geological Sciences, Binghamton University, State University of New York, Binghamton, New York 13902-6000

and

Lawrence A. Hardie, Department of Earth & Planetary Sciences, The Johns Hopkins University, Baltimore, Maryland 21218

SEPM Atlas Series No. 1

Peter A. Scholle, Editor of Special Publications

Tulsa, Oklahoma, U. S. A. *August 31, 1994*

COVER PHOTOGRAPHS: The color background photograph shows the cyclic, shallow marine carbonate deposits of the Middle Triassic (Ladinian) Latemar buildup, the Dolomites, northern Italy. View from the Latemar Towers looking north toward the Northern Alps in the background. Sedimentary structures and early diagenetic features played the key role in demonstrating that these strikingly bedded limestones constitute a vertical stack of over 500 submergence-emergence cycles. These submergence-emergence cycles were probably driven by meter-scale sea level oscillations in tune to Milankovitch astronomical rhythms. The inset photograph shows stromatolites nucleated on flat intraclasts of grainstone. This specimen is from the Upper Cambrian Conococheague Limestone of western Maryland.

To order additional copies contact:

S.E.P.M.
P. O. Box 4756
Tulsa, Oklahoma 74159-0756
(918) 743-9765

ISBN: 1-56576-013-1

PREFACE

Sedimentary structures remain a foundation for interpreting most ancient depositional environments. The product of physical, biological, or chemical processes either during or following deposition, they almost uniquely attest in the stratigraphic record to these processes. The application of sedimentary structures to paleoenvironmental reconstruction is not without pitfalls, however. First, because many processes are common to different environmental settings, the same structure can form in deposits of very different origin; few structures are truly diagnostic of a specific depositional facies. Second, dissimilar processes can generate remarkably similar structures; uncritical examination of a structure can mislead an interpretation. A thorough analysis of structures must include recognition of the subtlety of individual structures, the assemblage of structures and their spatial distribution and directional relationship within the architectural fabric of the deposit.

Most sedimentary structures are sufficiently complex as to defy communication by other than a visual representation. Our ability to effectively utilize structures depends in large part on the visual images that we retain. It is for this reason that outcrop experiences or photographs remain the primary mechanism whereby we learn to recognize structures and to discriminate between them. An atlas with high-quality photographs can be a fundamental resourcc.

The SEPM Atlas Series of sedimentary structures of important depositional environments is designed to provide a summary of the visual character of the individual structures that reside in specific facies and to relate that character to their origin. It is also intended to document the lateral and vertical distributions of structures relative to one another and in association with other lithologic features of a facies. The goal is a collection of volumes that will be of use to interpreters of rocks in outcrop and the subsurface.

H. Edward Clifton
Series Editor

TABLE OF CONTENTS

INTRODUCTION

THE IMPORTANCE OF SEDIMENTARY STRUCTURES AND EARLY DIAGENETIC FEATURES IN CARBONATE SEDIMENTOLOGY

The intensive studies in the 1960's and 1970's of modern shallow marine carbonate environments in the Persian Gulf (e.g., Shearman, 1963, 1966; Kinsman, 1966; Butler, 1970; Kendall and Skipwith, 1969; Purser, 1973), Florida and the Bahamas (e.g., Shinn and others, 1969; Hardie, 1977a; Enos and Perkins, 1979), and Western Australia (e.g., Logan and others, 1970, 1974b), led to spectacular advances in our understanding of the deposition and early diagenesis of carbonate rocks. These studies were part of a major revolution in sedimentology that saw a radical change from an approach based heavily on grain textures to one based on sedimentary structures and early diagenetic features. In this new approach, paleo-environments of sedimentary deposits are diagnosed from the vertical and lateral distribution of elemental rock units (subfacies and facies), characterized principally by their *assemblages of sedimentary structures and early diagenetic features* in combination with other properties such as sedimentary textures and biota, using analogs established from observations of processes and their sedimentary records in modern depositional environments (the "comparative sedimentology" method of Ginsburg, 1974).

Modern shallow marine carbonate environments carry a particularly rich inventory of primary sedimentary structures and early diagenetic features, such as current and wave bedforms, trough and tabular cross-stratification, "herringbone" cross-stratification, flat lamination, wavy and crinkled lamination, thin bedding, stromatolites, thrombolites, mudcracks, sheet cracks, prism cracks, flat pebble gravels, fenestrae, burrows and roots (and their casts and molds), evaporite minerals (and their casts and molds), early cements, hardgrounds, caliche crusts, tepee structures, dissolution cavities, and so on. And most significantly, the stratigraphic record back at least into the Proterozoic is replete with carbonate deposits that delicately preserve these primary and early diagenetic sedimentary features (see, for example, Ginsburg, 1975; Wilson, 1975; Hardie and Shinn, 1986; Grotzinger, 1989), demonstrating the existence through much of geologic time of environments and environmental processes analogous to those of modern shallow marine carbonate platforms and shelves.

There are other areas of study beyond reconstruction of paleo-environments where the environmental information preserved in the primary sedimentary structures and early diagenetic features of shallow water carbonates is of considerable value, for example, in the fields of paleo-oceanography, paleoclimatology, and cyclostratigraphy. In particular, tidal flat facies are unsurpassed "sea-level gauges," "tide gauges," and "climate recorders" (Hardie, 1977c, p. 188-189). An unambiguous record of the position of mean paleo-sea level is engraved within the intertidal subfacies of all ancient shallow marine carbonate deposits. The same intertidal subfacies carry a record, quantitatively determinable, of the tidal range in the depositional environment (cf. Klein, 1971; Terwindt, 1988). At the same time, the subtidal subfacies record the ambient and storm wave energy levels across the ancient platform, and this information in turn reflects, at least in a qualitative way, the prevailing weather patterns. In addition, the nature of the supratidal subfacies is a direct response to the prevailing climate in the region (Hardie and Shinn, 1986). If the climate is arid, like that of the modern Persian Gulf carbonate environments, then the supratidal subfacies will contain evaporite and aeolian features (Shinn, 1983a). If the climate is rainy, like that of the modern Andros Island tidal flats, then the supratidal deposits will carry freshwater marsh and lake features (Monty and Hardie, 1976; Hardie, 1977a; Shinn, 1983a).

The stratigraphic record abounds with shallow water carbonate deposits characterized by meter-scale vertical successions of subfacies that are organized into shallowing-upward "cycles," as revealed by analysis of primary sedimentary structures, grain types, sedimentary textures, biota, and early diagenetic features (Wilson, 1975; James, 1984; Hardie and Shinn, 1986). Vertical stacks of such shallowing-upward cycles record repeated fluctuations in relative sea level. In some cases, these sea-level fluctuations appear to be periodic oscillations, driven by Milankovitch astronomical rhythms (e.g., Goldhammer and others, 1987, 1990). Such cyclic shallow water carbonates take on a special significance as storehouses of information about global climatic and eustatic variations in the past.

Overall, an understanding of the origin and significance of primary sedimentary structures and early diagenetic features is vital to unraveling the origin and significance of carbonate deposits in the geologic record. Without such an understanding at the individual sedimentary structure scale, we cannot hope to accurately reconstruct the large scale accumulation history of carbonate deposits or to decipher the roles of sea-level changes, sedimentation rates, subsidence rates, and tectonics in determining the facies stratigraphy, cyclostratigraphy, and sequence stratigraphy of these buildups.

SOME SPECIAL ASPECTS OF CARBONATE DEPOSITION

In attempting to assess the origin and significance of primary sedimentary structures in shallow water carbonate deposits, it is necessary to recognize that carbonate sedimentation differs in a number of aspects from siliciclastic sedimentation. The more important of these special aspects of carbonate deposition are:

1) Shallow marine carbonates typically accumulate in the form of "buildups" (Wilson, 1975, p. 21) with a pattern of environments that is remarkably similar through time.
2) The carbonate sediment of which the major buildups are comprised is produced *in situ* in shallow water settings under warm tropical conditions.
3) Under compaction, carbonate muds behave in a manner that allows extensive preservation of primary sedimentary structures in even the finest grained of carbonate sediments.
4) Early diagenesis, an array of penecontemporaneous processes that includes cementation, is widespread in carbonate sediments, and its effects must be taken into account in interpreting the sedimentary structures of any ancient carbonate rock.

These aspects of carbonate deposition are briefly discussed below.

Carbonate Buildups and their Component Environments

Based on the results of the comparative sedimentology approach, it is now recognized that at least since the Proterozoic shallow marine carbonates have been deposited as *buildups* (Wilson, 1975, p. 21.) that took the form of: (1) isolated "hi-rise" banks with flat tops and walls that sloped steeply down to the surrounding deep ocean (a contemporary analog is the modern Bahama Banks), (2) continental shelves attached to the craton (analogous to the modern shelves of Belize and the Great

Barrier Reef), and (3) ramp-like shelves that sloped gently into shallow ocean basins (analogous to the southern shelf of the modern Arabian Gulf). Carbonate buildups are found in all types of tectonic basin settings. Isolated banks have nucleated on small upfaulted blocks related to the early opening of ocean basins that followed continental rupture and separation, as exemplified by the present day Bahama Banks and by the cluster of small buildups of middle Triassic age in the Dolomites of northern Italy (Bosellini and Rossi, 1974; Bosellini, 1984). The largest ancient buildups were carbonate continental shelves that formed Atlantic-type passive margins at the trailing edges of lithospheric plates during the long, stable, drift stage of the Wilson Cycle. The huge shallow carbonate shelf that rimmed the North American continent for over 100 million years during Cambro-Ordovician times (Palmer, 1971; Laporte, 1971; Dott and Batten, 1988) is an outstanding example. Smaller carbonate shelves and carbonate ramps most commonly are found in foreland basins (for example, the Devonian of the Appalachian Basin), strike-slip basins along transform boundaries (for example, the Pennsylvanian and Permian of the southwestern United States), and cratonic basin settings (for example, the Silurian of the Michigan Basin).

Across all these ancient carbonate buildups there appears to have existed, as there exists today on modern carbonate buildups, the following succession of depositional environments from the landward edges to the distal depths (Fig. 1):

1. shoreline environments (typically tidal flats),
2. very shallow inner shelf-lagoon,
3. deeper outer shelf-lagoon,
4. marginal sand shoals (typically, but not exclusively, oolitic),
5. marginal reefs (present only where conditions are favorable for development of barrier reef-building organisms),
6. foreslope,
7. proximal ocean basin floor, and
8. distal ocean basin floor.

It is remarkable that carbonate environments and their physical, chemical, and biological processes seemed to have changed so little across the span of at least 2 billion years (see, for example, Wilson, 1975; Ginsburg, 1975; Hardie and Shinn, 1986; Grotzinger, 1989). This is all the more impressive considering the extensive secular changes which have taken place in the marine organisms that built reefs (see James, 1983; James and Bourque, 1992) or were direct or indirect producers of carbonate sediment particles (see Wilkinson, 1979). As Robert Ginsburg (pers. commun.) puts it: "the play has remained the same, only the players have changed."

An explanation, or at least a partial explanation, for all this can be found in the very special characteristics of the *in situ* production and accumulation of carbonate sediment under shallow marine conditions, exemplified by the processes operating today on modern buildups such as the Bahama Banks, as outlined below.

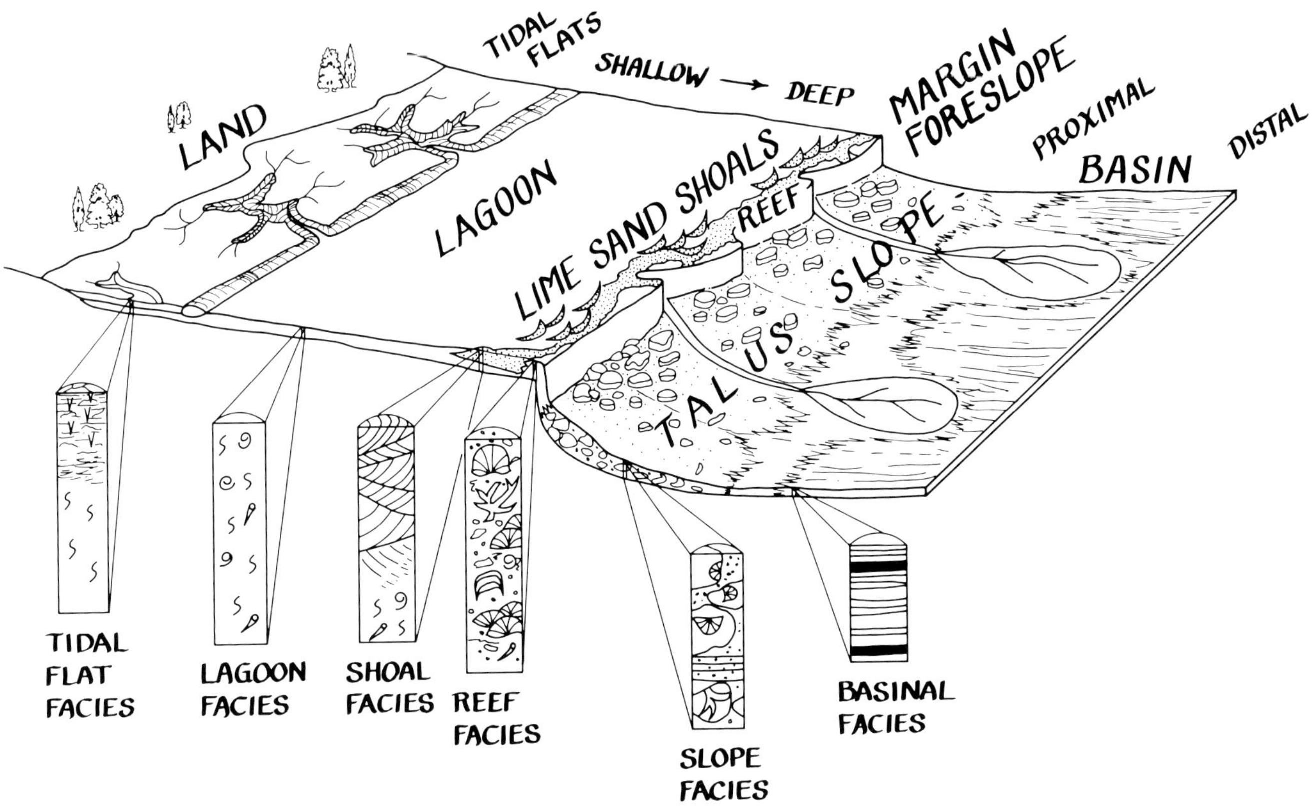

FIG. 1.-Depositional environments of carbonate buildups. Stylized cores depict typical deposits from each setting based on the modern Bahama Banks. See text for details.

The In Situ Production and Accumulation of Shallow Water Carbonates under Warm Tropical Conditions

The sand shoals on the margins of the modern Bahama Banks are composed of hardened aragonite mud peloids, aragonite ooids chemically precipitated *in situ*, the skeletal remains of benthic invertebrates, and lithoclasts (cemented aggregates of peloids, ooids, and skeletal grains). On those parts of the margins of the Bahamas (and other modern buildups) protected from muddy or other waters inimical to corals, barrier reefs are built by scleractinian corals (aragonite) and coralline algae (magnesian calcite). Cavities within the reef framework and the spaces between framework masses are filled by sediment made of the aragonitic plates of the marine green alga *Halimeda* mixed with reef fragments and benthic invertebrate skeletal debris. The carbonate sediment of the inner and outer lagoons and the tidal flats of the modern Bahama Banks is mainly peloidal *mud* made of aragonite needles produced *in situ* by marine benthic calcareous green algae (Stockman and others, 1967; Neumann and Land, 1975), by precipitation around the floating cells of phytoplankton (mainly 20-30 µm cells of marine cyanobacteria, Robbins and Blackwelder, 1992), and perhaps by direct chemical precipitation from sea water (Shinn and others, 1989). Skeletal remains of benthic invertebrates make up only a minor part of the sediment of the interior of the buildups (e.g., <10% on the Great Bahama Bank, see data of Purdy, 1963). In essence, then, the shallow water sediment production system of modern carbonate buildups is to a large extent dependent on *marine algae and cyanobacteria*, which require warm, shallow sunlit waters. This is true even of the reefs because the principal builders of modern reefs, the hermatypic scleractinian corals, are constrained by their symbiotic relationship with the zooxanthellae algae in their tissue to stay in warm sunlit waters. At the same time, shallow depths guarantee slow circulation of sea water across the buildup (long residence times) which in turn promotes further warming, evaporation, and ultimately a high degree of supersaturation with respect to aragonite (Smith, 1940; Cloud, 1962; Broecker and Takahashi, 1966). Thus, the conditions most favorable to carbonate production by, or mediated by, marine algae and cyanobacteria are also the conditions most favorable to inorganic precipitation of aragonitic needle muds and aragonitic ooids from seawater. All this explains why the optimum production of carbonate sediment on very shallow shelves and banks is to be found in and between the *tropics* (Wilson, 1975, Fig. 1-1) where winter seawater temperatures remain above the lower tolerance level for the algae.

On modern tropical shelves and banks such as those of Florida, the Bahamas, and the Caribbean, rates of *in situ* carbonate sediment production have been found to be depth dependent, falling off significantly below about 10 m (Schlager, 1981). In water less than 10 m deep, these rates are such that vertical accumulation can reach over 10 m/1000 yr for coral reefs (Schlager, 1981), up to 1 m/1000 yr for sand shoals (Schlager, 1981), and over 10 cm/1000 yr for lagoon mud (Hardie and Shinn, 1986, p. 64-65). When these shallow "carbonate factories" are in full production, net accumulation of sediment will easily exceed thermal subsidence of mature oceanic crust (typical thermal subsidence rates for passive margins in the drift stage are in the range of 1-10 cm/1000 yr). Therefore, as long as there is no intervention by external forces, the basic depositional system is *self-maintaining*, that is, the warm shallow water conditions guarantee adequate sediment production, and in turn adequate sediment production guarantees that shallow water conditions will be maintained. By keeping pace with subsidence, very thick deposits of shallow water carbonates build up vertically over long periods of geologic time, as, for example, the 5 km of Cretaceous to Holocene platform carbonates of the Bahama Banks, and the 3.5 km of Lower Cambrian to Middle Ordovician shelf carbonates of the Appalachians. Only a significant increase in subsidence rate due to onset of tectonism (faulting, subduction, etc.), or drift of continents to higher (colder) latitudes, can overcome the production rate of these tropical "carbonate factories" and therefore lead to the terminal drowning of the buildup. These tropical "carbonate factories" can be overwhelmed *temporarily* by high frequency eustatic sea level rises and falls such as has occurred on a 10^4 to 10^5 year scale on all existing carbonate shelves and banks around the world as a result of Plio-Pleistocene glaciation. However, these glacio-eustatic rises and falls of sea level do not destroy the basic self-regulating machinery of the buildup depositional system, they simply turn it off and on repeatedly. For the shallow platform core of a buildup this results in a "layer cake" stratigraphic record composed of a stack of meter-scale transgressive-regressive depositional cycles with soil caps (see, for example, Perkins, 1977), a cyclic stratigraphy that is easily read using sedimentary structures and the comparative sedimentology method (see below under Important Associations).

The Response of Carbonate Mud to Compaction

A special aspect of shallow water marine carbonates that makes them extraordinary as recorders of environmental information is the wealth of sedimentary structures preserved in even the muddiest of ancient deposits (see, for example, illustrations in Demicco and Mitchell, 1982; Mitchell, 1985). In this, they differ markedly from siliciclastic deposits in which the muddy units characteristically are highly compacted into fissile shales that preserve few if any primary sedimentary structures. There are two main factors that are instrumental in the preservation of sedimentary structures in shallow marine carbonate deposits: (1) aragonite needle muds compact far less than clay-rich muds, and (2) early cementation (pre-compaction lithification) is widespread in carbonate environments. The effects of compaction on primary sedimentary structures in shallow marine carbonates are discussed in this section. Early cementation is explored in the succeeding section.

Lime muds in modern shallow marine environments consist mainly of a loose aggregate of micron-sized aragonite needles with porosities in the range of 70-80%. Shinn and Robbin (1983) have shown in squeezing experiments that such muds cannot be condensed by mechanical compaction below a porosity of about 35% (see Fig. 7 of Shinn and Robbin, 1983), presumably because the packed needles behave as a grain-supported framework. A fully compacted lime mud, then, will have suffered a maximum of 50-55% reduction in volume and will have the porosity of a compacted sand but with a very low permeability because of the small pore throats. This means that many primary sedimentary structures, such as laminations, mudcracks, sediment-filled burrows, and so on, will survive compaction albeit somewhat distorted, as Shinn and Robbin (1983) convincingly demonstrated in their experimental compaction of Bahamian mudcracked laminites. Open burrows, vugs, and fenestrae, however, will be eliminated by this degree of compaction, organic matter will be reduced to "whispy seams" that resemble micro-stylolites, and soft peloids will be merged into a clotted muddy paste (to produce a texture like that of the "structure grumeleuse" of Cayeaux, see Bathurst, 1975). If, on the other hand, clay-rich siliciclastic mud were subjected to the same compaction loads, their porosity could be reduced to near zero because the platy clay particles will flatten like a stacked deck of cards. This major difference in the response of siliciclastic and carbonate muds to compaction most likely explains why carbonate mudstones preserve so many primary sedimentary structures whereas fissile shales preserve almost none. Nonetheless, compaction will have played a major role in modifying primary structures in carbonate mudstones and such modification must be identified and taken into account in assessing the origin and significance of the resulting distorted structures.

The Importance of Early Diagenesis in Carbonate Sedimentation

Studies in Bermuda and the Caribbean of Pleistocene reef, shoal, lagoonal, and aeolian limestones that were subaerially exposed during the glacio-eustatic sea-level lowstands demonstrated that meteoric diagenesis can effectively lithify significant volumes of shallow marine carbonates (see reviews in Bathurst, 1975, 1980a; James and Choquette, 1990c). This lithification was the result of precipitation of low magnesian calcite from meteoric groundwaters in both the vadose and phreatic zones, and was accompanied by partial to complete alteration of metastable aragonite and high magnesian calcite particles to stable low magnesian calcite (carbonate "stabilization"). At the same time, considerable dissolution of the original sediments occurred as the undersaturated meteoric groundwater percolated downwards ("karstification"). From these studies grew the notion that the early lithification of shallow marine carbonates throughout the geologic record reflected their frequent exposure to meteoric groundwater circulation. It was, therefore, a considerable surprise to most carbonate petrographers when it was discovered that Holocene subtidal carbonate sediments never exposed to meteoric waters have been contemporaneously cemented beneath the sea to varying degrees by aragonite and high magnesian calcite crystals precipitated *from seawater* (see Bricker, 1971; Bathurst, 1975, 1980a; James and Choquette, 1990a, b). Re-examination of ancient shallow marine carbonate rocks has confirmed that early marine cements are abundant throughout the record (James and Choquette, 1990a, b).

In modern marine carbonate environments, early marine cementation occurs at or near the sediment-seawater interface in most of the environments found across a buildup. The distribution of marine cements in the sediments of the present-day Bahama Banks serves as an important example (Fig. 2). Here, intergranular and intragranular cements occur in the sediments of the periplatform basins, in the foreslope debris, in the bank margin reefs, in the bank margin sands, in the shelf-lagoon peloidal muds, and in the tidal flat peloidal muds (Fig. 2). What the particular cementation sites in all these environments have in common are: (1) sedimentation and erosion rates are very low, (2) the host sediments have the same mineralogy as the cements and thus can act as seeds, and (3) large volumes of supersaturated seawater are pumped through the sediments. Nonetheless, specific conditions that promote precipitation of $CaCO_3$ are different in each of these environments, and have led to significant differences in the kind and degree of early marine cementation (Fig. 2), which in turn have potentially important effects on the kind and degree of preservation of the primary sedimentary structures, as briefly outlined below.

In many barrier and fringing reefs around the world, the intra-framework voids, the internal sediments of these voids, and the coarse skeletal gravels and sands between framework masses are extensively cemented by both aragonite and magnesian calcite (James and Choquette, 1990b). The coral colonies of these reefs are growing at rates as high as 1 cm/yr and in the process produce an open framework with a substantial network of connected voids through which immense volumes of seawater are flushed every day by tides and waves, providing ideal sites for extensive growth of void-lining cements unhindered by sedimentation or erosion. It is possible that at least some of the carbonate precipitation within these reef voids is mediated by microbial activity (Chafetz, 1986; MacIntyre, 1985). The sand shoals at the margins of the Bahama Banks are not as massively cemented as the neighboring reefs. Patchy cementation of the walls of crustacean burrows and along bedding as flat pods (centimeter- to decimeter-scale) occurs beneath the sediment-water interface in the interior of subtidal dunes protected from frequent grain agitation caused by tidal currents and wind waves. Eventually, however, slow migration of the dunes exhumes the patches of partially cemented sand which are broken up into rounded grainstone lithoclasts and distributed at the toes of dune avalanche slopes or worked into lenses of flat pebble lithoclast gravels by storm waves and currents. In this way, contemporaneous cementation is indirectly responsible for the construction of grainstone intraclasts that comprise flat pebble conglomerates, an important sedimentary feature characteristic of many ancient shallow marine carbonate deposits (see below under Physical Structures). Bankward of the actively migrating sand shoals, partial cementation at and near the sediment-water interface of the inactive lime sand sheets of the outer shelf-lagoon can be extensive, but major storms break up the loosely cemented crusts to produce lithoclast lag sands ("grapestone" sands belong in this category).

On the Bahama Banks, the vast bulk of the Holocene sediments are accumulating in the protected shallow interior "shelf-lagoon" (on the Great Bahama Bank this "lagoon" covers an area approaching 100,000 km^2). Here, the sediments consist predominantly of aragonite needle muds fashioned into fecal pellets mainly by polychaete worms and shrimp. Of considerable significance is the fact that in this environment individual pellets and other sand-sized mud aggregates (collectively referred to from here on in as *peloids*) become hardened by contemporaneous intragranular cementation. This combination of pelletization followed by intragranular cementation transforms soft mud deposited in a low current and wave energy environment into *hard* sand grains with significantly different hydrodynamic properties. As such, it is a remarkable process that flies in the face of conventional views of the origin, deposition, and environmental significance of sand and mud. On the Great Bahama Bank, the proportion of hard to soft peloids decreases substantially from the outer lagoon to the inner lagoon behind Andros Island, accompanied by a change in the sea bottom from firm, well sorted, wave rippled sand to soft peloidal mud. The hard-peloid sands of the outer lagoon are in essence storm lag sediments from which the smaller cemented peloids, the uncemented peloids, and the unaggregated needle mud have been winnowed. Onshore storms carry suspended soft mud, soft peloids, and smaller hardened peloids to the inner lagoon and onto the tidal flats (Hardie and Ginsburg, 1977) while offshore storms wash this muddy sediment off the banks and down into the periplatform deeps (Neumann and Land, 1975). This unusual selective cementation of peloids not only has drastically changed the hydraulic properties of the lagoonal sediments and their depositional destiny, but it also has affected their potential for preserving primary sedimentary structures. The outer lagoon sediments composed of hardened peloidal sands have a high potential for preservation of primary structures, whereas the soft peloidal muds of the inner lagoon have a high potential for compaction on shallow burial that will distort or destroy certain primary features (see below under Physical Structures).

One additional special aspect of the cementation of individual peloids is their role in the development of the ooid shoals accumulating at the bank margin. Hardened peloids provide

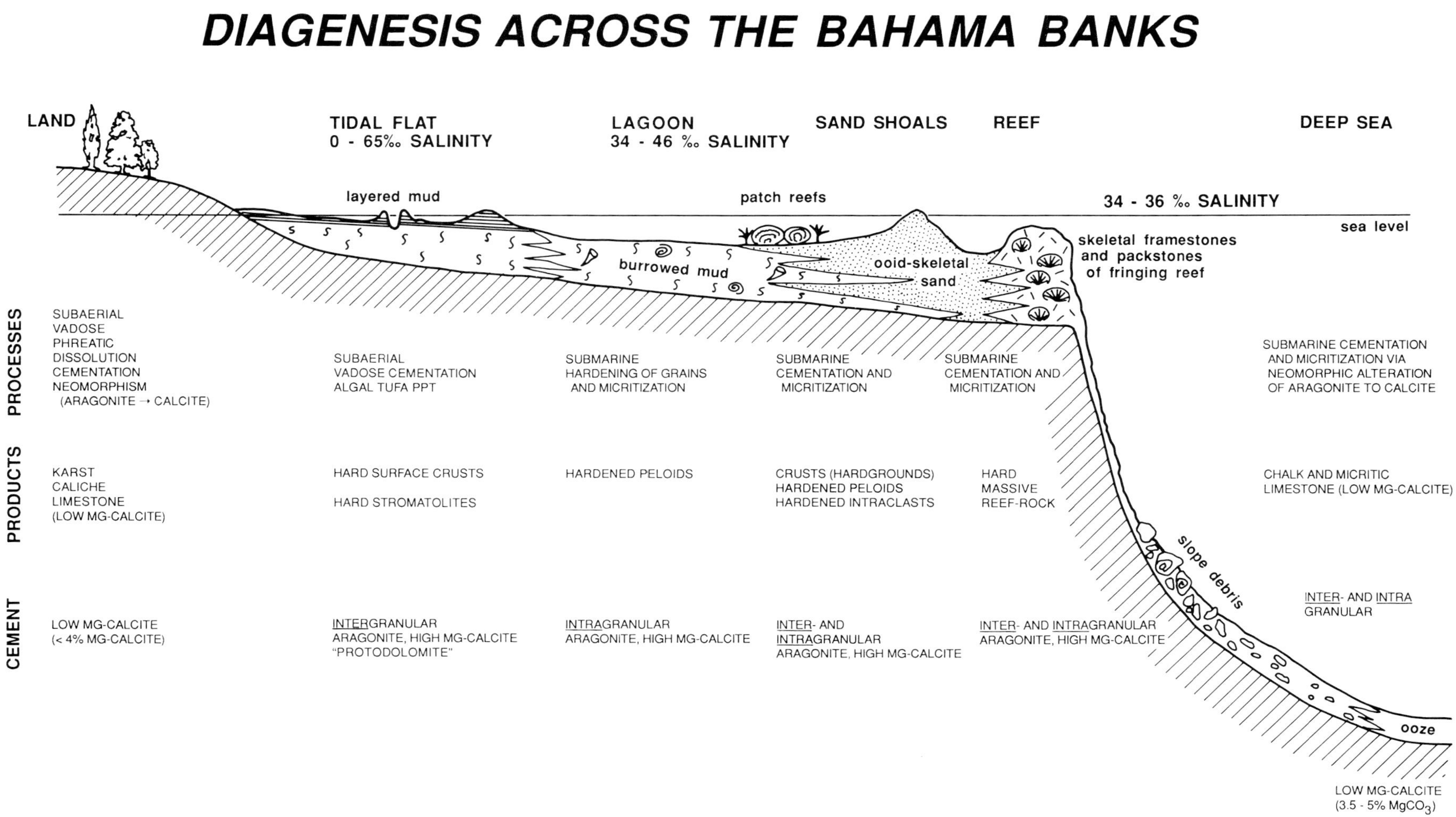

FIG. 2.-Distribution of cements and associated diagenetic features in the succession of laterally contiguous depositional environments across the Bahama Banks.

the nuclei for the bulk of these ooids. It is entirely likely that the formation of hard peloids is a necessary precursor process in the genesis of significant volumes of marine ooids (R. N. Ginsburg, pers. commun.). In this regard, it may be significant that the puzzling absence of ooid shoals from the modern Florida carbonate shelf is accompanied by a notable absence of hardened peloids. If hardened peloid sands are indeed a prerequisite for the development of ooid sands, then one of the most distinctive components of shallow marine carbonate deposits-cross-stratified ooid grainstones- will need to be considered as a by-product of selective early marine cementation of peloids. But there is yet another important consequence of the early lithification of the lagoonal sediments. Where locally on the Bahama Banks sedimentation has come to a halt (due in all probability to changes in bank water circulation as Holocene sea level has risen), the peloidal sediments are being cemented to produce local *hardgrounds* (centimeters to decimeters in thickness and tens to hundreds of square meters in area, see section on hardgrounds under Chemical Features). These hardgrounds act as substrates for growth of sponge-alcyonarian thickets and thus the distribution of early cemented surface crusts across the shelf-lagoon determines the distribution of patch reefs.

The tidal flat sediments at the landward margins of the shelf-lagoon are predominantly composed of soft peloidal muds deposited at the shoreline by onshore storms, as noted above. Contemporary cementation on the tidal flats occurs in those supratidal settings where sedimentation takes place only during hurricane-driven onshore surges spaced many years apart (Hardie, 1977b). In these areas thin (centimeter-scale), caliche-like surface crusts are forming by precipitation in the vadose zone of aragonite, high magnesian calcite, and poorly-ordered calcian dolomite cements (Shinn and others, 1965; Hardie, 1977b; Lasemi and others, 1989; see section on supratidal crusts under Chemical Features). The driving force is evaporative pumping which not only pumps seawater up from the marine phreatic zone (Hardie, 1977b) but also raises its salinity, a factor that seems to aid nucleation of calcian dolomite at surface temperatures (Hardie, 1987; but see Carballo and others, 1987, for a different view of the hydrologic processes involved in the formation of supratidal dolomite crusts). Lateral migration of the tidal channels concentrates gravel and sand-sized lithoclasts of these surface crusts in channel thalwegs so that contemporaneous cementation has a direct influence on the nature of the clastic components of the channels and hence on what sedimentary structures can be formed.

Despite the above observations of *partial* cementation of modern shallow marine carbonate sediments, many ancient shallow marine grainstones are *pervasively* cemented by early marine cements that have allowed delicate preservation of many primary sedimentary structures. One obvious explanation for this difference lies in the fact that Holocene shallow marine carbonates have been accumulating for only a very short time, a matter of a few thousand years (the latest Holocene sea level rise only flooded the interiors of the Bahama Banks about 5000 years ago), so that the cementation of the surface sediments we observe today represents just the preliminary stages. With time this kind of early marine cementation may lithify a greater and greater volume of Holocene surface sediments. Nonetheless, cementation related to the flow of seawater across the sediment-seawater interface on shallow carbonate banks and shelves is limited to the very near surface by the nature of the hydraulic driving force. This hardly seems sufficient to account for the pervasive marine cementation found in many shallow marine carbonates. But shallow marine carbonate deposits are characteristically cyclical on a meter scale (James, 1984; Hardie and Shinn, 1986). They consist of vertical stacks of shallowing-upward successions (subtidal lagoon sediments capped by tidal flat sediments) that record alternations of submergence and emergence of the buildup top with a mean frequency in the 10^4 to 10^5 year range (Hardie and Shinn, 1986). Under such conditions, it is not necessary to look for a lithification process that is deeply penetrative. The near surface early marine cementation that is occurring today could be effective if it operated as an integral part of the depositional machinery of each cycle. In other words, pervasive early marine cementation of a thick pile of shallow marine carbonates could be achieved one thin cycle at a time (a sort of "layer cake" diagenesis). Evidence of this piecemeal early marine lithification is found in the cyclical carbonates of Triassic buildups in the Dolomites of northern Italy (Goldhammer and others, 1987, 1990; Goldhammer, 1987; Goldhammer and Harris, 1989).

There are other possible marine cementation processes that need to be considered. For example, bank water could penetrate deeply downward (tens to hundreds of meters) into the underlying sediment pile if the density of the warm surface seawater was raised by evaporative concentration until it exceeded that of the cooler seawater occupying the pores of the subsurface sediments. Such conditions do exist on the Great Bahama Bank and have the potential to generate a significant downward flux of seawater (Simms and Hardie, 1983; Simms, 1984) that could result in pervasive cementation of those permeable subsurface sediments that acted as flow conduits (shallow burial marine cementation). Another process for driving large volumes of seawater supersaturated with respect to aragonite and magnesian calcite through large volumes of subsurface bank or shelf sediments is slope convection (Wilson and others, 1990; Phillips, 1991). This kind of massive lateral and upward flow through a carbonate buildup of the deep ocean water surrounding the bank has been documented for the present day Florida platform (Kohout and others, 1977; see also Simms, 1984). The cool seawater sucked into the center of the buildup is warmed by the higher geothermal gradient within the buildup, and this warming may increase the $CaCO_3$ supersaturation sufficiently to induce precipitation of pore cements. Like the early marine cementation of the surface sediments of modern buildups, any shallow burial marine cementation resulting from pervasive subsurface flow of seawater should occur preferentially in the more permeable sediments. Ooid-peloid-skeletal sands and intraclast gravels should be favored over lime muds. These ideas on shallow burial marine cementation are speculative and it remains to be demonstrated to what extent, if any, such mechanisms are responsible for pervasive cementation by aragonite and magnesian calcite found in ancient shallow marine carbonates.

In summary, then, the effects of the *extensive, but selective, early marine cementation must be taken into full account when assessing the significance of primary sedimentary structures in shallow marine carbonate deposits.* For example, the thick Cambro-Ordovician shelf carbonates of the central Appalachians carry a spectacular array of primary sedimentary structures, but, at the same time, selective early marine cementation has strongly influenced not only the final form of the structures but has also influenced the sites of later burial diagenetic alteration by such processes as compaction, dolomitization, and stylolitization (Wanless, 1979; Hardie and others, 1982; Goldhammer and others, 1985; Hardie, 1989). This is most commonly seen in the lagoonal subfacies where wave-rippled and hummocky cross-stratified storm lags are interlayered on a centimeter to decimeter scale with burrowed peloidal mudstones (e.g., Nguyen, 1986) and in those intertidal subfacies with heterolithic thin bedding analogous to that of the flaser, lenticular and wavy bedding of the modern North Sea tidal deposits (e.g., Demicco, 1983). Differential compaction has resulted in the fracture of the brittle sandy beds that were selectively cemented before compaction and has caused the ductile muddy interbeds to flow into the fractures and around isolated brittle "boudins." At some stage during post-compaction burial, the muddy layers were selectively dolomitized to produce "ribbon rocks" of alternating dolomite and limestone. Late bedding plane stylolites typically are confined to the early cemented units or concentrated at the boundaries between the sandy and muddy beds. Clearly, as long as such case histories are possible, it is essential in analyzing the sedimentary structures of shallow marine carbonates to fully consider the roles that early diagenesis might have played in modifying or even producing any given sedimentary structure.

SEDIMENTARY STRUCTURES AND DIAGENETIC FEATURES

In compiling this atlas, we were faced with an enormous array of features to describe and illustrate. Many of the features of shallow marine carbonates are common to the deposits of other depositional environments. We have chosen to present only those features that are pivotal to the recognition of those carbonate rocks that were deposited in shallow marine paleoenvironments. A significant exception to this is made in the case of metazoan reefs. The properties of these reefs, modern and ancient, are too extensive to be included in this atlas and the reader is referred to the excellent summaries by James (1983), James and MacIntyre (1985), and James and Bourque (1992).

PHYSICAL STRUCTURES

Stratification

Stratification or layering is the most characteristic and fundamental of all sedimentary structures. In outcrops, the physical resemblance of stratification to the pages of a book conjures up the idea that sedimentary rocks are natural history records, records that can be deciphered to yield a glimpse of the chronological succession of events that befell the surface environments of the earth. Stratification by its very nature reveals sedimentation as an episodic process in which periods of non-deposition are punctuated by bursts of sediment influx. The periods between sedimentation events may be as short as seconds (as occurs on the avalanche slopes of actively migrating ripples), may be months or years long (as occurs on the storm-built supratidal flats of the Bahamas), or they may occupy many millennia (as happened on the Florida platform during the Pleistocene glacio-eustatic sea level oscillations). In each case, the essential record of the non-depositional episode is a fundamental surface engraved within the sediments as a layer boundary, which is identified by distinctive changes in grain size, grain texture, grain composition, fossils, truncation of grains, truncation of primary sedimentary structures, or some combination of these features.

The nature of these layer boundaries and the characteristics of the sediments enclosed within such boundaries are the features on which our classification, terminology, and interpretation of the significance of stratification are based. However, the first problem encountered in describing layering in carbonate rocks is identification of primary sedimentary boundaries between layers and distinguishing them from boundaries produced by diagenetic processes (such as pressure solution) and other secondary processes. Figure 3 illustrates a common field problem. The shaly partings between limestone units here are obvious and these rocks might be referred to as thinly layered or thin bedded. However, within the limestone units between shaly partings are erosional sedimentary boundaries that sharply separate internal layers of notably different composition (Fig. 4). Shaly partings here actually divide complicated, composite sets of layers. Unambiguous sedimentary boundaries can be recognized between limestones in which primary grain fabrics can be identified. Stylolites commonly follow such boundaries but when traced laterally, stylolites may leave the boundary and move out into the rock. Limestone-shale and limestone-dolostone boundaries are the most ambiguous because of the effects of compaction and replacement diagenesis. We have devoted a section below to compaction features which includes a discussion on criteria for the recognition of diagenetic overprinting on primary layer boundaries. Finally, where there are gradational changes in internal features such as grain size, texture, and mineralogy, the choice of a layer boundary becomes an entirely arbitrary one.

We recognize three major groups of stratification types. These are: (1) strata of relatively well-sorted sediment which exhibit internal cross-stratification between horizontal layer boundaries (typically strata of either gravel, sand or coarse-grained silt), (2) strata of relatively homogeneous sediment which exhibit no internal cross-stratification between layer boundaries (layers of either sand or mud), and (3) strata in which sand layers alternate with mud layers (heterolithic stratification). The characteristics and origins of these three types of strata are briefly discussed below.

Cross-stratification.-

Terminology.-Cross-stratification comprises an important class of layering structures commonly found in gravel-sized, sand-sized, and coarse silt-sized, noncohesive materials. Cross-strata are textually or compositionally distinct layers of sediment that are more or less steeply inclined (up to the 30-34° angle of repose) to the principal surfaces of accumulation of the deposit in which they occur (Allen, 1984, p. 345). Cross-strata are organized into sets which are in turn commonly grouped into cosets (McKee and Weir, 1953). Where the upper boundary of a set has a ripple or megaripple cross-sectional shape, the internal cross-strata may be: (1) form-concordant and directly follow bedform morphology (a form set in the terminology of Imbrie and Buchanan, 1965); or (2) the internal cross-strata may be form-discordant. Set boundaries may be either sharp or gradational. Other than these terms, there is no generally accepted nomenclature or classification system for cross-stratification. Allen (1984, p. 346) provides a review of genetic and descriptive classifications (see also Reineck and Singh, 1980; Collinson and Thompson, 1989; and other atlases in this series). A voluminous literature exists on cross-stratification (see Reineck and Singh, 1980; Harms and others, 1975, 1982; Allen, 1984; Pettijohn and others, 1987; and Collinson and Thompson, 1989) that cannot be thoroughly reviewed here. However, because all cross-stratification types are commonly encountered in shallow marine carbonate deposits, the following discussion is intended as a basic starting point for more intensive inquiry by the reader.

It is important to note that although the following discussion focuses on internal layering structures, the complimentary bedforms (bedding surface sedimentary structures such as ripples, megaripples or dunes (Fig. 5), antidunes, wave ripples (Figs. 6, 7), and hummocks) are also common features of the sediments of modern carbonate shallow-water environments and their ancient equivalents. Furthermore, these bedforms in carbonate sediments are exactly like their siliciclastic counterparts.

FIG. 3.- "Thin-bedded" Middle Ordovician carbonates from central Pennsylvania.

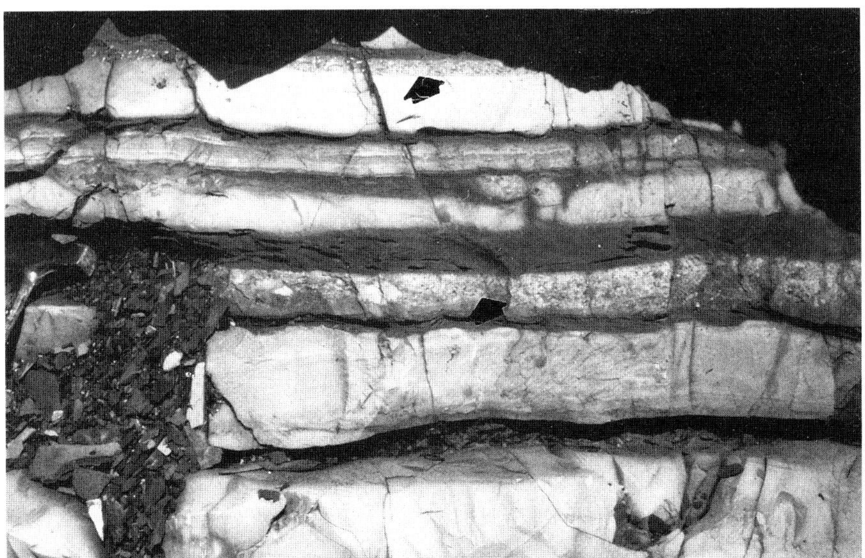

FIG. 4.- Detail of layers shown in Figure 3. The shale layers separate limestone layers that internally may consist of complex composite sets of layers (see arrows). Hammer at left for scale.

A

B

FIG. 5.-(A) Modern subaqueous dunes in carbonate sands. Aerial photograph of Cat Cay ooid sand belt, northeastern Great Bahama Bank (Plate 28, Pettijohn and Potter, 1964, see Ball, 1967). Photograph courtesy of Francis J. Pettijohn. This marine ooid sand belt is approximately 1.5 km wide and is composed of lobes and channels typical of tidal deltas. The lobes are covered with straight-crested, sinuous-crested, and discontinuous-crested dunes (Ball's "medium-scale ripples") with wavelengths of tens of meters and heights up to a meter. (B) Sinuous-crested dunes developed in skeletal sands on a tidal flat in Nilemah Embayment at the southern end of Hamelin Pool in Shark Bay. (C) (Following page) Sinuous-crested dunes from Ordovician grainstones, Brown County, Ohio (Plate 89A, Pettijohn and Potter, 1964). Photograph courtesy Francis J. Pettijohn. Crest-lines of dunes in the foreground were modified by erosion prior to burial. Erosion may have been due to a subordinate tidal current or by exposure during falling water.

C

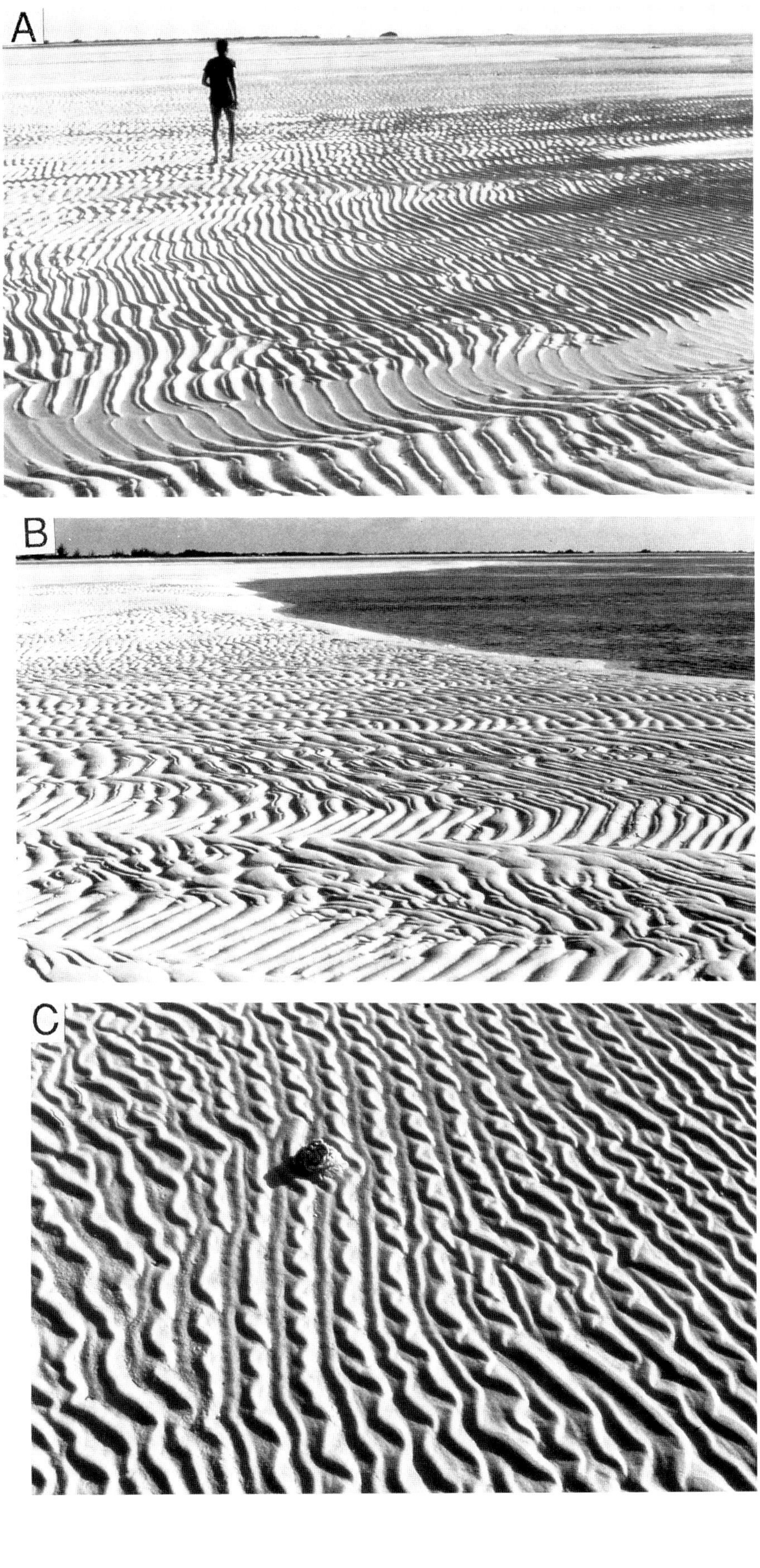

FIG. 6.-Complex ripples on Joulter's Cay ooid shoal, Great Bahama Banks. (A) Current-modified wave ripples with double crestlines (Halley and others, 1983, Fig. 33). Rounded crestlines (foreground) are a form modification produced during a falling-water stage in the tidal cycle. (B) Detail of ripples in another area of Joulter's Cay. Note the rounded crestlines and variable plan morphology of the ripples. Ripples are superimposed on low dunes oriented roughly perpendicular to ripple crest orientation. Ripple spacing approximately 100 mm. (C) Interfering wave ripples (Halley and others, 1983, Fig. 34). Fecal mound in center approximately 30 mm high. All photos courtesy of Paul M. Harris.

FIG. 7.-Current-modified wave ripple marks in a Precambrian ooid grainstone, Lower Proterozoic Wildbread Formation, Pethei Group, Northwest Territories, Canada. Geologic hammer for scale. Photograph courtesy of Paul Hoffman.

More illustrations of current ripples, megaripples (dunes), and wave ripples can be found in Reineck and Singh (1980), Scholle and others (1983), Allen (1984), Collinson and Thompson (1989), and the other atlases in this series. Cross-stratification types considered here are small-scale and large-scale cross-stratification, wave-ripple cross-stratification, and hummocky and swaley cross-stratification. Eolian ripple marks, dunes and their cross-stratification patterns are not considered here but details can be found in Hunter (1977) and the atlas on eolian structures in this series.

Distinguishing characteristics and origin.-The best known types of cross-stratification are small-scale cross-stratification (Fig. 8), where set thicknesses are less than 0.04 m, and large-scale cross-stratification (Fig. 9), where set thicknesses are greater than 0.04 m. These cross-strata are generally interpreted as the result of the downstream migration and vertical aggradation of ripples (small-scale cross-stratification) and dunes (large-scale cross-stratification) formed beneath steady, uniform, unidirectional currents. The 0.04 m boundary corresponds to the empirically observed minimum in frequency distribution of ripple and dune height (Allen, 1984, p. 347). However, where erosive set boundaries occur, dunes can leave behind truncated sets less than 0.04 m thick. A variety of geometric terms are available to describe the erosive set boundaries left where ripples or dunes have subcritical angles of climb (Allen, 1984, p. 351-352; see also Botvinkina, 1959; Allen, 1963, 1970, 1973; Elliot, 1964; Jacob, 1973), but the most common terms are tabular and trough cross-stratification (McKee and Weir, 1953). Tabular cross-sets are interpreted to have formed by the migration of straight-crested ripples (small-scale) or dunes (large-scale) whereas trough cross-sets record the migration of curved, discontinuous-crested ripples or dunes. Generally, three-dimensional exposure is necessary to accurately differentiate tabular from trough cross-stratified sets.

A supercritical angle of climb is uncommon for dunes but relatively common for ripples migrating during massive rain-out of suspended sediment and result in what Sorby (1908) termed "ripple-drift" (Fig. 8C). Climbing-ripple cross-lamination is another common term for this type of supercritical, small-scale cross-stratification which can form from the migration of both relatively straight-crested or sinuously-crested ripples (see Allen, 1984, p. 368). In coarse sand-sized and gravel-sized skeletal grainstones, large-scale cross-stratification may be difficult to see in outcrop or in core. In addition, stylolites may follow set boundaries and so make proper identification of cross-set geometries difficult (Fig. 10). Intraclast flat pebble conglomerates are typically cross-stratified but this is commonly misidentified as imbrication (see below under conglomerate structures). In both, unsteady unidirectional flows and periodically reversing tidal flows, cross-strata produced by dunes are modified by changes in flow direction, flow velocities, or both. Boersma (1969) referred to erosional surfaces caused by such changes as "structural diastems" but these features are now generally referred to as "reactivation surfaces" (Fig. 11). This term is derived from Collinson's (1970) study of the Tana River, Norway. Modification may occur on a tidal ebb-flood cycle basis, a spring-neap cycle basis, or a combination of both. Modifications include single (Boersma, 1969) and double (Nio and others, 1983) mud drapes from slack water suspension settle-out. Other common modifications are internal sigmoidally-shaped erosive surfaces, some of which are covered by upslope or along-slope directed small-scale cross-stratification (Boersma, 1969) (Figs. 10, 11). Large-scale "herring-bone" cross-stratification (Fig. 12) occurs where ebb and flood flows are roughly equal (Reineck and Singh, 1980). De Raaf and Boersma (1971) and Boersma and Terwindt (1981) have documented reactivation surfaces associated with spring to neap tidal cycles. Along these lines, Nio and others (1983) have reported systematic lateral width changes in "tidal bundles" which are groups of adjacent cross-strata bounded by mud drapes. In Figures 10 and 11, the mud drapes are selectively composed of dolomite, whereas stylolites follow set boundaries. Complicated cross-stratification patterns that resemble reactivation features are also commonly found in deposits that are interpreted to have been deposited as large-scale "sand waves" and "tidal sand ridges." These are discussed later under Important Associations.

Wave-ripple marks (Fig. 6) and wave-ripple cross-stratification (Fig. 13) in modern carbonate sediments and ancient carbonate rocks (see, for example, Demicco, 1983) are exactly analogous to their siliciclastic counterparts and carry the same environmental significance. The morphology of wave-ripple marks produced under oscillatory flows is well known (van Straaten, 1953; Reineck and Singh, 1980; Allen, 1984) and so offers the potential for quantitative reconstructions of the paleohydraulic conditions that existed during ripple growth (Komar, 1974, 1976; Clifton, 1976; Harms and others, 1982). However, the internal structure produced by sediment aggradation beneath oscillatory flows (wave-ripple cross-stratification) is not as well known as cross-stratification produced by unidirectional flows. De Raaf and others (1977), Reineck and Singh (1980), and Allen (1984) have summarized what is known about the internal structures of wave ripples, and the reader is referred to those works for further information.

Ripples and dunes formed under combined oscillatory and unidirectional flows (Fig. 14) have not been studied exten-sively, but some flume experiments have been carried out (Inman and Bowen, 1963; Harms, 1969; Arnott and Southard, 1990; Myrow and Southard, 1991). The internal structures of such combined flow ripples have been modeled by Arnott and Southard (1990) and Myrow and Southard (1991).

Hummocky cross-stratification (Fig. 15) (Harms and others, 1975, 1982; Walker and Plint, 1992) and swaley cross-stratification (Duke 1980, 1985; Leckie and Walker 1982; Walker and Plint, 1992) comprise poorly known and controversial cross-stratification types. The original definitions have been so greatly modified and misquoted that illustrations in current sedimentology textbooks bear little or no relationship to the originally defined sedimentary structure (e.g., Selley, 1988, p. 133). Allen (1984, p. 501) avoids using the term altogether. The original definition of hummocky cross-stratification includes the following essential elements: (1) lower bounding surfaces of sets are erosional and commonly slope at angles less than 10 degrees, though dips can reach 15 degrees; (2) laminae above these erosional set boundaries are parallel to the surface, or nearly so; (3) laminae can systematically thicken laterally in a set, so that their traces on a vertical surface are fan-like and dip diminishes regularly; and (4) the dip directions of erosional set boundaries and of the overlying laminae are scattered (Harms and others, 1975, p. 87). Swaley cross-stratification was formally defined by Leckie and Walker (1982, p. 143) and is characterized by: (1) a series of superimposed concave-upward shallow scours (swales) about 0.5 to 2 m wide and a few tens of centimeters deep, (2) stratification between scours that is gently curved and follows the basal erosive swale surface, (3) dips of the stratification at the edges of swales that rarely exceed 10 degrees, and (4) swales having the same geometry regardless of the orientation of a vertical cross-section. Clearly, three-dimensional exposure is necessary to insure proper differentiation of swaley cross-stratification from trough cross-stratification.

Most workers now would interpret hummocky cross-stratification and swaley cross-stratification as the result of bed

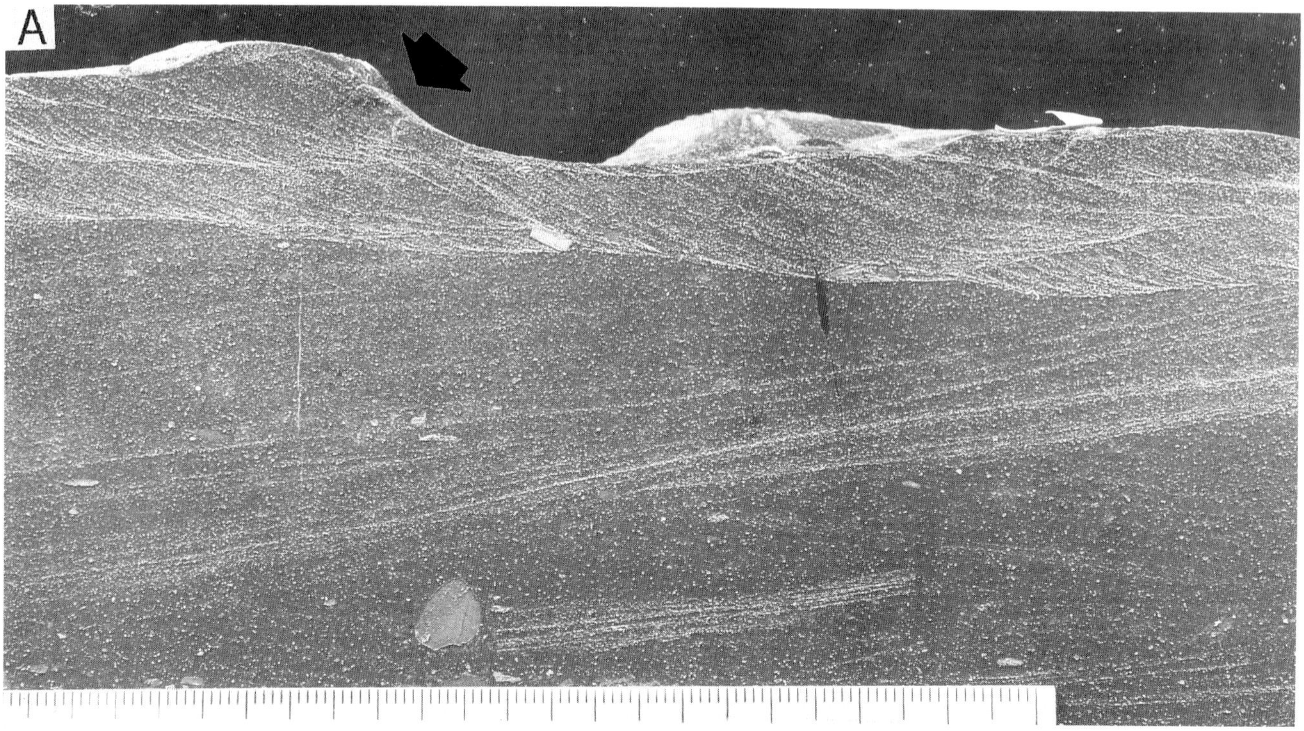

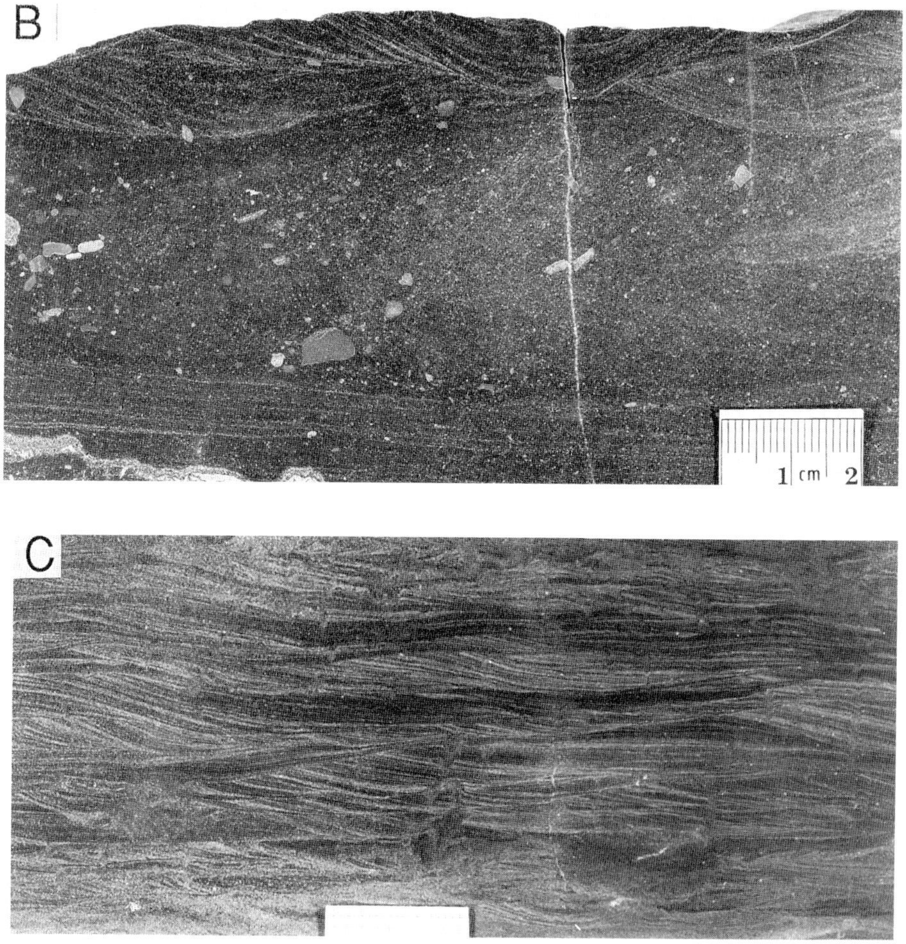

Fig. 8.-Small-scale cross-stratification in Lower Paleozoic grainstones. (A) Upper 30 mm of this specimen shows flow-parallel views of small-scale, trough cross-stratification. Lower part of the specimen shows large-scale, tabular cross-stratification. Note the hump-shaped profile typical of "classic" current-ripples at the upper left (arrow points to slipface; thin white layer draping the ripple is dolomitic mudstone) and internal form-concordant cross strata. Small divisions of the scale are mm, Upper Cambrian Conococheague Limestone, western Maryland. (B) Flow-perpendicular view of specimen shown in (A). (C) Flow-parallel cross-section of climbing small-scale cross-stratification. Scale bar is 20 mm long, Lower Devonian Manlius Formation, New York.

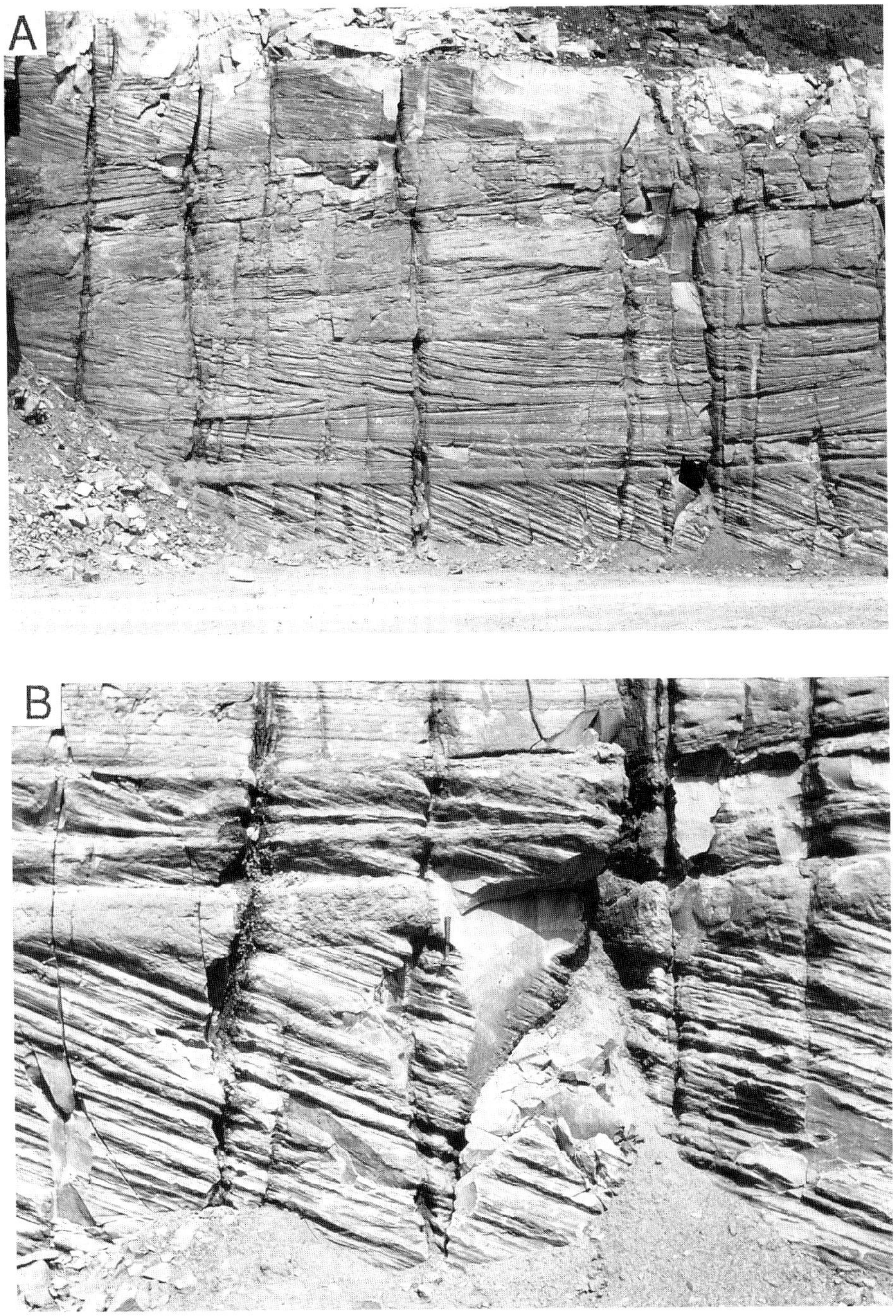

Fig. 9.-Large-scale cross-stratification in Upper Paleozoic grainstones. (A) Skeletal-peloidal grainstone unit in the Mississippian Loyalhanna Limestone from central Appalachians. Cliff face is 19.5 m high. Lower 3 m and upper 6 m mostly flow-parallel views. Central 10 m shows flow-perpendicular and oblique views of large troughs. (B) Detail of lower right portion of (A). Photographs courtesy of Robert W. Adams. Also see Adams (1970, Fig. 11, p. 94). (C) (Following page) Large-scale planar cross-stratification (geologist at lower left for scale) from an Upper Devonian skeletal grainstone found in siliciclastic coastal facies near Binghamton, New York. Photograph courtesy of J. S. Bridge. Also see Bridge and Droser (1985, Fig. 4D, p. 145).

Fig. 10.-Medium-grained to coarse-grained peloid-skeletal grainstones from the Upper Cambrian Conococheague Limestone, western Maryland, showing planar-stratification (lower portion of photograph) and large-scale cross-stratification (upper portion of photograph). Lichen-covered dolomitic mudstone layers (arrows) are interpreted as storm mud layers that settled as drapes over the sandy bedforms (seen in oblique view beneath the double arrows). Note the thin stylolitic seams in the upper cross-stratified grainstone that follow individual foresets and set contacts, highlighting the depositional boundaries. Metal frame for scale is 1 m on a side.

Fig. 11.-Reactivation surfaces in cross-stratified Paleozoic grainstones. (A) Composite set of large-scale cross-strata developed in quartz sand-bearing, skeletal-peloidal grainstones, Mississippian Loyalhanna Limestone, central Appalachians. The four flow-parallel sets of cross-strata that comprise the upper two-thirds of the outcrop include many reactivation surfaces, some of which have small-scale ripples directed up the foreset slope (arrows). Photograph courtesy of Robert W. Adams, hammer provides scale. Also see Pettijohn and Potter (1964, Plate 37). (B) The thin, dark, lichen-covered layers are dolomitic mudstones that drape reactivation surfaces in ooid-peloid grainstones, Upper Cambrian Conoco-cheague Limestone, western Maryland. Many of these reactivation surfaces are modified by current or wave ripples (e.g., the prominent wavy partings in the upper left quarter of the photograph). Lens cap for scale is 50 mm in diameter.

Fig. 12.-"Herring-bone" large-scale cross-stratification developed in ooid grainstone, Upper Cambrian Waterfowl Formation, Canadian Rocky Mountains, southwestern Alberta. Foreset true-dip measurements made on near-perpendicular faces of the outcrop demonstrate that the paleocurrent directions were bimodal, with the modes oriented at 180° to each other. Ruler is 150 mm long. Photograph courtesy of Brent B. Waters.

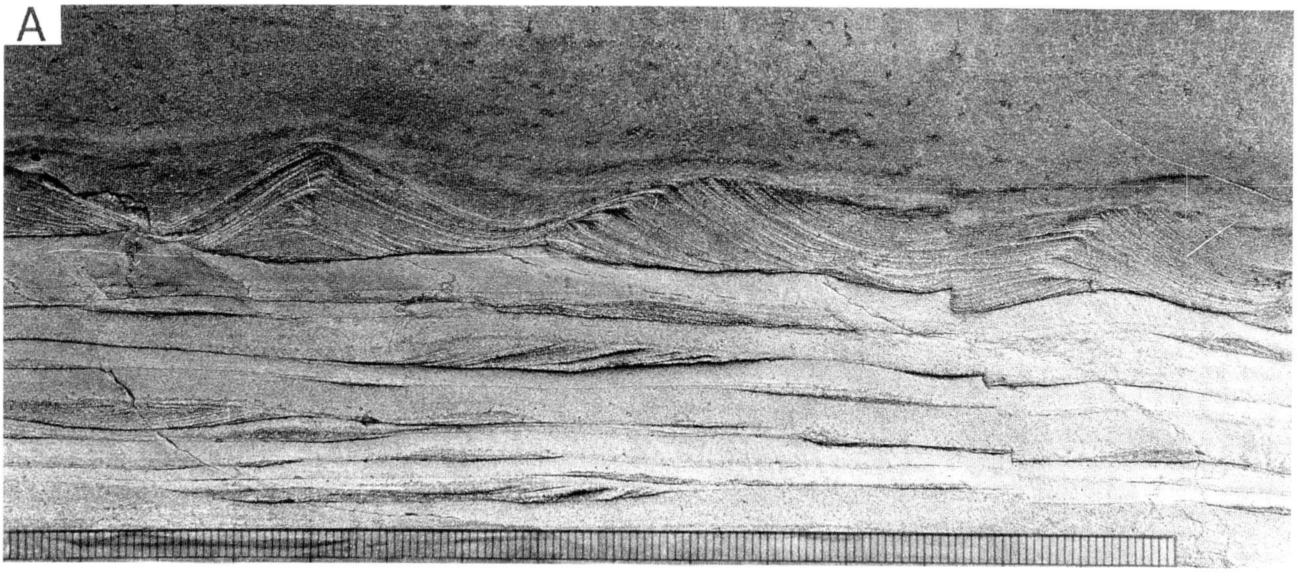

Fig. 13.-(A) Set of wave-ripple cross-stratification showing lateral variability from form-concordant to form-discordant geometry, "chevron structure", "bundled upbuilding", and draping lamination morphology (see de Raaf and others, 1977 for further details on wave-ripple cross-stratification). Lower half of photograph comprises wave-formed lenticular bedding (mostly upper flow regime). Scale in mm. Upper Cambrian Conococheague Limestone, central Appalachians. (B) Flow parallel view of wave-ripple form set (upper arrow) overlying flow oblique view of wave-ripple cross-stratification (lower arrow). Dolomite mud drapes both cross-stratal sets. Middle Cambrian Arctomys Formation, Canadian Rocky Mountains, southwestern Alberta. Coin is approximately 25 mm in diameter.

Fig. 13(continued).-(C) Flow parallel view of wave-ripple cross-stratification in a dolomite grainstone, Precambrian Stark Formation, Etthan Island, Canada. Scale in inches. Photograph courtesy of Paul Hoffman.

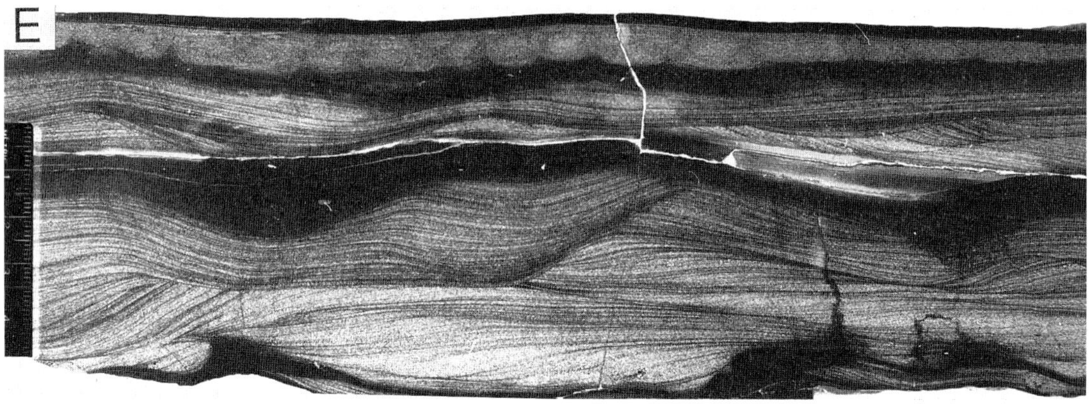

Fig. 13(continued).-(D), (E), and (F) Flow oblique cross-sections of wave-ripple cross-stratification. (D) and (F) from Upper Cambrian Waterfowl Formation, Canadian Rocky Mountains of southwestern Alberta. Coin in (F) approximately 20 mm in diameter. (E) from Upper Cambrian Conococheague "ribbon rocks", central Appalachians. Scale bar (lower left) is 50 mm long.

Fig. 14.-Nearly symmetrical form sets in isolated ripples (at hammer), Ordovician skeletal grainstones, Brown County, Ohio. The unidirectional internal large-scale cross-strata suggest that these are either combined flow (wave- and current-formed) features or produced by shoaling waves that generated enough asymmetry of flow to cause unidirectional ripple migration. Photograph courtesy of Francis J. Pettijohn (Pettijohn and Potter, 1964, Plate 90).

Fig. 15.-(A) and (B) Hummocky cross-stratification in peloidal grainstones from the Lower Ordovician Stonehenge Limestone, Beekmantown Group, western Maryland. Note wave-ripple form set at top (arrow) of hummocky cross-stratified set in (A). Photographs courtesy of Chau Nguyen. Lens cap in lower photograph is 50 mm in diameter.

aggradation under complicated, combined oscillatory-unidirectional flows generated by storms (references cited above; Kreisa, 1981; Craft and Bridge, 1987; Arnott and Southard, 1990; Myrow and Southard, 1991; and Southard and others, 1990). However, see Kreisa (1986) for an alternative perspective.

Occurrence and preservation.-In our experience, cross-stratification and planar stratification (described below) are common sedimentary structures of many modern carbonate gravels and ancient conglomerates, modern carbonate sands and ancient grainstones, and modern peloidal muds and ancient peloidal mudstones formed in shallow-water platform settings. However, these most important physical sedimentary structures have been given too little attention by carbonate sedimentologists. There are a number of reasons for this. Cross-stratification is commonly difficult to see in coarse, skeletal grainstones and intraclastic conglomerates where grains have irregular shapes. Compaction and stylolitization also commonly obscure cross-stratal boundaries, making identification difficult. In addition to these problems in identifying cross-stratification in ancient carbonates, a number of disruptive processes affect cross-stratified grainstones. Bioturbation commonly obscures cross-stratification, especially in post-Lower Ordovician rocks, and in many cases, all but obliterates it. In addition to bioturbation, grainstones can be disrupted by chemical processes such as caliche growth, tepee development, submarine hardground growth (see individual structures), etc., due to the inherent permeability of the original carbonate sands.

Significance.-It is impossible here to completely detail the significance of cross-stratification in carbonate rocks and the reader is referred to the references cited above and to other atlases in this series. Cross-strata are some of the most use-ful indicators providing stratigraphic "up direction" critera in deformed sequences. Cross-stratification patterns produced by unidirectional flows provide paleocurrent information which can be used to map regional paleoflow patterns (Pettijohn, 1962; Potter and Pettijohn, 1963, Pettijohn, 1975) and in detailed paleoenvironmental reconstructions (Cloyd and others, 1990). Finally, our ability to produce and study cross-stratification in flumes and to study it in natural flows holds the promise of detailed quantitative paleoflow reconstructions such as is now possible in channelized flows (Bridge and Gorden, 1985; Allen, 1985).

Strata without internal cross-stratification.-

Terminology.-Layering without internal cross-stratification is a class of stratification that is, in general, far less well understood than that with cross-stratification. We include in this catagory layering where cross-stratification may make up part of the layering but is, in general, a subordinate component of the rock. Layering without internal cross-stratification is most common in muddy sediments and mudstones, and mixed (heterolithic) sands and muds and sandstones and mudstones. Strata without internal cross-stratification constitute the major kind of layering found in carbonate sediments and rocks, particularly in modern and ancient peritidal carbonate deposits, but their analysis is problematical for a number of reasons. First, there is no common terminology and, although general thickness terms are available for classifying layering, other descriptive terms are not. Secondly, we now recognize that diagenesis can severely alter original sedimentary boundaries (Bathurst, 1987). In some cases, the primary depositional layer boundary is enhanced by a diagenetic process such as compaction. In other cases, diagenetic features such as stylolites may produce lithologic boundaries where none existed previously. Obviously, such boundaries of diagenetic origin if not identified and distinguished from primary depositional boundaries can lead to serious errors in the interpretation of stratification in carbonate rocks. Thirdly, there are many different depositional mechanisms that can produce layers without internal cross-stratification. The purposes of this section are: (1) to briefly review current terminology, (2) to draw attention to criteria which may help identify diagenetic modification of layer boundaries, and (3) to present a "key" for classifying and interpreting layering without internal cross-stratification.

The terms in use for stratification without internal cross-stratification include stratum, bed, lamina, division, and layer. Here we will basically follow McKee and Weir's (1953) approach of considering the terms stratum and layer as general terms. In McKee and Weir's terminology (1953, p. 382), a stratum "is a single layer of homogeneous or gradational lithology, deposited parallel to the original dip of the formation. It is separated from adjacent strata or cross-strata by surfaces of erosion, non-deposition, or abrupt change in character." In this definition, the interpretive terms erosion and non-deposition are explicitly used and strata apparently cannot have gradational boundaries. For these reasons we prefer to define a stratum or layer as a body of rock differentiated from those above and below by variations in grain composition, grain size, grain shape, grain sorting, grain orientation, or grain packing (Pettijohn and others, 1987, p. 98) where boundaries between adjacent layers are recognizable and are either sharp (and in some cases demonstrably erosional) or gradational (see also Elliot, 1965). McKee and Weir (1953) recognize two types of strata based on thickness. Laminae are strata less than 10 mm thick whereas beds are more than 10 mm thick. They further suggested modifiers for different thicknesses of laminae and beds (thin beds, thick beds, etc.). Ingram (1954) suggested different limits for thickness categories. Lamination is so common in carbonates that the term laminite is widely used. However, although thin bedding is also very common, especially in Precambrian and Lower Paleozoic carbonates, the term thinbeddite has not gained widespread use.

One serious problem commonly arises when attempting to apply McKee and Weir's nomenclature to sediments or sedimentary rocks. Many beds and laminae themselves contain "internal structures" or "internal laminae" or, indeed, "internal beds," whereas others apparently do not, demonstrating that thickness is not a parameter which easily and logically allows layering to be subdivided. This may pose no problem for one worker in one formation or a group of workers who understand their own rules for applying these terms. However, there is no uniform usage and hence terms such as "division" or "layer" were suggested by Blatt and others (1980, Fig. 5-1, p. 130) apparently following the suggestions of Elliot (1965). Bed thus becomes a rather subjective term and strongly scale dependent. Even further complications arise when there are sets of beds or sets of laminae. For example, nearly every "laminite" is a set (or coset). The terms set, coset (set of similar sets), and composite set (set of different types of sets and cosets) are useful in describing these layering types but have not been extensively employed. Finally, it should be noted that the division McKee and Weir place between laminae and beds at 10 mm is entirely arbitrary. There is apparently no observable groupings of layers into size categories with a natural break at this value, nor is there any established theoretical reason for picking this division as a break.

Campbell (1967), Reineck and Singh (1980), and Collinson and Thompson (1989) attempted to get around these problems by classifying and describing layering in terms of "sedimentation units." Otto (1938) defined a sedimentation unit as "that thickness of sediment which was deposited under essentially constant physical conditions" (Pettijohn, 1975, p. 103). Laminae arise within sedimentation units due to "momentary fluctuations in current velocities" (Pettijohn, 1975, p. 103). In Campbell's (1967, p. 8) usage, a bed "is the stratum that reveals the principal rock layering." In some cases, a genetic sedimentation unit can be identified in the field, such as a turbidite or a storm deposit ("tempestite"), but this is generally not the case.

In summary, no hard and fast rules exist for defining and classifying layering in beds that display no internal cross-stratification.

Distinguishing characteristics and origins.-Our approach here is to present a practical key to the identification and interpretation of layering without cross-stratification (Table 1). Table 1 is primarily intended as a field classification of layering without cross-stratification. It uses a practical rather than theoretical approach and is based on our experience with Paleozoic and Mesozoic platform carbonates. For each category of layering, there is: (1) a reference to *an illustration* of the layer type (Figs. 16 to 28), and (2) a list of *possible mechanisms* for producing the layer type. It is important to remember that slabbing, etching, petrographic study, and X-radiography may be necessary to fully reveal the critical elements of the layering.

The key is based on three criteria: layer thickness, layer geometry, and layer composition (Table 1). The key is entered in one of three thickness categories: layers less than 5 mm, layers 5 mm to 100 mm, and layers greater than 100 mm. The categories are arbitrary but in choosing 5 mm as the upper boundary for the thinnest category of layers we have taken into account the influence of a thin microbial mat upon deposition. Such mats play a major role in the origin of most of the thin laminae found on modern carbonate tidal flats (see, for example, Hardie, 1977a) and appear to have played this role on carbonate tidal flats since Archean time. Layers less than 5 mm influenced by microbial mats commonly occur in sets up to a few meters thick that cap peritidal cycles (see, for example, Demicco, 1983). The upper break of layer thickness at 100 mm is based on our experience in Upper Proterozoic and Lower Paleozoic carbonates. In these rocks, layer thicknesses are commonly between 5-100 mm and such thin-bedded carbonates may indeed be the most common rock type.

Each thickness class is divided into three categories based on layer boundary geometry: (1) planar, (2) undulatory, and (3) crinkled. Planar layers appear as flat, parallel-sided layers, although when traced laterally most are actually long, flat lenses. Crinkled layers are distinguished by boundaries that characteristically exceed 34 degrees, the angle of repose for cohesionless sediments, where at the same time the "crinkles" exhibit height/length ratios less than 10 and typically closer to 1. All gradations of undulatory geometry are possible between these end members. Layers can be dominantly mudstone (mud-sized material), dominantly grainstone (sand-sized material), wackestone (unsorted mixtures of mud-sized and sand-sized material), or graded (vertically sorted mixtures of sand-sized to mud-sized materials). In addition, layering may consist of couplets made of alternating grainstone and mudstone layers (see also heterolithic stratification below). A number of problems arise with the proposed compositional classes. First, virtually all sedimentary layers are graded to one degree or another, and indeed it is this grading that in many cases allows layering to be seen. Therefore, here we restrict the term grading to size gradations of at least an order of magnitude. Secondly, many "mudstones" are in reality composed of sand- to coarse silt-sized peloids (see discussion below under heterolithic stratification). As a general guideline, peloidal fine-sands and coarse silts are commonly carried and deposited as bed load whereas finer-sized peloids are deposited from suspension. However, it must be noted that in general the hydraulic behavior of silt-sized particles is not that well understood and varies with the flow. The hydraulic behavior of peloids in this size range is even less well understood.

Occurrence, preservation, and significance.-Strata without internal cross-stratification constitute the major kind of layering found in carbonate sediments and rocks, particularly in modern and ancient peritidal carbonate deposits.

One of the most important rock types found in ancient shallow marine carbonate deposits are laminated mudstones (commonly referred to as laminites) that contain: (1) exposure features (mudcracks, desiccation fenestrae, and the like), and (2) some kind of evidence that microbial mats covered the depositional surface at the time of deposition. Such cryptmicrobial laminites (also known as cryptalgal laminites, planar stromatolites, stromatolitic laminites, and so on) have long been recognized as recording deposition in periodically exposed mudflats in the intertidal and supratidal zones (see further discussion below under Biogenic Structures). In thick successions of carbonate rocks, this rock type fixes a general peritidal environmental setting and acts as a marker for recognizing the boundaries of the shallowing-upward cycles for which shallow marine carbonates are renowned (James, 1984; Hardie and Shinn, 1986; and see under Important Associations below). Hardie and Ginsburg (1977) referred to these types of rocks as the "laminite caps" that commonly overlie fossiliferous subtidal and lower intertidal deposits within shallowing-upward cycles.

It is a most common practice in identifying ancient peritidal deposits to use *any* laminated carbonate as evidence of deposition on cyanobacterial mat-covered tidal flats. Such an approach is fraught with problems. First, a number of different mechanisms not involving cyanobacterial mats can produce laminated deposits in any carbonate environment, as described in Table 1. Therefore, it is essential to look for independent evidence that a cyanobacterial mat covered the surface. Unfortunately, such evidence (e.g., "jelly roll" structures, cyanobacterial "tufas", filament molds and casts, and so on) is difficult to find (see discussion under Biogenic Structures). In this regard, it is important to recognize that the morphology of the laminae is not an unequivocal indicator of former cyanobacterial mats (see Table 1) as is well exemplified by modern microbial deposits which carry laminae that range from planar to highly crinkled (see, for example, Hardie and Ginsburg, 1977). These modern cryptmicrobial laminites also show a considerable range in the texture of the sediments making up the laminae, which may consist of sands, muds, or mixed sediment types. Secondly, even where it can be shown that cyanobacterial mats were involved in the construction of the laminae, their presence is not *a priori* evidence of a tidal flat setting. Cyanobacterial mats will thrive in any shallow water setting as long as browsing and deposit feeding organisms are kept at bay (Garrett, 1970, 1977). In modern tidal flats, it is the exclusion of cyanobacteria-consuming organisms due to salinity variations (e.g., the subtidal embayment of Hamelin Pool, Shark Bay) or to extended periods of subaerial exposure (e.g., the supratidal levees of Andros Island, Bahamas) that allows the mats to thrive. In the Precambrian before the evolution of invertebrate grazers, any and all shallow water depositional environments were accessible to the development of cryptmicrobial laminites (Garrett, 1970). All in all, the presence of laminites in a carbonate deposit, whether cryptmicrobial or not, is not sufficient evidence to identify an ancient tidal flat deposit. What is necessary is to show that the deposit underwent periodic subaerial exposure (see under Mudcracks below). It is the combined weight of (1) the internal features of the

TABLE 1

LAYER THICKNESS	LAYER GEOMETRY	COMPOSITION	POSSIBLE ORIGINS[1]	FIGURE[2]
LAMINAE < 5 mm	**planar fine laminae** (parallel boundaries to low flat lenses where wavelength (λ) = 100 laminae thickness (*h*): *must be observed in 3 dimensions*)	mud or mudstone	1) settle-out from suspension	**16**
		sand or grainstone	1) settle-out from suspension (lee-effect grainfall due to flow separation, Hunter, 1977) 2) unidirectional flows - upper plane bed -air or water - lower plane bed - coarse sand 3) oscillatory flows - upper plane bed 4) swash cross-stratification (Harms and others, 1975) 5) climbing wind ripples (Hunter, 1977) 6) flow perpendicular cuts of wave-ripple cross-strata	**17A** **17C** **17B**
		alternating sand and mud or grainstone-mudstone	1) gentle traction currents alternating with suspension settle-out of mud (*silt-streaked, sand-streaked, tidal bedding, distal storm deposits*, etc.) 2) as in 1 but with microbial mat influence 3) settle-out from alternating sources ("varves")	**18, 70C**
	wavy fine laminae (λ>> *h*, & starved ripple-forms)	mud or mudstone	1) settle-out from suspension draping irregularities 2) settle-out from suspension draping irregularities + agglutination by microbial mat	**19**
		sand or grainstone	1) small-scale cross-stratification - random cuts through small-scale cross-strata - critical to supercritical angle of climb 2) wave-ripple cross-stratification, various cuts 3) hummocky and swaley cross-stratification 4) antidune cross-stratification	**17C** **13D,E,F**
		alternating sand and mud or grainstone-mudstone	1) bedload deposition alternating with mud settle-out - sand-streaked (waves and storm deposits) - lenticular bedding (tidal bedding) unidirectional flows 2) as in 1 but with microbial mat influences	**21A,B**
	crinkled fine laminae (λ> *h*, gravity defying)	mud or mudstone	1) settle-out from suspension+/- microbial agglutination or baffling 2) intrasediment growth of evaporites and/or compaction around evaporite minerals	**22A**
		sand or grainstone	1) dewatering structures (*e.g. dish structures*) 2) soft sediment deformation 3) baffling by microbial mats 4) intrasediment growth of evaporites and/or compaction around evaporite minerals	**22B**
		alternating sand and mud or grainstone-mudstone	1) microbial agglutination +/- desiccation 2) baffling by microbial mats 3) soft sediment deformation 4) intrasediment growth of evaporites and/or compaction around evaporite minerals	**22C,D**

[1] An x-radiograph or acid etching may be needed to accurately assess structures present.

[2] Figures do not necessarily correlate to any particular origin. No photograph indicates no example. Please refer to figure captions.

TABLE 1 CONTINUED

LAYER THICKNESS	LAYER GEOMETRY	COMPOSITION	POSSIBLE ORIGINS	FIGURE
LAMINAE > 5 mm and THIN BEDS < 100mm *(Shaley and dolomitic partings common between layers and sets of layers)*	**planar thin beds** (parallel boundaries to long, flat lenses, to slight undulations of layer boundaries)	mud or mudstone (beware squashed peloids)	1) settle-out from suspension 2) bioturbation	**23**
		wackestone	1) bioturbation 2) liquification of graded layer 3) debris flow	
		sand, grainstone, or packstone	1) grain flow 2) bioturbation 3) liquification 4) climbing wind ripples (Hunter, 1977)	
		graded	1) decelerating (normally-graded) or accelerating (reverse-graded) flows: typically grainstone to mudstone, basal grainstone with planar to wavy internal laminae. a) truncated to complete Bouma sequences = turbidity deposits b) with hummocky cross-strata and wave ripples = storm deposits c) other = storm deposits on shallow shelf (?) 2) settle-out from suspension (cannot be too thick)	**24A,B**
		alternating sand and mud or grainstone-mudstone	1) gentle traction currents alternating with suspension settle-out of mud (*silt-streaked, sand-streaked, tidal bedding, distal storm deposits*, etc.) 2)settle-out from alternating sources ("varves")	**25A,B** **24B**
		mud or mudstone wackestone	1) compaction or pressure-solution modification of above	
	wavy thin beds	sand, grainstone or packstone	1) compaction or pressure solution modification of above 2) cross-strata too well sorted to see individual cross sets	
		alternating sand and mud or grainstone-mudstone	1) flaser, wavy and lenticulat bedding (Reineck and Singh, 1980): may be modified by burrowing, compaction and pressure solution	**26, 27A, 28**
	crinkled thin beds	all textures	1) soft sediment deformation 2) modification by burrowing, compaction and pressure solution (*nodular bedding, lumpy bedding*, etc.)	**27B,C**

TABLE 1 CONTINUED

LAYER THICKNESS	LAYER GEOMETRY	COMPOSITION	POSSIBLE ORIGINS	FIGURE
THICK TO VERY THICK BEDS (> 100 mm) (Generally parallel sided lenses, channels, or bedforms: lateral layer boundaries abrupt to gradational)	**planar to wavy thick beds** (parallel boundaries to long, flat lenses, to slight undulations of layer boundaries)	mud or mudstone (beware squashed peloids)	1) settle-out from suspension 2) bioturbation of layered sequence	
		wackestone	1) debris flow 2) bioturbation 3) direct sediment production by *in situ* organisms with associated burrowers	
		sand, grainstone, or packstone	1) grain flow 2) bioturbation 3) liquification and dewatering	
		graded	1) decelerating (normally-graded) or accelerating (reverse-graded) flows: typically grainstone to mudstone, basal grainstone with planar to wavy internal laminae. a) truncated to complete Bouma sequences = turbidity deposits b) with hummocky cross-strata and wave ripples = storm deposits c) other = storm deposits on shallow shelf (?) 2) settle-out from suspension (cannot be too thick)	
	crinkled thick beds	all textures	1) soft sediment deformation 2) modification by burrowing, compaction and pressure solution (*nodular bedding, lumpy bedding*, etc.)	

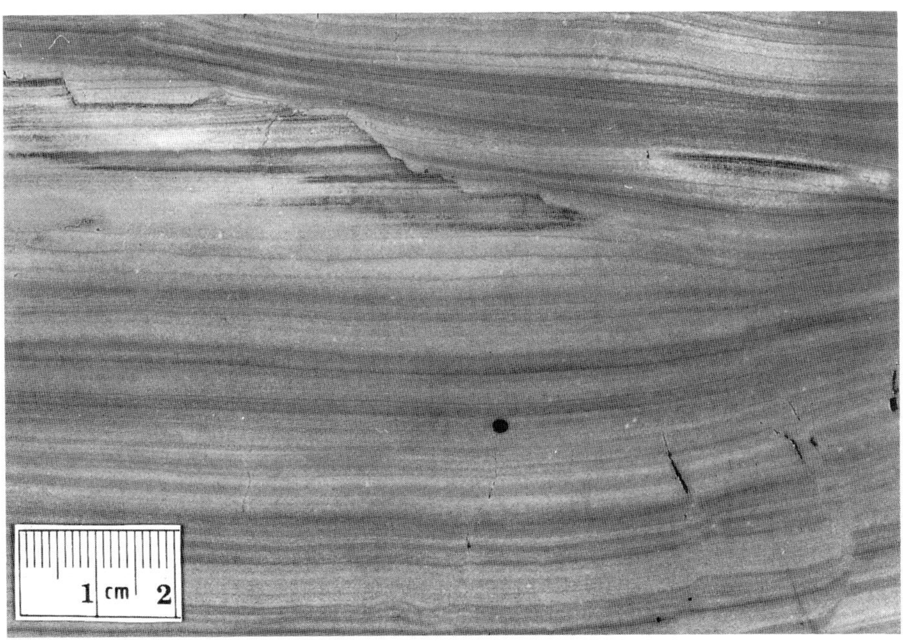

Fig. 16.-Parallel fine laminae in a dolomitic mudstone. Upper Cambrian, Grand Canyon, Arizona. Sample collected by Hal Wanless.

Fig. 17.-Parallel fine laminae in carbonate grainstones. (A) Skeletal-peloidal grainstone from Upper Cambrian Conococheague Formation, central Appalachians. Scale in mm. The discontinuous, lens-like geometry of the laminae suggest upper-stage "plane bed" deposition from unidirectional flows. (B) Ooid grainstone from lithified subrecent beach deposits, Joulters Cay, Great Bahama Banks. Gastropod shell is approximately 20 mm long. (C) Parallel fine laminated peloidal grainstone (at the base) grades up to wavy fine laminae, which in turn are capped by burrowed mudstone. This fining-upward set is interpreted as a storm deposit ("tempestite"). Lower Ordovician Stonehenge Limestone, central Appalachians. Metric scale bar in upper left corner.

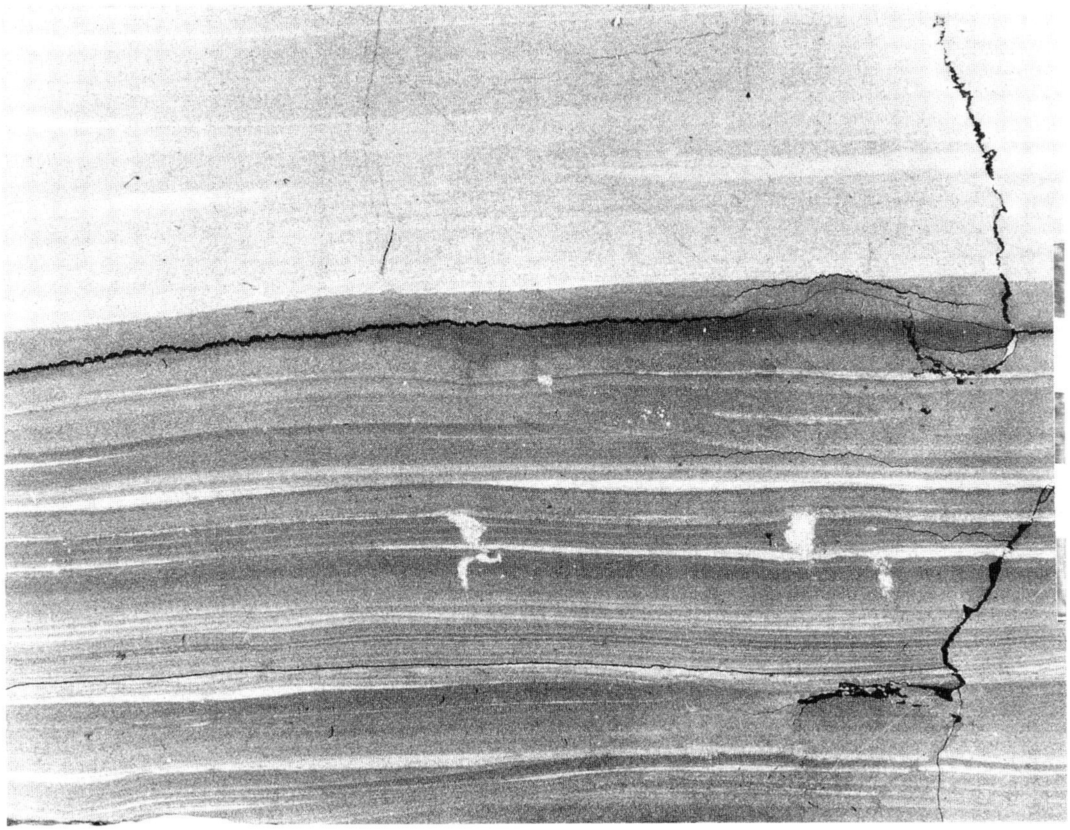

Fig. 18.-Planar fine laminae composed of couplets of very fine grainstone (light streaks and lenses) alternating with continuous mudstone (dark) layers. Middle Ordovician St. Paul Group, western Maryland. Note the mudcracks. See Mitchell (1985, Fig. 14, p. 23). Photograph courtesy of Ray Mitchell. Scale bar marked in centimeters.

Fig. 19.-Wavy fine laminae in carbonate mudstones. Note that the samples comprise composite sets of different lamination styles up to a few tens of millimeters thick. This interlayering of sets of different laminae geometries is common, if not typical, of laminites in ancient shallow marine carbonates. (A) Alternating sets of parallel fine laminae, wavy fine laminae, and crinkled fine laminae. Dolomitic mudstone from Upper Cambrian Waterfowl Formation, Canadian Rocky Mountains, southwestern Alberta. Scale bar is 50 mm long.

B

C

Fig. 19(continued).-(B) and (C) Sets of parallel and wavy fine laminae from dolomitic mudstones of the Upper Cambrian Conococheague Limestone, western Maryland. Note the mudcracks in (B), scale in mm. Scale bar in (C) is 50 mm long.

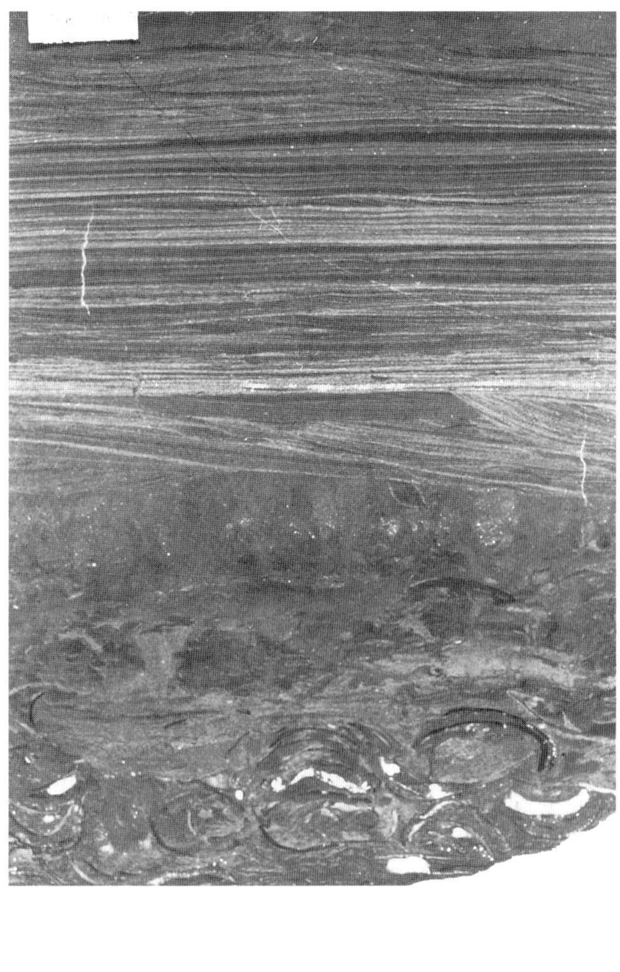

Fig. 20.-Wavy fine laminae in peloidal grainstone overlying burrow disrupted skeletal packstone, Lower Devonian Helderberg Group, central New York. Compare with the upper portions of Figure 17C. Scale bar is 20 mm long.

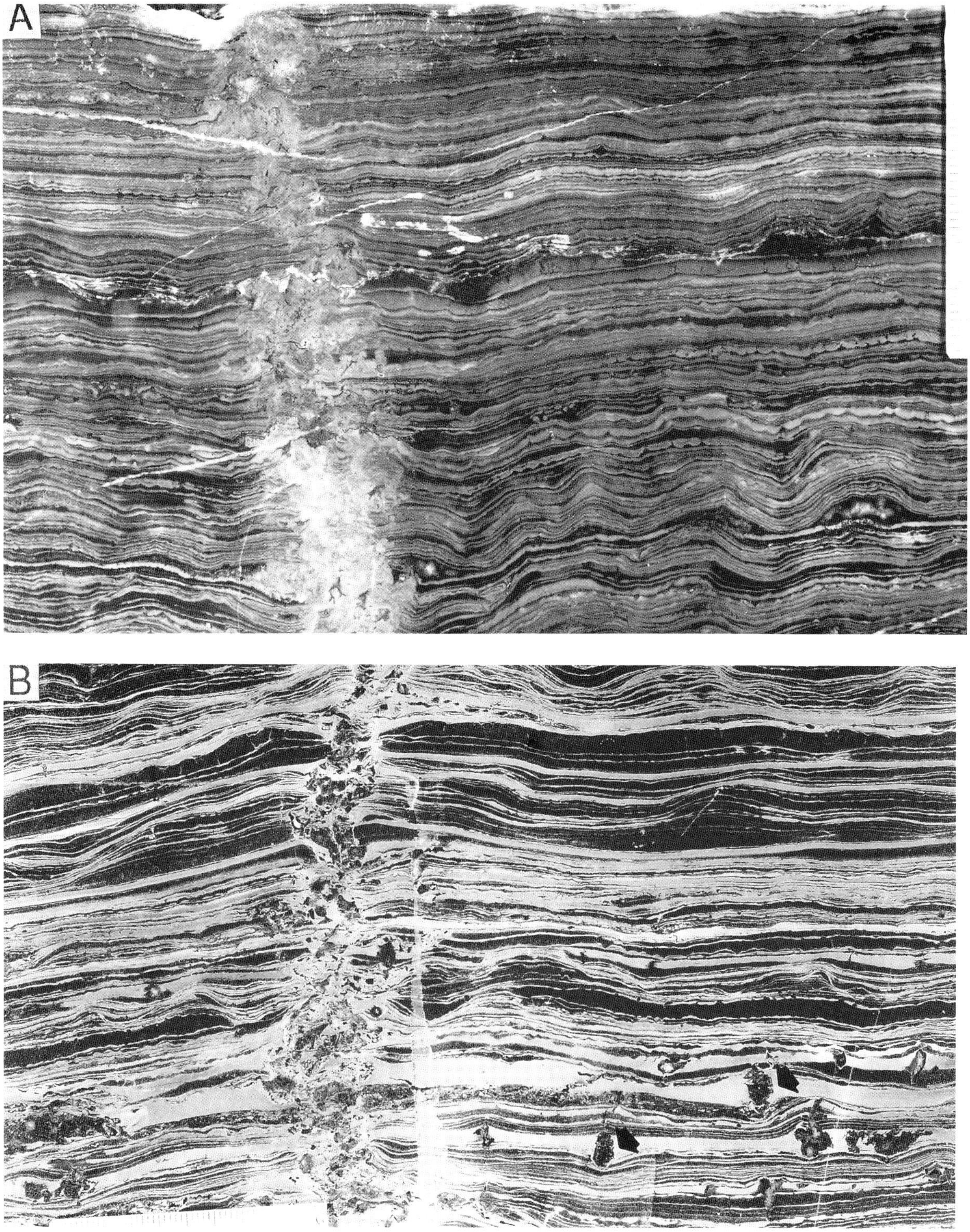

Fig. 21.-Wavy to crinkled fine laminae composed of couplets of grainstone layers (dark) and dolomitic mudstone layers (light). Both samples from the Upper Cambrian Conococheague Limestone of the central Appalachians, scale bar is 50 mm long in both examples. Note the deep prism cracks in both samples, and the smaller mudcracks (arrows in lower right-hand corner) in the thicker dolomitic mudstone laminae in (B).

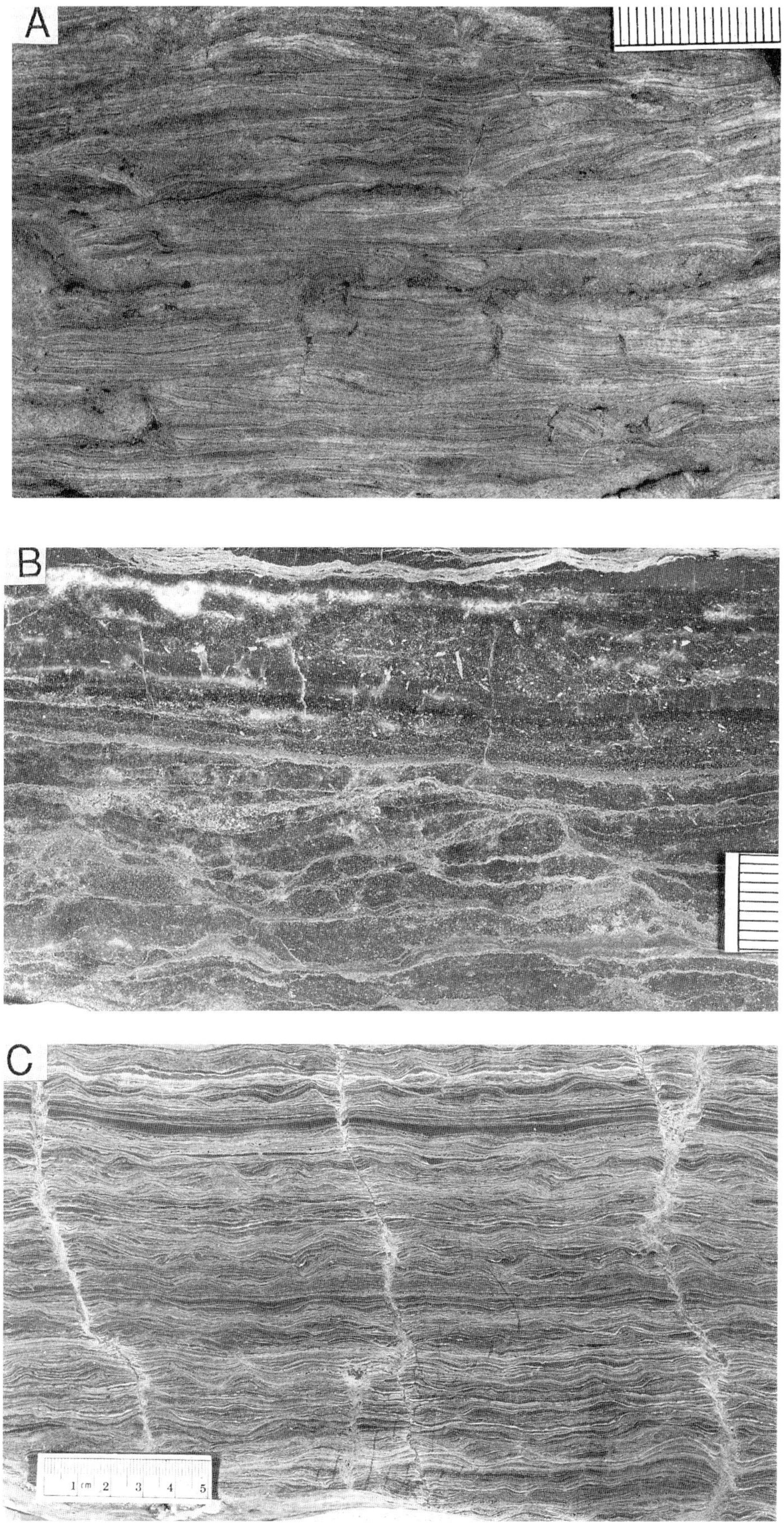
A
B
C
1 cm 2 3 4 5

Fig. 22.-Crinkled fine laminae in carbonates of various textures. (A) Crinkled fine laminae in carbonate mudstone from the Upper Cambrian of the Grand Canyon, Arizona. Note the small mudcracks. Small divisions in the scale bar are mm. Sample collected by Hal Wanless. (B) Crinkled fine laminae developed in a fine-grained peloidal grainstone. Layer boundaries in the lower and uppermost sets have been modified by differential compaction of the mudstone partings giving a stylolite-like morphology to the layering. Upper Cambrian of the Grand Canyon, Arizona. Sample collected by Hal Wanless. (C) and (D) Crinkled fine laminae composed of couplets of grainstones (dark) and dolomitic mudstones (light) from the Upper Cambrian Conococheague Limestone, western Maryland. Note the deep prism cracks in both examples. Scale bar in (C) is 50 mm long. Scale in (B) is in mm.

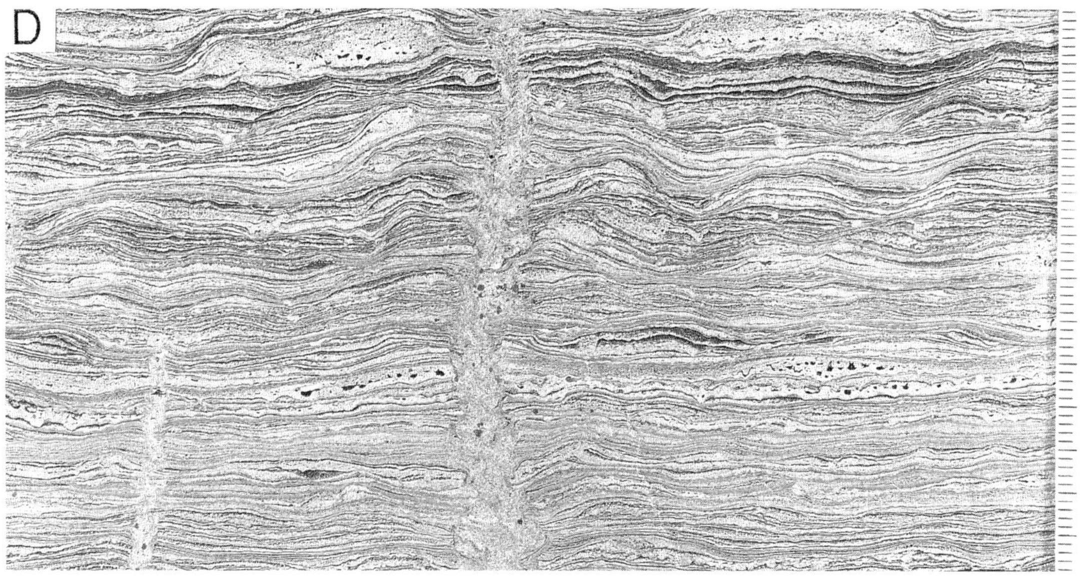

Fig. 23.-Planar laminae (greater than 5 mm thick) and thin beds developed in dolomite mudstone, Lower Proterozoic Wittenoom Dolomite, Hamersley Basin, Western Australia. Pocket knife, lower left center, is approximately 80 mm long.

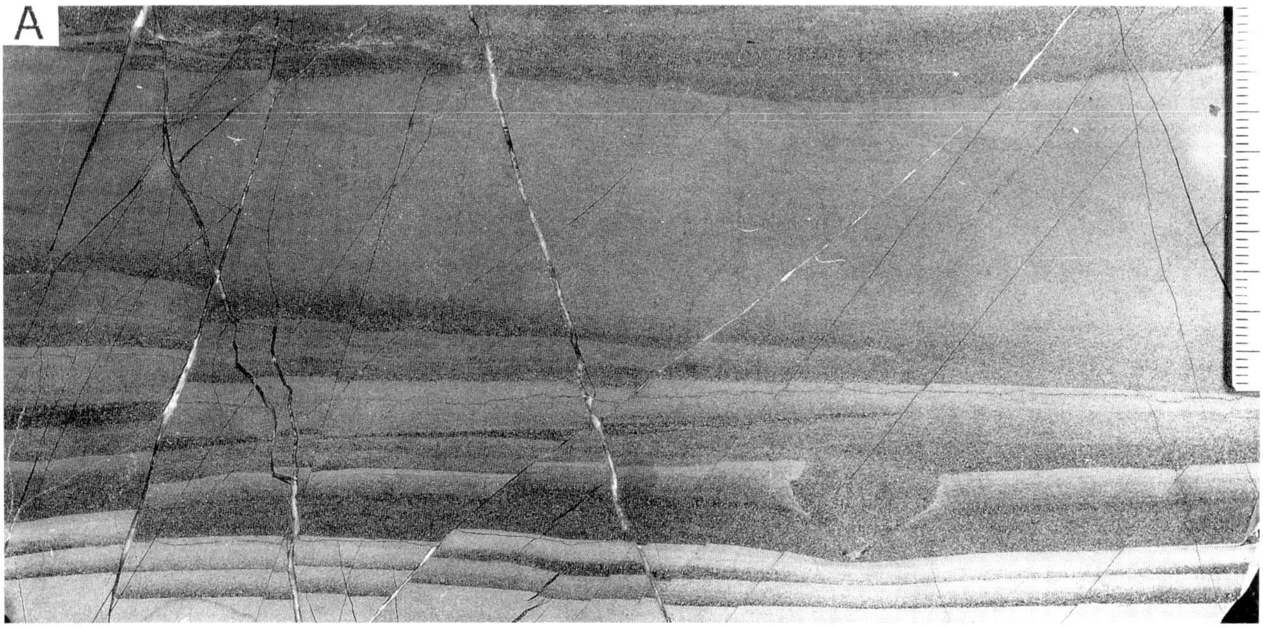

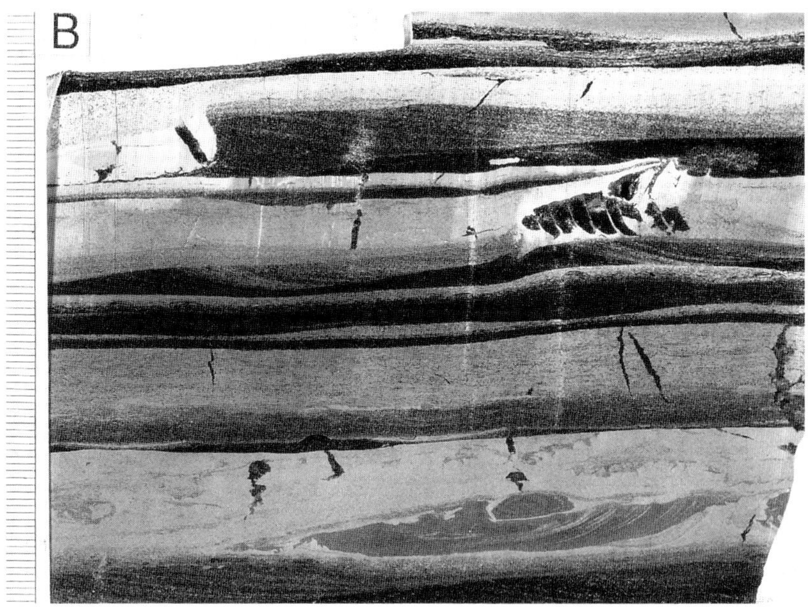

Fig. 24.-Graded planar laminae (greater than 5 mm thick) and thin beds. (A) From the Upper Cambrian Conococheague Limestone, central Appalachians. Fine peloidal grainstones (dark) grade up to light-colored dolomite. Note the erosional scour in the lower right. Scale bar in mm. Spar-filled fractures are late tectonic deformation features. (B) From the Upper Silurian Wills Creek Shale, central Appalachians. Medium- to fine-peloidal grainstones (dark) grade up to light-colored, mudcracked dolomite layers. Note the small-scale cross-stratification in some layers, and the load cast in the lower right. Scale in mm. Sample collected by Tom Tourek.

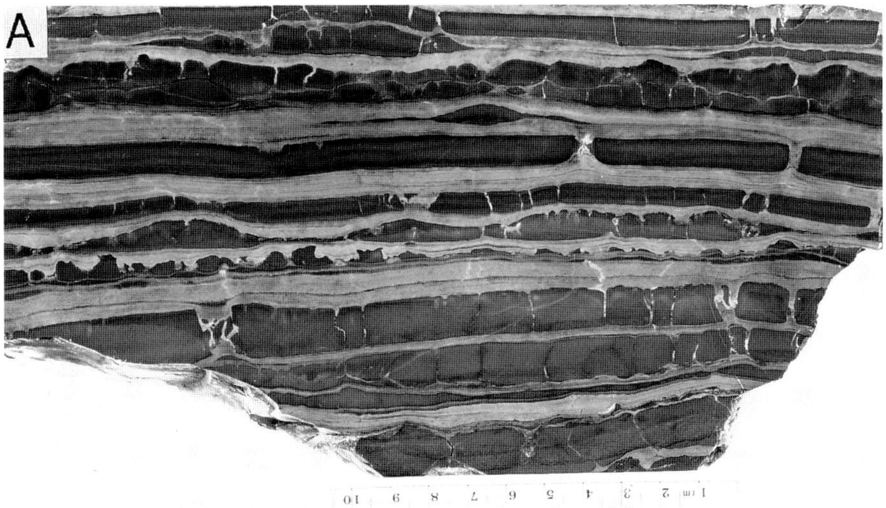

Fig. 25.-Planar and wavy laminae (greater than 5 mm thick) and thin beds. (A) Dark layers are fine peloidal grainstones whereas the light layers, some of which are laminae sets, are dolomite mudstones. Sample from the Middle Ordovician St. Paul Group, central Appalachians. Note the ripple-forms of some of the limestone layers. The numerous cement-filled fractures in the limestone layers that propagate from the tops and bottoms of layers, the fractures in the limestone layers filled by laminated dolomite from above and below, and the irregular upper surface of some of the limestone layers are mechanical compaction features. Scale is in cm. Sample collected by Ray Mitchell. (B) Outcrop view of planar and wavy laminae and thin beds of the Upper Cambrian Waterfowl Formation, Canadian Rocky Mountains, southwestern Alberta. Matchbook for scale is approximately 40 mm high. Photograph courtesy of Brent Waters.

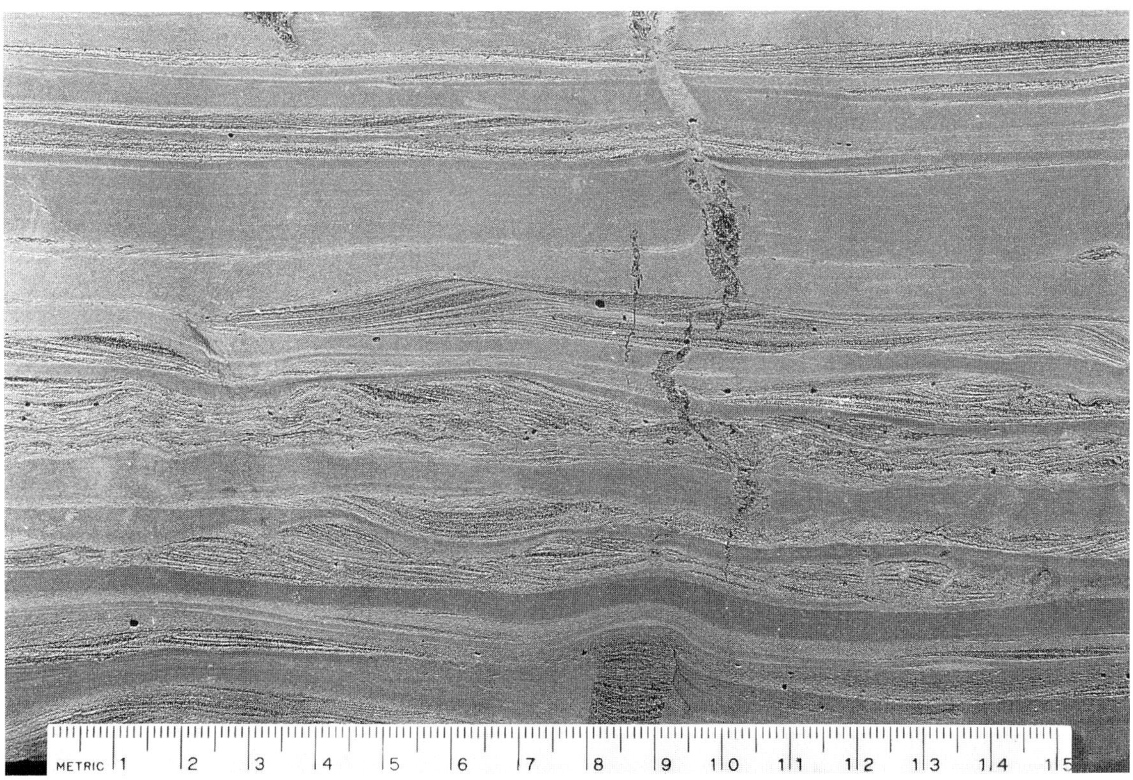

Fig. 26.-Flaser, wavy and lenticular bedded "ribbon rocks" of the Upper Cambrian Conococheague Limestone, western Maryland(see Demicco, 1983). Dark colored peloidal grainstones alternate with dolomitic mudstones. Note sediment-filled mud-cracks. Metric scale in cm and mm.

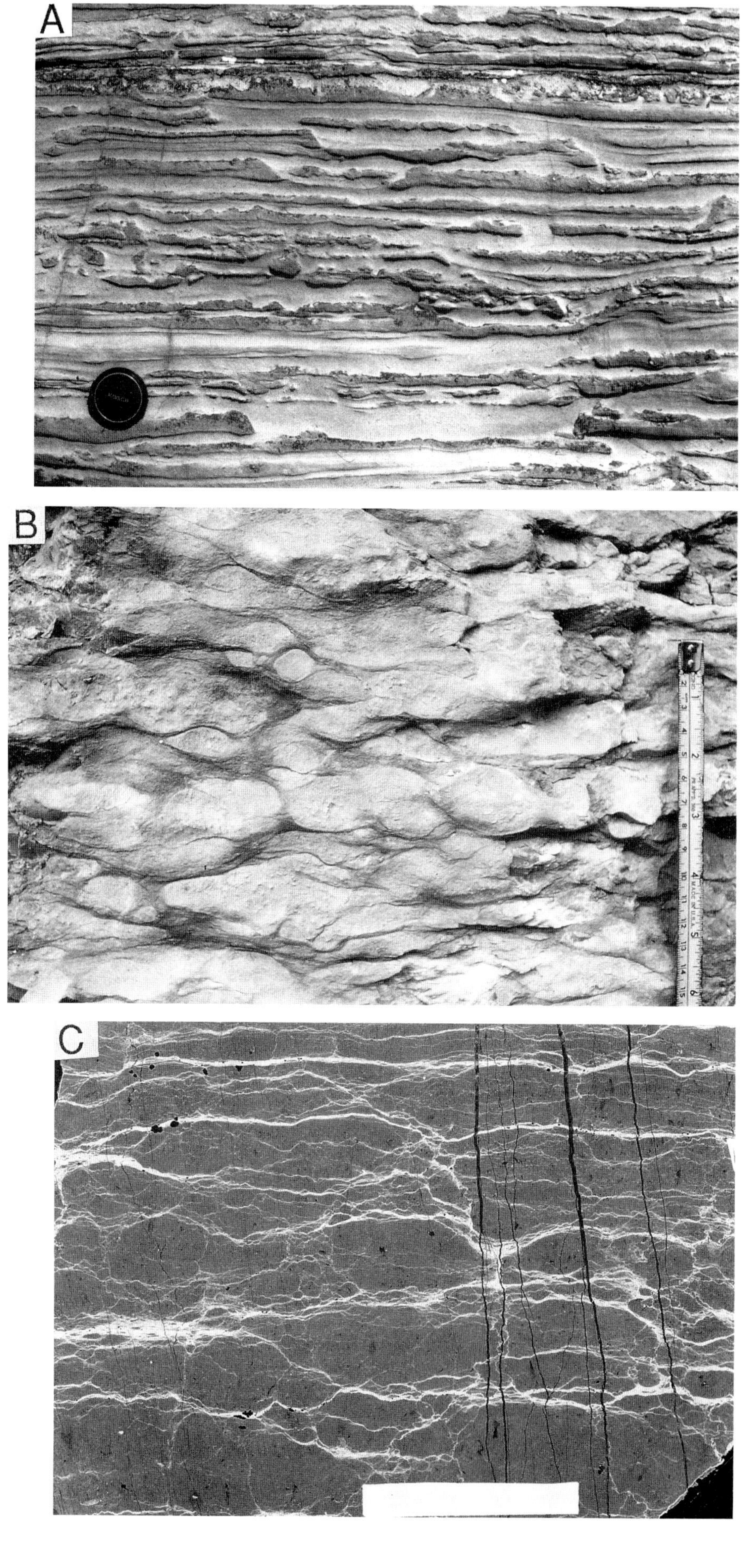

Fig. 27.-Discontinuous wavy to crinkled laminae (greater than 5 mm thick) and thin beds (a.k.a. "lumpy bedding", "nodular bedding", "sedimentary boudinage", etc., see text). (A) is from the Middle Ordovician St. Paul Group, central Appalachians. Lens cap 50 mm in diameter. Dark gray lime mudstone layers and pods are separated by lighter colored dolomitic layers. (B) is from the Jurassic Ammonitico Rosso of northeastern Italy. Light colored lime mudstone nodules are surrounded by dark colored siliciclastic-rich carbonate mudstone that is reddish on outcrop. (C) is from the Lower Devonian Helderberg Group, central New York. This sample comprises a fitted network of irregularly-shaped mudstone nodules separated by light gray dolomitic seams, and "horsetail" stylolites. Scale bar is 50 mm long. Sample courtesy James Kradyna.

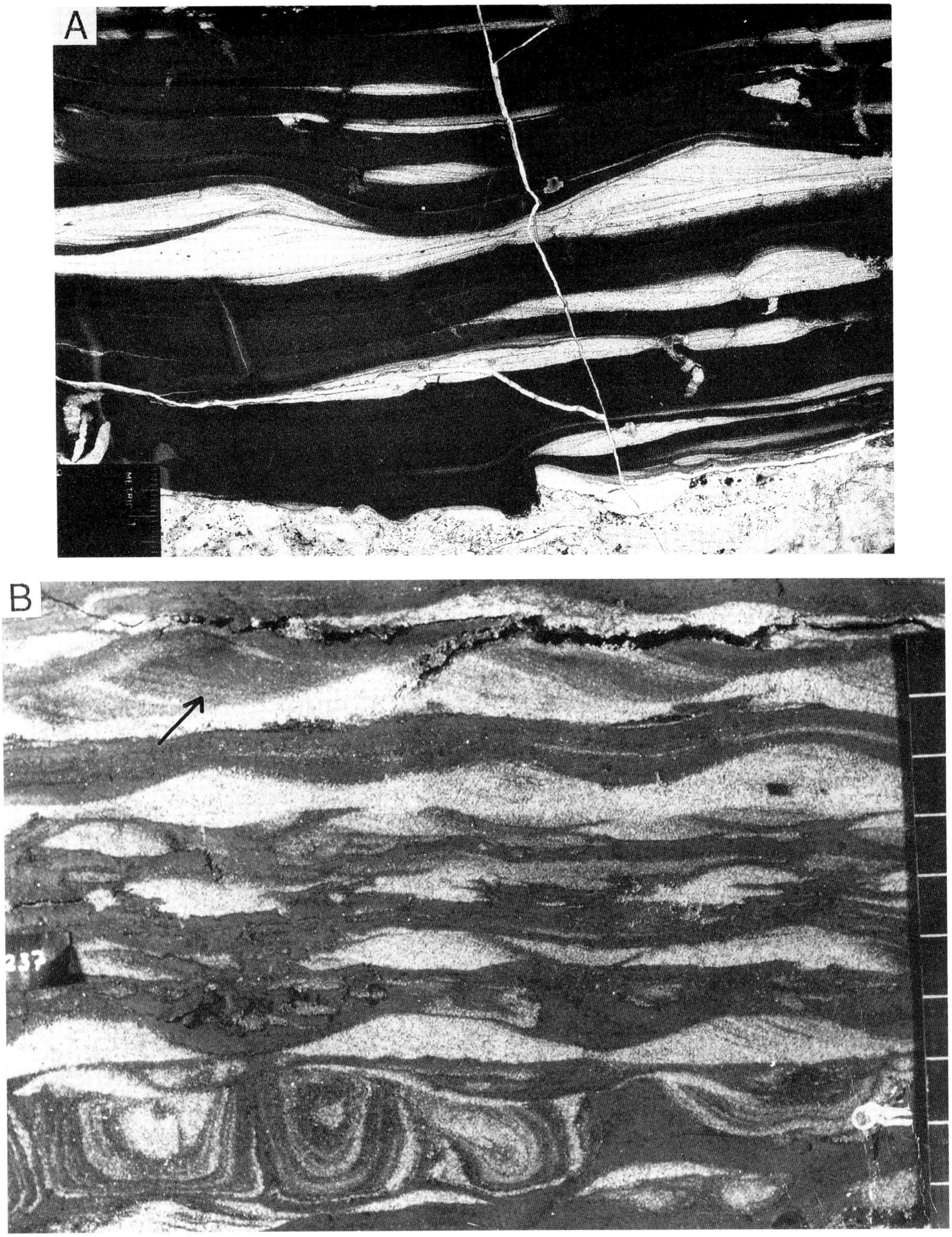

Fig. 28.-Wavy to lenticular bedding. "Ribbon rock" from Upper Cambrian Conococheague Limestone, central Appalachians, (A) (see Demicco, 1983), compared with modern siliciclastic wavy bedding from the North Sea, (B) (courtesy of Hans Erich Reineck, see Reineck and Singh, 1980, Fig. 195, p. 117). (A) Light lime grainstone layers have flat bottoms, ripple-form tops, and are draped by dark dolomitic mudstones which thin over ripple crests and thicken over ripple troughs. Dolomite layers are sets of planar to wavy fine laminae in some cases, and are mudcracked. Scale bar in lower right hand corner is 20 mm high. (B) Note the flat-topped lower set of ripple marks, sand grading into the mud drape at the arrow, and the convolute laminae at the base of the sample. Scale in cm.

laminites, (2) evidence that a cyanobacterial mat covered the surface, and (3) the demonstration that the sediments had undergone periodic subaerial exposure, that is critical to the recognition of ancient intertidal-supratidal deposits. Once recognized, such tidal flat deposits are important paleo-sea-level gauges and markers of depositional cycle boundaries (see later under Important Associations).

Thin-bedded carbonates are among the most commonly encountered rocks in the ancient carbonate platform deposits of Cambrian, Ordovician, and Silurian age in North America that we have studied. They are, however, the least well known of any bedding type. Part of this no doubt stems from lack of a clear descriptive terminology as discussed above. Another problem is the paucity of modern analogs, due in large part to rapid destruction of layering by the extensive bioturbation typical of most modern subtidal environments (see, for example, Garrett, 1977). Finally, diagenetic processes such as mechanical compaction, pressure solution, and dolomitization, can severely modify the primary sedimentary features of thin-bedded carbonates (see section on Mechanical Compaction Structures). In the deposits we have studied, planar and wavy thin beds predominate. In many of these deposits, limestone layers alternate with dolostone layers (Demicco, 1983) and this has led to the use of a variety of local field terms such as ribbon rocks, ribbon limestones, and tiger-striped limestones. Commonly, in these types of thin-bedded carbonates, the limestones are peloidal grainstones while the dolostones are selectively replaced mudstones so that such rocks properly belong in the heterolithic stratification category discussed in the next section.

Although some thin-bedded carbonates are mudcracked indicating deposition under intertidal-supratidal conditions, most types of thin beds carry no evidence of subaerial exposure and thus appear to be subtidal deposits. Modern supratidal thin-bedded carbonates in Florida and the Bahamas (Enos and Perkins, 1979; Hardie and Ginsburg, 1977) are demonstrably the products of strong hurricanes and it is extremely likely that most ancient thin-bedded carbonates in shallow marine facies, both supratidal and subtidal, originated as storm deposits ("tempestites," compare with Aigner, 1985). However, care must be exercised in the interpretation of thin-bedded carbonates because a number of other possible origins exist as outlined in Table 1. We hope that this table, together with the discussion of diagenetic overprinting of thin-bedded carbonates given in the section Mechanical Compaction Structures below, will help refocus the attention of sedimentologists on this abundant but long neglected suite of bedded carbonates.

Heterolithic stratification.-

Terminology.-Interlayering of sands and muds and their lithified equivalents (heterolithic bedding, heterolithic rocks) occurs on outcrop scale (meters to decimeters), hand sample scale (decimeter to millimeter), and thin section scale (millimeters to micron). Thick interbeds at the outcrop scale are probably more correctly regarded as subfacies (see below under Important Associations) than as individual sedimentary structures and should be dealt with accordingly. Here, we are mainly concerned with hand-sample and thin-section scale heterolithic stratification wherein layers are generally up to a few tens of millimeters thick. There is no commonly accepted classification for heterolithic layering. Reineck and Singh (1980) recognize flaser bedding, wavy bedding, lenticular bedding, coarsely interlayered bedding, and thinly interlayered bedding, which they also refer to as rhythmites. de Raaf and others (1977) introduced the terms silt-streaked and sand-streaked muds. Leeder (1982) uses the terms flaser lamination, wavy lamination, lenticular lamination, and streaky lamination.

Distinguishing characteristics and origin.-Flaser, wavy and lenticular bedding or lamination were terms introduced by Reineck (1960) (see also Reineck and Wunderlich, 1968) for interbedded sands and muds where the sands are current-ripple or wave-ripple cross-stratified on a small-scale (Figs. 26, 28A). Generally, the sand layers are composed of ripple form sets, and the muds drape the ripple form crests, thinning over the crests and thickening into the troughs (Fig. 28). However, many published examples of flaser, wavy and lenticular bedding have little, if any, discernible ripple form set geometry or internal small-scale cross-stratification (Figs. 29A, B) (see also Figs. 3 and 5 in Reineck, 1972, and Fig. 4B in Demicco, 1983). Disruption of the layering in these modern examples appears to be due to burrowing, soft sediment deformation, and mudcracking. We would interpret ancient examples in a similar manner (Fig. 30). Sand-streaked and silt-streaked muds are gradational into lenticular and wavy bedding but in this type of heterolithic stratification the sands layers have parallel to slightly undulatory upper and lower contacts and internal laminae that are parallel to undulatory. It is important to note that interlayered sands and muds must be examined in three-dimensions before they can be classified under these schemes because flow-perpendicular views of current and wave rippled sand layers give parallel sided layers of muds in sands that strongly resemble sand streaked muds. Sand streaked and silt streaked muds are roughly equivalent to Reineck and Singh's (1980) thinly interlayered rhythmites and are also dealt with under lamination above. Contacts between the interlayers of sands and muds in hand sample and thin section may be either sharp or gradational but most typically the bases of the sand layers are sharp and only rarely scour into the underlying muds (see below). Reineck and Singh (1980) use the term coarsely interlayered bedding where the coarser layers are not rippled. They recognize three categories: (1) sand and mud layers of subequal thickness, (2) thicker sand layers separated by thinner mud layers, and (3) thicker mud layers separating thinner sand layers. This categorization has not gained widespread usage.

Reineck (see Reineck, 1960; Reineck and Wunderlich, 1968; Reineck and Singh, 1980) explains flaser, lenticular and wavy bedding as follows. When tidal flats are covered during ebbing or flooding tides, currents or waves produce rippled sands and silts. During slack tides or calm seas, mud suspended in the water column settles out, burying the ripples to various degrees. The mud compacts and is thus not easily eroded during the next current cycle. McCave (1970, 1971) and Hawley (1981) have disputed this explanation based on the fact that prohibitively large volumes of mud would have to be suspended in the water column to produce a centimeter thick bed during a one-half to one hour slack tide period. Hawley's (1981) flume experiments further suggest that slack water periods do not allow enough time for adequate compaction of settle-out mud to successfully resist erosion of the next tidal cycle. Both McCave and Hawley suggest that longer period processes such as storm cycles or neap spring tidal fluctuations are responsible for flaser, wavy and lenticular bedding. However, the sediment dyeing experiments of Reineck and Wunderlich (1967a, b, 1969) clearly demonstrate that interbeds of sand and mud can accumulate over a tidal cycle (see also Allen, 1984, p. 511). We suggest that the answer lies in the peloidal nature of the mud fraction. Peloids as aggregates of clay-sized sediment behave hydraulically as silt- and sand-sized grains during transport and deposition but they are easily compacted after deposition to produce a clotted mud ("structure grumeleuse"). In a thin section of modern lenticular bedding supplied by Reineck (Fig. 10 in Demicco, 1983), the peloidal nature of the mud interlayers is evident (see also Reineck and Singh, 1980, p. 116-117). More-

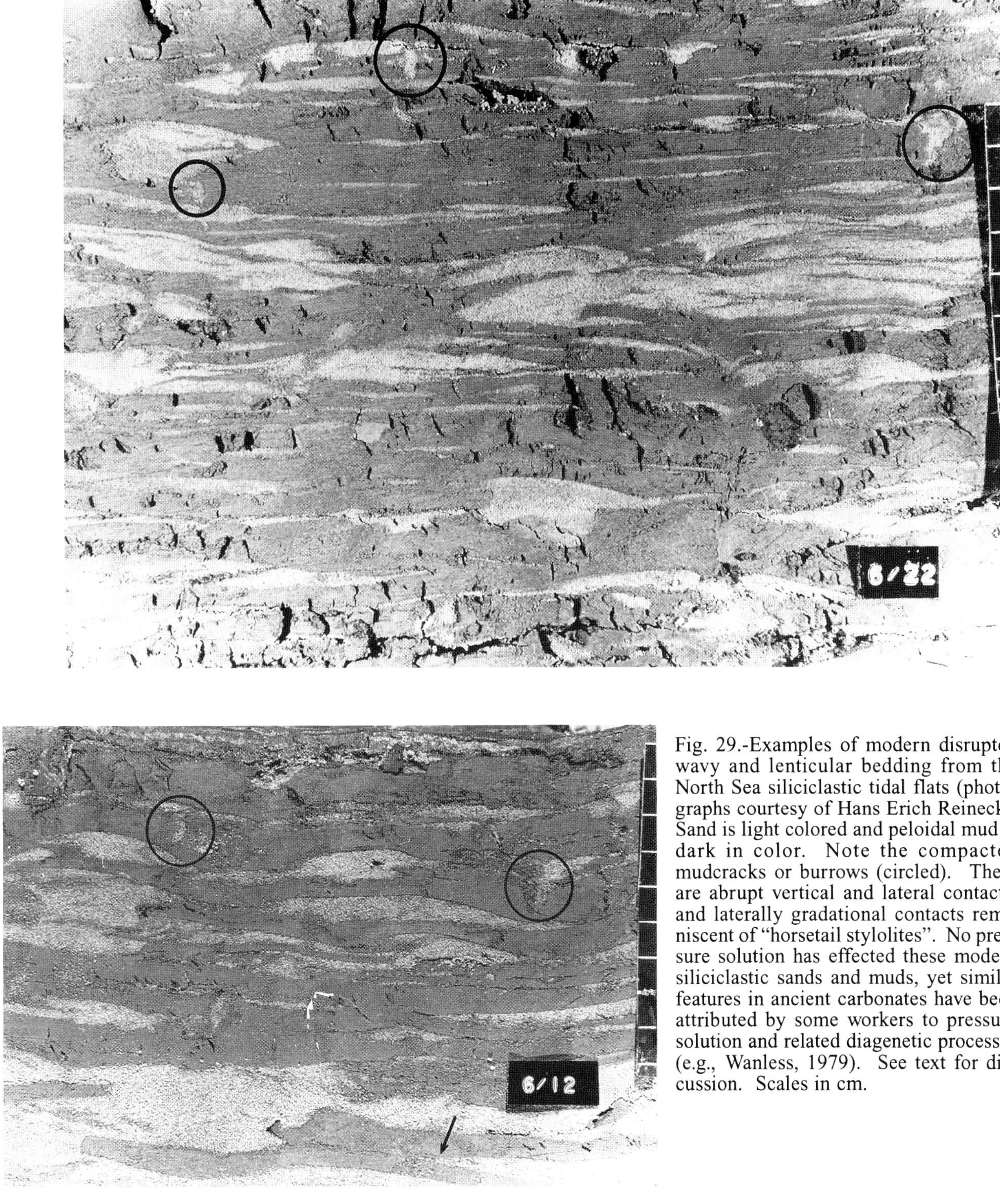

Fig. 29.-Examples of modern disrupted wavy and lenticular bedding from the North Sea siliciclastic tidal flats (photographs courtesy of Hans Erich Reineck). Sand is light colored and peloidal mud is dark in color. Note the compacted mudcracks or burrows (circled). There are abrupt vertical and lateral contacts, and laterally gradational contacts reminiscent of "horsetail stylolites". No pressure solution has effected these modern siliciclastic sands and muds, yet similar features in ancient carbonates have been attributed by some workers to pressure solution and related diagenetic processes (e.g., Wanless, 1979). See text for discussion. Scales in cm.

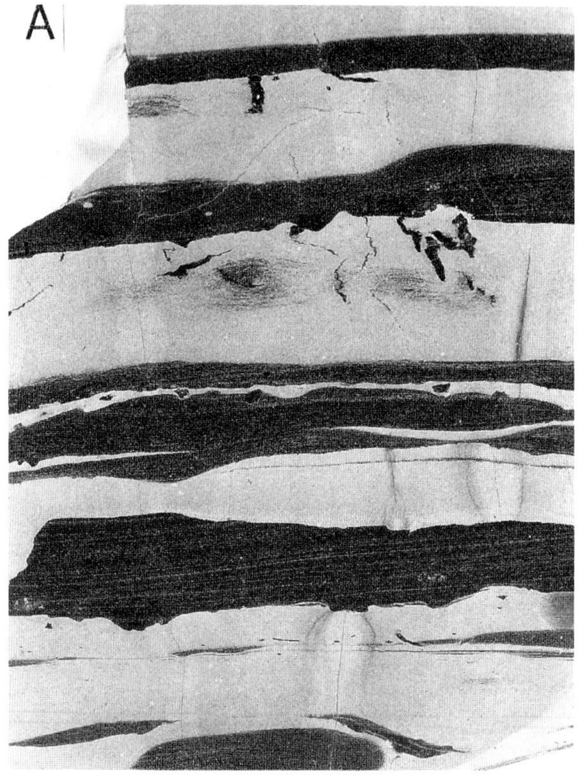

Fig. 30.-Examples of disrupted wavy bedded "ribbon rocks" from the Upper Cambrian Conococheague Limestone, central Appalachians (compare with Fig. 29). In (A) and (B) limestone layers are dark gray whereas dolomite layers are light gray. (C) is a negative print: limestone layers here are light gray whereas the dolomite layers are black. Note the dark, tubular, lime grainstone-filled disruptions interpreted as burrows in (B) and (C). Sample in (A) is 0.25 m high. Scale in (B) is in cm and mm. Dark scale bar in lower left hand corner of (C) is 45 mm high.

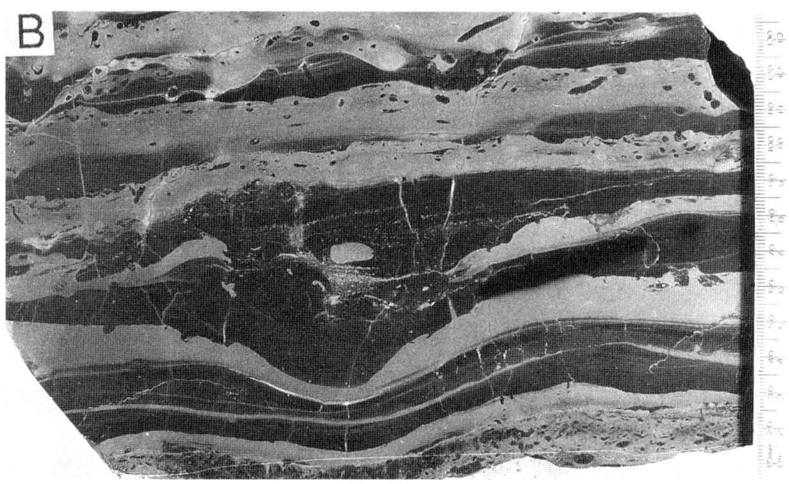

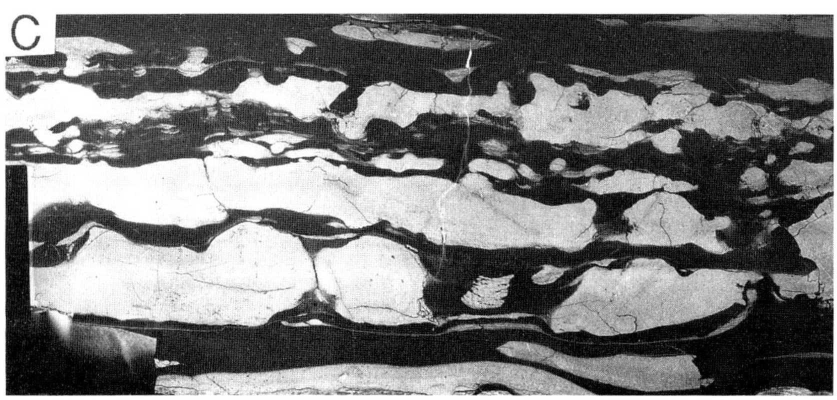

over, quartz grains in the mud layers are well sorted and only slightly smaller than quartz grains in the underlying rippled beds. The North Sea tidal flats are replete with mud peloids which are mainly the fecal and pseudofecal pellets of worms and molluscs. Allen (1987) documented the generation of sand-sized mud intraclasts in the Severn estuary due to reworking of previous muds. Both the pellets and the intraclasts are fragile and so it requires only very slight compaction to cause most of the pellets to merge into a uniform, cohesive mud. Ripples exposed on North Sea and Georgia (U. S. A) tidal flats commonly have collections of fecal pellets in the ripple troughs. Currents or waves produce rippled sands and as the currents or waves wane, slightly less dense fecal pellets of mud in the fine sand to coarse silt size range begin to be deposited with the sand. When the currents weaken further, the last foresets of the ripple are pellets that partially to completely fill the ripple trough (hence the drape), producing flaser and wavy bedding. This is illustrated in Figure 28B (Demicco, 1983, p. 1125) which shows the interlamination of sand and mud in the last foresets of the ripples in one of Reineck's box cores. An ancient carbonate equivalent is illustrated in Figure 13E. The origin of streaky bedding and finely interlayered bedding is not so well known. Parallel or undulatory internal laminae within sand layers and the erosive nature of the lower contacts of the sand layers suggest bed load deposition alternating with settle-out of peloidal muds. This in turn, is commonly associated with storm deposition (de Raaf and others, 1977). There are a number of descriptions of storm layers from modern shelves that resemble streaky muds and coarsely interlayered sands and muds (Reineck and Singh, 1980; Howard and Reineck, 1981; see also Allen, 1984, p. 493-505).

Interlaminated muds and sands on tidal flats can be produced by the preferential trapping of mud by microbial mats during storm flooding and then deposition of sand as the currents wane (Hardie and Ginsburg, 1977). This type of microbial heterolithic lamination is described in the section on cryptmicrobial structures under Biogenic Structures.

Other origins must be called on for interlayered sands and muds found in deep water settings where, in the absence of currents, settle-out from suspension is taking place. Examples of this type of interlayered sand and mud (rhythmites or "varves") are found in: (1) glacial and temperate lakes (Forstner and others, 1968); (2) fjords (Gross and others, 1963); (3) restricted embayments (Burky and others, 1970); and (4) oceanic abyssal plains (Bouma, 1968). In most of these examples, sand to silt layers of tests of planktonic organisms alternate with terrigenous mud layers. The classic interpretation is that the layers of planktonic debris are the result of seasonal "blooms" of the planktonic organisms (see Bradley, 1929a, 1931, for a discussion of "varves" in carbonate lacustrine deposits). However, in the Gulf of California, Calvert (1966) has shown that the rain of diatom tests is constant and the mud comes from episodic flooding of the Colorado River. In abyssal settings next to carbonate banks, planktonic oozes alternate with carbonate turbidites of varying thickness that contain the remains of benthic shallow-water invertebrates (Rusnack and Nesterhoff, 1964; Davies, 1968, 1972; Andrews and others, 1970; Huang and Pierce, 1971; Bornhold and Pilkey, 1971; Crevello and Schlager, 1980). For further discussion, the reader is referred to the sections on storm layers.

Occurrence and preservation.-Heterolithic stratification is quite common in carbonate platform deposits. However, the description and interpretation of interlayered sands and muds in carbonate rocks is complicated by the fact that diagenetic alteration of the original minerals commonly obscures the primary texture of the layers. Interlayers of limestone and dolostone are common in Paleozoic and Proterozoic carbonates where they are known as ribbon rocks, ribbon limestones, or tiger-striped limestones. Despite the fact that these types of carbonates are exceedingly common, their origin remains controversial (Sando, 1957; Aitken, 1967; Matter, 1967; Gebelein and Hoffman, 1973; Wanless, 1979; Read, 1980; Demicco, 1983). The role of mechanical compaction in altering such heterolithic strata is discussed at length in the section on Mechanical Compaction Structures. Dolostones can also have the sedimentary features of interlayered sands and muds (Figs. 13A, 26) but in such rocks the original textures of the layers are generally obscured by the replacement of the primary minerals by dolomite. Finally, interlayers of limestones and shales are also common in platform carbonate deposits. These are commonly interpreted as diagenetic rather than depositional features (for example, Bathurst, 1987) but a primary depositional origin cannot be ruled out.

Significance.-The genetic significance of the various types of heterolithic stratification has been touched on above. Paleocurrent measurements and stratigraphic younging determinations can be made where cross-stratification is preserved.

Breccias and Conglomerates

Terminology.-

Carbonate gravels and conglomerates can be composed of a variety of clast types but here we are mainly concerned with those in which non-skeletal components dominate. Walcott (1894, p. 192) defined intraformational conglomerates as "formed within a geologic formation of material derived and deposited within that formation." Folk (1959, p. 4) defined the term intraclast (adjectival form is intraclastic) as "fragments of penecontemporaneous, usually weakly consolidated carbonate sediment that have been eroded from adjoining parts of the sea bottom and redeposited to form a new sediment." A variety of older, mainly field-oriented, terms of obscure origin have been applied to intraclastic conglomerates. These are, in general, related to the shape or orientation of the clasts within an intraclastic layer (also see below). Flat, tabular clasts with angular edges have been called sharpstone conglomerates and breccias, flakestones, and flat pebble conglomerates, although the last term is also commonly used for conglomerates with rounded, discoidal clasts as well. Flat clasts in some cases stand on edge and are packed together to form fans of "edgewise" conglomerates. Clasts of roundstone conglomerates are subequant and rounded.

Distinguishing characteristics.-

Shallow-marine carbonate intraclastic gravels and conglomerates typically occur as individual layers up to a meter or so thick, most being a few decimeters thick. Layers range from discontinuous, meter-long lenses to continuous beds traceable in some cases for many tens of kilometers (e.g., up to 120 km, as reported by Grotzinger, 1986a, Fig. 10). Layers generally have a bimodal grain size distribution, with an interstitial, normally well-sorted, sand-sized component of peloids, ooids, or skeletal fragments. There is a complete range from clast-support fabrics to isolated clasts within grainstones. Large clasts may support shelter voids. These features of shallow marine carbonate conglomerates are in contrast to those of deep-marine carbonate slope and toe-of-slope conglomerates. These latter are either the result of direct gravity transport (e.g., rockfalls, talus piles, see James and Ginsburg, 1979) or sediment-gravity transport. Sediment-gravity transport mechanisms include debris flows, grain flows, fluidized flows and turbidity flows. Details on the deposits of these types of flows can be found in Crevello and Schlager (1980), Cook and Mullins (1983), Cook and others (1983), Enos and Moore (1983), Allen (1984), McIlreath and James (1984), and Coniglio and Dix (1992).

Carbonate intraclastic gravels and conglomerates from shallow-water settings are composed of a number of different clast types. Desiccated mud clasts from modern intertidal and supratidal settings (Fig. 31) are perhaps the best known intraclasts (see Hardie, 1977a; Shinn, 1983a; among many others). Desiccated intertidal and especially supratidal muds are commonly laminated. This imparts an initial tabular shape to the clast which is easily modified into a discoidal shape with transport and rounding off of corners. In modern examples of desiccated mud chips, evidence of their soft, uncemented nature is seen, for example, in the warping and bending of clasts about harder elements. Where a cyanobacterial mat impregnates the upper surface of the ripped-up layers, the chips will curl up, with secondary, small tension gashes and desiccation cracks opening on the convex side (Fig. 31A). Mudchip conglomerates are commonly found as pockets and interbeds within intertidal and supratidal laminites (Fig. 32). In a number of cases, beds can be found where there is a complete transition from in-place mudcracked laminites through a disturbed, ripped-up zone where clasts can be fitted back in place, up to a transported conglomerate as demonstrated by the organization of clasts into current-oriented, grain supported fabrics. This is in marked contrast to untransported "pseudobreccias" which may be found in similar settings (Bromley, 1968; Demicco and Mitchell, 1982, p. 196-197, and see sections on Mudcracks, Tepees, etc., elsewhere in this atlas). In examples of "pseudobrecciation" caused by invertebrates that burrow horizontally, the flat pebbles of the disrupted layer are surrounded by mudstone, and appear to fit back together. The superadjacent strata, however, are disrupted in a step-like fashion which gives the appearance of a stoped roof (Fig. 33) Finally, carbonate peloidal muds need not undergo exposure in order to be consolidated enough to form intraclasts. The best example of this comes from Andros Island where the channel lag of some tidal channels is composed of round clasts of burrowed, intertidal pond and shallow subtidal muds eroded from the walls of the channels during lateral migration of the channels. The round rather than tabular shape of these clasts of subtidal mud stems from the fact that bioturbation has destroyed all layering, leaving only a homogeneous peloidal mud that breaks into blocky lumps rather than tabular clasts.

In many modern and ancient examples, intraclasts have been derived from subaerial or submarine cemented hardgrounds and were rather firmly lithified by cementation before transport (see sections on Submarine Hardgrounds and Supratidal Crusts in this atlas). We would suggest that the term *lithoclast* be reserved for such clasts that show evidence of pre-transport cementation. Penecontemporaneously cemented carbonate peloidal muds may be sources for lithoclasts. These may be of either subaerial (e.g., Andros Island, Hardie, 1977b; Shinn, 1983a) or subaqueous (Shinn, 1969) origin. The Andros Island subaerial crusts are cemented by aragonite, microcrystalline magnesian calcite and poorly ordered calcian dolomite (protodolomite). Where exposed in eroding shoreline cliffs along tidal channels or beaches, this crust is a source of large, platy, peloidal-mud lithoclasts (Fig. 34) Another Quaternary example of well-cemented peloidal mudstone lithoclasts is provided by the blackened soil-crusts found on top of the Pleistocene "bedrock" in the Florida-Bahamas region (Shinn and Lidz, 1988). These have, in a few instances, been found in cores of Holocene carbonates at the boundary between the Pleistocene and the Holocene deposits (see below). Grainstone lithoclasts and recycled conglomerate lithoclasts (Fig. 35) are fairly common in modern carbonate sediments and in ancient carbonate rocks (Ball, 1967; Shinn, 1969; Harris, 1979; Markello and Read, 1981; Sepkowski, 1982; Halley and others, 1983; Demicco, 1985; Nguyen, 1986). Pre-transport cementation of grainstone clasts is indicated by: (1) cements which are truncated at clast margins; (2) ooids and skeletal fragments which are truncated at clast margins; or (3) the clasts are composed of ooids or skeletal grains. Clasts dominantly composed of peloids present somewhat of a problem as compaction and merging of peloids under shallow burial might generate a mud cohesive enough to be eroded and transported (Demicco, 1983). However, where peloids within clasts have an open, sedimentary packing (40% pre-cement porosity with one or less contacts per grain) early cementation is indicated. Modern subaqueous clasts of cemented grainstones are common on the shelf-margin shoals of the Bahama Banks (Fig. 35A, B) (as briefly discussed in the introduction to this Atlas). Erosion by waves of subaerial beachrocks exposed on windward lagoon beaches, such as those of Joulter's Cay, Bahamas, also produces intraclasts.

The hydrodynamics of transportation and deposition of shallow-marine gravels are, in general, poorly known, as is also the case for the resultant fabrics and sedimentary structures. This is especially true for tabular and discoidal-shaped clasts, the most common kind encountered in carbonate gravels and conglomerates. Allen (1984) gives a review of hydrodynamic aspects of depositional shape fabrics. Collinson and Thompson (1989) give a more general review of shape fabrics and sedimentary structures of conglomerates. Koster and Steel (1984) and a number of other atlases in this series contain more specific information on gravel shape fabrics and sedimentary structures. The compilation by Einsele and Seilacher (1982) contains 18 short papers dealing with "storm deposits" many of which have gravel-sized components.

The shape fabrics and sedimentary structures commonly encountered in carbonate intraclastic conglomerates include flat-lying clast orientation, imbrication, pseudo-imbrication, cross-stratification, and a variety of high-angle "edgewise" structures. Flat-lying orientation of tabular and discoidal clasts within carbonate conglomerate beds is the least well known clast fabric (Fig. 36A). Care must be taken to examine the bed in three dimensions to make sure it is not an imbricated or cross-stratified fabric viewed in the flow direction. Flat-lying orientation of clasts may also come about in part by compaction. Imbrication is the normal shape fabric of current-deposited gravels (Allen, 1984; Collinson and Thompson, 1989) with either the long or intermediate axis of the clast dipping roughly from 8 up to 20 degrees or so upstream. Imbricated gravels (Fig. 36B) are useful for paleocurrent determinations but care must be taken to make sure that the clasts are imbricated and not lying along the foresets of large-scale cross-stratification (Fig. 36C). Such "pseudo-imbrication" (Teisseyre, 1977; Allen, 1984, p. 217; Collinson and Thompson, 1989) may be quite common because cross-stratification in coarse sands and gravels is normally difficult to recognize. Paleocurrents determined from pseudo-imbricated conglomerates are 180 degrees in error. Evidence supporting a primary imbrication origin would be the partial infilling of interstices between gravel clasts by sand to form geopetal "shelter voids" that indicate infiltration of the sand after deposition of the gravel framework (Fig. 36B). In contrast, in cross-stratified gravelly sands, the inclined clasts will be incorporated within and between sandy avalanche layers of bedform foresets.

Edgewise fabrics are common in carbonate conglomerates and range from discrete lenses and pockets to laterally extensive conglomerate beds. Continuous beds of edgewise con-

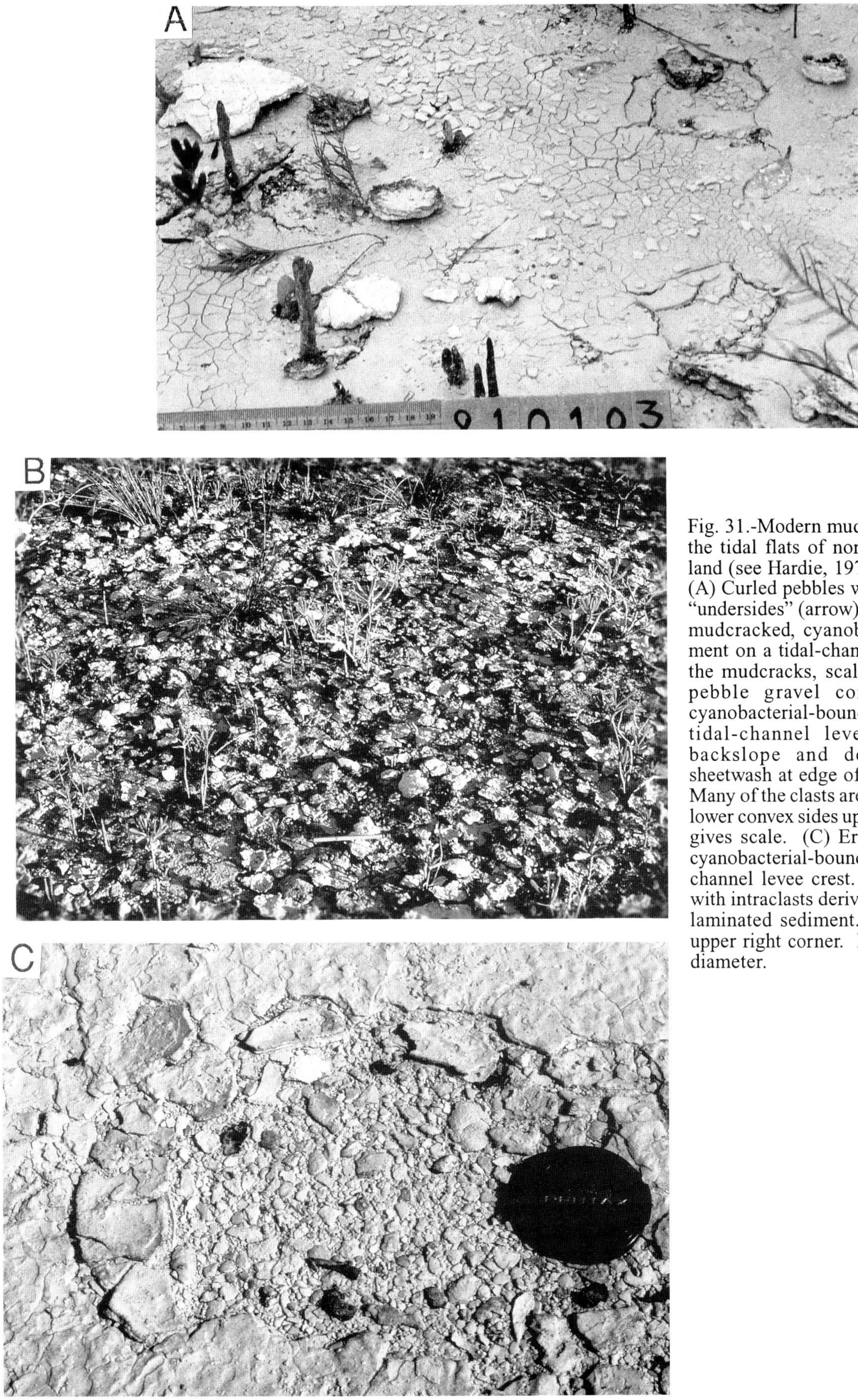

Fig. 31.-Modern mud chip intraclasts from the tidal flats of northwestern Andros Island (see Hardie, 1977 for further details). (A) Curled pebbles with mudcracks on the "undersides" (arrow) derived by erosion of mudcracked, cyanobacterial-bound sediment on a tidal-channel levee crest. Note the mudcracks, scale is in cm. (B) Flat pebble gravel composed of curled, cyanobacterial-bound clasts derived from tidal-channel levee crest and levee backslope and deposited by storm sheetwash at edge of intertidal pond zone. Many of the clasts are overturned with their lower convex sides up. Pencil in foreground gives scale. (C) Erosional depression in cyanobacterial-bound sediment on a tidal-channel levee crest. Depression is filled with intraclasts derived from erosion of the laminated sediment. Note mudcracks in upper right corner. Lens cap is 50 mm in diameter.

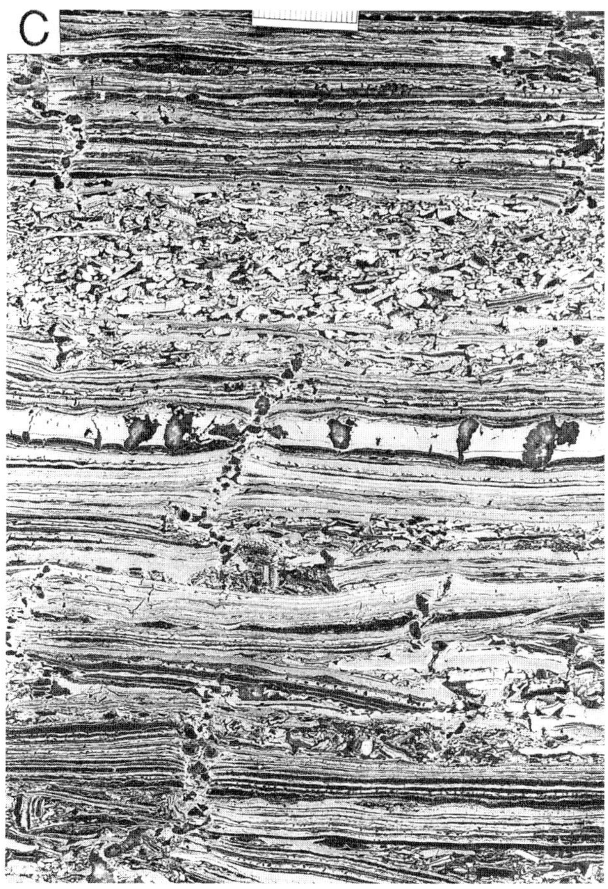

Fig. 32.-Mud chip intraclast conglomerates. (A) Detail of a modern Bahamian mudchip gravel of the type shown in Figure 31. Note that the mudchips, although rounded by erosion, are not cemented and when thoroughly wetted they become ductile and easily squashed. Penny in upper right for scale. (B) Bedding-plane view of dolomite mud chip conglomerate from the Middle Cambrian Arctomys Formation, Canadian Rocky Mountains, southwestern Alberta. Matchbook is approximately 40 mm long. Compare with mudchip gravel shown in (B). Photograph courtesy of Brent Waters. (C) Lenses and layers of intraclastic mud chips within a planar to wavy fine laminite composed of couplets of fine peloidal grainstone (dark) and dolomite mudstone (light). Note the deep prism cracks. Sample from the Upper Cambrian Conococheague Limestone, western Maryland, scale bar in mm.

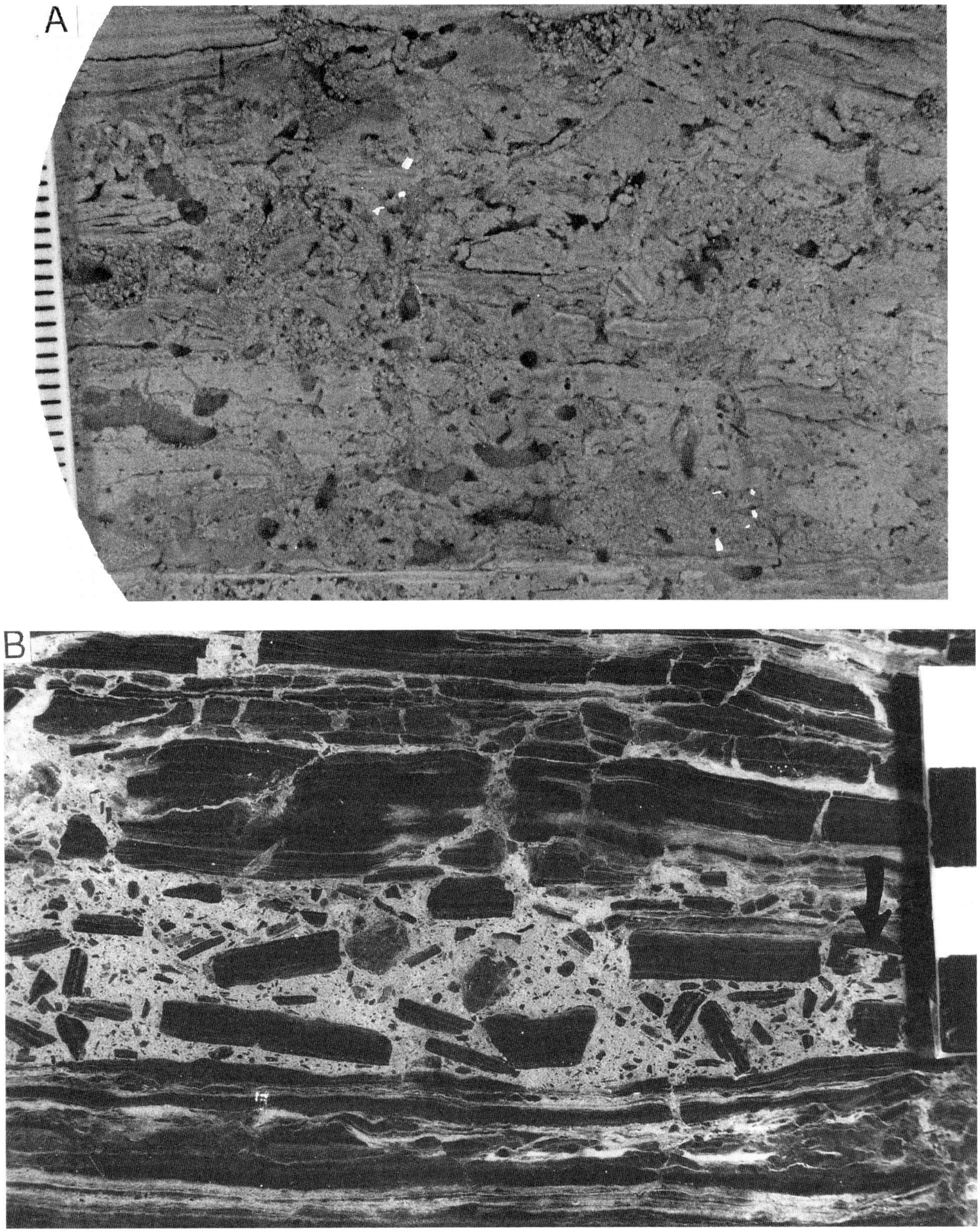

Fig. 33.-Untransported "pseudobreccias" produced by the horizontal burrowing of invertebrates such as worms. (A) is from modern tidal-channel levee crest deposits of northwest Andros Island, (B) from Middle Ordovician St. Paul Group laminites (photograph courtesy of Ray Mitchell). Note that the intraclasts are generally oriented parallel to layering, are poorly sorted, fit back together in some cases, and are disrupted by discrete tubular burrows (arrow in B). Superadjacent strata are disrupted in a step-like fashion and give the appearance of a stoped roof. Some of the brecciation in (B) may also be due to mudcracking, which can also generate in situ "pseudobreccias." Scale in (A) in mm, scale in (B) in cm.

Fig. 34.-Flat pebble gravels from the modern tidal flats of northwest Andros Island (see Hardie, 1977, for further details). Clasts are composed of cemented cyanobacterial crusts (see descriptions of these crusts in Hardie, 1977). (A) Upper beach levee zone where a sheet of buried cemented crust has been partly exposed by storm-wave erosion. To the right can be seen the scattered plates washed up across the outcropping crust sheet. (B) Cemented clasts on a low narrow levee at the edge of a tidal channel. This gravel has a notable imbricate fabric and sits atop an outcropping pavement of cemented cyanobacterial crust. Clasts were derived from the crust by storm-wave erosion. For scale, the mangrove pneumatophores in the middle ground are 15 to 20 cm high, and some of the very large clasts are 40 to 60 cm long.

Fig. 35.-Cemented intraclasts. (A) Intraclasts of modern cemented ooid sands from Joulters Cay, Bahamas. Photograph courtesy of Paul M. Harris. (B) Cemented ooid intraclasts from toe-sets of megaripples in subtidal ooid shoals, Berry Islands, Bahamas. Clasts are lightly cemented by an aragonite fringe cement. Scale in cm and mm. (C) Polycyclic conglomerate intraclasts from the Lower Ordovician Stonehenge Limestone, central Appalachians. Scale in mm. Some of the constituent clasts are truncated at clast margins (arrows). These features are interpreted as due to early cementation and re-erosion of intraclastic gravels. Photograph courtesy of Chau Nguyen.

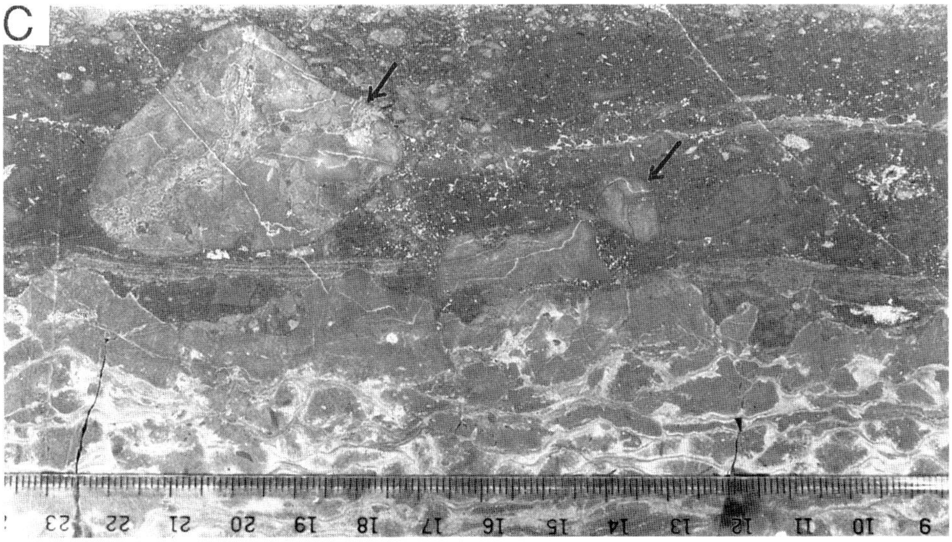

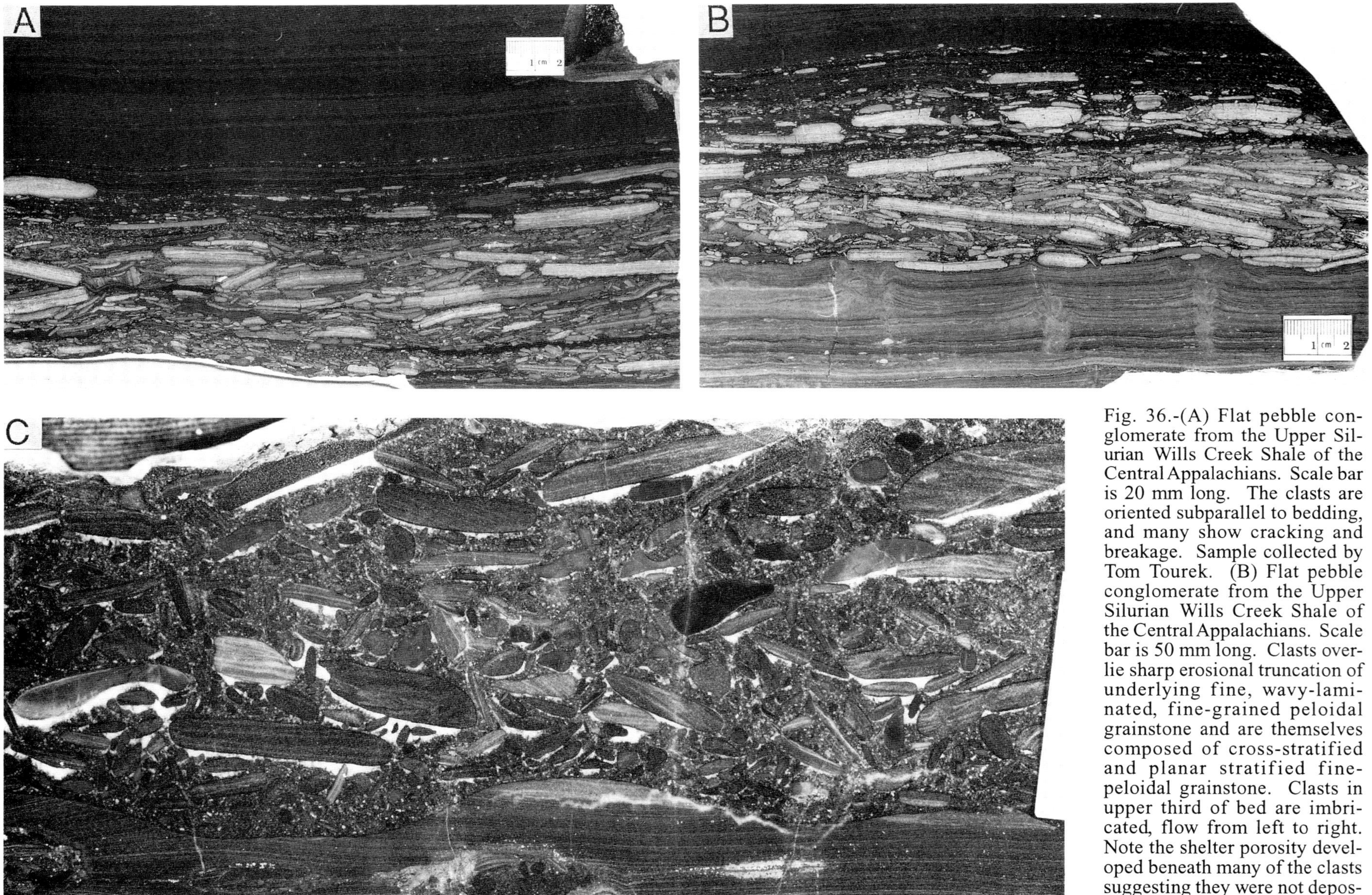

Fig. 36.-(A) Flat pebble conglomerate from the Upper Silurian Wills Creek Shale of the Central Appalachians. Scale bar is 20 mm long. The clasts are oriented subparallel to bedding, and many show cracking and breakage. Sample collected by Tom Tourek. (B) Flat pebble conglomerate from the Upper Silurian Wills Creek Shale of the Central Appalachians. Scale bar is 50 mm long. Clasts overlie sharp erosional truncation of underlying fine, wavy-laminated, fine-grained peloidal grainstone and are themselves composed of cross-stratified and planar stratified fine-peloidal grainstone. Clasts in upper third of bed are imbricated, flow from left to right. Note the shelter porosity developed beneath many of the clasts suggesting they were not deposited on the foresets of dunes (compare with C). Sample collected by Tom Tourek. (C) Set of large-scale, cross-stratified intraclastic conglomerate overlain by flat-lying clasts from the Upper Cambrian Conococheague Limestone of the central Appalachians, flow from left to right, scale bar is 20 mm long. Cross-stratified set sharply overlies wavy, finely laminated dolomite mudstone with prism cracks. Note that interpretation of the cross-stratified clasts as imbricated ("pseudo-imbrication") gives an paleocurrent direction incorrectly oriented by 180 degrees (see text for details). Tension fractures and close packing of the clasts indicate considerable mechanical compaction. Sample collected by Tom Tourek.

glomerates are commonly found in modern beach deposits as described by Sanderson and Donovan (1974), among many others. Where seen in plan view, tabular and discoidal edgewise clasts found on beaches are organized into decimeter scale clumps. These clumps have polygonal borders and within each clump the long axes of pebbles are subparallel. A wave origin is inferred for these types of edgewise fabrics. A modern and an ancient example of plan views of edgewise fabrics are shown in Figure 37. Discrete lenses and pockets of edgewise gravels are commonly reported from ancient shallow marine carbonate deposits (Fig. 38) but must be interpreted with care. Where clasts fan out within a pocket and are very tightly packed, a wave origin is inferred (Fig. 38C) (Allen, 1984, p. 229). Where plan views reveal clumps of pebbles with polygonal boundaries separating parallel-oriented clasts, a wave origin is probable (Fig. 38A). However, a number of other types of edgewise and near-edgewise gravels have been observed in the modern sediments of Andros Island (Hardie, 1977a; Shinn, 1983a). They occur: (1) as pockets in erosional scours on supratidal flats; and (2) as discontinuous low linear mounds on beach ridges along eroding shorelines (Fig. 39). Pockets of clasts are commonly more or less all inclined in the same direction (up to vertical) and may not be wave generated but instead may be pebble cluster bedforms generated by unidirectional flows where clasts stack up in front of an obstacle (see Brayshaw, 1984). Unidirectional flows may have stacked up pebbles in the small scour in Figure 40. The spaced ridges of imbricate intraclasts from Andros Island shown in Figure 39 may be transverse ribs; that is supercritical-flow bedforms developed in gravels (Allen, 1984, p. 387-390). Uniform spacing among imbricate pockets of conglomerates would be expected in ancient examples of transverse ribs.

Origin and significance.-
As outlined above, modern intraclastic carbonate gravels are found in a number of depositional environments in shallow marine carbonates. The flat pebble gravels made by erosional rip-up of intertidal and supratidal laminites during storm events on modern tidal flats have been widely used as analogs in the explanation of the origin of ancient flat pebble conglomerates composed of laminated mudstone clasts. The subtidal gravels and their ancient conglomeratic counterparts are much less familiar, but no less important. Three important categories are: (1) storm (or swell) lags, (2) lateral accretion lags found on the base of tidal channels, and (3) transgressive lags. Subtidal intraclastic conglomerates, both carbonate and siliciclastic, are commonly interpreted to be storm or swell lags deposited in shoreface environments (Einsele and Seilacher, 1982; Nemec and Steel, 1984; Aigner, 1985; Craft and Bridge, 1987). These may commonly form the basal parts of hummocky cross-stratified, or planar-stratified, graded beds that fine-upward into wave-ripple cross-stratified, bioturbated fine sands (Fig. 41). Many of these so-called "lag" deposits have been transported into place, as is clearly indicated by their composition (gravels in siliciclastic shoreface deposits; or exotic fossil content). Also, clast types are commonly indicative of local origin. Clasts of ooid grainstone are a very common component of intraclastic conglomerate beds within modern and ancient ooid shoal deposits (see references cited above). These conglomerates may be imbricated or large-scale tabular-planar cross-stratified. Shinn (1983a, 1986) has described modern tidal channel floor lag deposits from Andros Island. Very few ancient equivalents with well preserved lateral accretion bedding have been described (Bridge and Droser, 1985; Waters and others, 1989). One of the most commonly cited depositional interpretations of ancient conglomerates is that they are the transgressive lag formed by the rip-up of underlying older tidal flat sediments as sea level rose during the next cycle of platform drowning (see James, 1984; Grotzinger, 1986a). However, as mentioned above, in the Holocene of Florida and the Bahamas, the basal subtidal sediments formed during the latest glacio-eustatic sea level rise carry only a sparse scattering of blackened Pleistocene rock fragments in small karst holes on the Pleistocene surface. In these same Holocene subtidal sediments of the Bahamas, the prominent intraclast gravels are not transgressive lags but consist instead of fragments of penecontemporaneously cemented Holocene hardgrounds. In our view, most so-called transgressive lag breccias or conglomerates in ancient carbonate deposits are, like these modern intraclast gravels of the Bahamas, discontinuous lenses of ripped-up subtidally cemented hardgrounds (see elsewhere in this atlas) and are not the clasts of the underlying older "transgressed" deposit. Finally, care must be taken to distinguish between true imbricate structure and pseudo-imbricate (cross-stratification) structure when measuring paleocurrents in conglomerates of ancient shallow marine carbonates (see discussion above).

Occurrence and preservation.-
Intraclast breccias and conglomerates are a ubiquitous component of modern and ancient shallow marine carbonate deposits in both subaerial and subaqueous settings. As many appear to be storm deposits, they represent the extraordinary event and tend to be preserved.

Mudcracks and Other Disruption Structures

Open, polygonal mudcrack systems due to subaerial desiccation are one type of exposure feature common on modern carbonate (and siliciclastic) tidal flats (Fig. 42). Their ancient counterparts (Fig. 43) are commonly filled with sediment or, less commonly, with cement. Where such polygonal cracks are seen on bedding surfaces or in cross-section, they are considered one of the most useful sedimentary structures for identifying ancient carbonate tidal-flat deposits (Shinn, 1983a, p. 174, 1986, p. 8; James, 1984, p. 213). Yet, despite their importance, our understanding of mudcracks is incomplete. There are a number of studies on the plan morphology of subaerial desiccation cracks in siliciclastic and carbonate muds (see Plummer and Gostin, 1981 and Allen, 1984 for reviews). However, there are very few published photographs and virtually no detailed descriptions of modern mudcrack cross-sections even though most mudcracks in ancient rocks are seen in cross-section (Allen, 1984, p. 556, for examples). This is a particularly serious omission in view of the fact that there are a number of other mudcrack-forming mechanisms not involving subaerial drying that can operate at a sediment-water interface (subaqueously) or at some depth within buried deposits (substratally). These various non-subaerial desiccation mechanisms for mudcrack generation are far from fully understood, and, as a result, the plan and cross-section morphologies of non-subaerial desiccation mudcracks are exceedingly poorly known. Moreover, in the few cases where non-subaerial desiccation mudcracks have been experimentally produced, their plan and cross-section morphologies are similar to those of subaerial desiccation mudcracks. Although some authors prefer the term shrinkage crack over mudcrack due to their polygenetic nature, we believe that mudcrack is a suitably non-genetic term and have retained it. Also, in substratal cracks shrinkage may be intimately coupled with other mechanisms. Another serious problem arises where mudcrack-forming mechanisms that depend upon the special electrochemical properties of siliciclastic clays are used to interpret mudcracks in carbonate mudrocks that contain little or no siliciclastic clays. For these reasons,

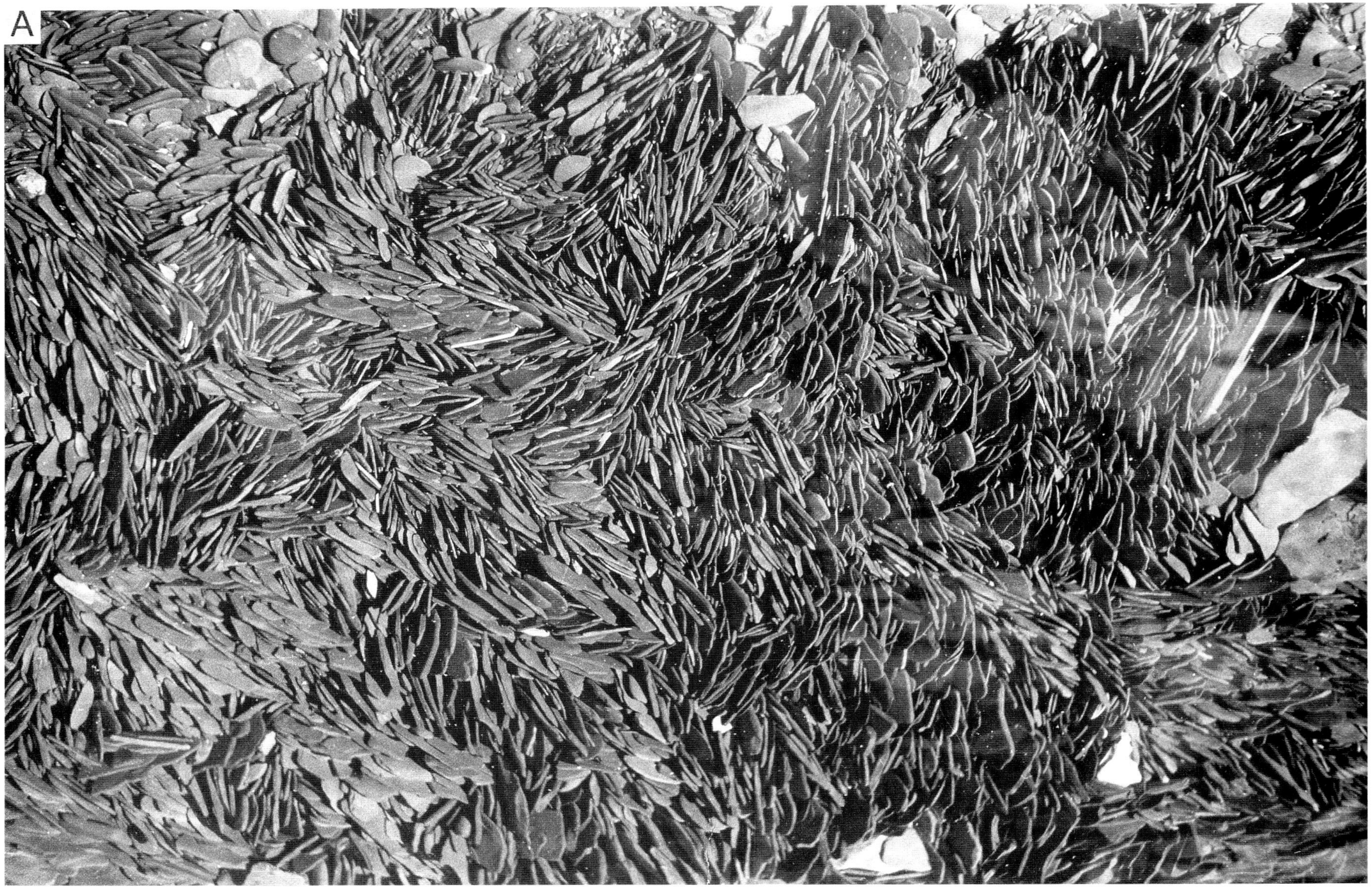

Fig. 37.-(A) Surface view of a modern edgewise gravel of shale chips on a beach at the west end of McLeod Bay, Great Slave Lake, Northwest Territories, Canada. Chips are a few tens of millimeters long. Note how the chips are packed into decimeter-scale polygonal domains. Within domains, chips are subparallel. (B) Bedding surface of carbonate edgewise conglomerate from the Lower Proterozoic Wildbread Formation, Pethei Group, Northwest Territories, Canada. Compare with (A). Scale in tenths of inches (2.54 mm). Both photographs courtesy of Paul Hoffman.

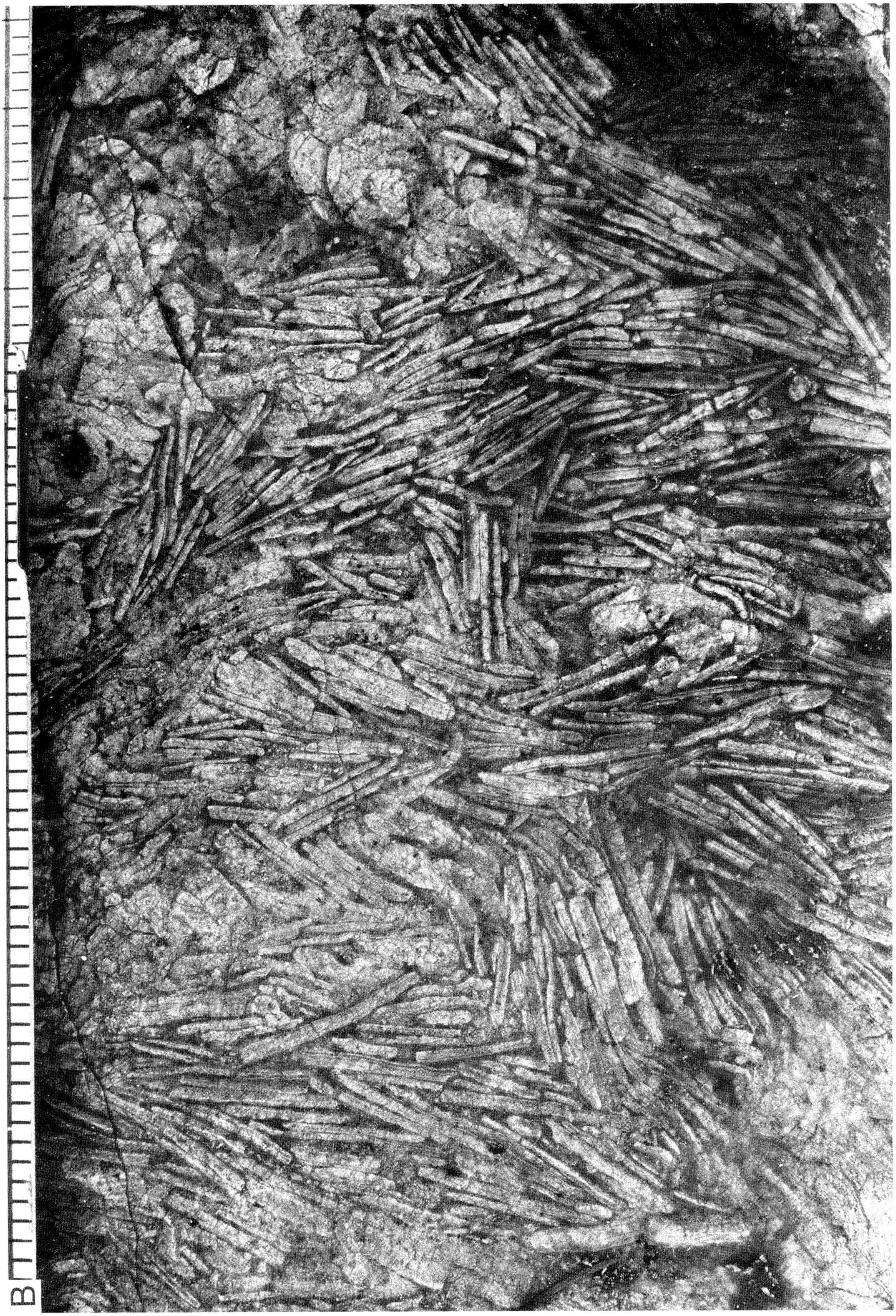
B

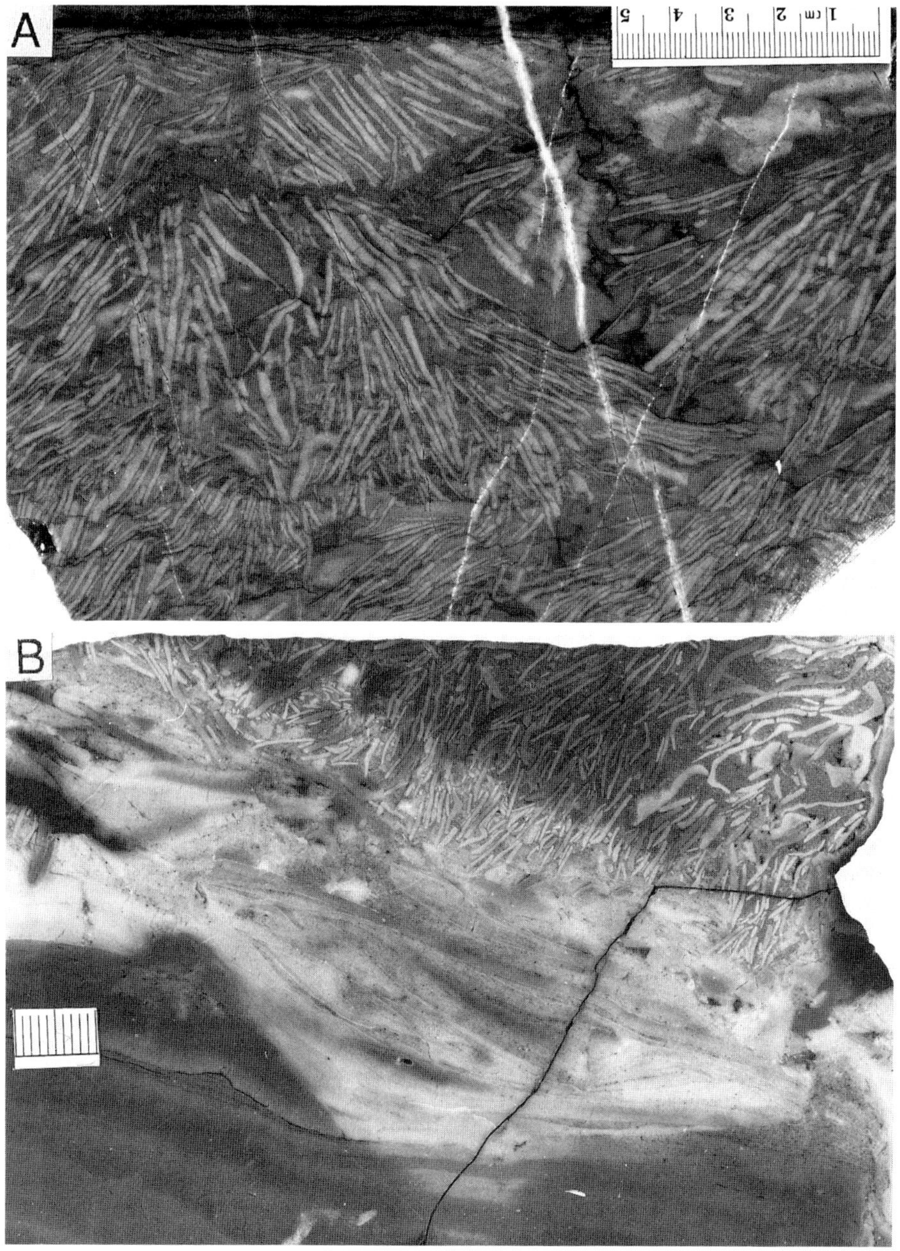

Fig. 38.-(A) plan view, and (B) cross-section of pocket of edgewise conglomerate from Precambrian wavy laminated dolomite mudstone. Note the polygonal fabric of (A) (compare with Fig. 37A). Scale bar in (B) is 10 mm. Samples collected by Paul Hoffman. (C) (Following page) Cross-section of a lens of edgewise clasts from "ribbon rocks" of the Upper Cambrian Conococheague Limestone, western Maryland. Clasts are composed of planar laminated peloidal grainstones. Scale bar in mm.

C

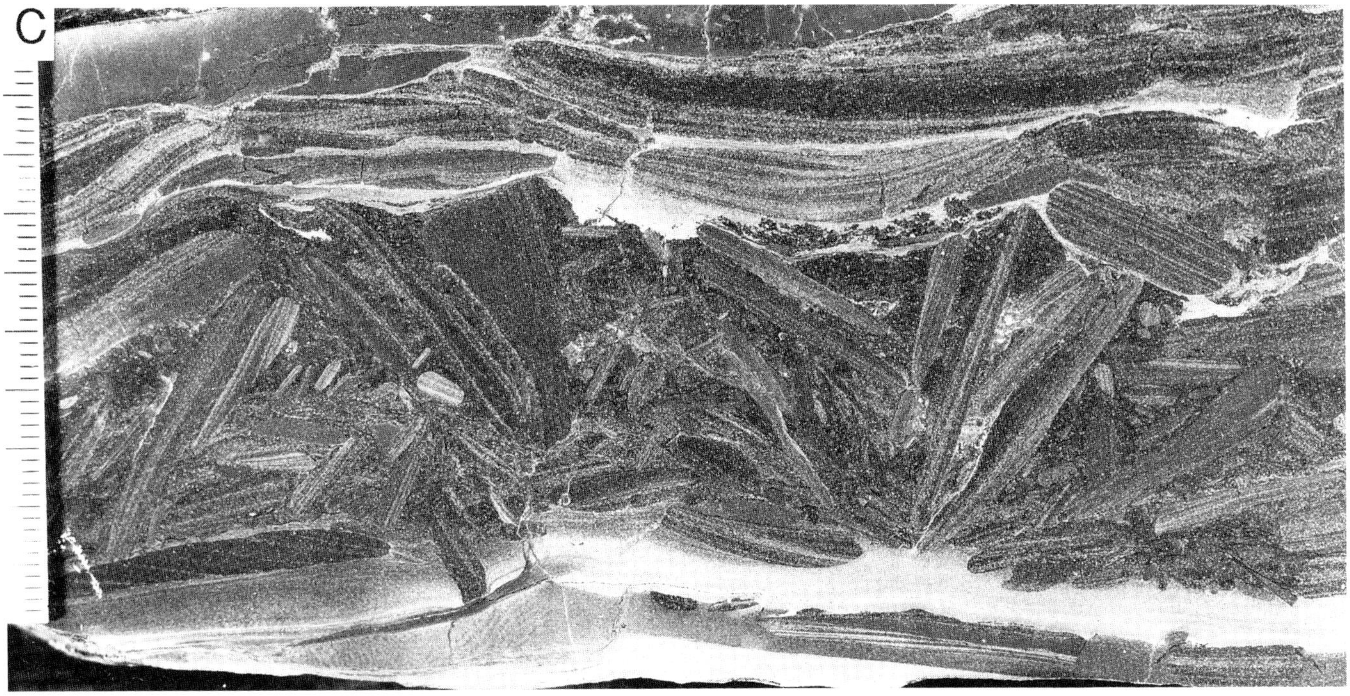

Fig. 39.-Transverse ribs composed of imbricated flat pebbles deposited on beach-ridge washover crest on the modern carbonate tidal flats of northwest Andros Island. Clasts are composed of cyanobacterial-bound laminites (layering in couplets of fine peloidal sand and mud). Current flow from upper left to lower right.

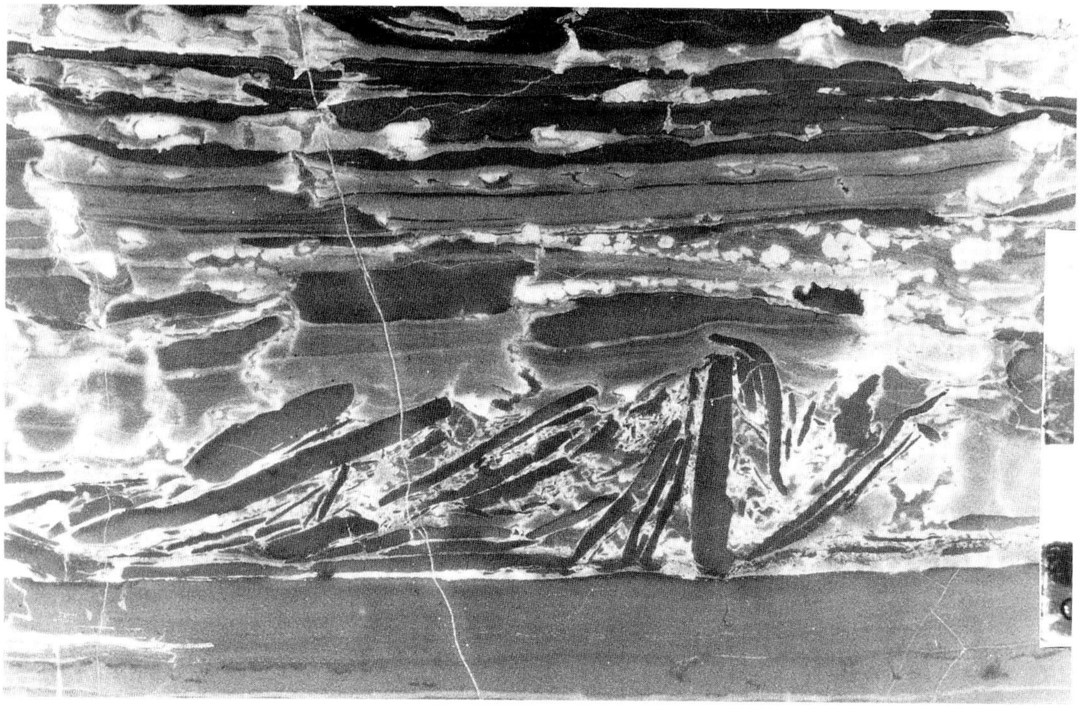

Fig. 40.-"Edgewise" conglomerate pocket in fine planar laminites of the Middle Ordovician St. Paul Group, western Maryland. With few exceptions, clasts dip to the right. Possible origins of this fabric include oscillation wave packing, unidirectional current pebble clustering, or deposition of clasts in a transverse rib bedform (see text for details). Note bent chips, indicative of microbial binding of some of the chips. Also see Mitchell (1985, Fig. 17, p. 26). Photograph courtesy of Ray Mitchell, scale bar in cm.

Fig. 41.-Skeletal grainstone conglomerate at the base of a fining-upward unit interpreted as a subtidal storm deposit ("tempestite"). Conglomerate grades up into planar to wavy laminated peloidal grainstone (plane-bed and hummocky cross stratification) which in turn is capped by small-scale cross-stratified grainstone. Scale bar is 10 mm long. Lower Devonian Helderberg Group of Pennsylvania.

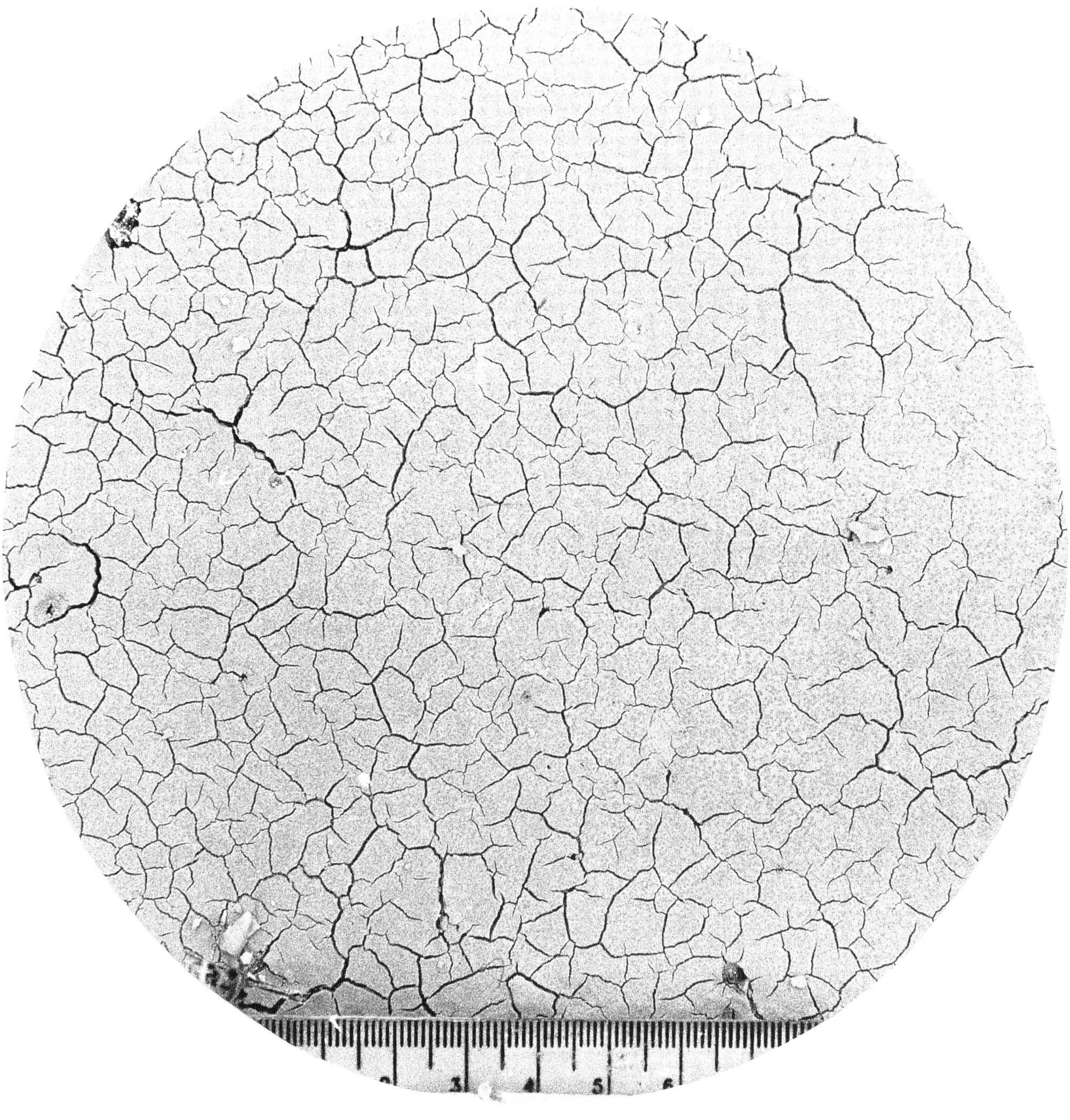

Fig. 42.-Surface of mudcracked cyanobacterial-bound sediment on the backslope of tidal channel levees of the modern tidal flats of northwest Andros Island. Note the orthogonal junctions of most cracks and the spindle-shaped incomplete cracks on the right side of the photograph. Small divisions of the scale are mm.

Fig. 43.-Bedding plane view of mudcracks in argillaceous limestones of the Wills Creek Formation (Silurian) of western Maryland. Also see Pettijohn and Potter (1964, Plate 97). Photograph courtesy of Francis J. Pettijohn.

the unambiguous interpretation of mudcracks as subaerial desiccation features is not a trivial problem. The commonly cited criteria used to distinguish subaerial desiccation mudcracks from other types of mudcracks are equivocal at present. This is a particular problem where mudcracks in ancient sequences are first interpreted as non-subaerial desiccation features based on these inadequate criteria and then used as pivotal evidence in an environmental reconstruction.

The purposes of this section on mudcracks are: (1) to review mudcrack forming processes and products in general, and (2) to evaluate criteria for differentiating among the various kinds of mudcracks. We shall consider both siliciclastic muds and carbonate muds here as there are considerable fundamental similarities between the two systems but also a number of important differences.

Subaerial desiccation cracks.-

Terminology.-We recognize three different types of desiccation cracks formed at subaerially exposed surfaces by drying out of newly deposited mud: (1) mudcracks, (2) prism-cracks, and (3) sheet-cracks. Their properties and origins are outlined below.

Distinguishing characteristics and origin.-The tensional stresses that cause *mudcracks* in siliciclastic and carbonate muds which are losing water directly by evaporation into the atmosphere are created by surface tension films of water that bridge the mineral grains at water contents less than complete saturation (Allen, 1984, p. 550). Analysis by Allen (1985, p. 155-156) shows that internal tensile stress due to surface tension varies with grain size, and can be on the order of 10^4 to 10^5 Nm^{-2} for mud-sized particles. At these stresses, the haphazardly arranged siliciclastic clays and carbonate muds collapse on themselves generating the tensional cracks. In general, the initial rupture is localized by some inhomogeneity. Initial crack patterns as shown in a number of published photographs from both carbonate and non-carbonate settings include simple open gashes that are spindle-shaped in plan, three-armed stellate patterns much like the spreading arms at a triple-junction, and curvilinear open gashes (Fig. 42) (also see Plummer and Gostin, 1981, Fig. 4). With continued water loss, a connected polygonal set of fractures, usually with orthogonal crack junctures develops. Commonly, the largest (or first order) fracture pattern is broken up by a second and in some cases third order pattern (Fig. 43). Plummer and Gostin (1981) and Allen (1984) give useful reviews of the plan geometry of subaerial desiccation mudcracks. A number of nonsystematic observations have been made about the possible environmental influence on the geometry of mudcracks (e.g., orthogonal junctures vs. polygonal junctures; convex up vs. concave up curling of the desiccating layer). Allen (1984) maintains that these are apparent differences due to sampling bias and that there seem to be no consistent, detectable geometrical differences between subaerial desiccation mudcracks developed in various environments, principally marine versus nonmarine. There does, however, appear to be a direct relationship between scale of first-order polygons and thickness of the desiccating layer (Allen, 1984, Figs. 13-27). Cross-sectional views of mudcracks are commonly considered to be simple "v"-shapes which taper downward to nothing or to a minimum width at the top of a underlying layer not involved in the cracking (Allen, 1984, p. 547). In some cases, however, where the desiccating layer is thin, the mudcrack walls may not taper very much and the cracks are relatively parallel sided or may, in fact, increase in width downward (Fig. 44).

Subaerial desiccation mudcracks in sedimentary rocks are usually completely filled from above by detrital sediment, but cements can also occur as partial or complete crack fillings. Crack fillings made of sediment may be simple, consisting of a homogeneous, single sediment type. In such cases, the sediment is typically coarse grained and the entire bedload may be trapped within a crack (Fig. 45) (the "floating mudcracks" of Smoot, 1983, or Allen's, 1984, "unbridged" mudcracks). More commonly, though, the sediment from the layer above a mudcrack can be traced directly down into the mudcrack fill (Fig. 45). Desiccation mudcracks in succeeding layers may recrack old mudcrack fills, that is, the locations of the cracks stay relatively constant through successive sedimentation events because the old cracks provided the surface inhomogeneities at which the new mudcracks formed. These types of cracks have compound, multiple fills where the stratigraphy of the overlying layers is reflected in the stratigraphy of the fills. Finally, in cross-section, subaerial desiccation mudcracks commonly start at the upper boundaries of layers.

Fischer (1964) defined *prism cracks* as a polygonal network of cracks, essentially normal to bedding, that breaks the sediment up into vertical prisms. Fischer originally used the term to refer to all the vertical mudcracks of the Late Triassic Dachstein Limestone (Fig. 46A) of the Northern Limestone Alps to distinguish them from the horizontal cracks that he called sheetcracks (Fischer, 1964, Fig. 8). However, the term is now commonly used for vertical desiccation mudcracks with the following features (Figs. 22B, C, 46B-D). First, the polygons are typically a few decimeters in diameter and the cracks a few decimeters to a meter deep. These cracks plot well off the trend of crack depth versus polygon diameter (e.g., Fig. 13-27 in Allen, 1984). Second, the cracks are relatively narrow, usually no more than a few tens of millimeters wide. Third, prism cracks tend to be developed in cryptmicrobially laminated sediments. In these cases, the laminae that comprise the prisms are generally concave up but may show complex cyanobacterial "stick-on" morphologies at crack edges. Some laminae may "bridge" the crack filling, recording places where the surface microbial mat temporarily overgrew the mudcrack. Finally, prism crack fills tend to be a complex assortment of brecciated fragments derived from the walls of the cracks. Shrock (1948, p. 159) called attention to prism cracks without specifically naming them. Shrock thought that the columnar structure of these mudcracks was maintained by repeated cracking over the same polygonal cracks after deposition of each lamina. This interpretation is consistent with the characteristics listed above and with our own observations on the spectacular prism cracks of the modern siliciclastic tidal flats of the Colorado River delta, Baja California (Fig. 46B-D; see also the illustrations in Weimer and others, 1982, Fig. 27, p. 209). In ancient rocks, the presence or absence of multiple cracking and filling features, bridging laminae, and so on, should allow differentiation between prism cracks and "one-shot" drying cracks due to the extended exposure of laminated sediment brought on by a relative drop in sea level. These latter cracks should have normal crack depth to polygon diameter ratios, should all come from one surface, should be wide, have no "bridges" within crack fills, and should be overprinted with a vadose diagenetic cap (a carbonate "soil").

Prism cracks associated with cyanobacterially laminated deposits of modern carbonate tidal flats are found in Shark Bay, Western Australia (see Logan and others, 1974a) and the Persian Gulf (Kinsman and Park, 1976). Spectacular prism cracked laminated muds (Fig. 46B-D) comprise subrecent and recent deposits found in the siliciclastic tidal flats of Baja California along the northwest coast of the Gulf of California (Thompson, 1968; Castens-Seidell, 1984). Other significantly different "prism cracks" were reported from the modern carbonate tidal flats of northwest Andros Island by Hardie and Ginsburg

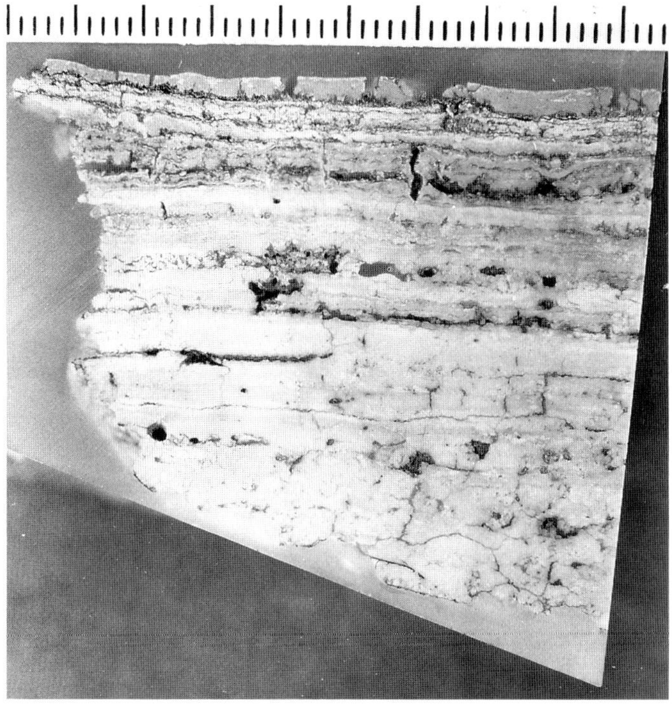

Fig. 44.-Plastic impregnated core of modern sediments from backslope of a tidal channel levee on northwest Andros Island. The upper lamina buries a mat of coarsely filamentous *Scytonema* cyanobacteria. Mudcracks in this lamina have a variety of cross-sectional shapes. Remnants of older buried *Scytonema* mats are seen as the dark wavy partings between light-colored peloidal mud laminae. Dark circular to tubular voids are open burrows of oligochaete worms. Scale in mm.

Fig. 45.-"Unbridged" or "floating" mudcracks from planar laminated dolomite mudstone in the Upper Cambrian Conococheague Limestone of the central Appalachians. Mudcracks in center of photograph are filled with intraclastic conglomerate. This coarse infill is strikingly different from the fine material making up the host laminated mudstone and presumably was imported to the depositional site by storm flooding. Also note the complicated cross-sectional shapes of the mudcrack fills. Scale bar on right side of photograph is 50 mm long.

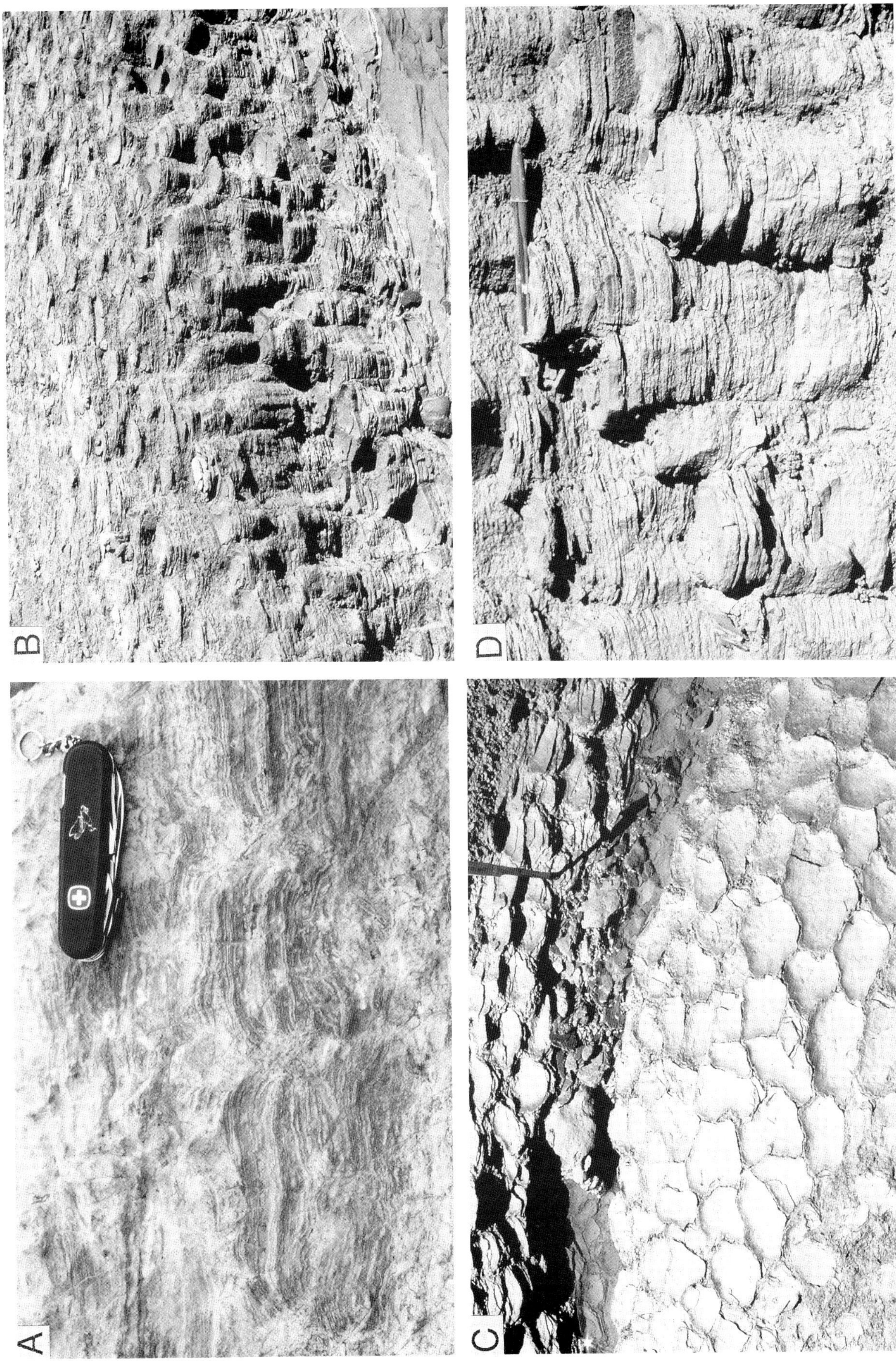

Fig. 46.-Prism cracks. (A) Prism cracks from the Lofer facies of the Triassic Dachsteinkalk at Dachstein section, Austria (see Fischer, 1965, Fig. 3). (B), (C) and (D) Prism cracks in modern siliciclastic tidal flats of the Colorado River Delta, Baja California (see Castens-Seidell, 1984, and Hardie and Shinn, 1986, p.46-49).

(1977). These prism cracks are developed in essentially homogeneous, bioturbated peloidal muds in two settings: (1) on periodically exposed banks and midchannel bars of tidal creeks (Fig. 47A); and (2) on the floors of intertidal ponds (Fig. 47B). These cracks developed in areas exposed 10-60% of the time. Intertidal pond mudcrack polygons are 0.2 to 0.3 m in diameter and 0.15 to 0.3 m deep. Channel bar prism cracks have similar diameters and depths of "many tens of centimeters" (Hardie and Ginsburg, 1977). Hardie and Ginsburg (pers. commun.) noted that the cracks form as soon as the pond drains but before it begins to dry out. Cracks may close again upon the rising tide. These observations led them to interpret these prism cracks as a result of draining of the water out of the sediment as the sea water level drops on the falling tide. It is not known whether any of the prism cracks in ancient cryptmicrobially laminated deposits carry the same message.

Fischer (1964) in his classic study of the Upper Triassic Dachstein Limestone of the Austrian Alps defined *sheetcracks* as planar to gently undulatory cracks, commonly parallel to bedding, filled by spar or mudstone (Fig. 48) In many cases, sheetcracks radiate out from vertical mudcracks and clearly formed as the desiccating layers curled up. Other sheetcracks are not apparently connected to vertical mudcracks. Fischer attributed most sheetcracks to subaerial desiccation. Sheetcracks can be differentiated from burrows by their sheet-like geometry in three dimensions.

Hardie and Ginsburg (1977) showed a number of examples of sheetcracks in laminated sediments from the modern tidal-flat deposits of northwest Andros Island. These are 50-100 mm long fractures a millimeter or so high that separate layers along bedding planes. Some of these sheetcracks propagate from vertical mudcracks causing the upper layers to lift and curl but other sheet cracks are isolated and appear to have developed independently of the polygonal network of vertical mudcracks. Small vertical fractures may connect sheet cracks in overlying and underlying layers making a "stair-step" crack. Smoot (pers. commun., and see Smoot and Lowenstein, 1991) has studied cross-sections of mudcracks developed on dry mudflats in semiarid closed basins of the American southwest. His epoxy impregnated box cores show in detail how complex subaerial desiccation mudcracks and their associated sheetcracks appear in cross-section. Complex mudcrack cross-section patterns (Fig. 48C) arise from: (1) interconnection of vertical mudcracks with a sheet-crack network producing a brecciated fabric; (2) intersection of vertical and horizontal cracks with desiccation fenestrae producing wavy, irregular crack outlines; and (3) spalling off of the corners of the desiccated sediment at the intersections of crack patterns. Smoot reports that these modern mudcracks may have linings of aligned, illuviated clays ("cutans"). Sheetcracks are commonly partially to completely filled with sediment where they are connected to vertical crack networks. Cements are more common in isolated sheet cracks.

In both form and origin, some laminoid fenestrae (see section below on fenestrae) belong in the category of sheetcracks. However, sheetcracks differ morphologically from stromatactis structure, which commonly has an irregular, digitate upper surface (see under stromatactis in Chemical Features).

In summary, the major features of subaerial desiccation cracks are as follows. In plan mudcracks can vary from simple, spindle-shaped, stellate or curvilinear gashes where desiccation is incomplete through to completely interconnected orthogonal polygon patterns. Cross-sections can vary from simple "v" shapes (that scale with the diameter of the surface polygon and thickness of cracked layer) to relatively wide, parallel walled fissures through to very deep, narrow forms. Where interconnecting sheetcracks develop off of vertical mudcracks, complex, brecciated cross-sections can develop. Crack walls need not be smooth and can "wiggle." Cracks typically are generated from common surfaces that mark the upper boundaries of mud layers. Fills vary from simple to complex. Commonly a distinct stratigraphy of a multiply filled crack can be related to the stratigraphy of overlying laminae or thin beds which ultimately bury the crack. Debris can be spalled off crack walls and cements can be found as crack fills. Cryptmicrobial laminae may bridge mudcracks.

Occurrence and preservation.-Mudcracks are one of the most common sedimentary structures of muds deposited in modern carbonate peritidal settings. Their ancient counterparts are commonly filled with sediment or, less commonly, with cement. It is mainly due to such infillings that mudcracks are so readily preserved and easily recognized in outcrops and rock slabs. Mudcrack infills composed of sandy sediment have a particularly high potential for preservation even when the host mud has been subjected to severe mechanical compaction. Such compaction will considerably deform the mudcracks into ptygmatic folds (cf. molar-tooth structure) but will not obliterate them.

Significance.-Where seen on bedding surfaces or in cross-section, mudcracks are considered one of the most useful sedimentary structures for identifying subaerial exposure in particular and ancient carbonate tidal-flat paleoenvironments in general (Shinn, 1983a, p. 174, 1986, p. 8; James, 1984, p. 213). However, mudcracks can also form by substratal and subaqueous processes. Subaerial desiccation mudcracks can be mistaken for "syneresis cracks" and vice versa and the critical problem of distinguishing between them is discussed below under syneresis cracks. Clearly, much care must be taken to correctly interpret mudcracks in ancient carbonates.

Fenestrae.-

Terminology.-Tebbutt and others (1965, p. 4) introduced the term *fenestra* (Fig. 49) defining it as: "a primary or penecontemporaneous gap in rock framework, larger than grain-supported interstices. A fenestrae may be an open space in the rock, or it may be completely or partially filled by secondarily introduced sediment or cement. The distinguishing characteristic of fenestrae is the spaces have no apparent support in the framework of primary grains composing the sediment." The adjectival form "fenestral" is now a common descriptive modifier. *Birdseye* is an older American usage describing a circular to irregularly-shaped spar-filled void. Birdseye limestone refers to a rock containing birdseyes. Fischer (1964) referred to these features as *shrinkage pores* which he defined (Fischer, 1964, p. 148) as: "irregular pores formed in muddy sediments by shrinkage (desiccation). Partly synonymous with 'birdseye' when spar filled. Grades into microsheet cracks and small prism cracks (fenestra of Tebbutt, Conley and Boyd)." Fischer (1964, p. 148) also defined the rock term *loferite* (Fig. 50) as a: "carbonate sediment riddled by shrinkage pores. Partly synonymous with 'birdseye limestone' " (see also the discussion in Shinn, 1983b). Stromatactis is generally used to refer to larger (generally centimeter-scale) irregularly-shaped cavities that are commonly elongate and have a partial filling of internal sediment that has a flat upper surface. The remainder of the void is filled by spar which preserves an irregular, digitate upper cavity surface (see discussion elsewhere). The term fenestra is now used for any kind of a hole of any shape in a rock or unconsolidated sediment that is not of an intergranular nature. Note, however, that Choquette and Pray (1970) suggested that the term fenestra be restricted to fabric selective cavities and that the term vug be applied to cavities that are not fabric selective.

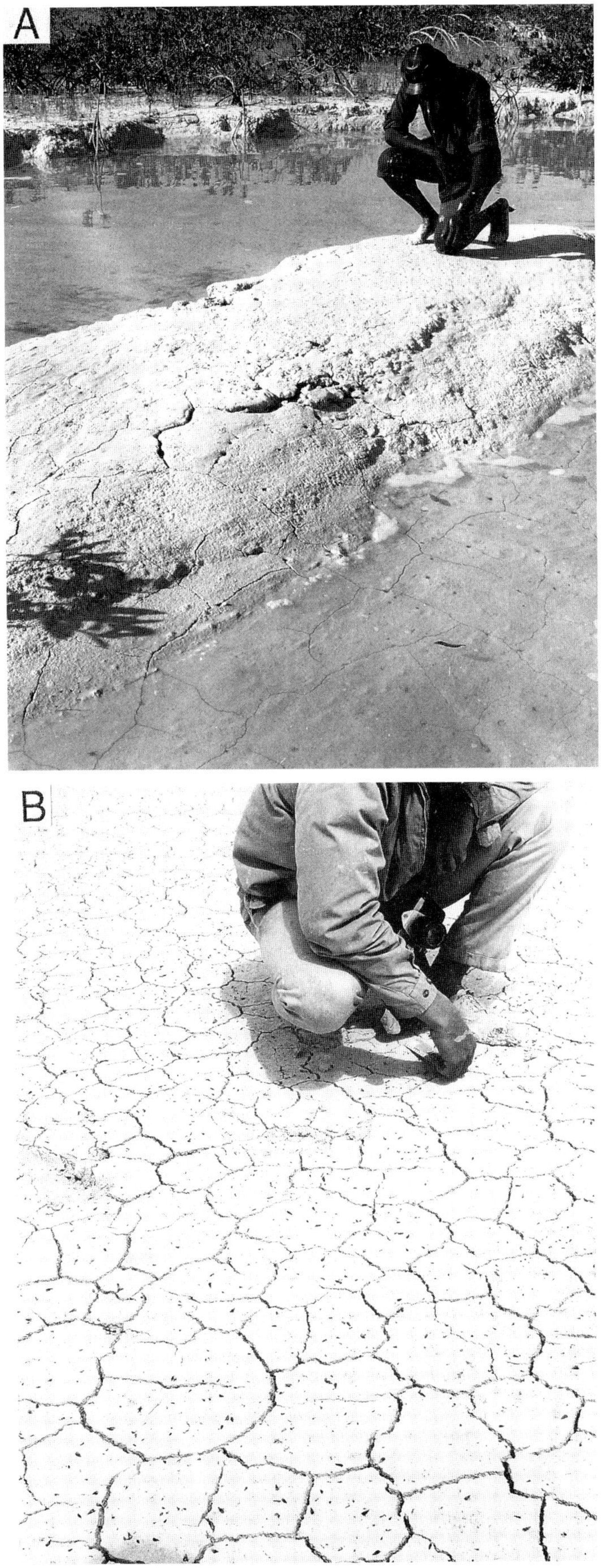

Fig. 47.-Prism cracks in bioturbated peloidal mud on the modern carbonate tidal flats of northwest Andros Island. (A) shows large, deep prism cracks on the surface of a bar in the axis of a tidal channel. (B) shows the prism-cracked surface of intertidal "pond" at low tide. Hardie and Garrett (1977) interpreted these cracks as due to draining of seawater from the sediment during ebbing tides. Dark spots in (B) are cerithid gastropods.

Fig. 48.-Sheet cracks. (A) Lower half of photograph shows spar-filled sheet cracks (light-colored horizontal lenses) from Upper Proterozoic Belt Group carbonates, Glacier National Park, Montana. Note the contorted mudcracks "molar tooth structure" (see elsewhere in this atlas) in the dark layer in the upper third of the photograph. Lens cap is 50 mm in diameter. (B) Sediment-filled sheet cracks (arrows) developed in an Upper Devonian fresh-water marl in the Oneonta Formation of the Catskill "delta." Width of specimen is approximately 15 cm. (C) Sediment-filled sheet cracks (dark layers) bridging complicated mudcracks from the Upper Silurian Wills Creek Formation of western Maryland. Scale bar in mm. Sample collected by Tom Tourek.

Fig. 49.-Fenestral carbonates from the Middle Triassic (Ladinian) Esino Formation, the Lombardic Alps, northern Italy. These tidal flat carbonates are made up of thin shallowing-upward cycles consisting of subtidal burrowed, peloidal mudstone (dark gray in photograph, thickest one beneath the coin which is approximately 35 mm in diameter) capped by intertidal-supratidal fenestral, wavy to crinkled laminites (light colored sets). All portions of the cycle are riddled by fenestrae of various types shown in more detail in Figures 51 and 53.

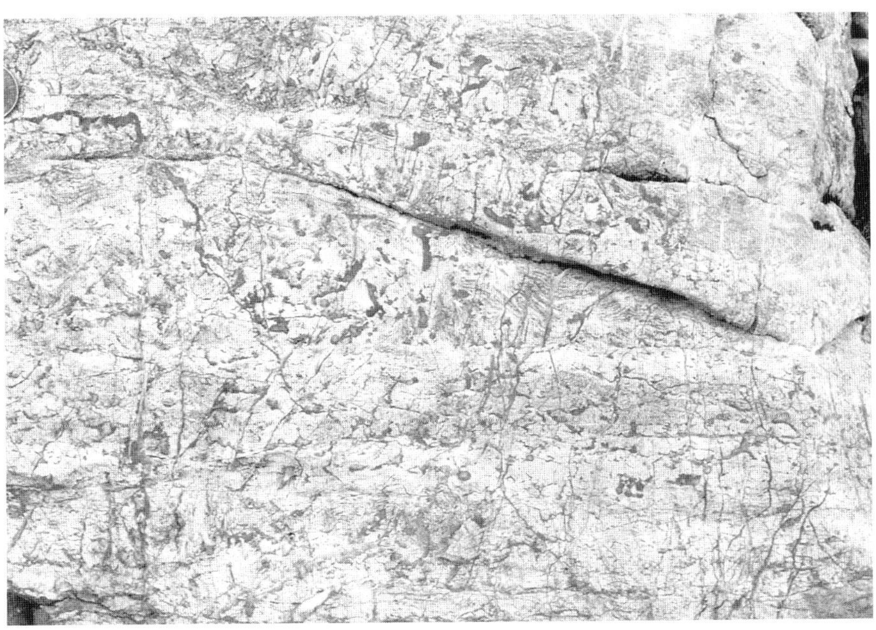

Fig. 50.-A fenestrae-riddled "Loferite" from the Lofer facies of the Triassic Dachsteinkalk at the Dachstein section, Austria (Fischer, 1964). The fenestrae are spar-filled and appear as dark irregular blebs in this outcrop photograph. Coin in upper left is approximately 25 mm in diameter.

Distinguishing characteristics and origin.-Fenestrae are generally millimeter-scale features (Fig. 49). Equant forms have diameters of a few millimeters whereas elongate forms are generally a few millimeters high and up to a few tens of millimeters long. Fenestrae can be open, or filled by any combination of surface-derived internal sediment, diagenetic internal sediment (i.e., the crystal silt of Dunham, 1969a), or cement. Fenestrae are most common in modern peloidal carbonate muds and in ancient carbonate mudstones that are either peloidal or have a clotted texture ("structure grumeleuse" of Cayeaux, see Bathurst, 1975, p. 511). However, fenestrae are also common in peloidal and intraclastic grainstones. Useful descriptions and reviews of ancient fenestrae are found in Choquette and Pray (1970), Read (1973), Wilson (1975), Grover and Read (1978), and Demicco and Mitchell (1982).

Fenestrae have variable shapes which are intergradational. In general, four shape categories are recognized: (1) laminoid, (2) tubular, (3) spherical to subspherical, and (4) irregular. The cements and sediments that fill fenestrae are not taken into account in this classification. Careful observations of three-dimensional samples or serial sections are needed to properly describe fenestrae shapes, particularly to distinguish mudcracks, sheetcracks, and spherical-subspherical fenestrae from tubular fenestrae. Two-dimensional vertical cuts down sub-vertical tubes give crack-like cross-sections, whereas perpendicular and oblique cuts through tubes yield spherical and subspherical shapes. If the tubes are dominantly horizontal, they can give an apparent laminoid fenestral geometry in longitudinal cuts.

Laminoid fenestrae (Figs. 51A-C) are perhaps the most distinctive fenestral style but are polygenetic and need to be interpreted with care. Desiccation is perhaps the most commonly cited mechanism for the generation of laminoid fenestrae, and indeed almost all fenestrae have been interpreted in this way. Horizontal sheet cracks that propagate from mudcracks are common in modern supratidal sediments of Andros Island (Hardie and Ginsburg, 1977) where they have roughly parallel sides with outlines that match each other. These cracks commonly follow bedding but need not, being concentrically disposed about burrows in some cases. In a widely quoted study, Shinn (1968a, see his Fig. 7) produced subspherical, irregular, and laminoid fenestrae by experimentally wetting and drying homogeneous aragonite mud. The development of sub-parallel laminoid fenestrae gave rise to a distinctly laminated appearing fabric in what was previously massive mud. Shinn compared these experimentally produced fenestrae to modern and ancient examples. Whereas some of the laminoid fenestrae Shinn produced experimentally have matching sides (and are therefore sheet cracks), others do not. The experimentally produced fenestrae tend to have both smooth and irregular walls.

A second commonly cited origin of laminoid fenestrae is that they are the molds of fleshy cyanobacterial mats buried by sediment. In addition, drying out of surface cyanobacterial mats can wrinkle, lift, and separate them from the adjacent sediment. Fenestrae produced in this way are illustrated by Hardie and Ginsburg (1977), Davies (1970), and Logan and others (1974a). These laminoid fenestrae may have flat bottoms and irregular tops, especially where the mat has been dried out before being buried by a layer of sediment. In such cases, the upper boundary of the fenestrae has a distinctive "pendant" structure. Gases generated by organic decay can also lift subaqueous mats or buried mats to produce laminoid fenestrae. Lower boundaries of fenestrae produced in this way need not be planar, especially where coccoid cyanobacterial mats are involved. Ancient laminoid fenestral fabrics interpreted to have had a microbial mat origin are commonly given the same environmental significance as fenestrae thought to be of subaerial desiccation origin, namely formation under intertidal to supratidal conditions. However, fenestrae related to microbial mats need not be solely restricted to these subaerially exposed settings because such mats can develop in a wide variety of subaqueous environments both marine and non-marine (see discussion of stromatolites under Biogenic Structures). Laminoid fenestrae may also be produced by the compaction of subspherical gas bubbles. Indeed, some of the modern gas-generated fenestrae apparently start out with a laminoid shape. Also, horizontal burrows produced by organisms that preferentially "mine" certain layers can produce laminoid fenestrae (see Garrett, 1977).

Tubular fenestrae (Fig. 51C) in modern carbonate sediments originate as burrows and as root holes, but they can also be formed by the upward escape of gas bubbles biogenically produced within the sediment (see next paragraph and also under Bioturbation Structures in Part II, Biogenic Structures).

Spherical and subspherical fenestrae are produced by air and gas bubbles trapped during deposition of the host sediment or generated by post-depositional decay of organic matter. Slight compaction of gas bubble fenestrae leads to subspherical fenestrae with their long axes aligned parallel to bedding. Cloud (1960, Fig. 1) pictured subspherical, irregular, and tubular millimeter-scale voids that formed in a sealed core of Bahamian aragonitic mud from a subtidal setting. Cloud attributed these to gas evolution and migration during decay of organic material in the sediment. These features are common in modern siliciclastic muds. Gas bubbles in cores from the Bodensee are illustrated by Reineck and Singh (1980, Fig. 95). They are also very common in modern muds from the bottom of Chesapeake Bay (Reinharz and others, 1982). The small (1-2 mm) bubble fenestrae in the Bodensee are subspherical whereas the larger (2-3 mm) ones tend to be flattened and more irregular. Bubble fenestrae in the Chesapeake Bay mud are notably irregular and many are subvertical irregular tubes apparently caused by upward migration of gas. A special expression of spherical and subspherical fenestrae are "*keystone vugs*" which are most commonly seen in modern beach sands and in tidal flat and playa storm layers (Figs. 52A-C) (Shinn, 1968a; Deelman, 1972; Reineck and Singh, 1980, Fig. 94). Such keystone vugs are produced by trapping of air bubbles (particularly organic-rich froths) during storm deposition in the swash zone (on beaches) and in the sheetwash zone (on tidal flats and playas). Such keystone vugs, if they could be identified unequivocally, would be useful indicators of deposition from very shallow sheets of turbulent water. However, there are no reliable criteria to separate fenestrae formed by *in situ* generation of gas due to decay (a process that is essentially ubiquitous and has no depth or environmental significance) from those formed by sedimentation of bubbles along with detrital grains at the sediment-air interface of a shallow water environment. There is a further complication in that trapped air bubbles apparently can be also be transported and aggregated internally within a porous sediment by rising groundwater. DeBoer (1979) published photographs of bubble-rich sands associated with convolute bedding from a modern, sandy intertidal-bar. Based on experimental work, he interpreted these bubbles to have formed during a rising tide stage. After the bars are submerged, trapped capillary air is compressed as the groundwater table rises. The trapped air is eventually converted into bubbles which separate sand grains.

Irregular fenestrae (Figs. 53A-C) are the most problematic of the four types of fenestrae because they can be formed in a variety of ways, and, as yet, no good criteria exist to distinguish among them. They can form by desiccation, as has been demonstrated experimentally by Shinn (1968a). They can form by soft-sediment deformation of other, originally more regular shaped fenestrae, such as gas bubbles, evaporite molds,

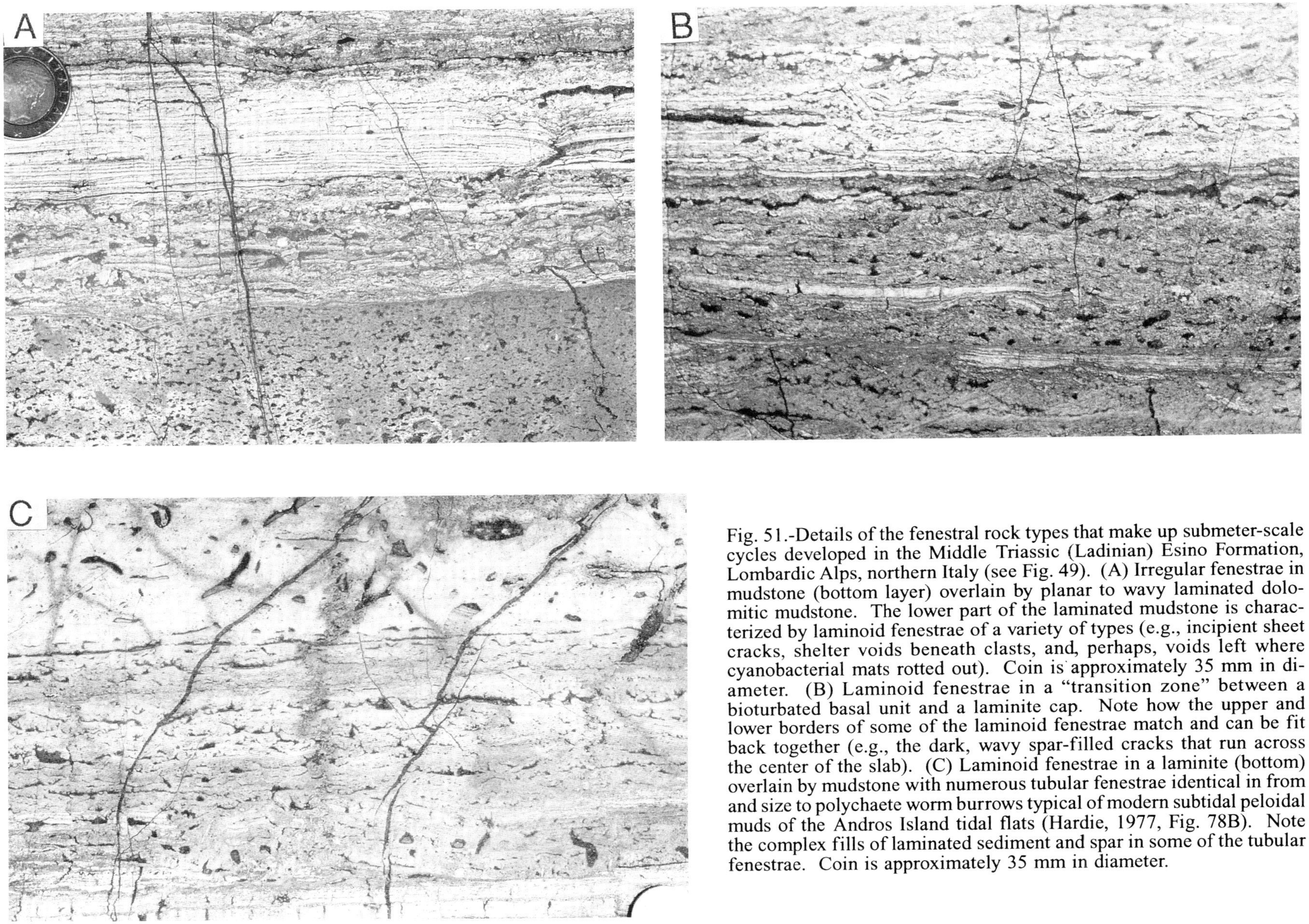

Fig. 51.-Details of the fenestral rock types that make up submeter-scale cycles developed in the Middle Triassic (Ladinian) Esino Formation, Lombardic Alps, northern Italy (see Fig. 49). (A) Irregular fenestrae in mudstone (bottom layer) overlain by planar to wavy laminated dolomitic mudstone. The lower part of the laminated mudstone is characterized by laminoid fenestrae of a variety of types (e.g., incipient sheet cracks, shelter voids beneath clasts, and, perhaps, voids left where cyanobacterial mats rotted out). Coin is approximately 35 mm in diameter. (B) Laminoid fenestrae in a "transition zone" between a bioturbated basal unit and a laminite cap. Note how the upper and lower borders of some of the laminoid fenestrae match and can be fit back together (e.g., the dark, wavy spar-filled cracks that run across the center of the slab). (C) Laminoid fenestrae in a laminite (bottom) overlain by mudstone with numerous tubular fenestrae identical in from and size to polychaete worm burrows typical of modern subtidal peloidal muds of the Andros Island tidal flats (Hardie, 1977, Fig. 78B). Note the complex fills of laminated sediment and spar in some of the tubular fenestrae. Coin is approximately 35 mm in diameter.

Fig. 52.-Keystone vugs (air bubble molds) from modern and ancient beach deposits. Lens caps in all photographs are approximately 50 mm in diameter. (A) Surface view of air-bubble molds in ooid sand on a modern beach at Joulter's Cay, northern Andros Island, Bahamas. The limestone block at the top of the photo is a lightly cemented Holocene ooid grainstone with abundant keystone vugs. This 1000-year old grainstone, typical of the young bedrock that underlies Joulter's Cay, is characterized by the same very low-angle cross-lamination found in the modern foreshore (swash zone) ooid sands on which the sample in the photo is lying. (B) Keystone vugs from modern siliciclastic sands of Wachepreague Island, Delmarva Peninsula, Virginia. (C) Cross-section of keystone vugs developed in Pleistocene grainstone in South Rock Sound Quarry, Eleuthera Island, Bahamas. Photograph courtesy of Ray Mitchell.

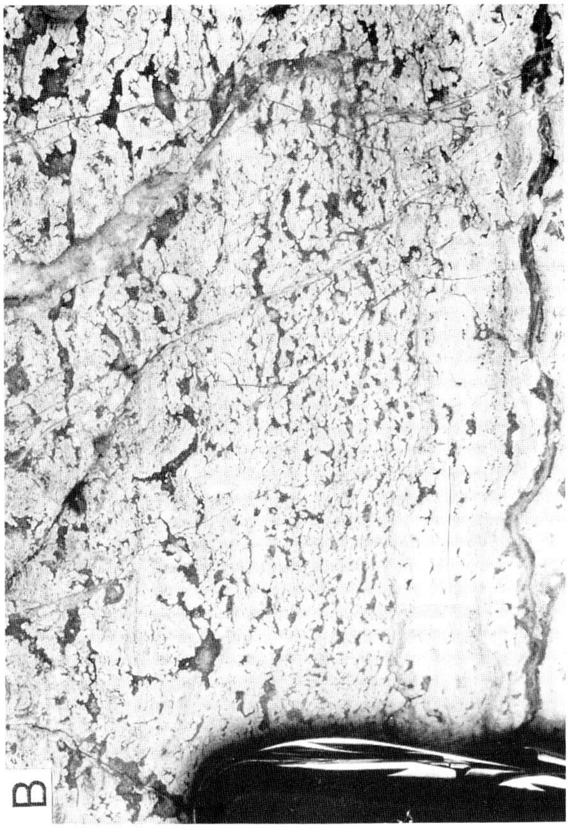

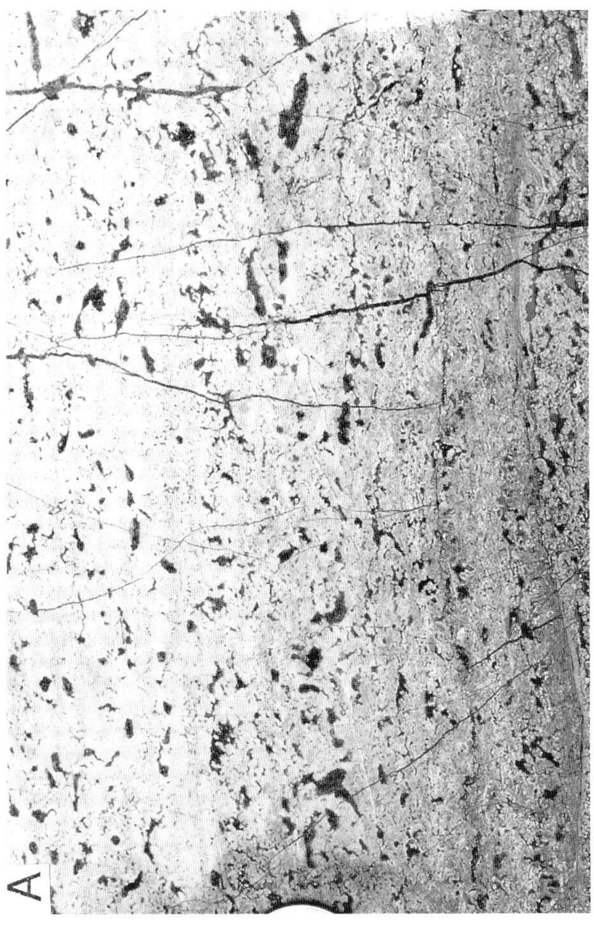

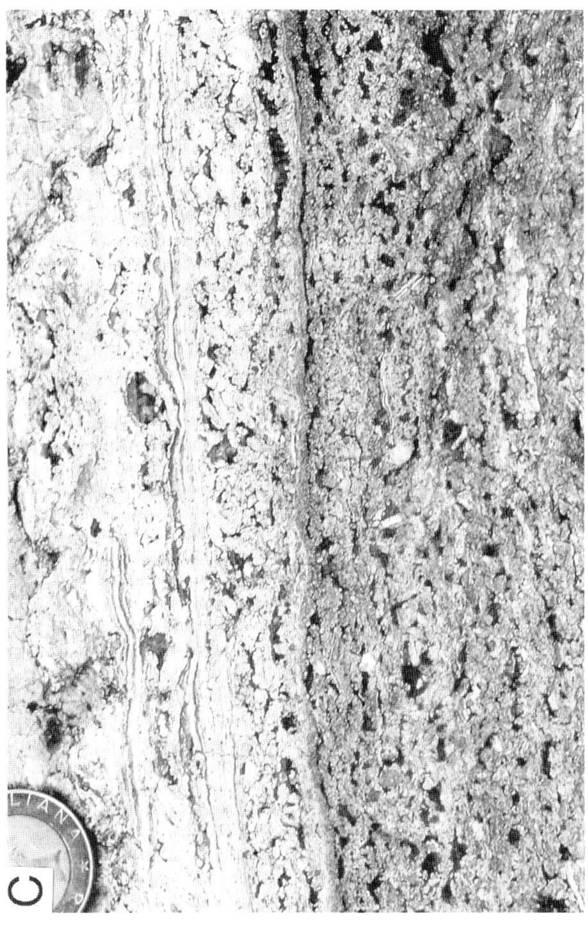

Fig. 53.-Irregular and tubular fenestrae from the Middle Triassic (Ladinian) Esino Formation, the Lombardic Alps, northern Italy. In many cases, the irregular fenestrae are simply large intergranular pores or shelter voids beneath mudstone intraclasts (see especially the fenestral fabric in B and in the thin layer in middle of C; cf. Shinn, 1983b). In some of the more elongated fenestrae, the opposing walls match in shape, suggesting a shrinkage origin. Fenestrae with circular, oval and tubular outlines are most likely burrows (as, for example, the smooth-walled fenestrae in A). Coins in top and bottom photographs are approximately 35 mm in diameter; portion of knife in (B) is approximately 40 mm high.

burrows, and so on. Logan and others (1974a) show examples formed by the burial of "pustular mats" of coccoid cyanobacteria. Most significant is that many irregular fenestrae in grainstones and packstones are clearly nothing more than intergranular pores, as is particularly evident in bioturbated grainstones composed of relatively poorly packed lithoclasts (Fig. 53C). Shinn (1983b) reported laminoid to irregular *shelter pores* developed in subtidal hardgrounds on the Bahama Banks. These pores are found in peloidal grainstones and occur beneath intraclasts and "grapestone" clasts that apparently were derived from reworking of hardground layers. As these are grain-supported interstices, Shinn (1983b) proposed the name *pseudofenestrae.* Goldhammer (1987), Goldhammer and others (1987), and Hardie and others (1986) have reported this type of "shelter pore" fenestrae from the subtidal lithoclast grainstones of the Middle Triassic Latemar buildup in the Dolomites of northern Italy. These kinds of large, irregular intergranular pores and shelter pores in some cases can strongly resemble vuggy laminoid fenestrae found in mudstones and have been mistakenly identified as such in the past and interpreted as due to drying on subaerial exposure (e.g., Scoffin, 1987, p. 271, Fig. 18.3). Other origins that have been suggested for irregular fenestrae include dewatering of gel-like carbonate muds (see discussion under syneresis cracks), replacement of evaporites, and replacement of allochems.

Occurrence and preservation.-Fenestrae are very common in both carbonate grainstones and mudstones. The experiments with modern carbonate mud carried out by Shinn and others (1977) and Shinn and Robbin (1983) clearly showed that open fenestrae are obliterated by mechanical compaction so that the preservation of fenestrae of all types in mudstones signifies that the host sediments were cemented before even shallow burial.

Significance.-Fenestrae are commonly considered one of the most useful sedimentary structures for identifying subaerial exposure. However, despite this widespread use of fenestrae in shallow marine carbonate rocks as indicators of desiccation, it should be clear from the foregoing discussion that fenestrae are polygenetic. Special care must to taken to consider all possible origins before using any given set of fenestrae in a paleoenvironmental interpretation.

Syneresis cracks.-

Terminology.-There are five commonly cited mechanisms for forming crack patterns in mud not subject to subaerial desiccation. These are: (1) contraction due to flocculation of a settled clay sediment, (2) contraction due to syndepositional compaction of a settling clay layer, (3) contraction of a smectitic clay sediment due to loss of interlayer water as a result of salinity changes in the surrounding seawater or lakewater, (4) compaction dewatering under a sediment load causing either injection from below or foundering from above, and 5) tensional gashes opened up due to downslope creep of a surface mud layer. Mudcracks formed by these processes are generally lumped by sedimentologists under the blanket term of syneresis cracks. However, syneresis is a term used by colloid chemists and biochemists to refer to the expulsion of water out of a contracting colloidal mass such as coagulating blood. Furthermore, mechanisms 1 and 3 above depend on special electrochemical properties of siliciclastic clay minerals, namely flocculation in electrolytic solutions and exchange of loosely-bound interlayer water in smectites (swelling clays). These last two mechanisms clearly are not appropriate for relatively pure carbonate sediments. We believe that the blanket use of the term syneresis to cover these various processes is unwarranted and should be discontinued.

Distinguishing characteristics and origin.-There have been a number of reports, beginning with the short note by Moore (1914), of open, polygonal mudcrack systems apparently developed under a standing cover of water (Fig. 54). Eardley (1938) reported a rough, hexagonal pattern developed on newly exposed sediments of the Great Salt Lake. These polygons developed in carbonate-rich mud and were approximately 1.5-6 m in diameter and 0.07-0.10 m wide. This pattern extended out under the surface of the lake for a considerable distance. The cracks were filled with a "soupy clay" and were probed to depths of about 3 m. The cracks "appeared like small fissure eruptions and indicated rising or upwelling of clay." The most widely quoted reference on "modern" subaqueous mudcracks is van Straaten's (1954a) work on the tidal flat sediments of the Netherlands coast. van Straaten reported seeing mud cracks "in recently deposited mud on the floors of (tidal) channels which are permanently covered with water" (van Straaten,

Fig. 54.-Mudcracks in modern carbonate mud open under water in a roadside ditch on Big Pine Key, Florida Keys. Remains of an old sneaker in right center (arrow) gives scale.

1954a, p. 70). However, the description and photograph he provided to illustrate these features (Van Straaten, 1954a, p. 75-76, and his Fig. 12) are of a filled mudcrack pattern found in an excavation 13.55 m below sea level in sediments he interprets to have been tidal channel deposits. The age of these deposits is bracketed between 2200-6900 years BP (van Straaten, 1954b, c). van Straaten's description of these features is that they are: "less regular (than desiccation mudcracks) and very often of the incomplete type ... commonly of small dimension, not more than a few millimeters wide ... not more than 1 to 3 centimeters deep, and at most a few decimeters in length. Normally they have straight to gently curved forms, with smooth walls ... The fillings are sometimes folded by differential compaction and may resemble ptygmatic veins in crystalline rocks." Thus, the most widely quoted report of "modern" subaqueous mudcracks is not an eyewitness account of cracks forming contemporaneously below water in an active tidal channel but rather a description of fossil mudcracks interpreted to be of subaqueous origin. Moreover, there is nothing about the morphology of these subrecent cracks that is inconsistent with their being desiccation crack patterns.

A number of workers have produced subaqueous mudcracks experimentally, albeit in a number of cases serendipitously. Jungst (1934), who was one of the first to try to produce subaqueous mudcracks, failed to make cracks and produced instead millimeter-scale mounds with central craters. He was the first to apply the term "synaeresis" to these experimentally produced structures which he attributed to the flocculation and settling of clays in electrolytic solutions. The mound structure with its central crater indicates that these are water escape tubes. Jungst suggested that subaqueous mudcracks could be made in this way but he never confirmed this experimentally. White (1961), in poorly constrained and poorly described experiments, produced small mudcracks in clays he had leached in a strong electrolytic solution and then apparently washed in distilled water. He attributed the cracks to contraction as the flocculated gels settled under their own weight. White figures two sets of cracks in plan view (White, 1961, Plate 1). In one petri dish are displayed two or three weakly curvilinear cracks. In a second dish are shown a number of very short, randomly oriented gashes that have a few orthogonal junctures. It should be noted that the scale of the cracking is restricted by the size of the dish. White figures one crack pattern in cross-section. This pattern (White, 1961, Plate 2) consists of a dominant set of open fissures (tubes?) connected by shorter subhorizontal open fissures (tubes?). These features do not reach the surface so their relationship to the figured surface cracks is unclear. Furthermore, this pattern could equally well be interpreted as water escape structures (see Kuenen, 1965). In this same paper, White (1961, Plate 3 and 4) illustrates two slightly curvilinear, complete to incomplete mudcrack patterns preserved as molds on the soles of sandstone beds. He interprets these as having formed subaqueously based on the following criteria (White, 1961, p. 566): (1) no evidence of curling at the surface, (2) no gradation in grain size from bottom to top of the layer containing the mudcracks, (3) cracks are not wedge-shaped but of fairly uniform width from bottom to top (note that the figured cracks are at most a few millimeters deep), (4) the cracks do not extend into the fissile shales beneath (although the figure captions make clear that the cracks shown are from "a thin shale layer between two sandstone layers"), and (5) there are no clay aggregates (intraclasts) in the sandstones above. Clearly, none of these criteria are unequivocal evidence for a subaqueous origin and in fact all are perfectly consistent with the characteristics of mudcracks of subaerial desiccation origin, as described earlier. Kuenen (1965, Fig. 3; see also Reineck and Singh, 1980, Fig. 75) accidently produced a polygonal network of cracks in mud settling in the bottom of a tank. He attributed the crack pattern to simple compaction of the settling clay layers. Few details of the experimental conditions are given.

In the experiments of Jungst, White, and Kuenen, cracks and water escape structures formed as a flocculated gel of clay settled, compacted, and apparently contracted. It is not at all clear to what extent the flocculation of siliciclastic clays was the driving mechanism in each of these experiments. If so, then this mechanism would not be applicable to the production of subaqueous disruption structures in relatively pure carbonate muds. Even if settling of fine particles is sufficient for contraction to take place, cracks produced in this way would only affect that sediment layer. Thus, if a single layer is cracked a number of times by cracks emanating from different surfaces, then this mechanism can be ruled out.

The best documented and most widely quoted series of experiments that produced subaqueous mudcracks were those of Burst (1965). Burst formed mudcracks by leaching siliciclastic clays with distilled water and then filtering a 3.5% NaCl solution through them. The cracks he produced were small but showed incomplete, slightly curvilinear patterns to complete orthogonal polygonal fractures. Mudcracks were only produced where 10-25% of the sample was composed of "bentonite" (smectite). Shrinkage apparently took place by the expulsion of interlayer waters in the expandable clays driven by the H_2O chemical potential gradient imposed by immersing distilled water treated smectite in a saline solution with a lower activity of H_2O. In actual sedimentary systems, this mechanism is dependent not only upon a significant proportion of swelling clays being present but also requires an environment where abrupt and significant changes in salinity can occur.

Donovan and Foster (1972) accidently produced short gash-like and stellate incomplete mudcracks subaqueously. These cracks were produced in a "fine silt-clay grade of sediment composed dominantly of quartz and calcite...[with]... small proportions of montmorillonite [amount unspecified], muscovite, kaolinite, chlorite and mixed-layer clay" (Donovan and Foster, 1972, p. 314). The cracks appeared in three month old settled muds when a bleaching solution was added to the tank to "clean the water" (presumably of bacteria and cyanobacteria). There are two possible mechanisms that might explain the formation of these cracks. First, there may have been enough swelling clays in the sample to shrink the sediment in the manner of Burst's experiments. This is the mechanism advocated by Donovan and Foster. However, many, if not most, of the cracks (Donovan and Foster, 1972, Fig. 8) are clearly associated with small domal uplifts of sediment. These may represent trapped bubbles of gas produced by oxidation of organic matter within the sediment. These cracks could then simply be "micrograbens" produced at the tops of domes. In either case, this accidental experiment is poorly constrained.

One final aspect concerning the morphology of these experimentally produced subaqueous cracks needs to be stressed. A number of workers have used the complexity of cracks as a criterion for "syneresis" (see below). However, the shrinkage cracks in the experiments discussed above are identical in plan morphology to incomplete and complete mudcracks produced by subaerial desiccation. Also, where cross-sections of subaqueously produced cracks that are connected to surface fissures are shown (e.g., Burst, 1965) they have simple "v" shapes and are no more complex than subaerial desiccation cracks. The small scale of the cracks is most likely due to the small sizes of the containers used in the experiments. The complex cross-section figured by White is not connected to the surface and may not be a crack pattern but a set of interconnected water-escape tubes instead. When the simple morphology of the experimentally produced subaqueous mudcracks is combined with the observation that modern subaerial desiccation cracks in playa muds can be extremely intricate (Smoot, 1983, and see under mudcracks above), it follows that crack complexity is not a valid criterion of a "syneresis" origin for ancient mudcracks.

Many ancient examples of mudcracks interpreted as having a non-desiccation origin have been reported from both siliciclastic and carbonate mudstones. Unfortunately, in a number of cases, these are simply cited as "syneresis" cracks with few, if any, criteria given (see, for example, Rich, 1951; Picard and High, 1972; Pratt and James, 1986). Where interpretations of ancient mudcracks as non-subaerial desiccation features are more explicitly discussed, three mechanisms are most commonly evoked: (I) "syneresis" shrinkage of a mud due to variable salinity (this mechanism is based on the experimental works cited above) (see van Straaten, 1954a, b, c; White, 1961; Donovan and Foster, 1972; Fairchild, 1980); (II) downslope creep of semiconsolidated surface layers resulting in "linear shrinkage cracks" (Picard, 1966, 1969; Pfeil and Read, 1980); and (III) substratal "compaction dewatering" either leading to injection from below (Van Houten, 1964; Hoffman, 1975; Hesse and Reading, 1978; Grotzinger, 1985, 1986a, b) or foundering from above (Plummer and Gostin, 1981). Compactional dewatering cracks formed during burial are also commonly termed syneresis cracks. Criteria used by workers who evoke mechanism I above include: (1) incomplete, simple gash to stellate patterns in plan; (2) curvilinear plan patterns; (3) small size (both in plan and cross-section); (4) parallel-sided cross-sections (Fairchild, 1980); and (5) lack of mudchip (intraclast) breccias. The last two, cited by Fairchild, are not in keeping with his own photos. Fairchild's Figure 6 (Fairchild, 1980, p. 492) clearly shows strongly folded cracks becoming narrower downward, and his Figure 8 shows mudflake breccias in intimately interbedded facies. Even in cases where the criteria are met they are clearly not unequivocal. Strongly curvilinear to tightly spiraling curlicue markings found in the troughs between oscillation ripple-marks on sandstone bedding planes are commonly interpreted as "syneresis cracks" (Picard, 1966, Fig. 4; and Plummer and Gostin, 1981, Fig. 4). These can be common in Upper Proterozoic rocks where their "merger into well-formed polygonal desiccation cracks led to the realization they had an inorganic origin" (Plummer and Gostin, 1981). The "figure-8" surface feature shown by Plummer and Gostin (their Fig. 5) is clearly not a mudcrack but rather it is a furrow gouged into the upper surface of a ripple-marked sandstone bed; most of these features are probably best interpreted as surface tracks and trails. Such surface features could easily serve as the nucleation points of desiccation cracks hence the merger of one into the other. Cracks attributed to mechanism II are typically spindle-shaped in plan and are commonly found in wave ripple troughs where their long axes are perpendicular to the wave impingement direction (Picard, 1966, 1969; Donovan and Foster, 1972). Picard attributes these "linear shrinkage cracks" to downslope creep of a surface layer, interpreting the ripple trend to be nearly that of the paleoshoreline. However, the strong alignment between trough trend and mudcrack trend may just reflect the confinement of mudcracks to mud drapes in the ripple troughs. Pfeil and Read (1980, p. 106, Fig. 11) show "pseudo-mudcracks" they interpret to be the result of downslope creep. The features have an orthogonal pattern in plan. In cross-section, the polygonal blocks are seen as slabs with cleanly broken edges (i.e., they have deformed in a brittle manner). The gaps between rigid plates were filled from both above and below by sediment that was soft enough to deform plastically and flow into the gaps, suggesting that these disruption structures may have been the result of compaction deformation (see under Mechanical Compaction below).

One of the most commonly cited mechanisms for the non-subaerial desiccation origin of mudcracks is substratal intrusion of sheet-like sedimentary dikes (injection structures, see Allen, 1984, p. 554-556). These are planar to irregular, discordant to concordant sheets that commonly branch or send out short apophyses (Allen, 1984, p. 555). These appear to have been emplaced relatively forcefully from below while the material composing them was deformable. Allen (1984) states that these features form where there is a state of horizontal tension in the layer(s) to be intruded and where the material to be injected is suitably plastic, namely fine-grained mud or sand and gravel with a muddy matrix or with a water pressure in excess of lithostatic pressure. In plan, sedimentary intrusions or dikes commonly have an en-echelon pattern. However, Eardley (1938) interpreted the cracks on the floor of Great Salt Lake as sedimentary "upwellings". Anketell and others (1970) described polygonal deformation patterns in experiments where layers with density inversions were deformed. Criteria suggested by Hesse and Reading (1978) for distinguishing injection features include: (1) crack filled from an underlying layer, and (2) top of the crack has a mound of sediment (i.e., sand or mud volcano) which may show internal strata dipping away from the crack. Where these features are not exposed however, such an injection origin will be difficult to recognize. Other features which suggest an injection mechanism would be a wall-parallel orientation of platy grains within the fill and transverse textural grading with the coarsest material in the center of the dike. These two features arise through shear sorting during the laminar flow emplacement of the material.

Complex mudcrack patterns interpreted to be of the upward injection type are reported from the Mesozoic Lockatong Formation (van Houten, 1964) and the Precambrian Rocknest Formation of Canada (Hoffman, 1975; Grotzinger, 1985, 1986a). In these studies, the following criteria were used in support of an injection origin for the cracks: (1) complex pattern of cracking in plan (i.e., not simple polygons), (2) wavy to irregular crack cross-sections that commonly branch, (3) brecciated crack fill composed of host sediment broken apart by the forceful injection of the surrounding material, (4) common mud fillings, (5) protrusions up into an overlying layer at the top of a crack, (6) cracks "opening upward" ("veeing" upward?) with the fill coming from below, (7) lack of intraclastic conglomerates, and (8) cracks originating from many different surfaces when viewed in cross-section. Clearly, the single most convincing criterion is documentation of the source of the crack fill from a layer below. Cracks that contain several generations of sediment fills with micro-stratigraphy relatable to overlying layers cannot be formed by this mechanism. Both Hoffman (1975) and Grotzinger (1985, 1986a) interpreted mudcracks in the Precambrian Rocknest Formation as due to compaction dewatering involving injection. However, in their published photographs of crack cross-sections the cracks are clearly filled from above and show multiple generations of fillings. The other criteria cited above are equivocal. A complex surface pattern is common in modern subaerially mudcracked sediments and results from the superimposition of cracks from overlying layers and from repeated wetting and drying of a sediment surface exposed for many years (see Smoot, 1983). Waviness, branching, and brecciated fills are common in many modern desiccation mudcracks. Where mudcracks are compacted, waviness has no genetic significance. Also, compaction can produce intrusion of a noncompactable fill up into muddy layers at the top of a crack (see Pettijohn, 1975, p. 123). Thus, care must be taken in applying this criterion. Finally, in a number of published photographs of "syneresis cracks," the claim has been made that the cracks arise from many different surfaces (e.g.,

Grotzinger, 1986a, Fig. 13-b). However, Grotzinger's photographs of "syneresis cracks" (see also Hofman, 1975, Fig. 30-6) clearly show mud cracks originating at a few stratification surfaces. Plummer and Gostin (1981) show a number of examples of mudcracks radiating from small load structures and covering larger ball and pillow structures. They maintain that the majority of "syneresis" cracks in the geologic record were the result of sediment compaction or foundering induced by deposition of overlying sediment. Cracks generated in this way would be simple with only one generation of fill. Laminae overlying the fill should be deformed.

In deciding on the origin of a mudcrack pattern, the important step is to determine whether the cracks are filled from above or from below. If the source can be identified as below, or if mud or sand volcanoes can be identified at the top of a crack, then a compaction injection origin is a viable interpretation. Cracks filled from above and that show multiple generations of cracking and filling with a fill stratigraphy relatable to overlying layers were surely open at either the sediment-air interface or the sediment-water interface (Fig. 55). It seems most likely that such cracks in carbonate rocks are of a subaerial desiccation origin considering the typical paucity of swelling clays in most shallow marine carbonates and the lack of any field evidence from modern carbonate environments of subaqueous mudcracks. Small, incomplete mudcracks with simple fills from above are difficult to interpret. If associated with load structures or if there are deformed laminae over the crack fill, they may be due to loading. However, if such evidence does not exist, it is most probable that such cracks are of subaerial desiccation origin. Again, this is especially true in carbonates which contain at most only small proportions of siliciclastic clay minerals. No reliable criteria exist to separate simple incomplete desiccation cracks from subaqueous cracks.

Occurrence and preservation.-As noted above, mudcracks in a number of shallow-water carbonate platform deposits have been interpreted to be of a "syneresis" origin. However, as outlined in the discussion above, until more reliable criteria to unequivocally identify "syneresis" cracks are developed, little can be said with certainty about the occurrence and preservation of "syneresis" cracks in the shallow marine carbonate record.

Significance.-It is our view that the criteria commonly used to interpret cracks as being of non-subaerial desiccation origin, that is, "syneresis cracks," are inadequate and untrustworthy. We must stress that to assign a "syneresis" origin to mudcracks based on the nature of the crack plan or on the cross-sectional morphology is quite unwarranted in view of the evidence and arguments presented here and in the section on subaerial desiccation cracks above. Clearly, much more work is needed in this area. It must be emphasized that this is not a trivial matter. The "classic" syneresis (compaction dewatering) cracks of the Lockatong Formation have been convincingly reinterpreted as desiccation mudcracks formed on a playa mudflat (Smoot and Katz, 1982). Likewise, the "syneresis" origin of mudcracks of "subtidal" parts of the peritidal cycles from the Rocknest Formation is open to serious question in our opinion. Based on the published photographs (Hoffman, 1975, Fig. 30-6; Grotzinger, 1986a, Fig. 13b), these cracks have the attributes of subaerial desiccation cracks, an origin that would be in keeping with the halite casts characteristic of these mudcracked mudstones. Such a reinterpretation of the mudcracks would necessitate a complete reinterpretation of the Rocknest cycles. This, in turn, would obviously raise serious problems with the analysis of the sea-level oscillation control of the cycles proposed by Grotzinger (1985, 1986a, b).

Neptunian Dikes.-

Terminology.-Fischer (1964, p. 133) applied the term neptunian dike to discordant dikes of limestone within the Triassic Dachstein Formation of the Northern Limestone Alps (Fig. 56). Playford (1984) introduced the term neptunian sill for similar fissure fills that follow bedding planes (see also sheet cracks).

Distinguishing characteristics.-The neptunian dikes of the Dachstein are up to a meter wide and traceable laterally and vertically for up to tens of meters. Significantly, the dikes are filled with both sediment and radiaxial fibrous calcite cement. The dikes at the top of the Dachstein are filled with crinoidal grainstones and ammonite-bearing mudstones of early Jurassic age. However, in other portions of the formation, dike fills are composed of peloidal mudstones similar to one of the subfacies of the Dachstein. Rare mass occurrences of Triassic rhynchonellid brachiopods are found in the peloidal mudstones within the dikes. Dikes rarely show stratigraphic offset of bedding across them. Some of the meter-wide neptunian dike fills in the Dachstein are composed of mauve-colored mudstones cut by an extensive system of sheet cracks developed normal to the dike walls. These sheet cracks are filled by internal sediment and radiaxial spar. Fischer (1964, p. 135) introduced the term zebra limestone to describe this fill. However, the term zebra structure or zebroid structure is now commonly used to describe breccias cemented by saddle dolomite (see Beales and Hardy, 1980; Morrow and others, 1986). This latter variety of breccia is common in Mississippi Valley Type hydrothermal deposits such as the Presqu'ile and Manatoe "facies" of the Devonian of Canada (Aulstead and Spencer, 1985; Aulstead and others, 1988).

Playford (1984) describes spectacular examples of neptunian dikes and "neptunian sills" within Famennian shelf margin shoal and reef deposits in the Oscar Range of the Canning Basin, Western Australia. The dikes range from 100 m long, 50 mm wide and 20 m deep, up to 6 km long, 20 m wide, and 80 m deep. Dikes are parallel to the reef front but a set of fractures normal to the reef front, only rarely has fills. The dikes pinch out at depth but some pass into sills in underlying slope deposits. Other sills in the slope deposits are commonly less than 20 mm thick. Canning Basin neptunian dikes and sills show a complex history of crack development, filling, cementation, and recracking. Crack walls are encrusted by *Renalcis* and stromatolites. Crack-filling detrital sediments include skeletal, ooid, and peloid grainstones, siliciclastic sandstones, and lime mudstones. Cements are composed of radiaxial calcite. In addition, Playford (1984) describes two sediment grain types, unique to the fissures, which he interprets as internal sediments: cavity peloids and spar balls. Cavity peloids are distinctly colored peloids that are commonly interlayered with and cemented by radiaxial calcite. Spar balls are subspherical masses up to 0.25 m in diameter composed of a number of layers of radially-oriented, radiaxial spar nucleated around fragments of dike-fill material. Neptunian dikes and sills are also found in shelf-margin facies of the Proterozoic Rocknest Formation of Canada (Grotzinger and Read, 1983; Grotzinger, 1986a, b). These contain dolomite and silica pseudomorphs after botryoidal fibrous aragonite (up to 50 mm in diameter) that lines fissure walls.

Origin and significance.-The observations above suggest that neptunian dikes are fractures that developed in penecontemporaneously cemented platform and shelf-margin facies. These fractures may have been caused by local tectonic movements. In addition, in the Canning Basin, fracturing, slippage,

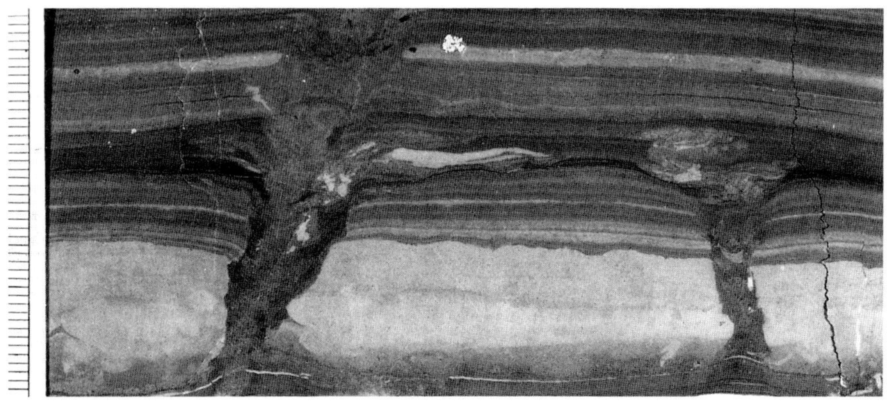

Fig. 55.-Cross-sections of mudcrack fills from the Middle Ordovician St. Paul Group of the central Appalachians. Note that the cracks contain several generations of sediment fills with a micro-stratigraphy relatable to overlying layers. These observations demonstrate that these cracks were open at a sediment-air or sediment-water surface a number of times and that these cracks could not have formed from upward "injection" of sediment. See text for further details. Scale is in mm.

Fig. 56.-Laminated area below pocket knife is the fill of a Neptunian dike from the Lofer facies of the Triassic Dachsteinkalk at Dachstein section, Austria (see Fischer, 1964).

and wholesale collapse of the shelf margin apparently took place along the neptunian dike and sill fracture system as uncemented muds in front of the margin compacted. Cracks were open at the surface and filled from above by marine sediments in many cases (Smart and others, 1988). The cements and the marine internal sediments interlayered with the cements provide evidence that these were fractures filled from above and are not injection structures. Internal sediments such as spar balls and encrusting organisms within the cracks suggest that there were significant flows of marine waters through these conduits, especially those at shelf margins. Neptunian dikes might also represent important pathways for diagenetic fluids other than seawater (Playford, 1984).

Occurrence and preservation.-Neptunian dikes and sills are most commonly reported from areas where large exposures are available (see examples cited above). However, in small outcrops, these features might be mistaken for tepee fills (see below).

Injection dikes.-
These have been discussed above under syneresis cracks.

Soft Sediment Deformation Structures

Convolute folds.-
Terminology.-Convolute bedding and convolute lamination (see Kuenen, 1952, p. 31; Kuenen, 1953, p. 1056; Pettijohn and Potter, 1964, p. 292; and Allen, 1984, p. 349) are soft-sediment deformation structures that are confined within one layer (set or coset) in otherwise small-scale cross-stratified, large-scale cross-stratified, or planar stratified fine-sands to coarse silts. There are no generally accepted definitions for distinguishing between convolute bedding and convolute lamination (see, however, Collinson and Thompson, 1989).

Distinguishing characteristics.-A set of convolute folds consists of a laterally extensive series of more or less regular folds that are confined between two undeformed layers (Fig. 57). The folds, which may range in height from centimeters to meters, generally comprise upright, cuspate to conjugate, tight anticlines separated by broader, more open synclines. Anticlines amplify (i.e., increase in amplitude and sinuosity) up-

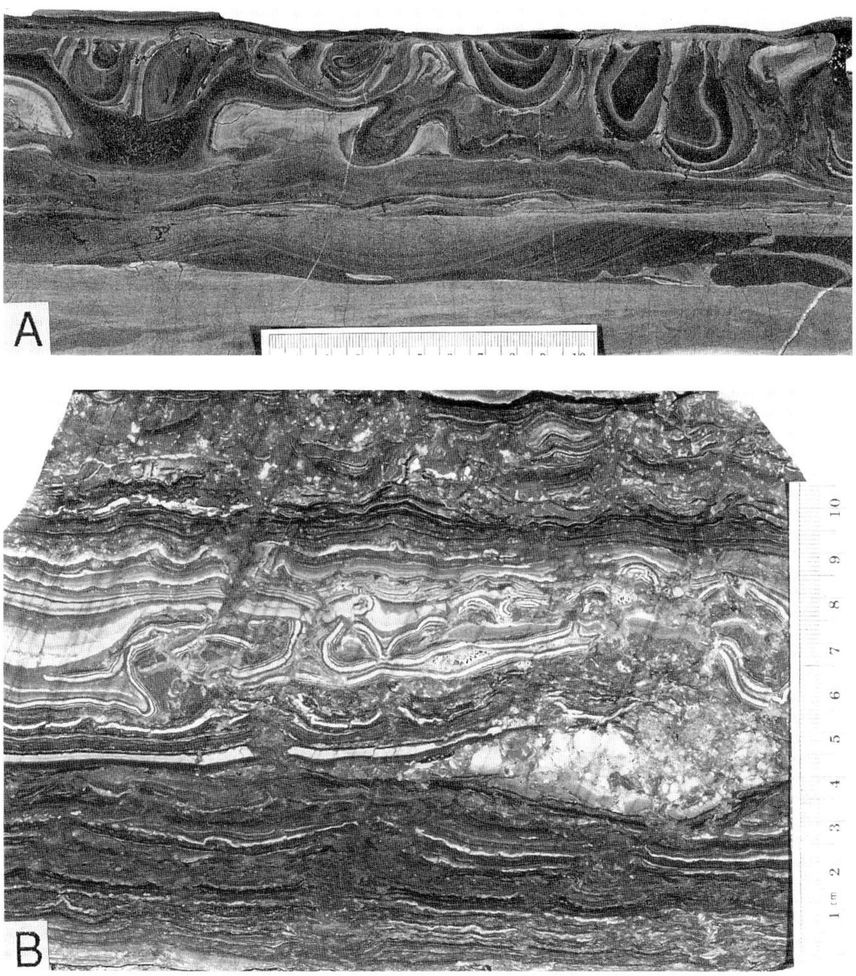

Fig. 57.-Convolute folds. (A) Thin bedded heterolithic subfacies from the Upper Silurian Wills Creek Formation, western Maryland, that includes a zone of convolute folds. Note the internal erosion in some of the synclines and the truncation of the entire folded layer. Scale in mm. (B) Convolute folding in crinkled laminated mudstones from the Upper Silurian Tonoloway Formation, western Maryland. Scale in mm. Sample courtsey of Tom Tourek. (C) (Following page) Thin zone of convolute folds (a little below the ruler) in a Precambrian wave-ripple cross-stratified dolomite from the Stark Formation, Christie Bay, Northwest Territories. The layering above and below the convolute folds is undeformed and consists of a composite set of planar laminated, wavy laminated, and small-scale cross-stratified dolomite. Photograph courtesy of Paul Hoffman.

C

ward and may have secondary, smaller folds in their cores. In some instances, anticline cores are completely disrupted and structureless. Where folding dies out within a layer, the synclines amplify downward. Commonly, however, the folds are erosionally truncated at the top of the bed. Fold axes are mostly symmetrical to moderately asymmetrical; overturned limbs are uncommon. Axial surfaces of the rare asymmetric folds on average lean in the paleocurrent direction (Allen, 1984, p. 351, and references cited therein) and may be of limited use as paleocurrent indicators. In bedding plane views, the folds are seen as basins and domes. Rare cylindrical folds in bedding plane views have axes either parallel to or at an angle to the paleocurrent direction (Allen, 1984). Laminae are commonly traceable through a number of folds. However, laminae geometry allows three types of convolute bedding or lamination to be distinguished. Allen (1984, p. 351) calls these: (1) post depositional, (2) metadepositional, and (3) syndepositional convolute bedding or lamination. In "post depositional" convolute bedding and lamination, folds die away upward and downward into undeformed laminae and discordances between laminae are confined to recognizable, preserved remnants of cross-stratification. In metadepositional convolute bedding and lamination, folds are all sharply truncated by erosion along the top of the layer. Synsedimentary convolute bedding and lamination (Fig. 57B) is characterized by complicated internal erosion of anticline cores and contains lensoidal sets of laminae that drape up onto anticlines from adjacent synclines. Thinning of the sets is commonly erosional. The internal erosion and deformation during deposition and wholesale erosion of the folds at the top of the layer set the timing of deformation (Kuenen, 1953; Crowell and others, 1966; Allen, 1984, p. 353).

Origin and significance.-Allen (1984) has given a detailed, critical discussion of the possible origins of these structures and other soft sediment deformation features. He attributes as the primary mechanism of convolute bedding, dish-and-pillar structures, load casts, ball-and-pillow structures, and "sand balls" the synsedimentary to post depositional liquefaction of sediments with a gravitationally unstable bulk-density gradient. Liquefaction can be induced in cohesionless sands and silts by (1) changes in pore fluid pressure brought on by, for example, pore fluid seepage (static liquefaction), (2) vibratory pressure changes caused by turbulent currents passing over the unstable sediment (dynamic liquefaction), and (3) earthquake-induced shaking (cyclic liquefaction). With regard to the latter origin, Allen (1984, p. 298-304), in his detailed discussion of dynamic and cyclic liquefaction, notes that seismically-induced liquefaction is mainly associated with earthquakes of magnitude 6 or more and may lead to deformation of cohesionless sediments over areas covering hundreds to thousands of square kilometers. Clearly, then, convolute bedding and other related soft sediment deformation features are potentially of considerable significance as recorders of major earthquakes. However, it has yet to be established how we can determine with any certainty that any given convoluted bed resulted from seismic activity rather than from some more commonplace phenomenon such as pore fluid seepage or turbulent current bed stress fluctuations. In this regard, the scale of the deformation, the suite of associated soft-sediment deformation features, and the paleoenvironmental setting will need to be taken into account. In shallow water carbonate environments, unstable depositional slopes are typically absent (except perhaps along the banks of tidal channels) so that seepage-induced failure can be given a low probability. Associated deformation structures, if present, could possibly aid in distinguishing between a seismic origin and, say, a storm-induced current origin. Both may produce soft-sediment deformation over an area of many tens or hundreds of square kilometers but if the convolute folding is accompanied by other liquefaction features such as sand dikes, dish-and-pillar structures, etc., then serious consideration should be given to a seismic origin. For example, near the top of the Lower Ordovician Beekmantown Group of the central Appalachians in western Maryland, a 5 meter thick unit within the tidal flat facies of the Pinesburg Station Dolomite that carries beds with convolute-like folds, injection dikes coupled to delaminated layers, and breccia pods, can be traced in scattered outcrops over an area of at least several tens of square kilometers (A. Freeman and L. Hardie, unpublished study). The complex nature of the soft-sediment deformation, particularly the presence of breccia pods that contain blocks of tidal flat laminites up to a meter in length, would seem to warrant stresses more in keeping with shocks from a major earthquake or succession of earthquakes than those provided by tidal or storm currents.

Occurrence and preservation.-Convolute folds are not commonly reported from shallow-water platform carbonate deposits, giving perhaps some weight to an origin by exceptional events such as major earthquakes. It is possible that convolute lamination and similar soft-sediment folding has been overlooked, or in some cases even mistaken for other structures such as stromatolites, but if our own observations are any gauge then this soft sediment deformational structure is simply not common in ancient shallow water carbonate deposits.

Load casts and flame structures.-

Terminology.-Load casts were first recognized by Fuchs (1895) who called them "Fleisswulste." Sorby's (1908) illustration of these structures was reproduced by Shrock (1948, p. 159, Fig. 118c) who called them "flow casts." Pettijohn and Potter (1964, p. 318) attribute the term "load cast" to Kuenen (1953) and the term "flame structure" to Walton (1956). Many other older synonyms and ancillary terms are listed by Pettijohn and Potter (1964).

Distinguishing characteristics.-Load casts are downward-oriented "swellings in the shape of slight bulges, deep or rounded sacks, knobby excrescences, or highly irregular protuberances" (Kuenen, 1953, p. 1048) commonly found on the base of sand, sandstone, and grainstone layers that overlie muds or mudstone (Fig. 58). Load casts are commonly millimeters up to decimeters in diameter. Where seen on bedding surfaces, the protrusions may be more or less equant with "a tendency toward polygonal form" (Allen, 1984, p. 537) or highly irregular. In cross section, the coarser material is commonly in rounded pockets with mudstone plumes in sharper, diapir-like "flame structures." Pockets are relatively symmetric about more or less vertical axes. In other cases, however, the axes of the pockets are all inclined normal to the paleocurrent direction (Allen, 1984, p. 356). Where sand pockets are attached to the overlying bed by a narrow constriction, they are known as "load pouches," and where completely detached, "load balls" (Pettijohn, Potter and Siever, 1987, p. 113). Layering within load casts may be preserved whereas layering within the muds commonly is concordant to the shapes of the diapirs and "flames."

Origin and significance.-Allen (1984) following Dzulynski and his coworkers (Dzulynski and Walton, 1963, 1965; Dzulynski, 1965, 1966; Anketell and others, 1970) attributes

Fig. 58.-Outcrop view of load casts and associated flame structures on the base of fine peloidal grainstones from the Ordovician of Adams County, Ohio. See Pettijohn and Potter (1964, Plate 102). Photograph courtesy of Francis J. Pettijohn.

load casts and flame structures to liquefaction of layers in a gravitationally unstable system where a denser layer overlies a lighter layer. Irregularities along that surface (e.g., ripples, flute marks, burrows, etc.) may localize the load casts, but are not necessary for their initiation and growth.

Occurrence and preservation.-Load casts and flame structures are not commonly reported from shallow-water carbonate platform deposits.

Mechanical Compaction Structures

Terminology.-

Bedding plane stylolites and interpenetrating or sutured allochems in grainstones (Meyers and Hill, 1983) are familiar volume reduction features of carbonates that result from chemical dissolution driven by overburden pressure (the "chemical compaction" of Shinn and Robbin, 1983). On the other hand, mechanical compaction structures in carbonate mudstones are usually overlooked, even though they may be quite common (for example, the compacted mudcracks of Pettijohn, 1975, Fig. 4-26; and other, more extreme "molar tooth" structures of Horodyski, 1976). *Mechanical compaction here refers to volume reduction in unlithified sediment by simple mechanical rearrangement of cohesionless grains due to pressure produced by the weight of overlying material.* It has become a paradigm of carbonate sedimentology that carbonate mudstones rarely show mechanical compaction features. For example, Scoffin (1987, p. 114) states that "mechanical compaction is rarely manifested in shallow-water limestones, because they are generally partially cemented (and thereby made rigid) early on... Very few limestones with micrite matrix show crushed grains." Shinn and others (1977) cite a paper by Weller (1959) and an abstract by Pray (1960) as the main sources of the "early cementation- no mechanical compaction" view of carbonate mudstones. In addition to the much cited observations that delicate fossil grains are rarely crushed or broken in carbonate mudstones, Pray (1960, p. 1946) offered a number of other lines of evidence for lack of mechanical compaction in such rocks, as follows: "supporting data are the scarcity of drag or penetration effects where rigid clasts or fossils occur in a calcilutite matrix; and the general similarity in fabric of sediment within a noncompacted cavity, such as a shell, and sediment of the enclosing matrix. Larger-scale evidence is provided by observations of attitude, thickness, lithology, and geopetal fabrics of calcilutites which are laterally adjacent to other rocks for which minor compaction can be demonstrated." Another aspect of the "early cementation- no mechanical compaction" view of carbonates has been raised by Logan and Semeniuk (1976) and Wanless (1979, 1982). Wanless (1979, 1982) especially has used volume reduction features in dolostones and interlayered limestones-dolostones to argue for pervasive post-lithification pressure solution. In particular, Wanless claims that pervasive dolomitization is a result of massive pressure solution, and that together the pressure solution and dolomitization processes remove over 70% of the original rock and thereby alter the primary sedimentary features, such as cross stratification, beyond recognition. Many authors, however, accept neither the "early cementation- no mechanical compaction" view nor the pervasive pressure solution alteration ideas of Logan and Semeniuk (1976) and Wanless (1979, 1982). Shinn and others (1977) point out that the "early cementation- no mechanical compaction" view leads to the problem of the sources for the huge amounts of cements needed to transform an uncompacted lime mud with 70-80% porosity into a carbonate mudstone with porosity of only a few percent. Shinn and others (1977) and Shinn and Robbin (1983) have argued that a major amount of pre-lithification mechanical compaction may be common in carbonate mud-rich rocks. Their major lines of evidence come from simple experiments where cores of lime mud from modern carbonate environments were artificially compacted in a device that allowed for pore water escape. The compacted material was then compared to an uncompacted control sample. In compacted cores, fossils and other platy fragments commonly were flattened and reoriented perpendicular to the pressure, but showed little evidence of breakage. Open burrows and fenestrae collapsed (but filled burrows were flattened), and soft peloids merged into mud whereas hard, cemented peloids remained intact. Also, in the compacted cores, primary layering structures were preserved and in some cases accentuated. Most significantly, seagrass leaves and concentrations of other organic matter produced wispy "stylolite-like" seams which wrapped around grains and produced a crude layering in what were originally homogeneous, bioturbated muds. Shinn and others (1977) and Shinn and Robbin (1983) made direct comparisons of their experimentally produced features with similar features in ancient carbonate mudstones. Most controversial has been their comparisons of the wispy organic seams of the compacted modern lime mud with the concentrations of insoluble residues commonly found in ancient carbonates and referred to as micro-stylolites and horse-tail stylolites (Fig. 59).

Distinguishing characteristics, occurrence, and preservation.-

The Cambro-Ordovician carbonates of the central Appalachians comprise over 3.5 km of peritidal and shallow-platform carbonates that contain a large number of well-preserved sedimentary and diagenetic features (Reinhardt and Hardie, 1976; Demicco and Mitchell, 1982; Hardie and Shinn, 1986; Nguyen, 1986). Volume reduction features are very common in these rocks, especially in the mudstones and wackestones (Hardie and others, 1982; Goldhammer and others, 1985; Hardie, 1989), and the characteristics of these features can be used to shed more light on the question of mechanical compaction in carbonates. In this section, we: (1) describe the volume reduction features and associated vertical and lateral changes of rock type, rock fabric, and rock mineralogy commonly found in the Cambro-Ordovician carbonates of the central Appalachians; and (2) examine the roles of pre-lithification mechanical compaction, fabric-selective differential cementation, and dolomitization in the origin of these features. While the following discussion focuses on Cambro-Ordovician carbonates of the central Appalachians, we would stress that we have found these features to be common to one degree or another in virtually all ancient shallow-water muddy platform carbonates that we have examined.

The common suites of volume reduction features in the Cambro-Ordovician carbonates described here are associated with: (1) thrombolites; (2) limestone patches in dolostone; (3) alternating, millimeter- to centimeter-thick layers of limestone and dolostone ("ribbon-rocks"); and (4) dolomite mudstones. Observations are drawn from outcrops, hand samples, and thin sections.

Thrombolites are decimeter- to meter-scale bioherms commonly interpreted as in-place biohermal mounds produced by the problematic framework-building fossils *Girvanella*, *Renalcis*, *Epiphyton*, and *Nuia*, and by the calcification of coccoid microbial communities. Thrombolites of various scales are common in the Cambro-Ordovician carbonates of the central Appalachians and carry abundant evidence that they were rigid when deposited (see discussion of thrombolites under Biogenic Structures). In many examples (Fig. 60), these bioherms are encased in thin beds of carbonate mudstone which bend

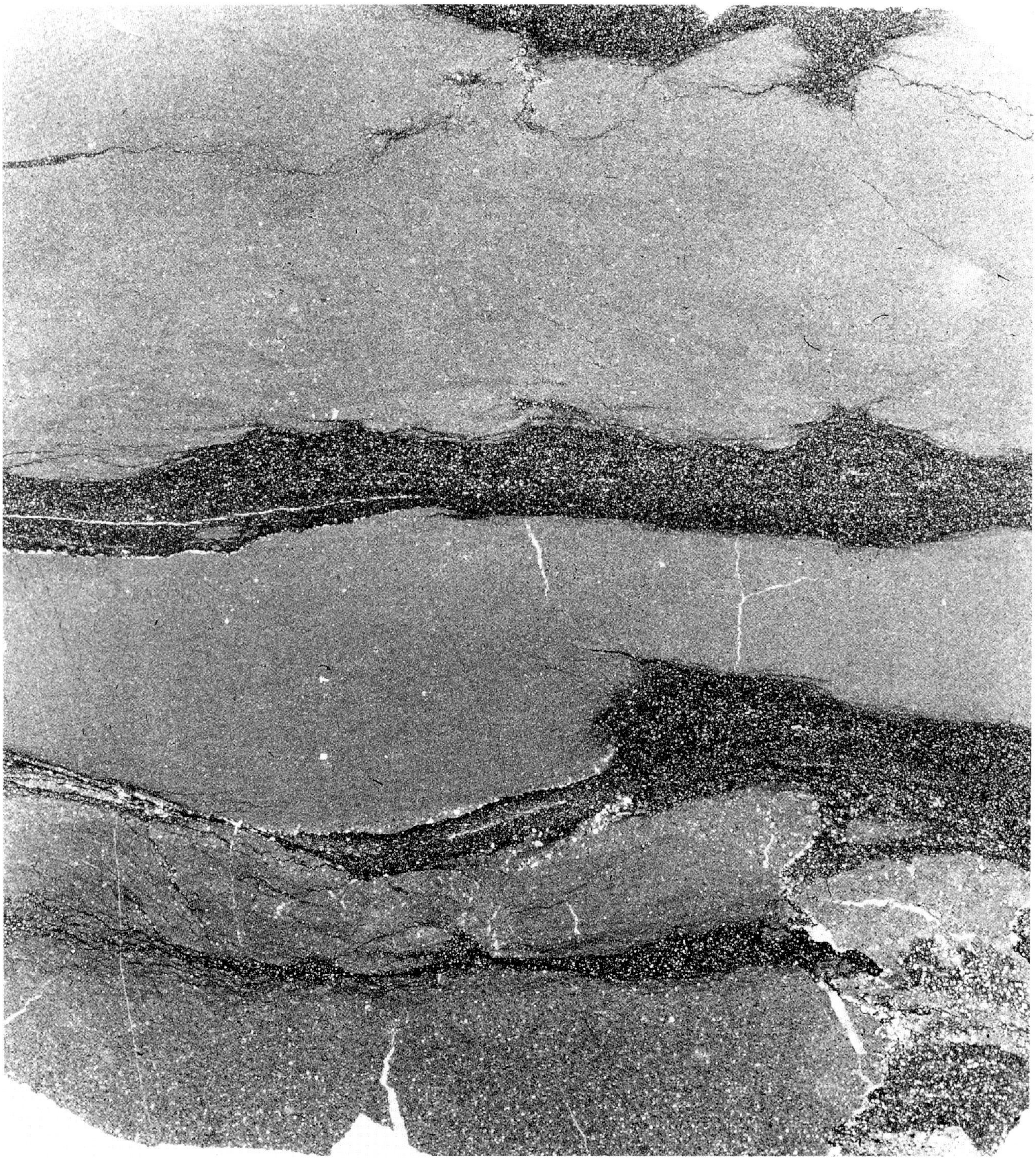

Fig. 59.-Thin-section photomicrograph of "horsetail stylolites" from wavy laminites of the Upper Cambrian Conococheague Formation of western Maryland. Field of view is approximately 25 mm high. Light gray layers are lime mudstones (note the fine-sand to silt-sized peloidal textures in lowest layer). Dark layers comprise dolomite rhombohedrons approximately 30 microns in diameter floating in a dolomite mudstone. Thin section courtesy of Miner Long.

Fig. 60.-Small stromatolitic bioherm encased in thin beds of carbonate mudstone that bend around the bioherm from both above and below. Affected beds thin laterally above and below the bioherm and show increasing dips and thicknesses down the sides of the bioherm. Beds are thickest along the flanks of the bioherm and thin laterally as well. Upper Cambrian Conococheague Limestone, western Maryland.

around the thrombolite from both above and below (drag or penetration effects of Pray, 1960). The affected beds thin laterally directly above and below the bioherm and show increasing dips and thicknesses down the sides of the bioherm. Significantly, beds are thickest along the flanks of the bioherm and thin not only over and under the bioherm but laterally away from the bioherm as well.

Limestone patches partially to completely surrounded by dolostone (Figs. 61, 62) are fairly common in the Cambro-Ordovician carbonates of the central Appalachians. It is important to note that there is a complete gradation between these features and the interlayered limestones and dolostones described below and that they share many features. Significantly, many of these features are similar to nodular, irregular, "flaser," and lumpy bedding described by Wilson (1975), Wilson and Jordan (1983), Matter (1967), and Schwarz (1975), among others. Figure 62 shows a dark, centimeter-scale limestone column partially surrounded by lighter dolostone. Note how the dolostone laminae plunge down and around the dark limestone column. Especially note how some of the laminae in the dolostone both to the right and left of the column are deformed into folds directed away from the column (Fig. 62B). This offers clear evidence that volume reduction took place while the material that is now dolomite was free to flow, i.e., was ductile. The limestone in the column is a fine sand-sized peloidal grainstone with a high pre-cement porosity and with well preserved, unbroken ostracod shells (Fig. 62C). The surrounding dolostone is composed of microcrystalline dolomite with isolated euhedral rhombohedrons several tens of microns in diameter. The microcrystalline dolomite has coarse silt-sized, rounded clotted areas that probably represent relict peloids (structure grumeleuse, Bathurst 1975, p. 511). Ostracod shells in the dolomite are preserved both intact and crushed depending on whether they possess an internal fill (intact shells) or not (broken shells) (Fig. 62D). Slabs cut parallel to the bedding plane clearly show that the limestone patches are following mudcracks (Fig. 63) and burrows. Figure 63B is a cross-section of the mudcracks of Figure 63A. Note here that the dolomite-calcite boundaries cut across sedimentary layering. Laminae in dolostone above and below a limestone patch show drag effects laterally as they pass around the patches. Important features of these limestone patches in dolostone are: (1) delicate shells are preserved in the limestone but crushed in the dolostone, (2) spar-filled fenestrae and intergranular pores are preserved in the limestone but not in the adjacent dolostone, and (3) cylindrical burrows in limestone become attenuated and elliptical in cross-section where they pass into dolostone (Demicco and Mitchell, 1982, p. 177).

Interlayering of limestone and dolostone on a millimeter- to centimeter-scale is a common and striking feature of the Cambro-Ordovician carbonates of the central Appalachians. These interlayered rocks grade from the lumpy, irregular and nodular varieties (Fig. 27) to wavy- and lenticular-bedded ribbon-rocks (Fig. 28, see Demicco, 1983). In Figure 28A, note how flat-bottomed, small-scale cross-stratified grainstone beds with ripple-form tops alternate with dolostone layers which thin over the crests of underlying ripples and thicken over their troughs. Also note how the mudcrack fills are deformed (compacted) in the dolostone but not in the limestone. In many examples of interbedded limestones and dolostones, there is ample petrographic evidence that the limestones were sands and the dolostones were muds when deposited (Demicco, 1983). Many of the limestone layers carry small-scale, spar-filled fractures that penetrate from both the top and the bottom of the layers (Fig. 25A). The layers themselves may be broken into trains of isolated plates around which the dolomite is draped in the style of boudinage structure. These features unequivocally indicate that the limestone layers underwent brittle deformation whereas the dolostones deformed ductilely by flowage around the limestone boudins and, in many cases, into the cracks.

Dolomitic mudstones carry evidence of volume reduction in the form of mudcrack fills that are telescoped and burrows that are flattened. Platy grains and shell fragments are generally parallel to bedding. Anastomosing, wispy argillaceous seams are very common. These "microstylolite swarms" *do not cut through larger allochems, but instead drape around them* (Fig. 59), demonstrating that they are not dissolution features but instead are the result of mechanical compaction.

Origin.-

Clearly, many of the finer-grained dolomitic mudstones of the Cambro-Ordovician carbonates of the central Appalachians

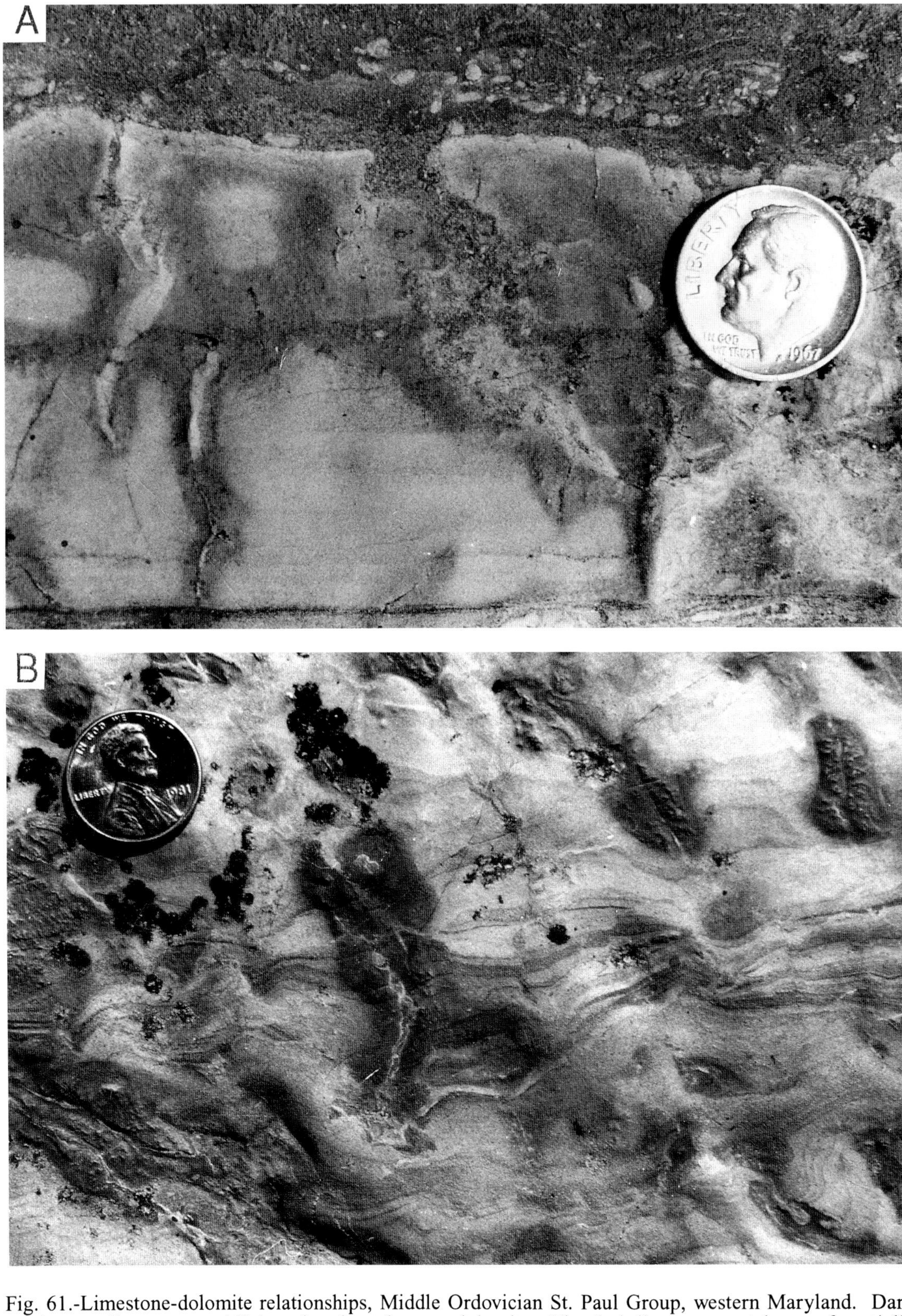

Fig. 61.-Limestone-dolomite relationships, Middle Ordovician St. Paul Group, western Maryland. Dark patches are limestone and light-colored patches are dolomite. The dark limestone patches preferentially surround mudcracks (A; see also Fig. 63) or burrows (B). Note how the limestone-dolomite boundaries cut across primary layering, indicating post-depositional alteration (see text). In (B), note the differential compaction around the limestone patches.

A

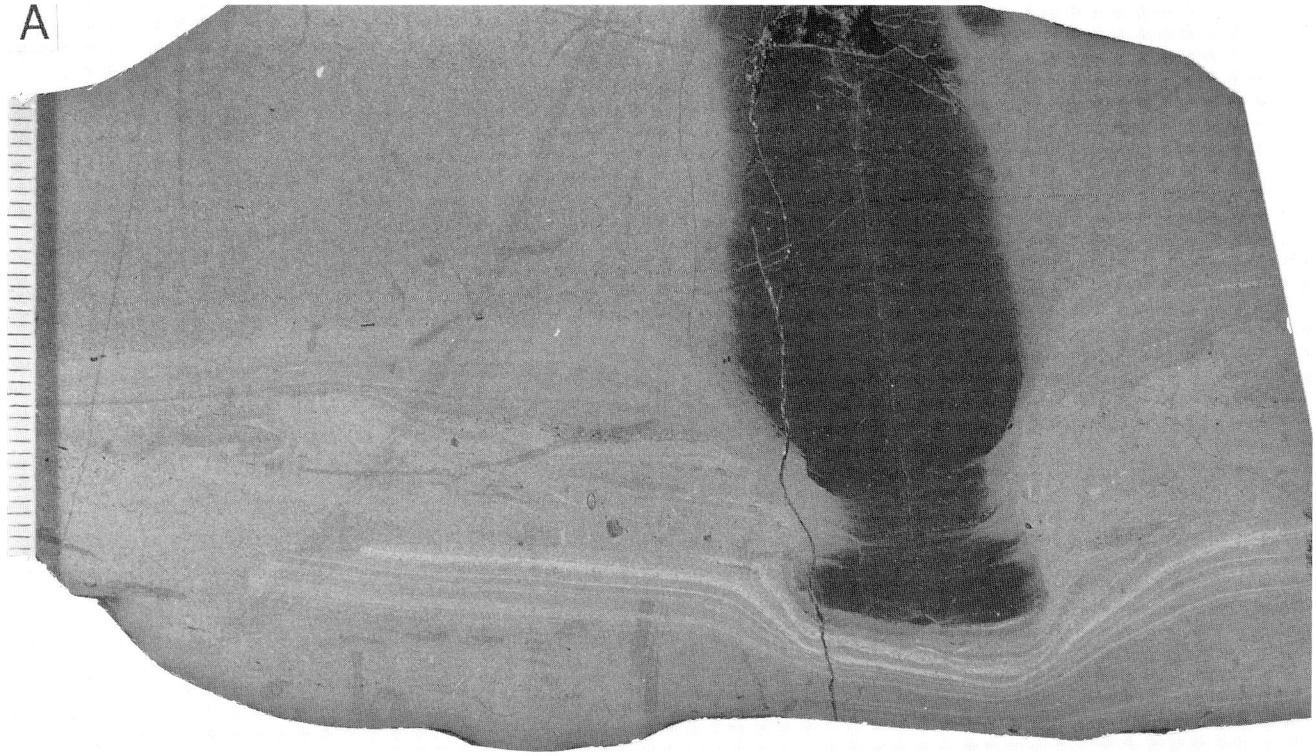

B

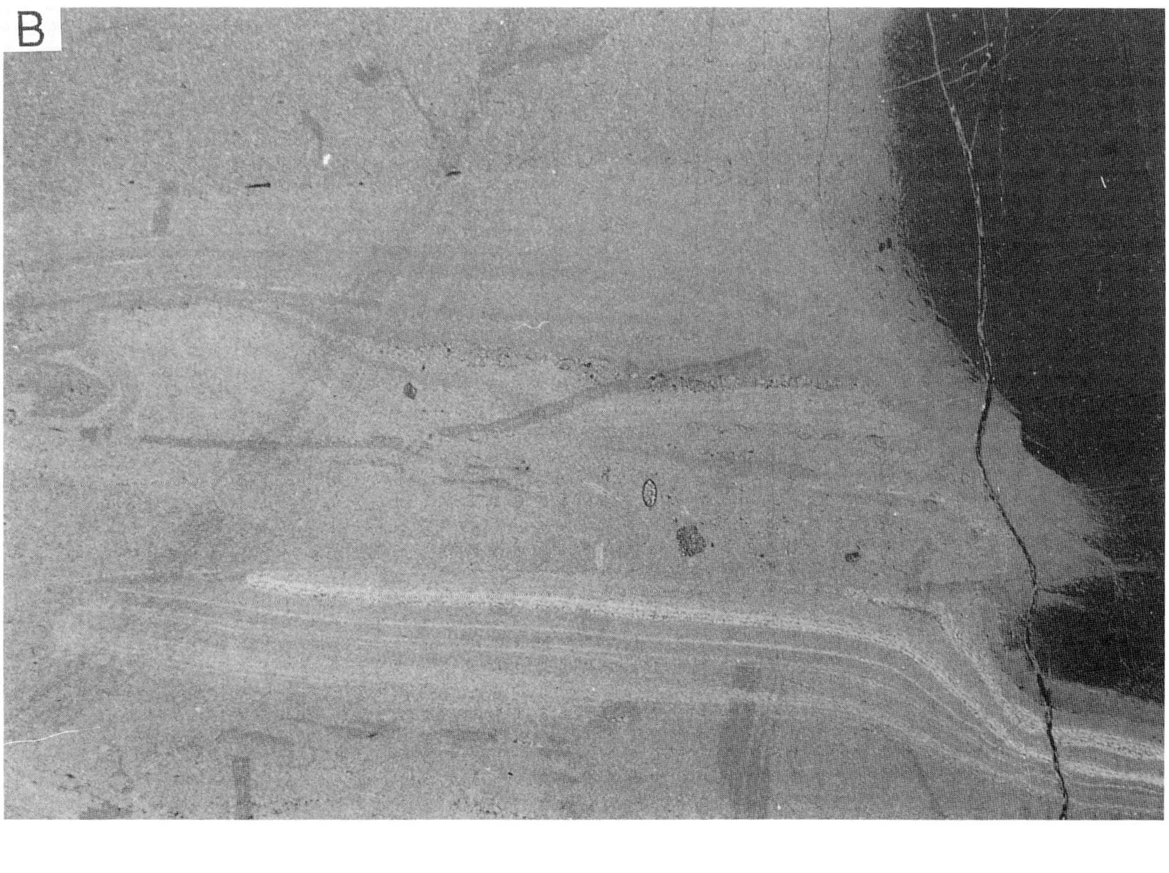

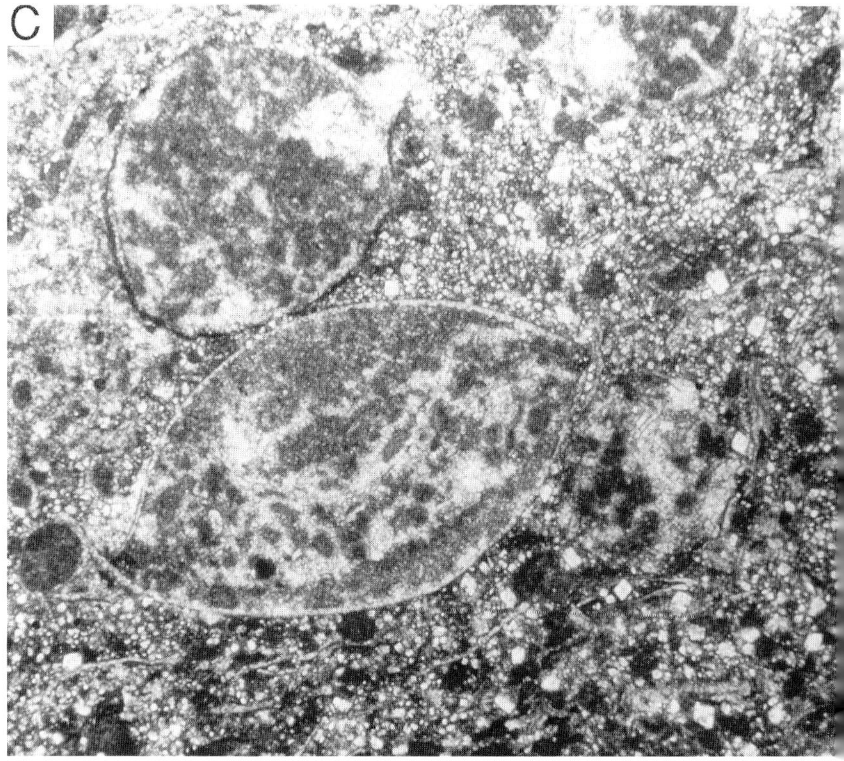

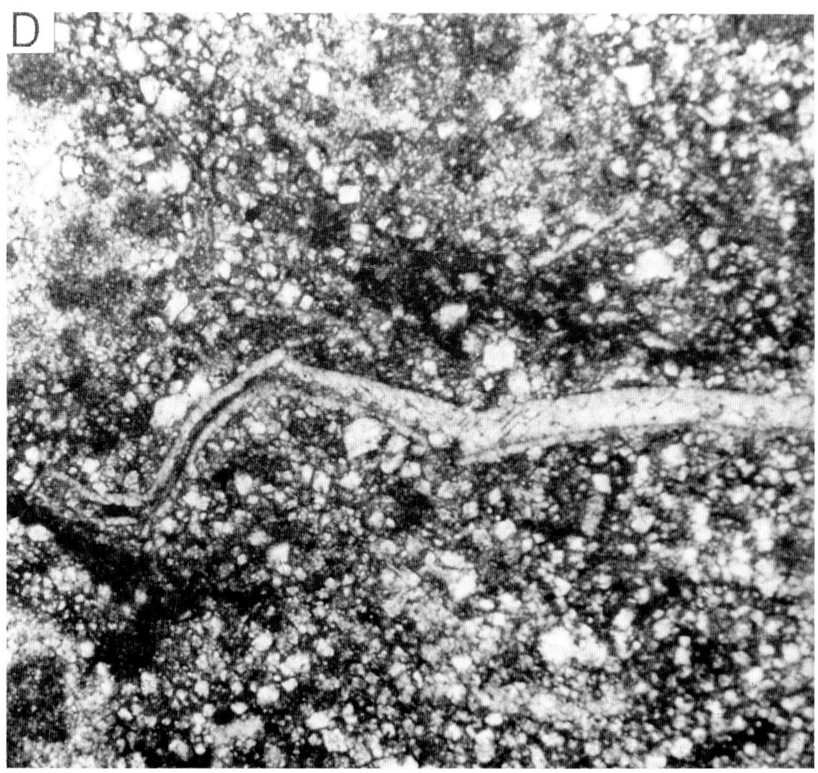

Fig. 62.-Details of limestone-dolomite relationships, Middle Ordovician St. Paul Group, western Maryland. (A) and (B) (Preceeding page) Dark limestone column partially surrounded by lighter dolostone, scale in mm. Note: (1) how the dolostone laminae plunge down and around the dark limestone column, and (2) how some of the laminae in the dolostone both to the right and left of the column are deformed into folds directed away from the column (shown in detail in B). This offers clear evidence that volume reduction took place while the limestone was rigid but the material that is now dolomite was free to flow, i.e. was ductile. (C) Thin-section photomicrograph of limestone in the column showing fine sand-sized peloidal grainstone texture with a high pre-cement porosity and with well-preserved, unbroken ostracod shells. Small light-colored rhombs are dolomite crystals which in places make up to 30-40% of the limestone columns. Width of field of view is approximately 5 mm. (D) Thin-section photomicrograph of dolomite surrounding limestone column; field of view is approximately 5 mm. Dolostone is composed of microcrystalline dolomite in which are embedded isolated euhedral rhombohedrons several tens of microns in diameter. The microcrystalline dolomite has coarse silt-sized, rounded, clotted limestone areas that probably represent relict peloids. Ostracod shells in the dolomite that do not have internal sediment fills are notably crushed (e.g., the collapsed, broken shell in center of the photograph), providing additional evidence of preferential mechanical compaction of the material now dolomite.

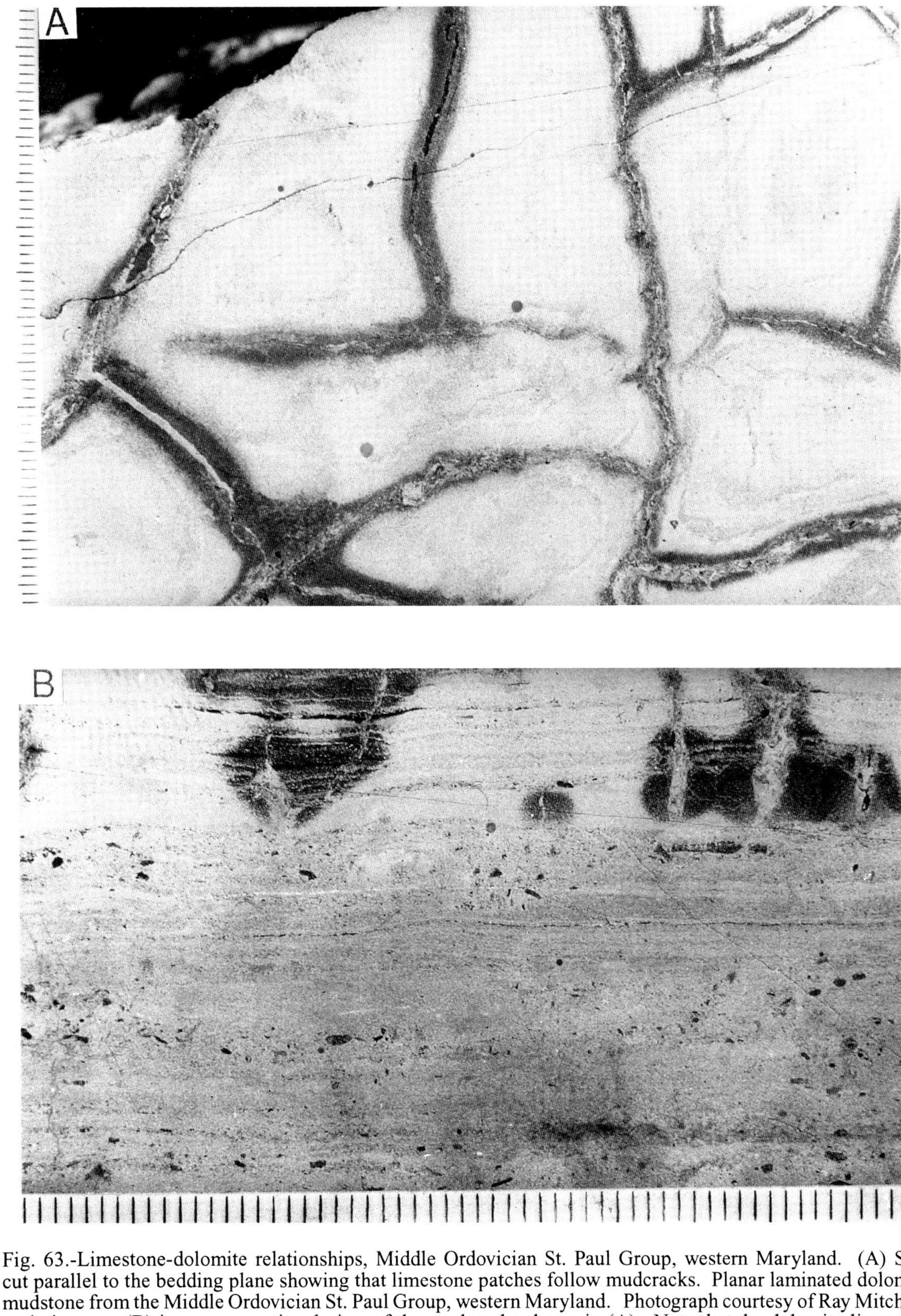

Fig. 63.-Limestone-dolomite relationships, Middle Ordovician St. Paul Group, western Maryland. (A) Slab cut parallel to the bedding plane showing that limestone patches follow mudcracks. Planar laminated dolomite mudstone from the Middle Ordovician St. Paul Group, western Maryland. Photograph courtesy of Ray Mitchell; scale in mm. (B) is a cross-sectional view of the mudcracks shown in (A). Note that the dolomite-limestone boundaries cut across sedimentary layering. Laminae in dolostone above and below a limestone patch show compaction effects laterally as they pass around the patches.

have been deformed ductilely. They must have been plastic enough to flow at the time they were subjected to the deforming forces. This is particularly well demonstrated by the wraparound and other volume reduction features in the interlayered limestone-dolostone rocks (as described above). We interpret this to mean that the volume reduction features common to these rocks were due to pre-lithification mechanical compaction of soft, deformable muds and not due to pressure solution of a cemented rock (for an opposing view see Wanless, 1979). In contrast, the limestones interbedded with the dolomitic mudrocks have been deformed brittlely by fracturing and rotating to form sedimentary boudins. Without doubt, the lime grainstones were cemented and hard whereas the dolostone mudrocks were soft and ductile before deformation. The preservation of open grain packing, open burrows, and open fenestrae within the limestones is clear supporting evidence that the limestones were cemented early, prior to the compaction which plastically deformed the dolomitic mudstones. It is our view that the differential early marine cementation of permeable sand layers in what were originally heterolithic sand and mud strata produced a stack of alternating brittle and ductile layers that on burial responded differentially to mechanical compaction to produce the complex volume reduction features described above. But the problem has a further level of complexity, because dolomitization, like compaction, has been differential, confined principally to the mudstones (Hardie and others, 1982; Demicco and Mitchell, 1982; Hardie, 1989). Most significant are patches of limestone that are partly to completely surrounded by dolostone where the limestone-dolostone boundaries cross-cut primary sedimentary layering. This is unequivocal evidence that the dolomite is secondary. What is more, in thin sections, this dolomite can be seen to replace spar-filled compaction cracks in limestone boudins (unpublished observations of Chau Nguyen, see Hardie, 1989, p. 77). This demonstrates the post-compactional origin of the dolomite. It remains then to explain the differential dolomitization of the mudstone interbeds that makes these "ribbon rocks" or "tiger stripe" rocks so distinctive in outcrops in the central Appalachians. The petrographic observations suggest that the dolomitization was controlled by the early cementation, namely that the original sedimentary elements that were not cemented prior to compaction were later dolomitized (Hardie and others, 1982). As has been described above, in the overwhelming majority of cases, it was the grainstones that were cemented early and deformed in a brittle fashion during mechanical compaction whereas the mudstones were ductilely deformed during mechanical compaction by flowage and were later preferentially dolomitized. Nguyen and Hardie (Hardie, 1989, p. 77) have offered the following explanation of the differing susceptibilities of the grainstones and mudstones to dolomitization: "the grainstones appear to be the result of preferential early cementation of porous sands, as evidenced by high pre-cement porosities and abundance of visible early cements. The mudstones, like modern lime muds, must have been effectively impermeable to the low pressure circulation of seawater at the sediment-water interface of the shallow platform and so remained uncemented before burial. Early syndepositional and early burial cementation occluded most major connected pores in the grainstones, which left mainly the micro-pores in the compacted mudstones as pathways for the later dolomitizing fluids. At the elevated fluid pressures of burial, groundwater could have flowed, however slowly, through these compacted mudstones in sufficient volume and at high enough temperatures to effect extensive dolomitization." We cannot establish the exact timing of compaction and dolomitization except that they were post early cementation but pre-stylolitization. In Figure 10, bedding plane stylolites cut across limestone-dolostone boundaries, demonstrating that these stylolites are younger than the dolomitization. In the thin-section photomicrograph shown in Figure 64, a bedding plane stylolite follows a limestone-dolostone boundary for a short distance before cutting off into the limestone. Furthermore, dolomite rhombohedrons along the stylolite show evidence of dissolution. These observations demonstrate that the bedding plane stylolites in these rocks are post-dolomitization and post-mechanical compaction features.

In summary, we propose the following scenario to account for the volume reduction features of the carbonates of the Cambro-Ordovician of the central Appalachians. Deposition was closely followed by early marine cementation of grainstones whereas mudstones were left uncemented. For the most part, this differential cementation was controlled by the differences in the primary grain textures of the sand and mud layers, but early cementation also occurred preferentially around other primary sedimentary features such as mudcracks and burrows that offered permeable pathways through the sediments for flow of the seawater from which the cement precipitated (see Fig. 63, which shows preferential cementation of the walls of mudcracks that must have acted as fluid flow conduits; see also Mitchell, 1981). This differential, fabric selective cementation is different from the better-known ovoid, early cemented *nodules* commonly found in shales (Fig. 65). Upon burial, the cemented grainstone layers and the differentially cemented halos around mudcracks and burrows behaved brittlely whereas the uncemented muds deformed plastically. At some point prior to the initiation of bedding plane stylolites, the uncemented but compacted muds were preferentially dolomitized.

Significance.-

The scenario outlined above in the discussion of the origin of differential compactional features in the Cambro-Ordovician carbonates of the central Appalachians emphasizes two processes which we believe have far reaching effects on the nature of primary sedimentary structures in muddy carbonate rocks of all kinds and heterolithic strata in particular. These processes are: (1) differential early cementation, and (2) mechanical compaction. If the Cambro-Ordovician rocks of the central Appalachians are any guide, then simple mechanical compaction due to overburden pressure plays an important role in the diagenesis of carbonate mudstones. It is important to note that all of the modifications of primary sedimentary textures and features produced experimentally by Shinn and his coworkers can be found in the Cambro-Ordovician dolomitic mudstones of the Appalachians. This lends strong support to the contention of Shinn and his coworkers that these experimentally produced features may actually be criteria whereby mechanical compaction can be identified in ancient carbonates. This is especially true of the enigmatic microstylolites and horsetail stylolite seams so common in ancient carbonate mudrocks. The second implication of this scenario is its applicability to the origin of carbonate structures commonly referred to as "nodular bedding," "irregular bedding," "lumpy bedding," "flaser bedding," or "sedimentary boudinage" (Fig. 27) (Wilson, 1975; Matter, 1967; Schwarz, 1975; Wilson and Jordan, 1983; Scoffin, 1987). Patchy early cementation followed by differential compaction of uncemented sediment can generate these bedding features. However, it is also possible that some of the complex geometric relations between the sandy and muddy elements of the rocks displaying these particular structures could have been inherited from the original primary sedimentary structures. For example, Figure 29 shows modern

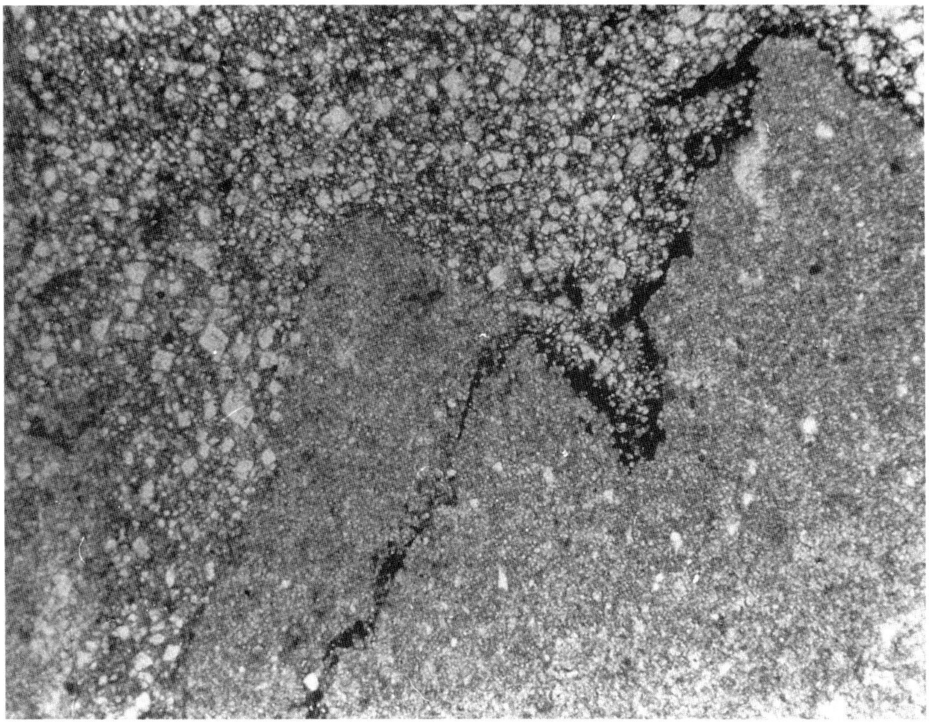

Fig. 64.-Thin-section photomicrograph of boundary between limestone mudstone (lower right) and dolomite mudstone with rhombohedrons of dolomite. Width of field is approximately 2 mm. Note how the stylolite follows the limestone-dolomite boundary for a portion of its length but then moves out into the limestone. Cambro-Ordovician carbonates of western Maryland.

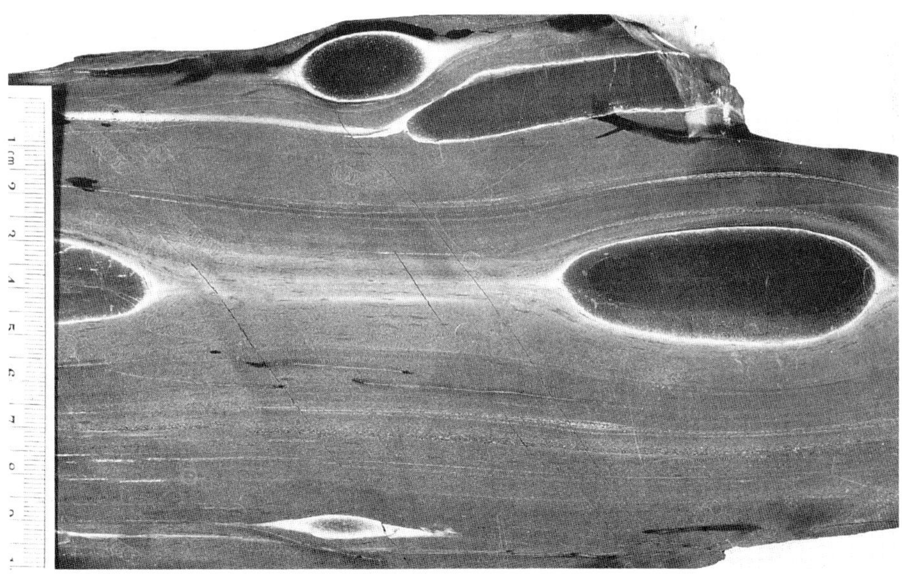

Fig. 65.-Limestone concretions in a laminated dolomitic mudstone, Silurian of western New York. Note how the laminae bend over and under the concretions, demonstrating preferential volume reduction of the material now dolomite. Scale in mm.

interbedded sands and muds from North Sea tidal and subtidal areas. Note the irregular outlines of the sand and the apparent lateral gradation of sand to mud (see Demicco, 1983, for further information on this style of heterolithic stratification).

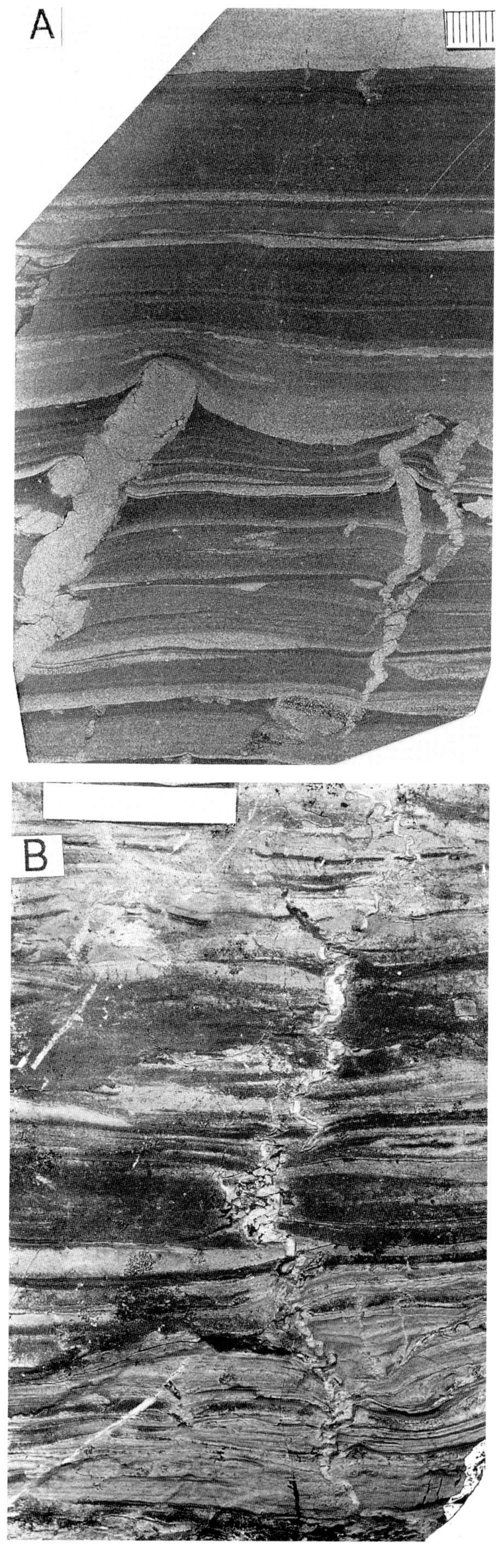

Miscellaneous Physical Structures

Ptygmatically folded mudcracks ("molar tooth structure").-

Terminology.-Mudcracks and sedimentary dikes in ancient sedimentary rocks are commonly folded (Fig. 66), reflecting compaction of the host sediment around the more competent crack filling (see Allen, 1984, p. 556). Intensely-folded horizontal and vertical cracks filled with calcite spar in a dolomitic host rock are relatively common in some carbonates of ern United States. Such structures were termed "molar tooth structures" by Bauermann (1885) after their resemblance on bedding plane surfaces to the intricate folds of enamel seen on an elephant molar tooth. Discussions of the structure are found in Fenton and Fenton (1937), O'Conner (1972), Horodyski (1976), and Young and Long (1977), among others. O'Conner (1972) provides a classification.

Distinguishing characteristics and origin.-Where seen on bedding planes, molar tooth structure can be quite complex, although well defined polygonal patterns do occur (see Horodyski, 1976, Fig. 10F). As with mudcracks, this complexity could simply reflect superimposition of a number of crack patterns. In cross-section (Fig. 67), molar tooth structures are commonly described as "spindle-shaped" (note this different use of spindle-shaped to describe the cross-sectional shape of the cracks, not the plan shape). Here the cracks are closed on both ends and widest in the middle. Load casts and flat-pebbles of dolomite that contain molar tooth structures and intraclast conglomerates composed dominantly of calcite spar similar to that comprising molar tooth fills attest to the very early origin of this structure. The spindle-shaped cross-sections of the cracks (i.e., they do not "v" upwards) and their reputed confinement to single sedimentation units has led to their common interpretation as "syneresis" cracks (mechanism unspecified). Their development in relatively pure carbonates makes this interpretation unlikely (see discussion on syneresis cracks above). Instead, these molar tooth structures of the Belt-Purcell would seem to be most compatible with an origin as severely compacted mudcracks (for a very different view see O'Connor (1972), who interprets them as organic structures).

Occurrence and preservation.-While mudcracks folded by compaction to one degree or another are very common in carbonates, the extreme folding represented by molar tooth structure is uncommon. The best known examples are found in the Upper Proterozoic Belt-Purcell Supergroup of the Rocky Mountains, as noted above.

Significance.-Compacted mudcracks allow measurement of the amount of compaction. Donovan and Foster (1972) and Allen (1984), among others, provide formulas for decompacting sequences with folded mudcracks. We agree with Allen (1984) that the use of a special term to designate folded mudcracks is unwarranted.

Erosional scour structures.-

The full range of erosional scour structures and tool marks known from siliciclastic rocks also occurs in carbonates. These include obstacle marks, flutes, longitudinal ridges and furrows, gutter casts, grove casts, chevron structures, prod-marks, skip-

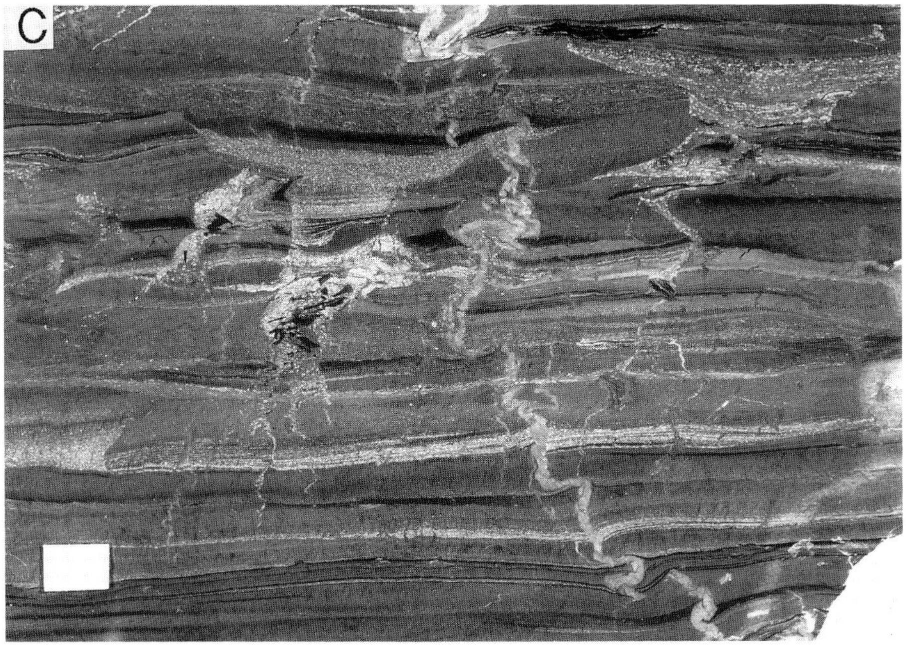

Fig. 66.-Ptygmatically folded mudcracks. (A) (Preceeding page) Upper Cambrian Conococheague Formation, central Pennsylvania. Scale in mm. Sample courtesy of Francis J. Pettijohn (also see Pettijohn, 1975, Figure 4-26, p. 122). (B) (Preceeding page) and (C) Middle-Upper Cambrian Waterfowl Formation, southern Canadian Rocky Mountains, southwestern Alberta. Scale in (B) is 50 mm long, scale in (C) is 10 mm long.

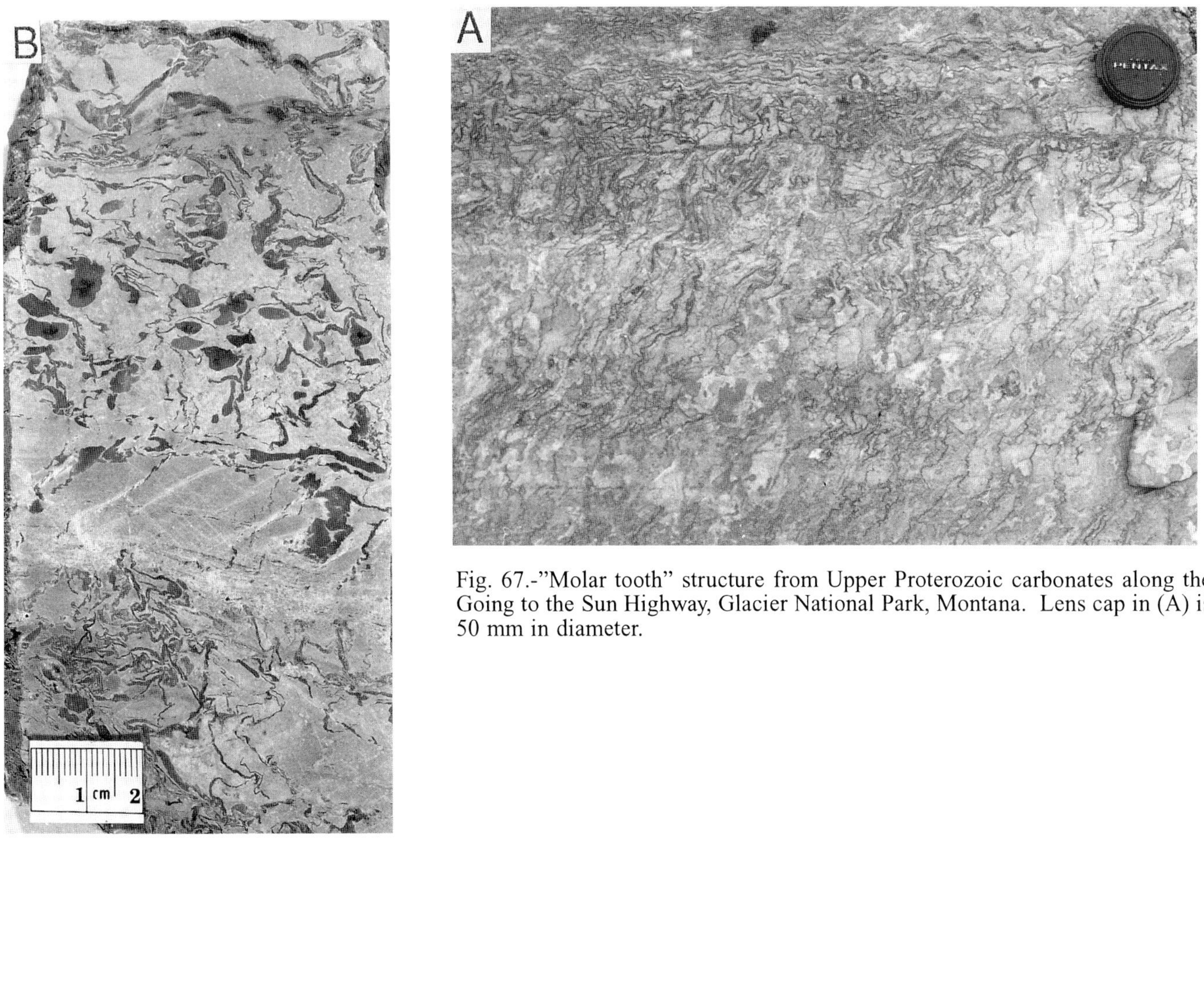

Fig. 67.-"Molar tooth" structure from Upper Proterozoic carbonates along the Going to the Sun Highway, Glacier National Park, Montana. Lens cap in (A) is 50 mm in diameter.

marks, bounce-marks, and channels. These are exactly like their siliciclastic counterparts and will not be dealt with here. The reader is referred to Pettijohn and Potter (1964), Reineck and Singh (1980), Collinson and Thompson (1989), Leeder (1982), Allen (1984) and other atlases in this series for details.

BIOGENIC STRUCTURES

Cryptmicrobial Structures

In a paper that stands as a benchmark in carbonate sedimentology, Maurice Black (1933) demonstrated the important role that surface mats of filamentous cyanobacteria ("blue-green algae") played in the construction of modern laminated carbonate sediments analogous to the common, but problematic, laminated domal sedimentary structures known since the time of Kalkowsky (1908) as *stromatolites*. Black's discovery, that sticky cyanobacterial mats on the supratidal flats of Andros Island, Bahamas, trap and bind sediment washed over the flats by storms to produce laminated sediments, was used by Young (1935) to explain the stromatolites of the Precambrian Transvaal Dolomite of South Africa, but otherwise received little attention. It was Robert N. Ginsburg (Ginsburg and others, 1954; Ginsburg, 1955, 1960,1967) who revived Black's ideas on "algal sediments," so successfully in fact that almost all ancient stromatolites and wavy laminites in carbonate rocks have since then been interpreted to be of "algal" origin (Walter, 1976a). Indeed, even where no direct evidence for cyanobacteria exists in stromatolites and laminites, these structures are called "algal stromatolites" and "algal laminites," or, in recognition of the lack of fossil cyanobacterial cells and filaments, "cryptalgal stromatolites" and "cryptalgal laminites." Aitken (1967, p. 1163) proposed the following definition for the term *cryptalgal*: "cryptalgal sedimentary rocks or rock structures may be defined as those *believed* (our italics) to originate through the sediment-binding and/or carbonate-precipitating activities of nonskeletal algae." Aitken considered the nonskeletal "algae" to include both the "blue-green algae" (Cyanobacteria) and "green algae" (Chlorophyta). The prefix "crypt" was derived from the Greek root "kryptos" meaning hidden or secret. The adjectives algal and cryptalgal have long been used by geologists. In the light of developments in understanding of the biology of algae, these terms are now inappropriate and their use should be discontinued. In modern environments where biogenic benthonic mats influence sedimentation, the mats are generally a complex, multispecific community dominated by cyanobacteria (or cyanophytes). Cyanobacteria are single-celled prokaryotes (i.e., chromosomes not organized into a nucleus, and lacking other organelles) and in the past have been referred to inappropriately as blue-green algae. Algae are more complex eukaryotes and indeed are classified in an entirely different kingdom than the cyanobacteria. Moreover, modern benthonic mats may also contain other unicellular microbes (most notably bacteria), multicellular fungi, true unicellular and multicellular algae, and diatoms (Golubic, 1976a, b; Bauld, 1981; Krumbein, 1983; Gerdes and Krumbein, 1987). For these reasons, "microbial" and "cryptmicrobial" (Kennard and James, 1986, p. 496) are more appropriate terms than "algal" or "cryptalgal."

Cryptmicrobial structures considered here are cryptmicrobial laminites, stromatolites, and "jelly roll" structures. Common to, as well as definitive of, all three of these organo-sedimentary structures is *lamination*, which in ancient examples is composed of detrital sediment, crystalline precipitates, or both. Our approach to the understanding of these structures is that of comparative sedimentology and so we begin below by looking at mechanisms known to produce laminated cryptmicrobial structures in modern carbonate environments.

Oncoids (unattached stromatolites) and "algal biscuits" (Ginsburg, 1960; Toomey and others, 1988) are considered as grains rather than sedimentary structures and will not be treated here (refer to Peryt, 1983; Jones and Goodbody, 1985). Thrombolites are discussed separately in a following section.

Microbial mats and the origin of organo-sedimentary lamination.-
The laminated structure of modern analogs of stromatolites, cryptmicrobial laminites, and "jelly roll" structures can be produced by one or more of a large number of processes which can be grouped into four categories: (1) mechanical sedimentation not influenced by microbes, (2) microbially-influenced sedimentation, (3) chemical precipitation of carbonate, and (4) episodic growth of a microbial mat with little or no sedimentation or precipitation. We are immediately faced with two problems. First, the number of modern examples is small, so we may not be seeing all the processes that are capable of building stromatolitic laminae in any given environment. Obviously, we can only evaluate how serious this problem is by further studies. Secondly, and more importantly, it is clear that in many cases the laminae and their internal fabrics produced by different mechanisms cannot always be told apart with ease, and serious difficulties arise when stromatolites contain laminae produced by more than one mechanism.

*Sedimentation not influenced by microbial mats.-*Wherever the sediment surface does not exceed the angle of repose, sediment deposition of material from suspension or bedload can take place whether or not there is a microbial mat present. Mechanical deposition from bed load can preferentially fill small depressions or irregularities on domal or flat microbial mat surfaces with carbonate sand grains, carbonate mud or even siliciclastic grains. Modern examples are figured in Purser and Loreau (1973, Figs. 17, 18), Mendelsohn (1976, Fig. 3), and Hardie and Ginsburg (1977, Figs. 37, 38, 63). An ancient example of depression-fills in a wavy laminite is shown in Figure 68. Small-scale ripples and ripple-like starved lenses of sediment may also be deposited from traction on the surface of domal or flat microbial mats which simply act as passive substrates. Some of the surface and internal lenses of fine-grained peloidal-skeletal material from unlithified Bahamian stromatolites described by Hardie and Ginsburg (1977, p. 55 and Fig. 34) originated in this way and are found interstratified with laminae that clearly show the influence of a sticky microbial surface mat (see below). Figure 69 shows a composite set of planar, wavy, and crinkled fine-laminae from a Precambrian carbonate. Although original textures have been obliterated by later diagenesis, the primary layering is delicately preserved. The gray laminae, which are interlocking mosaics of calcite crystals, are interpreted as originally having been layers of fine carbonate sand based on the ripple-forms in the thicker lamina. These rippled layers clearly were mechanically deposited. In contrast, the whitish-tan laminae, now microcrystalline mosaics of dolomite, are interpreted as originally having been mud agglutinated to surface microbial mats by analogy with modern microbial laminites (see below). Note that in a number of places, *thinner* light gray laminae (interpreted as sand) clearly exceed the angle of repose of sand-sized noncohesive grains, indicating that at least some of the sand layer deposition was affected by microbial mats. Thus, this Precambrian laminite records the same alternation of purely mechanical deposition with deposition influenced by surface microbial mats as do modern laminites from the Bahamian carbonate tidal flats (see below).

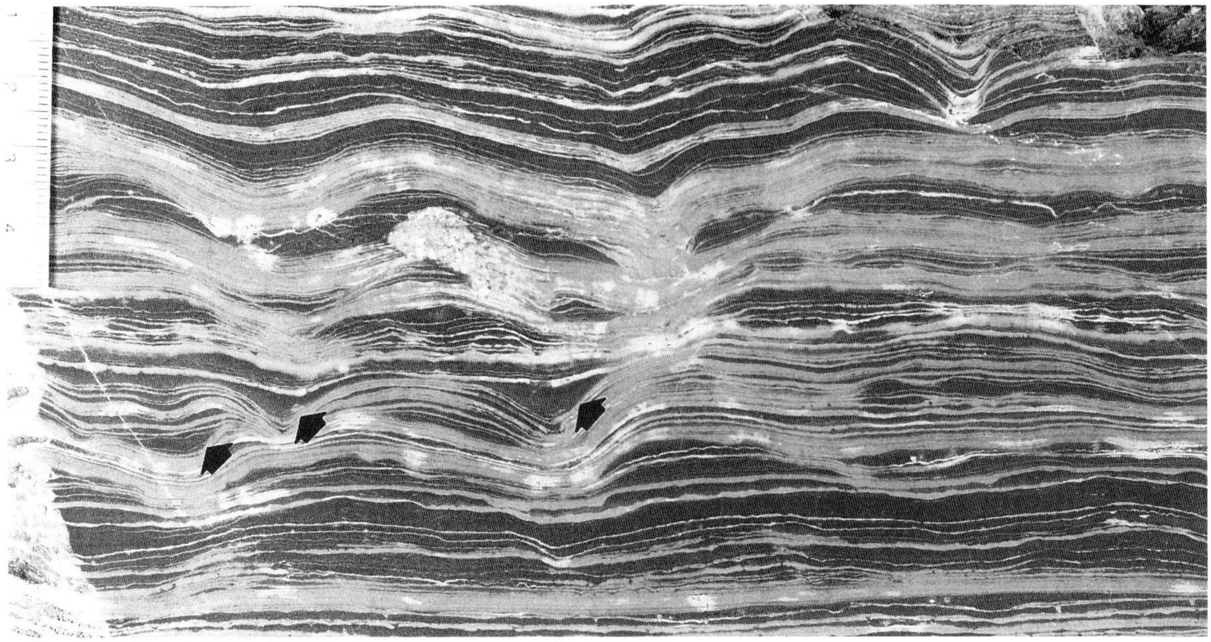

Fig. 68.-Wavy to crinkled laminite composed of couplets of peloidal lime grainstone (dark) and dolomite mudstone (light). The lamination is interpreted as due, in part, to the sediment trapping and binding of cyanobacterial surface mats. However, note the three depression-fills composed of fine sand-sized peloidal grainstone at left center (arrows). These are interpreted as trapped bedload and not the result of cyanobacterial trapping. Scale bar is 50 mm long. Specimen from the Upper Cambrian Conococheague Limestone, western Maryland.

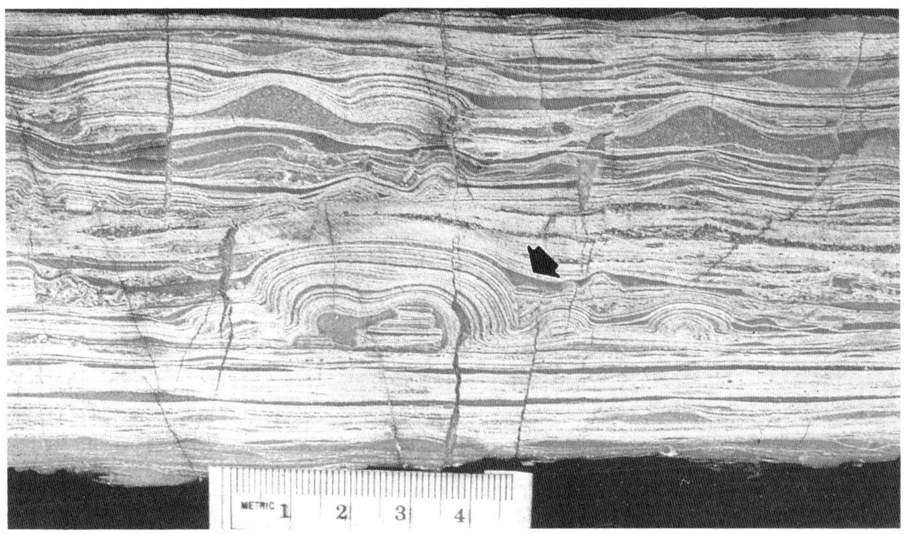

Fig. 69.-Composite set of planar, wavy, and crinkled laminae and small LLH-stromatolitic mounds from the Precambrian of the Great Slave Lake area, NWT, Canada. Sample courtesy of Paul Hoffman. Dark gray lenticular layers are composed of a microsparry mosaic of calcite while the thin, light-colored, continuous laminae consist of interlocking dolomite anhedra. Note: (1) the ripple-form morphology in the dark layers in the upper half of the specimen, and (2) the wedge-shaped calcite layers at the flanks of the stromatolitic domes have the morphological characteristics of lee deposits (e.g., dark lense at the arrow in the middle of the specimen). These features point to mechanical, bedload deposition of sand-sized sediment for the original material of the calcite lenticular layers. In contrast, the very thin but laterally continuous dolomitic laminae (light-colored) are analogous to the very thin continuous muddy laminae agglutinated by cyanobacterial mats on supratidal levees of the modern Andros Island tidal flats (see Figs. 70, 71; see also description and illustrations in Hardie and Ginsburg, 1977). This "stick-on" origin for the dolomitic layers is strongly supported by the way in which the laminae drape over laminated intraclasts at inclinations that far exceed the normal angle of repose to produce small LLH-stromatolitic mounds (see particularly the mound above the scale in the slab photograph). Note that some of the dark calcitic "sand" layers interlaminated with the dolomite "muddy" layers also drape over the stromatolitic mounds as thin "gravity-defying" laminae, adding weight to the case for the former presence of cyanobacterial mats capable of trapping and binding not only mud but sand-sized grains.

Microbially influenced sedimentation.-Microbial mats can either agglutinate sediment particles onto their sticky mucilaginous sheaths or act as baffles that trap sediment washed into the tangle of filaments. Which process dominates in a given situation in part depends on the type of microbe that dominates the mat, as briefly outlined below.

Mats composed mainly of small Oscillatoriacea cyanobacteria agglutinate mud and silt-sized to very fine sand-sized peloidal-skeletal grains into sub-millimeter to millimeter thick sediment laminae. Oscillatoriacea are **motile filamentous cyanobacteria**, such as *Phormidium*, with filament widths of a few microns and lengths of a few tens to a few hundreds of microns. On the modern carbonate tidal flats of northwest Andros Island, mats composed of motile filamentous cyanobacteria occur in two settings: (1) on the smooth flat crests of tidal channel levees and beach levees (Fig. 70); and (2) on the upper edges of channel banks where they occur as knobby encrustations (unlithified domal stromatolites) between fiddler crab burrows (Fig. 71). In both of these settings, agglutinated muddy laminae with uniform thickness and a typical clotted texture alternate with mechanically deposited peloidal sand laminae, which typically occur as discontinuous lenses (Figs. 70D, 71C). These microbially influenced muddy laminae and films are continuous across the unlithified domal stromatolites (see Hardie and Ginsburg, 1977, Figs. 34, 35), and will even drape vertical and overhanging edges, testifying to their origin as organically agglutinated layers (Fig. 71C). Oscillatoriacea not only agglutinate sediment onto their sticky sheaths, but they also have the ability to move up through the newly agglutinated sediment layer and re-establish themselves on the surface, leaving only their mucilage as evidence that they were once present in the underlying layers. For more information on the motile behavior of the Oscillatoriacea and their "fly-paper" role in agglutinating sediment on a modern Bahamian tidal flat, the reader is referred to the descriptions and discussions in Hardie and Ginsburg (1977) and other chapters in Hardie (1977a).

Mats dominated by larger, **immotile filamentous cyanobacteria**, such as *Scytonema*, *Rivularia*, etc., are found on the modern tidal flats of northwest Andros Island on the backslope of levees and in an extensive, freshwater "algal marsh" (Fig. 72). These mats are made up of filaments a few tens of microns in diameter and up to a few millimeters in length which are typically concentrated into small tufts and bundles (Fig. 72B). This "forest" of large filaments acts physically as a sediment baffle, allowing trapping of mud to sand-sized grains within the tangle of upright filaments at the surface of the mat. Laminae produced in this way tend to be up to a few millimeters thick, have irregular bounding surfaces, and consist of very porous, poorly-sorted peloidal sediment that contains bundles and tufts of buried filaments (Fig. 72C, D), or, where the filaments have decomposed, their tubular molds (Hardie and Ginsburg, 1977, Fig. 51). In some cases, the bulbous shape of the cyanobacterial columns are preserved upon burial, lending a distinctive crinkled geometry to the baffled laminae (Fig. 73).

Mats dominated by **non-filamentous coccoid cyanobacteria**, such as *Entophysalis*, generally produce internally unlaminated organic-rich layers with fenestral fabrics (Fig. 74) (also see descriptions in Logan and others, 1974a). The sediments found below crinkled coccoid bacterial mats of the Persian Gulf and Shark Bay are generally described as only vaguely laminated, clotted, or massive (cinder mats of Kendall and Skipwith, 1968; tufted and convoluted mats of Davies, 1970; blister, pustular, and pinnacle mats of Kinsman and Park, 1976).

The sediment layers agglutinated or trapped by microbial mats on the modern carbonate tidal flats of Andros Island, Bahamas, and Caicos Platform, British West Indies, record individual storms that washed suspended and bedload carbonate sediment from the shelf-lagoon onto the flats (Hardie and Ginsburg, 1977; Wanless and others, 1988). This finding contradicts the widely held view that stromatolitic layering is a register of daily tidal deposition or of daily growth cycles of the cyanobacteria (e.g., Pannella and others, 1968).

Precipitation fabrics.-Carbonate can be precipitated at the surface of or within a microbial mat by a number of mechanisms, both biogenic and abiogenic. Basically, two end member cases can be recognized. First, the entire water mass may be supersaturated with carbonate, and precipitation of crystalline layers of alkaline earth carbonate (aragonite, magnesian calcite, calcian dolomite, monohydrocalcite, etc.) takes place regardless of the activities of the organisms. This, for example, is the situation where upwelling springwaters degas CO_2 as they reach the surface or in evaporating bodies of standing surface water such as closed lakes or marginal marine lagoons in semi-arid and arid climates. The second case is where local extraction of CO_2 by photosynthetic organisms or by heterotrophic bacteria induces precipitation of alkaline earth carbonates around their cells (see under Tufas elsewhere in this atlas). Combinations of these two cases produce laminae with a variety of internal fabrics, but three basic types can be recognized: (1) micritic, (2) radial crystalline, and (3) micritic or peloidal containing calcified molds of organic structures (such as filaments).

Micritic laminae vary in thickness from a few microns up to a centimeter or so (Fig. 75). Boundaries between laminae may be sharp and marked by impurities, or may be diffuse, typically being defined by color-banding. Some very finely-laminated units have a microcolumnar structure that cross-cuts laminae boundaries (Fig. 75). In a number of cases, the micrite has a distinctive clotted texture, and this is especially true of thicker laminae (a few millimeters up to centimeter) which also tend to have irregularly-shaped fenestrae. In thin section, such fabrics may be impossible to separate from those produced by purely mechanical deposition of detrital mud and peloidal laminae.

Oriented crystal growth fabrics comprise either competing growth "druse-fans" (Fig. 76) or acicular needles. In the competitive growth druse-fan fabric, the seed crystals with c-axis orientations more or less perpendicular to the substrate surface outgrow the horizontally oriented seed crystals to produce a radial crystalline fabric in which the crystals may reach lengths measured in centimeters. In these radial crystalline precipitates, the lamination boundaries are marked by nucleation of a new set of seed crystals or by the inclusion of a layer of impurities at a growth face which is subsequently buried by continued crystal elongation (syntaxial outgrowth). A particular form of radial druse-fan is composed of millimeter-diameter branched clusters of subspherical nodules ("cave popcorn" of Thrailkill, 1976). These can have fine internal laminae or coarse radial druse-fan growth. Commonly, they contain complete spheroids in which the outer laminae join to make compound structures. For further description, the reader is referred to the discussion on chemical lamination below.

Micritic or peloidal laminae may contain the calcified remains of both filamentous and coccoid cyanobacteria (Monty, 1967, 1972, 1976; Monty and Hardie, 1976; Hardie, 1977a), coccoid bacteria (and other microbes), and higher plants, particularly mosses and charophytes. Light calcification and subsequent breakdown of the organic material and release of car-

Fig. 70.-Surface features and deposits of crests of tidal channel levees on the modern carbonate tidal flats of northwest Andros Island, Great Bahama Bank. The peloidal surface sediment of these highest and driest portions of these tidal flats is impregnated by a mat of motile cyanobacteria. (A) General view of the surface showing flow-parallel longitudinal ridges a few millimeters high composed of peloidal sand. Pencil for scale. (B) Smooth, flat, cyanobacterial mat-covered surface typical of the levee crest subenvironment. Note that there are few mudcracks even though the levee crest remains subaerially exposed for over 90% of the time. The binding of the surface sediment by the mat prevents mudcracking. 15 cm ruler at the base of stake in the center of the photograph. (C) (Following page) Epoxy-impregnated core slab of levee crest sediments showing typical uniformly thin laminae. Scale is in mm. (D) (Following page) Thin-section photomicrograph of laminae shown in (C). Scale bar is 1 mm long. Layering consists of alternations of relatively thick, lenticular laminae made of well-sorted peloidal fine sands (upper flow regime bedload deposits) and very thin, continuous laminae made of clotted mud (cyanobacterial mat "stick-on" layers). Dark grains are hard, cemented peloids whereas gray sand-sized grains are soft, uncemented peloids. In places, the soft peloids merge into a clotted mud with a "structure grumeleuse" texture. Note the cyanobacterial filament molds in a number of the laminae.

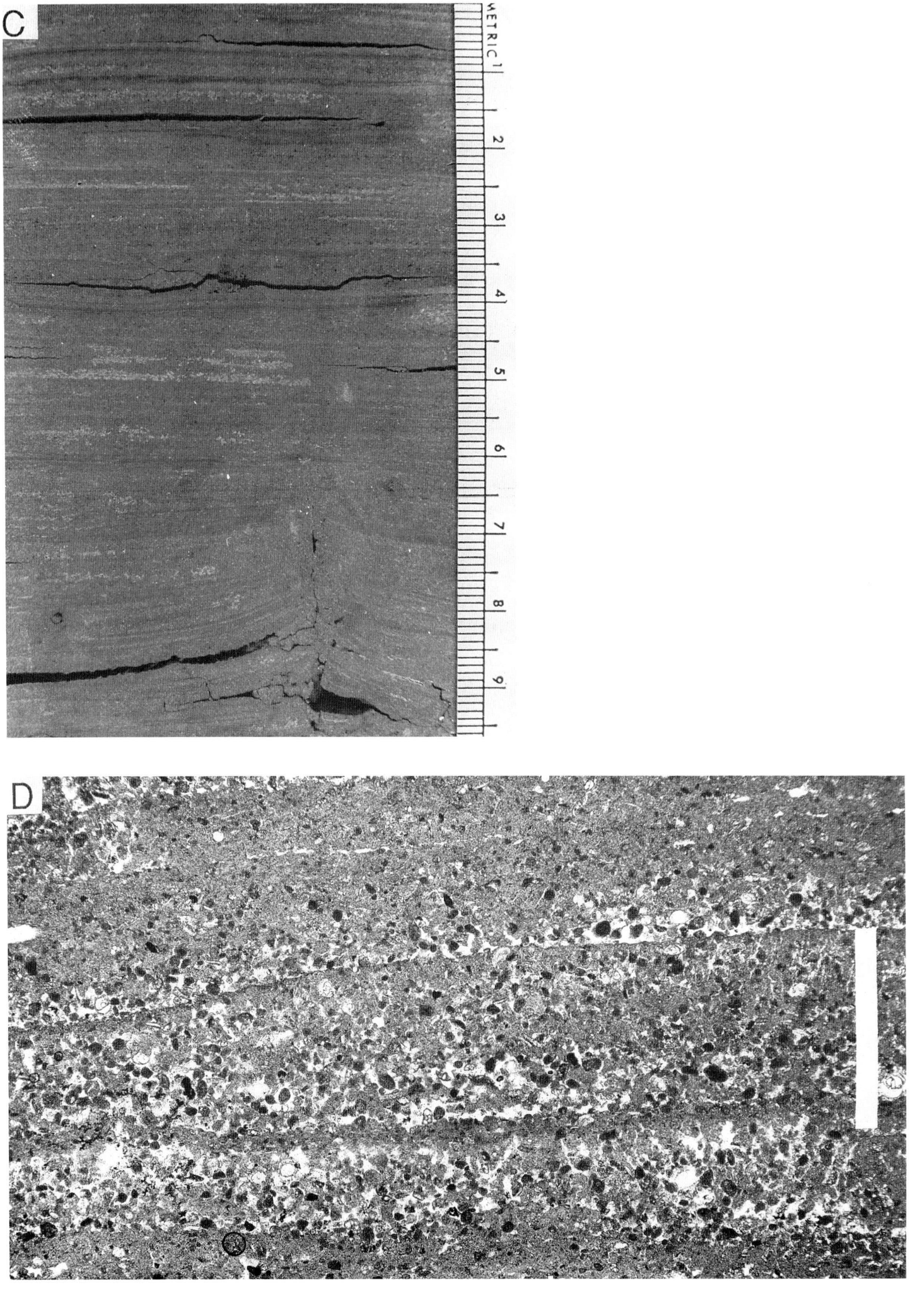
C
METRIC 1
2
3
4
5
6
7
8
9
D

Fig. 71.-"Protostromatolites" from tidal flats of northwest Andros Island (Hardie and Ginsburg, 1977). (A) Photograph of the bank of a tidal channel showing the cyanobacterial mat-draped knobs between fiddler-crab burrow entrances. Note the oblique ridges across the knobs and the coarse, dark *Scytonema* filament stubble on the sides of each knob. (B) (Following page) Vertical cross-section through an epoxy-impregnated knob. Scale in mm. This knob has a laminated internal structure and accreted over a vertically oriented clast of laminated sediment. (C) (pg. 95) Thin-section photomicrograph of (B). Scale bar is 1 mm. Mud layers with a clotted fabric alternate with peloid-rich laminae. Note the wavy, dark stringers that mark organic remnants of the mat of the motile cyanobacterium *Schizothrix* . Thread-like voids at right angles to laminae are filament molds of the non-motile cyanobacterium *Rivularia* .

B

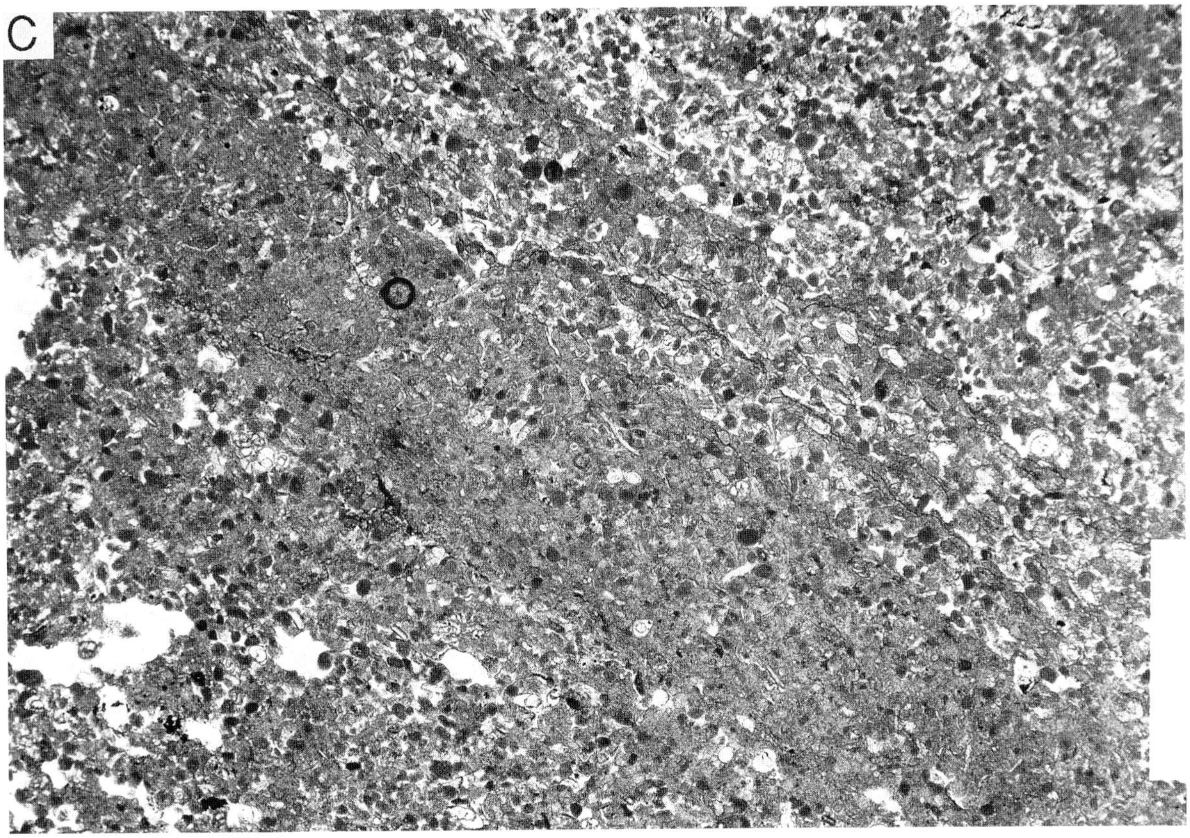

bonate debris is an important source of carbonate mud in some settings (Monty and Hardie, 1976; Mitchell, 1985). With heavier calcification, a cast of the organism is produced (see, for example, the "algal tufa" stromatolitic structures characteristic of the supratidal marshes of the Andros Island tidal flats, Hardie, 1977b, and the appropriate sections in CHEMICAL FEATURES below). Large calcified filaments of cyanobacteria and calcified clots of coccoid cyanobacteria may represent forms transitional to the problematic fossil forms *Epiphyton*, *Girvanella*, and *Renalcis*. Recently, MacIntyre (1985) and Chafetz (1986) have suggested that calcification of coccoid bacteria (generally much smaller forms than coccoid cyanobacteria) produce both isolated peloids (up to a few tens of microns in diameter) and arborescent "shrubs" (composed of peloids) which may have an overall size of a few centimeters. Once initial calcification takes place, void space may become filled with detrital material or micritic cements which themselves can have clotted fabrics. It is not known whether these clots are related to calcification of organic remains. Radial druse-cements, acicular fringe cements, and blocky mosaic cements can also be found between calcified microbes. A note on filament molds is in order here. All of the sedimentary environments in which modern stromatolites are found have bacteria and cyanobacteria living on and in the surface sediments. The calcified remains of these microbes may have nothing to do with the depositional processes. The microbial organisms may be, in effect, innocent bystanders in a dominantly chemically precipitated sediment or stromatolite. Stretching the cautionary tale further, on the tidal flats of Andros Island, coarse filamentous non-motile *Scytonema* coexists with small motile oscillatoracean cyanobacteria. The main sediment agglutinator is the oscillatoriacean that after deposition of a lamina moves up through the sediment leaving no trace of its former presence and leaves behind the buried *Scytonema* to take the blame.

Mats with little or no sedimentation or precipitation.-Laminae up to a few centimeters thick comprised of filamentous and coccoid cyanobacteria with little or no detrital or precipitated components are common in a number of modern carbonate environments. They are typically interbedded with storm-deposited sediment layers (e.g., Davies, 1970, Figs. 10, 11; Hardie and Ginsburg, 1977, Fig. 62A), and record long periods of microbial mat growth in a subenvironment such as the supratidal marshes of Andros Island that are inundated by sediment laden seawater only during hurricanes (Hardie, 1977a). On burial, differential compaction of the cyanobacterial layers coupled with decay of the organic matter will result in a laminated or thin bedded rock with organic-rich partings between the storm layers.

Summary.-Table 2 is an attempt to summarize the characteristics of various types of modern "cryptmicrobial" layers reported from a variety of sources. Note that many of these layers exhibit strictly chemically precipitated fabrics which may have nothing to do with organisms which happen to be living upon the surface.

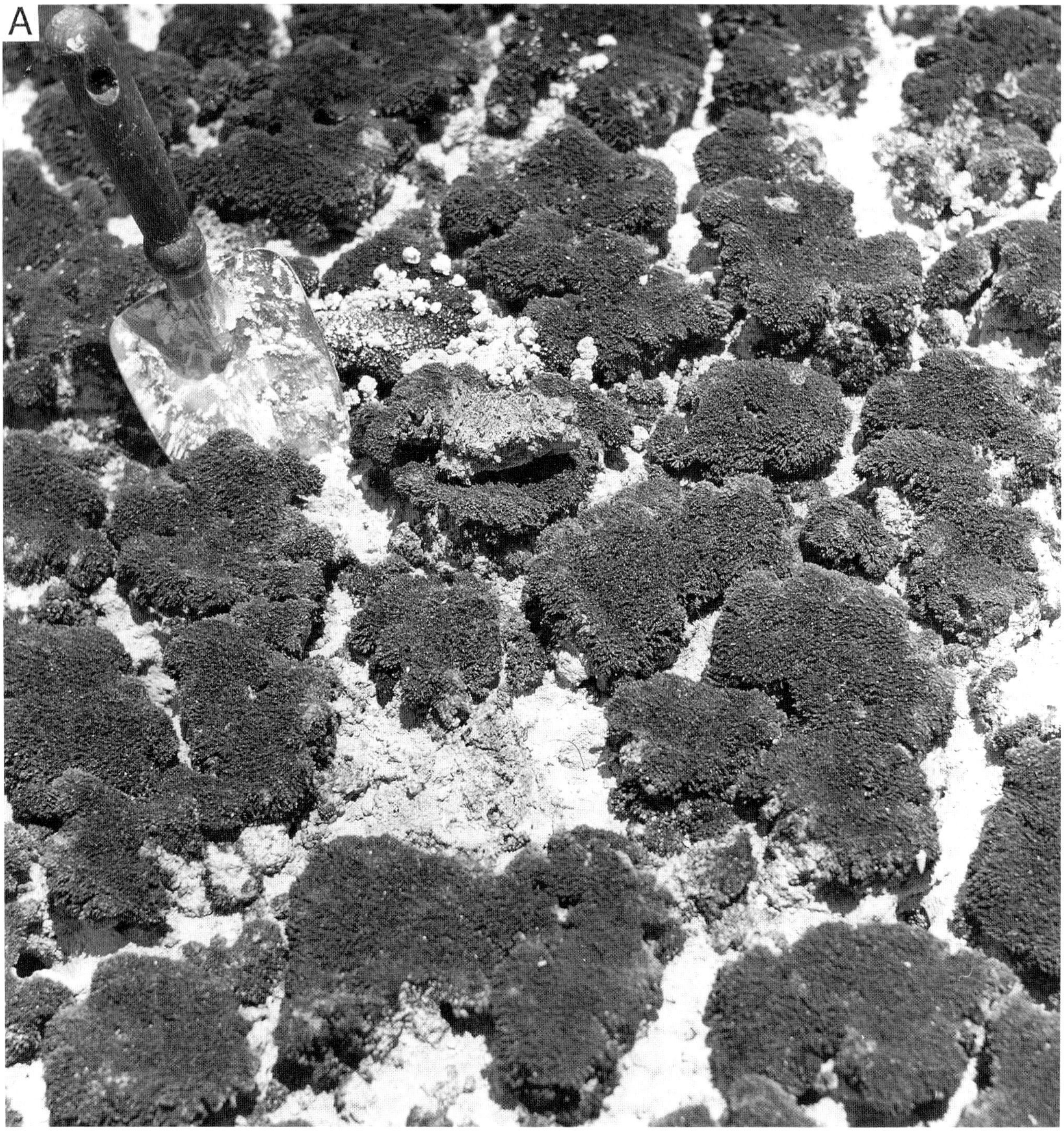

Fig. 72.-Surface form of, and deposits beneath, *Scytonema* cyanobacterial mats on the modern carbonate tidal flats of northwest Andros Island. (A) Surface of luxuriant "pincushions" of *Scytonema* growing at edges of intertidal ponds during the rainy summer season (see Hardie, 1977). (B) (Following page) View looking down onto the surface of a living *Scytonema* mat showing the honeycomb-like structure of the filament tufts. Scale is in cm. (C) (pg. 98) Epoxy-impregnated slab of deposits beneath *Scytonema* mats. Scale is in mm. Crinkled white sediment laminae alternate with dark organic partings that are the remnants of *Scytonema* mats. In upper portions of the slab, the cyanobacterial mat is still visible whereas lower in the sample it has rotted away and left laminoid fenestrae. Scale is in mm. (D) (pg. 98) Thin-section photomicrograph of laminae in (C). Scale bar is 0.5 mm. Note the clotted peloidal fabric of the sediment and the twisting mass of vertical filaments and filament molds smothered by sediment. White areas are plastic-filled intergranular voids. The thin dark bifurcating layers that have the appearance of "horse-tail" stylolites are the compressed organic remnants of *Scytonema* mats (see dark layers in C).

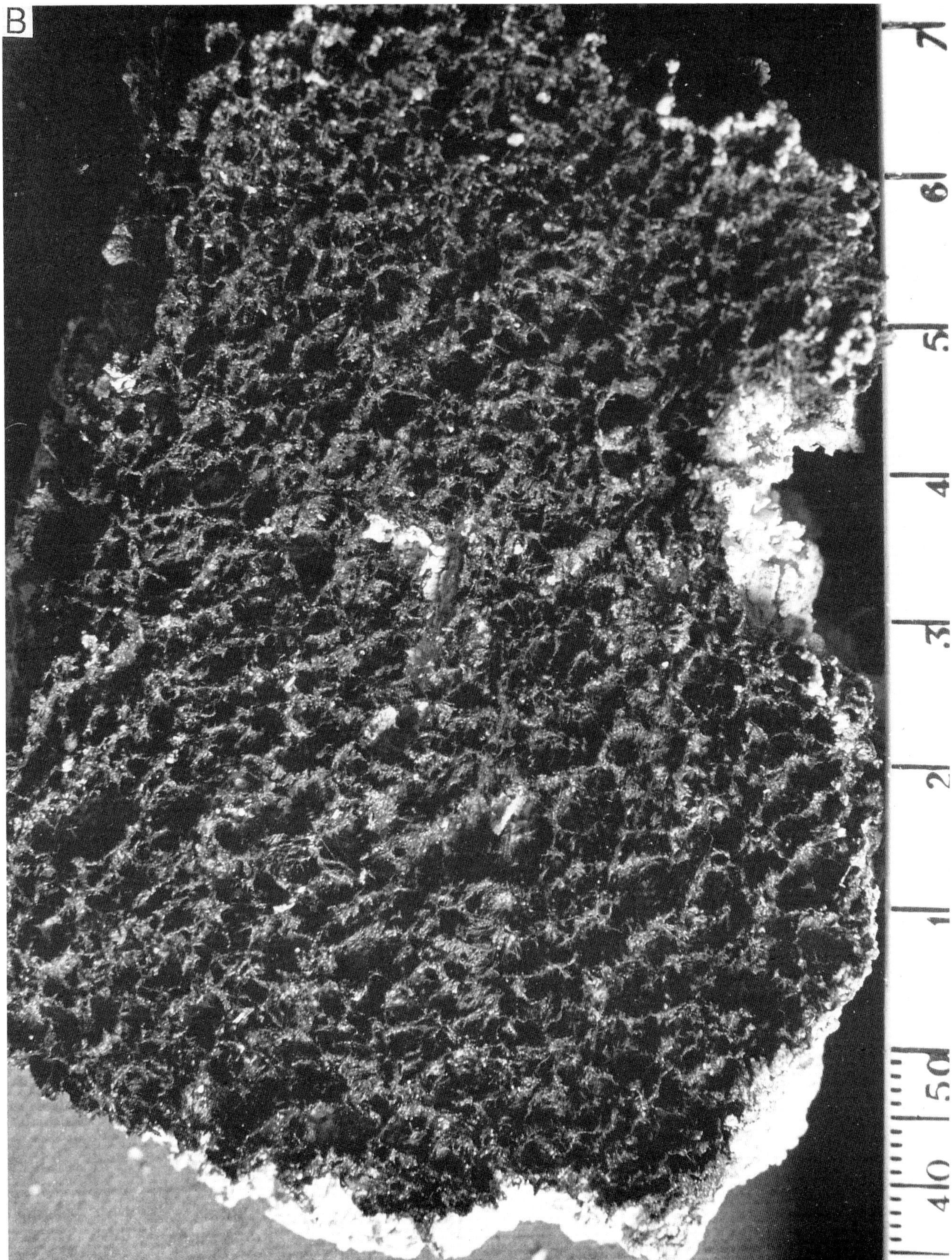
B

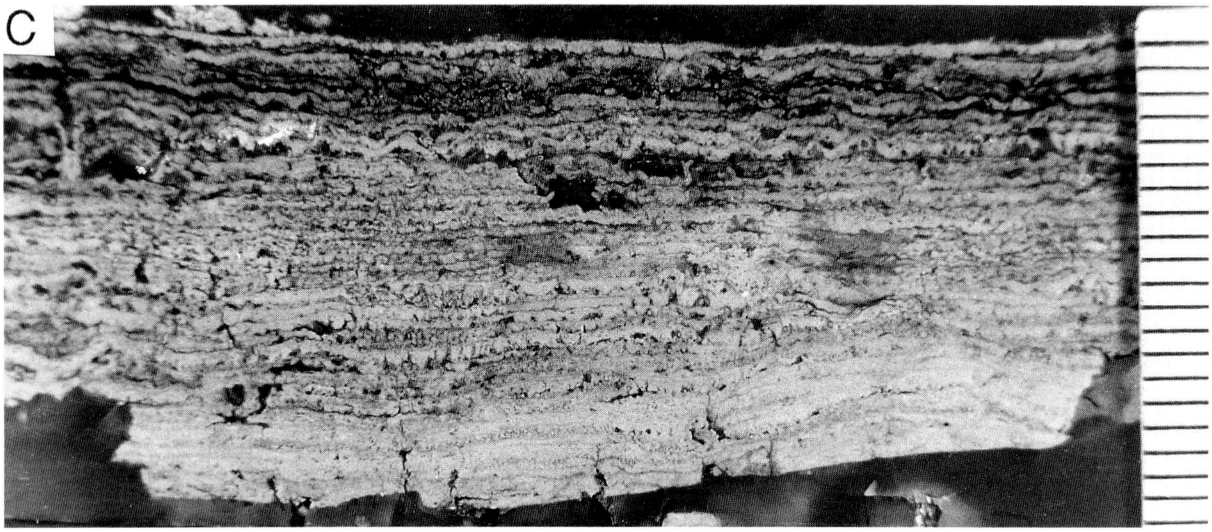

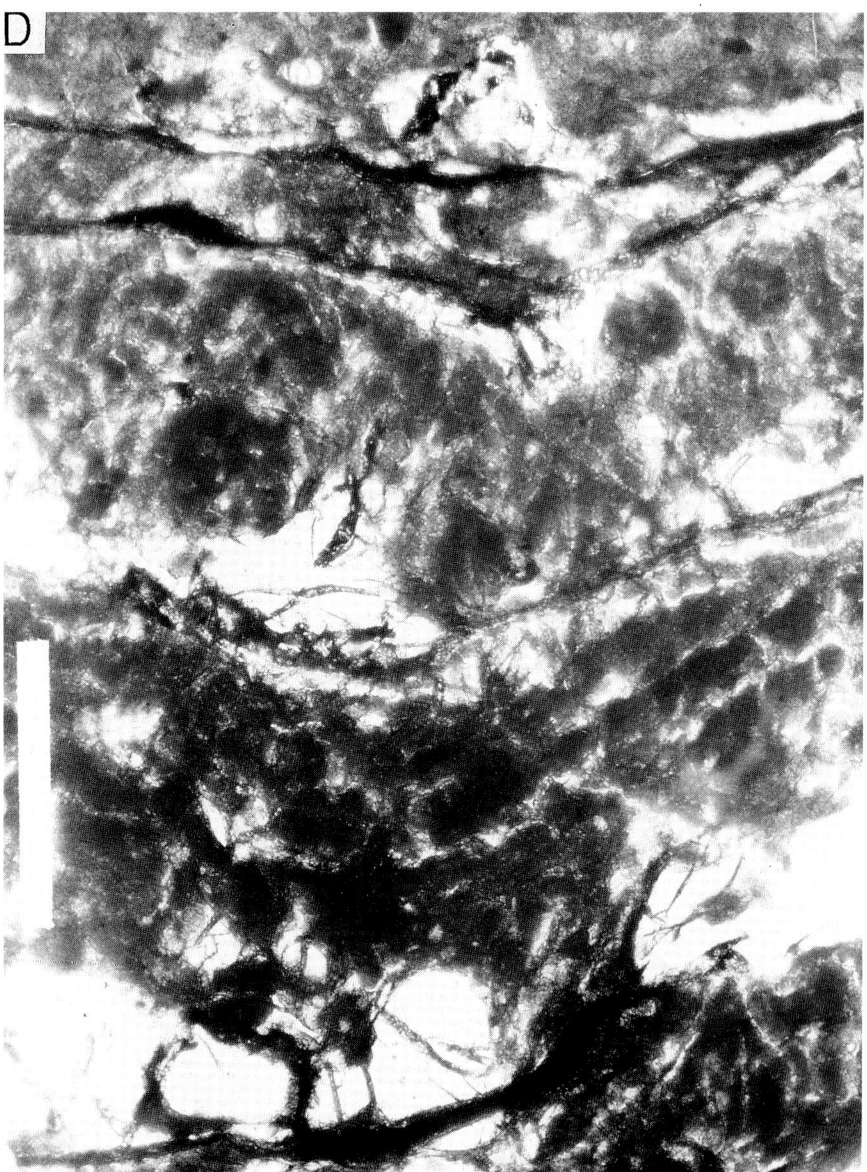

A
B

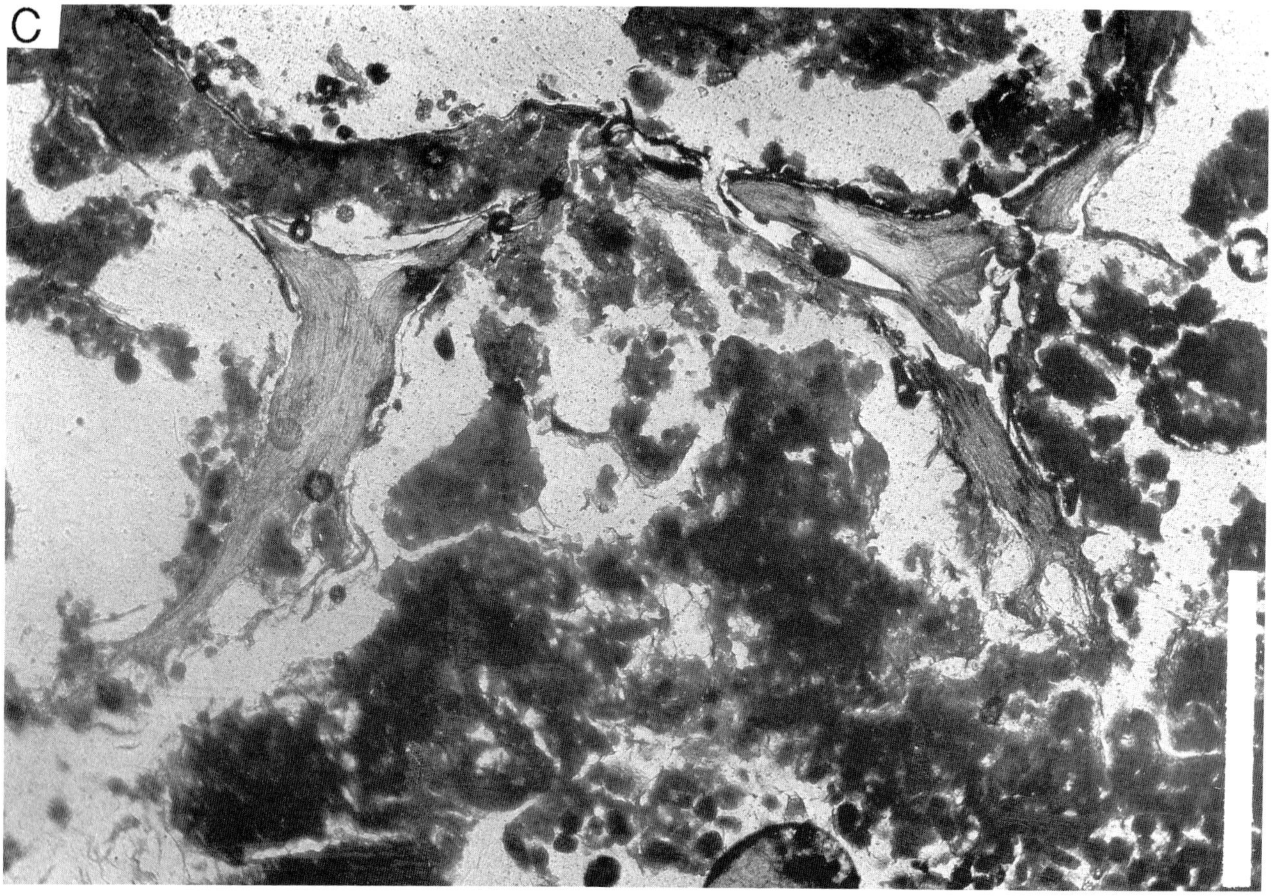

Fig. 73.-Deposits with preserved "pincushions" of *Scytonema* from backslopes of tidal channel levees on the modern carbonate tidal flats of northwest Andros Island. (A) and (B) (Preceeding page) are epoxy-impregnated slabs, scales in mm. The two sediment domes in (A) are where *Scytonema* tufts have trapped peloidal sediment by the baffling action of the filaments. Note the new growth of *Scytonema* on the crests of the domes. (B) Draping of sediment laminae over *Scytonema*-entrained tufts (center of photograph). Dark horizontal stripes are cracks produced during sample preparation. (C) Thin-section photomicrograph of sediment-smothered tufted *Scytonema* mat, showing the vertical columnar structure of the cyanobacterial tufts (bundles of filaments). White areas are plastic-filled voids. Scale bar is 1 mm.

It should be noted that modern microbial layering can be modified by early diagenetic processes. Early cementation can line or fill primary voids between detrital grains and in and around microbially-produced framework elements in any of the laminae types outlined above. These early cements typically are acicular aragonitic fringe cements, and very finely-crystalline, magnesian calcite cements (micrite). In vadose and other exposed settings, these cements can have meniscus and gravitational fabrics. In vadose settings, cementation may be part of calichification of the surface sediments that involves the development of pisoids, intergranular micrite "matrix," grain brecciation, laminar crusts, and the formation of tepee and sheet fracture structures (Esteban and Klappa, 1983). The hard surfaces of cemented or precipitated stromatolites may serve as the substrate for microscopic and macroscopic boring organisms. Here, boring by bacteria, cyanobacteria, fungi and sponges may lead to the formation of filament molds and grain micritization (Klappa, 1979). In uncemented microbial layers and stromatolitic structures, the decay of the organic matter can produce gas bubble fenestrae. This short list highlights some of the early diagenetic modifications possible in microbial structures. Very little has been done in this area and more work is clearly needed to generate criteria for differentiating secondary diagenetic from primary features.

With the above descriptions and discussions of the mechanisms that can produce cryptmicrobial lamination as a foundation, we are now in a position to tackle the principal sedimentary structures that have been widely ascribed to the sediment accretion activities of microbes, that is, stromatolites, cryptmicrobial laminites, and "jelly-roll" structures.

Stromatolites.-

Terminology.-More has probably been written about stromatolites than any other sedimentary structure. Nonetheless, there remains a great deal of confusion about stromatolites. Indeed, there are many fundamentally different definitions of the term, in addition to many lesser terminological problems associated with the name. Our purposes here are to: (1) review the definitions of the term stromatolite and address associated terminological problems; (2) provide field-oriented geologists with criteria to aid them in interpreting stromatolites; and (3) provide a selected bibliography that will ease the reader into

Fig. 74.-(A) Surface view of pustular mat composed of coccoid cyanobacteria, intertidal flats of Hutchinson Embayment, Shark Bay, Western Australia. Lens cap is 50 mm in diameter. (B) Cross-section of sediment below pustular mat. This coarse sand of skeletal fragments shows no internal layering. Tip of ball-point pen for scale.

Fig. 75.-Continuous, isopachous mudstone laminae with radial crystal sprays separated by detrital peloidal laminae. The isopachous mudstone laminae may be very finely crystalline chemical precipitates. Stromatolites from the Cambrian Waterfowl Formation, Canmore, Alberta. Field of view approximately 2 mm. Photograph courtesy of Brent Waters.

TABLE 2 - CHARACTERISTICS OF MODERN CRYPTMICROBIAL LAYERS

Process(es)		Laminae Fabrics	muddy lagoon subtidal	ooid shoal subtidal	peritidal	hypersaline (Shark Bay)	hypersaline (Arabian Gulf)	fresh temperate lakes	fresh tropical lakes	hypersaline lakes	cool springs, rivers waterfalls	warm springs	caves	soil
Non-Microbially Influenced Sedimentation	1	pockets of sand +/- mud; lenses (ripples) of sand	?	?	?	?	C	?	?	?	?	?	?	?
Microbially Influenced Sedimentation → agglutination	2	mud	C	R	C	?	?	?	?	?	?	?	?	R
Microbially Influenced Sedimentation → agglutination	3	sand	C	C	C	C	R	?	?	?	?	?	?	R
Microbially Influenced Sedimentation → agglutination	4	sand-mud couplets	?	?	C			?	?	?	?	?	?	R
Microbially Influenced Sedimentation → baffling	5	mud +/- sand	?	?	C			?	C	?	?	?	?	R
Microbe Growth	6	organic film μ-100's μ thick	C		C			C	?	C	C	C	C	C
Precipitated → fine grained (+/-microbial influence?)	7	laminated (μ-10 μ's) mudstone	R		?		C	C	C	C	C	C	C	C
Precipitated → fine grained (+/-microbial influence?)	8	diffuse, color-banded mudstone, spheroidal vugs	R		?		C	C	?	C	C	C	C	C
Precipitated → fine grained (+/-microbial influence?)	9	10's-100μ; microcolumnar	R		?			C	?					C
Precipitated → fine grained (+/-microbial influence?); microbial influence	10	clotted laminae	?				C	C	C					C
Precipitated → microbial influence	11	clotted shrubs or zones; 2 scales: bacteria (μ), cyanobacteria	?					C	?	C	C	C		R
Precipitated → microbial influence	12	vertical-horizontal filament molds; fill = 0, mud, sand	?		C			C	?	C	C	C		C
Precipitated → radial fabric (little or no microbe influence)	13	competing "druse fans"	?	?	?		C	C	?	C	C	C	C	C
Precipitated → radial fabric (little or no microbe influence)	14	radial needles	?	?	?	C ?	C	C	?	C	C	C	C	C
Precipitated → radial fabric (little or no microbe influence)	15	cave popcorn	?	?	?		C	C	?	C	C	C	C	?
Neomorphism	16	neospar mosaic					?		?					?

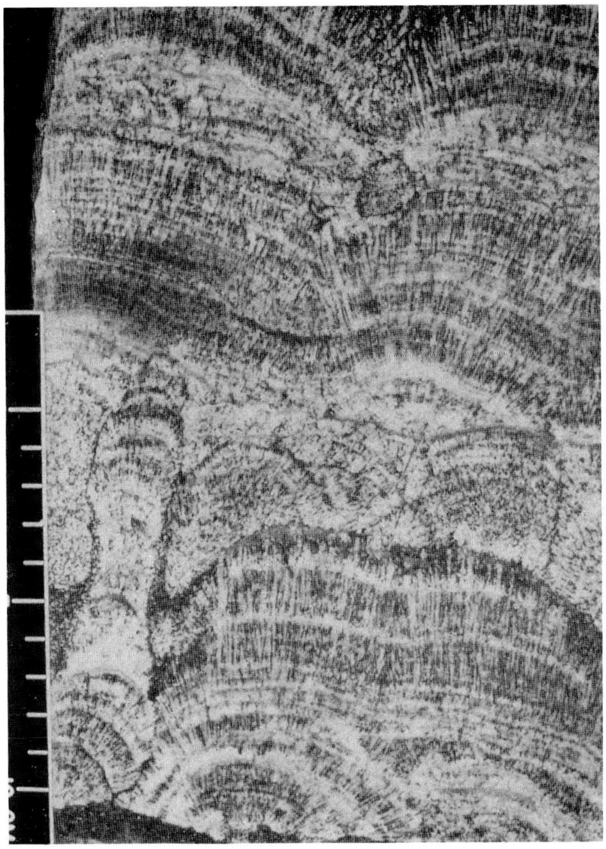

Fig. 76.-Stromatolites composed of radiating crystal sprays from the Lower Proterozoic Beechey Formation of the Kilohigok foreland basin, Northwest Territories, Canada. Photograph courtesy of John P. Grotzinger, scale in cm.

the enormous literature on this subject, which cannot be adequately or thoroughly reviewed here.

There are three widely used types of definitions of the term stromatolite (from the Greek, *stroma* = mattress or bed; *lith* = rock): (1) purely descriptive definitions, (2) purely genetic definitions, and (3) hybrid genetic-descriptive definitions. Purely descriptive definitions have been given by Logan and others (1964), Semikhatov and others (1979) and Buick and others (1981), among many others. These definitions contain three basic elements: (1) stromatolites are discrete, in-place structures that possess recognizable boundaries; (2) stromatolites possess fine, generally millimeter-thick, internal laminae; and (3) these laminae are usually convex-up and generally correspond to the shape of the underlying laminae, implying episodic addition of material to some identifiable surface oriented with respect to the earth's gravitational field (i.e., "growth-bedding" in the sense of Pettijohn, 1975). We would add a fourth element that we consider to be fundamental to the description of a stromatolite, that is, somewhere in the structure there are laminae that "*defy gravity*," in other words, they exceed the normal angle of repose of cohesionless sediment in the earth's gravitational field. This fourth element applies even in those rare stromatolites with concave-upward laminae (see Preiss, 1976, Fig. 2). The significance of this fourth element is that it implies that normal settle-out or traction deposition cannot be solely responsible for growth of the structure, thereby taking stromatolites out of the category of purely physical structures. Compare this definition to that of Semikhatov and others (1979, p. 993): "a stromatolite is an attached, laminated, lithified, sedimentary growth structure, accretionary away from a point of limited surface of initiation....it may be of any origin, composition, shape, size, or age." The most commonly cited wholly genetic definition of the term stromatolite is that given by Walter (1976b, p. 1, see also Krumbein, 1983, p. 493): "stromatolites are organo-sedimentary structures, produced by sediment trapping, binding and/or precipitation as a result of the growth and metabolic activity of micro-organisms, principally cyanophytes." Hybrid, descriptive-genetic definitions of stromatolites are also common. For example, Collinson and Thompson (1989, p. 137) define stromatolites as: "structures that show fine lamination caused by the trapping and binding of material by algae" (see also Pettijohn, 1975, p. 127).

In our view, a purely descriptive definition of stromatolites is best for a number of reasons. First, and most importantly, it is difficult (if not impossible) in many cases to prove that a structure was produced by microbial activity. This is especially true when confronted with stromatolites in the field. All of the authors cited above (and many others) point out that stromatolites (as defined descriptively) are produced by a number of mechanisms, some of which are not primarily influenced by microbial communities. Soil crusts, caliche profiles, splash-zone evaporitic crusts, cave deposits, travertines, and tufas all contain stromatolites. According to Semikhatov and others (1979, p. 994) "if a biological origin had to be demonstrated before a geological object could be called a stromatolite, the term would in most instances be inapplicable (or at least provisional) even after a careful search for associated microorganisms had been made." Although good descriptions of all types of stromatolites are widely available in the literature, there are only a few papers in which criteria are discussed in detail which allow the origins of different stromatolites to be sorted out. One of our major goals is to review and amplify these criteria. A second good reason to employ a purely descriptive definition of the term stromatolite is that nearly all modern examples of stromatolites of different origins in different environments have microbial communities living on or in them. However, in many, if not most of them, the main carbonate production was physio-chemical. Thus, filaments or coccoid forms could easily be encrusted and preserved in the deposit even though they had little, if anything, to do with deposition (see also discussion below). Finally, a purely descriptive definition is probably more akin to what Kalkowsky (1908) originally intended. Kalkowsky (1908) introduced the term *stromatolith* to describe sedimentary features associated with ooid grainstones in the Bund Sandstone of northern Germany. It is clear from Kalkowsky's descriptions, line drawings and photographic illustrations that he was describing discrete structures (masses) with recognizable boundaries that possess a fine (millimeter-thick) "more or less even layered structure" (ebene Lagenstruktur, Kalkowsky, 1908, p. 69), that is, laminae. Furthermore, Kalkowsky specifically mentions the tendency for laminae to be convex up. Monty (1977) and Krumbein (1983) provide partial translations of much of Kalkowsky's paper. However, both authors use the English word "planar" as a translation of the German "ebene" which is probably more correctly rendered in English as "even." Kalkowsky probably did not intend to include the flat laminated continuous beds that we would now refer to as "cryptalgal laminites, stratiform stromatolites, or planar stromatolites" as both Monty and Krumbein maintain. Indeed, Monty (1977, p. 17) translates "... fur Stromatolith mit mehr planem Aufbau" (Kalkowsky, 1908, p. 117) as "... for the flat-lying stromatolites" whereas Kalkowsky's original words were "... for stromatolites with more planar structure." Finally, Kalkowsky clearly says that he will interpret the stromatolites (and the ooids) of the Bund Sandstone as organic structures. In section 6, page 70, Kalkowsky begins with the sentence "It is the proposition of the following treatise to show that ooids and stromatolites of the north German Bund Sandstone have been built by plants." Compare this to Krumbein's translation of the same passage (Krumbein, 1983, p. 388): "Oolites and stromatolites of the North German Bundsandstein are produced by plants... ." Without the introductory clause, it would seem that Kalkowsky intended a genetic definition of stromatolites, but, as also suggested by Logan and others (1964), Kalkowsky almost surely intended stromatolite to be a purely descriptive term. In summary then, it is proposed that a purely descriptive definition of stromatolite be used. Thus, ***stromatolites are discrete, in-place structures with recognizable boundaries that are characterized by "gravity-defying" internal laminae reflecting addition of material to a discrete surface***. Note that continuous beds ("cryptalgal laminite," "planar stromatolites," or "stratiform stromatolites") are specifically excluded from this definition. The main point to be stressed here is that it is very difficult (and indeed often impossible) to assign a specific origin to a stromatolite, especially in the field. However, if application of the criteria discussed below does allow a more specific diagnosis of origin to be made, then this can be simply done by adding an appropriate adjective as suggested by Logan and others (1964). For example, a useful field term for an abiogenic stromatolite is "chemical stromatolite" but a more specific genetic term, such as "speleothemic stromatolite" could be used if the evidence so warrants (see below). For those stromatolites diagnosed to have formed by microbes, the terms "microbial" and "cryptmicrobial" (Kennard and James, 1986) as discussed above should replace Aitken's (1967) "algal" and "cryptalgal" terms long employed by geologists. A minor terminological problem arises from the fact that stromatolite is clearly a rock term (see also Krumbein, 1983). Thus, it should properly exclude modern unlithified examples. Krumbein recommended that such modern forms be called "potential stromatolites," whereas Hardie and Garrett (1977, p. 21) employed the term "proto-stromatolite." We would like to see a terminology that allows modern unlithified structures to be included in the stromatolite family but which at the same time remains descriptive. A term like "unlithified stromatolite" seems an appropriate descriptive name but any similar usage would be satisfactory.

The following problems with a purely descriptive definition of stromatolite are commonly cited: (1) it excludes biogenic laminites ("algal" laminites); (2) it is apparently incompatible with the binomial nomenclature (i.e., "form genera") of stromatolites; (3) it excludes some of the famous stromatolites of Hamelin Pool, Shark Bay; and (4) it unnecessarily increases jargon (Semikhatov and others, 1979; Buick and others, 1981; Walter, 1983; among many others). Even though most modern settings, where microbial mats are affecting sedimentation, are actually broad, flat surfaces (see review in Semikhatov and others, 1979), it is appropriate to exclude their ancient counterparts (i.e., "cryptalgal laminites", etc.) from stromatolites, because it is the lateral restriction of the microbial mats to specific topographically high surfaces that is the essence of the stromatolite morphogenesis problem. This is also true even where discrete structures "intergrade" with flat laminated beds. Microbial mats that spread out over huge, flat surfaces deserve attention in their own right and we have dealt with them below under "cryptmicrobial laminites."

There is nothing inherently wrong with using a "binomial nomenclature" to describe microbial stromatolites if: (1) the criteria which allow that diagnosis (i.e., that the stromatolite was of microbial origin) are specifically stated, and (2) it is clearly understood that it is a complex community of microbes (and perhaps multicellular organisms) that are responsible and that physical processes may also play a major role in stromatolite morphology. The "stromatolites" of Hamelin Pool, in Shark Bay are a very problematic group of structures. Although Logan and others (1974a) claim them to be actively accreting today, this view is disputed by Playford and Cockbain (1976). Thus, it is not at all clear just what is the true mechanism(s) by which these structures were formed. Also, Shark Bay "stromatolites" are commonly only crudely laminated and even unlaminated in some areas, apparently contain smaller-scale "finger structures" on the centimeter-scale, are fenestral, and are composed of sand-sized skeletal grains, peloids and ooids, as opposed to the generally mudstone composition of ancient stromatolites. These Shark Bay "stromatolites" are in many ways much more closely related to ancient thrombolites than to most ancient stromatolites (see discussion below). Finally, the adjective-noun system for ancient stromatolites is concise and does not increase, but rather decreases jargon. It eliminates the need for the term "stromatoloid" (Buick and others, 1981). It eliminates the need for confusing terms such as "pseudostromatolite" and "stromatolite-like" (Walter, 1976a). Finally, it eliminates the need for terms such as "stiriolites" and "coniatolites" (Walter, 1983) as proposed by advocates of genetic definitions of stromatolites.

Distinguishing characteristics.-In keeping with our overall comparative sedimentology approach, we first treat modern stromatolites and then tackle ancient stromatolites.

1. MODERN STROMATOLITES. The discussion below is based on studies of modern lithified and unlithified stromatolites found in a number of active depositional environments, both marine and nonmarine. Modern stromatolites will be very briefly described environment by environment, with emphasis on fabrics, geometry of laminae, and overall morphology (see Table 2 and the references for more complete details).

<u>Normal Marine Subtidal Environments</u>. Microbial mats are fairly common on and in carbonate lagoon floors where they stabilize the bottom (Neumann and others, 1970, see also Garrett, 1977). However, modern normal-marine subtidal stromatolites have only been reported from a few muddy lagoon settings, such as those of eastern Andros Island (Monty, 1967, 1976) and Bermuda (Gebelein, 1969, 1976; Hofmann, 1973, Fig. 10). These are unlithified with a maximum diameter of 10 cm and height of 5 cm. They tend to be solitary, vertically stacked hemispheroids with variable basal radius (terminology of Logan and others, 1964). Laminae are millimeter-thick and comprise alternations of organic material (principally Oscillatoracean sheaths) and well-sorted peloids of silt to very fine-sand size. Laminae tend to pinch out down the sides and are notably discontinuous resulting in small unconformities within the stromatolite. These unlithified stromatolites are found in association with oncoids (the "algal biscuits" of Ginsburg, 1960).

Large (up to 2 m high) solitary stromatolites and clusters of coalesced stromatolitic mounds are found in actively migrating areas of Bahamian ooid shoals (Dravis, 1983; Dill and others, 1986, 1989; Reid and Browne, 1991). Laminae are composed of ooid and skeletal sand grains with maximum sizes in the fine sand size range. Two types of stacking of laminae appear common. The most abundant are small, centimeter-high, vertically-oriented "fingers" that contain some concentric laminae, but are for the most part unlaminated. Less commonly, laminae span most of the way across a head generally pinching out down the sides; the stromatolite sides are commonly bored and micritized, obscuring detail. Again, laminae seem to be notably discontinuous. Both the "fingers" and the laminae contain large fenestrae that represent entombed algae (e.g., the non-calcareous green alga *Batophora*) and invertebrate borings. Surface layers are unconsolidated but within a few centimeters of the surface the stromatolites become cemented by fine-grained aragonite and magnesium calcite cements. These large stromatolites are the subject of current research.

Normal Marine Peritidal Environments. Black (1933) was first to recognize the importance of agglutination of detrital sediment on to the organic sheath of cyanobacteria in the formation of modern unlithified stromatolites on tidal flats (see earlier). Many of Black's stromatolites represent desiccation-modified features of more continuous cyanobacterial mats. Hardie and Ginsburg (1977, p. 53-57) have described and illustrated unlithified, well-laminated stromatolites up to 12 cm in diameter and 6 cm high built between the entrances of fiddler crab burrows along tidal creeks of northwest Andros Island (Fig. 71). Laminae (millimeter to sub-millimeter in thickness) composed of clotted mud alternate with thin lenses of silt-sized peloids (see Hardie and Ginsburg, 1977). The detrital lenses are more abundant on the landward sides of the knobs, making the stromatolites asymmetrical. Although most of the laminae pinch out down the sides of the stromatolites, some layers maintain their thickness down to the very base of the stromatolite (see, for example, Hardie and Ginsburg, 1977, Figure 34C). Monty (1967, Fig. 16-3) shows a somewhat similar unlithified stromatolite composed of alternating peloidal laminae and thin organic mats stacked in such a way that they make upward branching columns. Small, isolated colonies of the non-motile *Scytonema* also produce both unlithified and lithified stromatolites as described by Monty (1967, 1972, 1976), Monty and Hardie (1976), and Hardie and Ginsburg (1977). These thick-filamented cyanobacteria can baffle suspended and traction load sediment, producing crude, concentrically-laminated structures composed of variable amounts of organic matter and detrital peloids and mud. The concentric laminae of these *Scytonema* "heads" tend to be a few millimeters thick and have irregular boundaries. They are riddled with shelter fenestrae which are a consequence of the complex tangle of filaments and filament bundles into which the detrital sediment filtered. In the supratidal freshwater cyanobacterial marsh landward of the channelled belt of the northwest Andros tidal flats, the *Scytonema* mats not only baffle storm-derived sediment but also become calcified to produce lithified stromatolites (the "algal tufa" of Hardie and Ginsburg, 1977; Hardie, 1977b). Luxuriant growths of *Scytonema* flourish in the marsh during the rainy season, only to calcify to various degrees during the dry season. This process, which involves preferential precipitation of low magnesian and high magnesian calcite around both living and dead filaments is, on rare occasions, punctuated by hurricane floods which bury the cyanobacterial tufas with up to 5 cm of bank-derived peloidal-skeletal sediment. Upon draining and desiccation, this hurricane layer cracks and develops a very thin, cemented crust on its upper surface. The nonmotile *Scytonema* eventually recolonizes the surface by growing up the mudcracks and spreading out from there A striking ancient example is shown in Figure 77. (Compare Figure 77 with Figure 114 in the section on CHEMICAL FEATURES below).

Hypersaline Marine Peritidal Environments. A wide variety of stromatolites occur in hypersaline lagoons in Shark Bay, Western Australia (Logan and others, 1970, 1974a; Hoffman, 1976a). The most famous are indurated, decimeter high examples from Carbla Point and other areas (Fig. 78). These are found from the supratidal down into the subtidal to a water depth of at least 3.5 m. They occur as isolated, upward-widening columns up to 0.75 m high, as coalesced columns, and as elongate, isolated ridges oriented subparallel to the wave approach direction (subperpendicular to the shoreline). Internally these columnar structures range from laminated to unlaminated (Fig. 78B). Laminae are composed of *Fragum* shell fragments, peloids and ooids and are millimeters thick. Laminae are noticeably lensoidal with unconformities between packages of laminae being a common feature. Laminae are indistinct toward stromatolite edges. Particles are cemented by fine-grained and acicular aragonite cements. Laminae are disrupted by numerous fenestrae, many of which clearly have been enlarged by dissolution at some stage in their development (Fig. 78B). Logan and others (1974a) reported that these structures are currently accreting by addition of sediment layers at rates of up to 10 mm/yr, but this conclusion is vigorously disputed by Playford and Cockbain (1976), who maintain that markers hammered into the structures show that there has been no addition of sediment for many years. This crucial point of disagreement remains unresolved at this time.

Laminated aragonite layers and stromatolites up to 2 cm thick encrust beach rock surfaces in certain upper intertidal to supratidal areas of the arid Trucial Coast tidal flats along the Arabian Sea (Purser, 1972; Purser and Evans, 1973; Purser and Loreau, 1973; Scholle and Kinsman, 1974). These stromatolitic crusts are apparently the result of the abiogenic precipitation of aragonite from degassing or evaporating sea water in a splash zone setting. The crusts have a range of internal laminae including: (1) very thin mudstone laminae (microns to tens of microns thick); (2) millimeter-thick mudstone laminae with undulatory boundaries and patches of neomorphic (?) spar (Purser and Loreau, 1973, Fig. 9.3); and (3) coarse, clotted-peloidal laminae that contain ovoid to irregular fenestrae. Stacking geometry of laminae is variable, but in general the overlying laminae tend to "smooth out" depressions in the underlying laminae. The fine laminae appear to be quite continuous although some finely laminated crusts contain discontinuous laminae and unconformities. In one example figured by Purser and Loreau (1973, Fig. 9C), where two small stromatolites abut each other, the laminae meet in 90° "kink-fold" geometry at the boundary between the stromatolites. Small irregularities on the surface of the layers trap detrital grains (including quartz grains) which are then coated, encrusted and incorporated into the fabric. Scholle and Kinsman (1974) point out that the crusts are closely associated with concentrically coated pisoid grains presumed to have been chemically precipitated. In addition, they describe laminae composed of chemically precipitated radial druse-fans and grains completely coated with isopachous laminae. Overhangs and internal cavities commonly have pendant cements, both drusy and finely-crystalline. All these features point to a chemical precipitation origin for the stromatolitic crusts.

Lacustrine Environments. Stromatolites and unlithified stromatolites are common in lakes, where they range in size from centimeter-scale up to mounds meters to tens of meters in height and diameter (Dunn, 1953; Scholl, 1960; Scholl and Taft, 1964). They are found in both "fresh" (low salinity) and saline lakes. Virtually all of the processes described above, and the lamination fabrics so produced, have been reported. Indeed, heterogeneity seems to be the watchword of these deposits. Three fabrics appear to be most common: (1) very finely laminated micrite layers; (2) small complex protuberances shaped like popcorn; and (3) "shrubs" of clotted peloids. The peloidal shrubs appear to come in two sizes, peloids a few tens of microns across (similar to the "bacterial shrubs" of Chafetz and Folk, 1984) and clots a few hundreds of microns in size. The "popcorn" shows both radial crystalline fabric and concentrically laminated micrite; it is similar in every way to the "cave popcorn" described by Thrailkill (1976). There is an almost complete gradation between the two fabrics, both of which ap-

Fig. 77.-Bedding-plane (A) and vertical cross-section (B) views of laminated cryptmicrobial mounds covering mudcracks in a tidal flat carbonates of the Middle Ordovician St. Paul Group of western Maryland. Scale in mm. Compare with Figure 114.

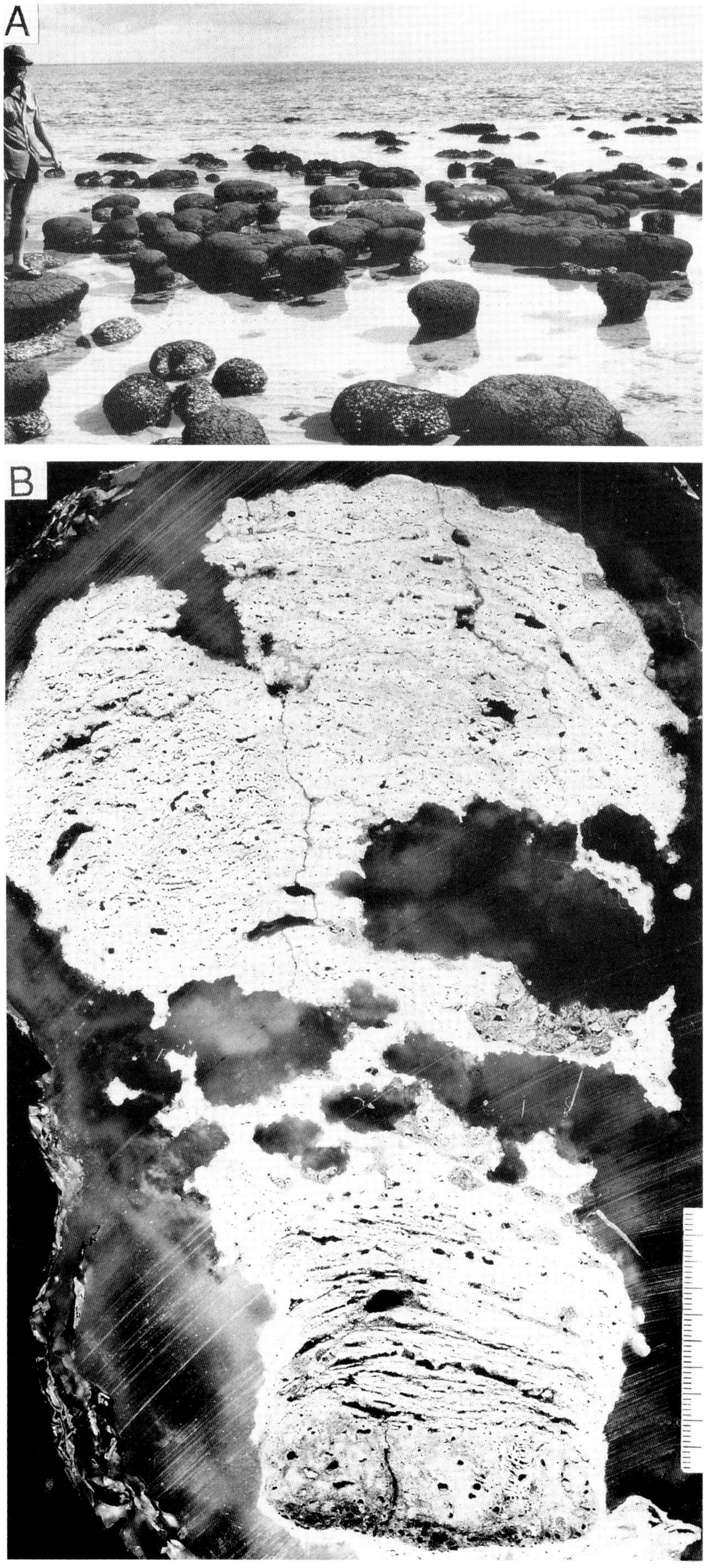

Fig. 78.-(A) Stromatolites from Carbla Point, Hamelin Pool embayment of Shark Bay, Western Australia. In the epoxy impregnated slab in (B), note the irregular laminae, fenestrae, large dissolution vugs, and lithoclast upon which the stromatolite nucleated. Unlike most ancient examples, the Hamelin Pool stromatolites are composed of medium to fine sand-sized skeletal grains and are pitted by dissolution vugs and cavities. Scale of lower photograph in mm. Sample shown in (B) courtesy of Brian Logan.

pear to be the result of *in situ* precipitation of low magnesian calcite in low salinity lakes, and aragonite, magnesian calcite, or monohydrocalcite in saline lakes. It is not clear to what extent microbes influence precipitation.

Unlithified stromatolites occur in the freshwater lakes immediately landward of the tidal flats of Andros Island (Monty, 1972; Monty and Hardie, 1976). In these very shallow lakes, fed entirely by rainwater, mats of *Scytonema* drape mudcrack polygons in underlying hurricane layers to produce laterally-linked, hemispherical stromatolites made of continuous wavy laminae of white calcite mud with dark organic partings (see Monty and Hardie, 1976, and Mitchell, 1985). The low magnesian calcite mud is produced initially by precipitation around the *Scytonema* filaments and then redistributed within the lakes by storm agitation. Freshwater calcite mud is produced in the same way in the Everglades of Florida (see Gleason and Spackman, 1974, on the "periphyton" marshes of the Everglades; also Monty and Hardie, 1976).

Fluvial Environments. Stromatolites found in rivers, springs and waterfalls are, for the most part, similar to those found in lakes. The spring settings range from cold to hot springs with the deposits of the latter commonly comprising quite impressive edifices (e.g., Mammoth Hot Springs in Yellowstone National Park, Wyoming) and may be siliceous (Walter, 1976c,d). Much of the deposition of calcium carbonate from upwelling spring waters actually takes place in the surrounding spring-fed pools rather than in the spring orifice. However, stromatolites composed of calcareous coatings of higher plants, particularly mosses, seem to be more common in spring orifices.

Cave Environments. Speleothems (flowstones) commonly have stromatolitic layering in which the laminae are composed of either radial crystalline carbonate (in some cases of coarse druse) or color-banded micrite. The latter type of laminae are typically isopachous in subaqueous settings such as cave pools but in the subaerial parts of the cave they show gravitational fabrics (e.g., pendants). Most typical, as might be expected, is "cave popcorn" (Thrailkill, 1976) (see above). Low magnesian calcite is the predominant mineral of speleothems but aragonite and calcian dolomite have been reported from some cave deposits.

Soil Environments. Non-marine caliches with stromatolitic layering are discussed under Chemical Features below.

2. ANCIENT STROMATOLITES. Ancient stromatolites have attracted interest from a diverse group of scientists, including paleontologists, carbonate sedimentologists, stratigraphers, biologists, geochemists, and cosmologists, because: (1) stromatolites are generally regarded as direct evidence of the earliest life in the form of microbial populations (circa 3.5 Ga), and (2) chert stromatolites and microbial laminites from Archean and Proterozoic formations commonly carry identifiable, well-preserved microfossils. In addition, stromatolites have been used by carbonate sedimentologists as environmental indicators (see Monty, 1977), and by Russian and Australian stratigraphers as guide fossils to subdivide the Upper Proterozoic stratigraphic column.

Ancient stromatolites come in a vast array of sizes and shapes ranging from simple club-shaped columns a few centimeters in diameter and a few centimeters in height up to complex masses tens of meters high and hundreds to thousands of meters in extent. In general, the larger a stromatolite is, the more complex its structure, with many possible levels of organization. Useful descriptive terminologies are provided by Logan and others (1964) and Preiss (1976). It is impossible to illustrate here all the morphologic complexities of stromatolites but Figures 79 to 91 give some small idea of the range of sizes and morphologies of carbonate stromatolites found in the stratigraphic record. Sources in which ancient stromatolites are well-illustrated and described include: Aitken (1967), Walter (1972, 1976a, 1983), Hofman (1973), Hoffman (1976b), Horodyski (1976), Krumbein (1983), Toomey and Babcock (1983), and Riding (1990).

The range of fabrics of ancient stromatolites include all those described for modern organo-sedimentary layering in Table 2 and present the same problems encountered in their interpretation. In making comparisons between modern and ancient stromatolite fabrics, it should be recognized that it is very common for the ancient stromatolites, particularly Precambrian examples, to have undergone neomorphic alteration. The original detrital nature of stromatolitic laminae in many neomorphosed examples is difficult or impossible to identify petrographically (Fig. 69). In other examples, laminae microstructure are preserved. Figure 92 shows preserved detrital fabrics where peloids comprise laminae that exceed the angle of repose of detrital sediment. These fabrics are remarkably similar to those of the modern stromatolites from Andros Island illustrated in Figure 71C. Other ancient stromatolite laminae include chemical mud laminae (Fig. 75), and radial druse fans (Fig. 76). In some cases, laminae that comprise stromatolites show a variety of interlaminated fabrics (Fig. 93). In view of these difficulties, and those outlined above, few attempts have been made to classify the microstructures of ancient stromatolites (see, as one example, Bertrand-Sarfati, 1976).

Occurrence and preservation.-Stromatolites occur in many types of sedimentary rocks of all ages, however, they are most common and most spectacular in their size and complexity in carbonate rocks from approximately 1.65 - 0.65 Ga. Stromatolites suffered an abrupt decline after the Early Ordovician (Fischer, 1965; Garrett, 1970; Ginsburg, 1971b; Awramik, 1971, 1984). The decline in stromatolites corresponds to the expansion of bottom grazing and burrowing invertebrates (Garrett, 1970). Carbonate sedimentary rocks are the hosts for the vast majority of stromatolites although examples in chert are not uncommon, especially in Archean and Lower Proterozoic carbonate and banded ironstone deposits. Rarer rock types that contain stromatolites include deep sea manganese nodules, phosphorites, barites, quartz sandstones, and siliciclastic mudstones. In the early development of stromatolite studies, it was commonly assumed that stromatolites were diagnostic of an intertidal setting. However, studies from both modern environments (see above) and ancient rocks quickly undercut this simplistic notion. Precambrian stromatolites probably grew in freshwater to marine settings, and, in marine basins, from basinal to supratidal environments (Hoffman, 1974). It is difficult to assess whether Precambrian stromatolites are characteristic of any particular carbonate platform environment, although, platform margin buildups appear to have the most diverse and largest forms (see Walter, 1976a). Playford and Cockbain (1969) were among the first to draw attention to non-intertidal stromatolites in Phanerozoic rocks.

Origin and significance.-Table 2 shows the distribution in modern stromatolites of laminae styles and internal fabrics, and types of depositional processes that produced them. It must be recognized that several problems arise with the use of this checklist to interpret the origin of ancient stromatolites and their depositional environments. First, the laminae of most ancient stromatolites are composed of micrite. Micritic laminae can be either agglutinated or precipitated. Both agglutination and precipitation are common processes that form stromatolites in almost all modern carbonate environments. Thus, micritic stromatolitic laminae are not diagnostic of any particular environment. Second, agglutinated peloidal fabrics are difficult to distinguish from precipitated clotted fabrics, even in modern examples. Third, the fabrics of the laminae of many ancient stromatolites are coarse-crystalline neomorphic spar or dolomite mosaics that preserve no evidence of the original depositional fabrics. Thus, care needs to be exercised in using the characteristics of modern stromatolites to interpret ancient stromatolites.

Fig. 79.-Bedding plane view of elongate giant stromatolite mounds. Note airplane at beach in lower right (arrow) for scale. Glaciers have plucked out the inter-mound fill so that the view today is nearly the depositional relief. Beechy Formation, Lower Proterozoic ramp deposits from the Kilohigok foreland basin, Northwest Territories, Canada. Photograph courtesy of John P. Grotzinger.

Fig. 80.-Dolomitic columnar stromatolites from the Lower Proterozoic Taltheilei Formation, Pethei Group, Northwest Territories, Canada. Photograph courtesy of Paul Hoffman.

Fig. 81.-Dolomitic columnar stromatolites from the Lower Proterozoic Taltheilei Formation, Pethei Group, Northwest Territories. Scale in inches. Photograph courtesy of Paul Hoffman.

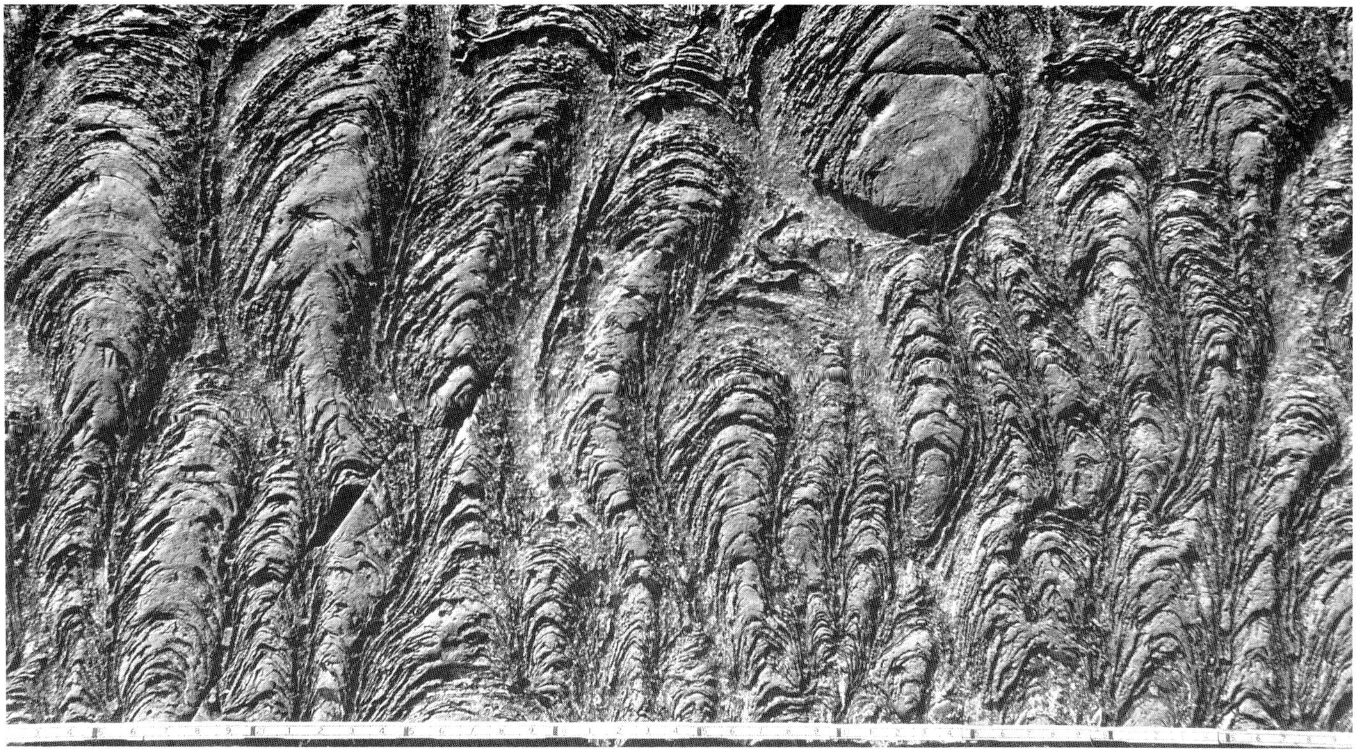

Fig. 82.-Dolomitic columnar stromatolites from the Lower Proterozoic Taltheilei Formation, Pethei Group, Northwest Territories. Scale in inches. Photograph courtesy of Paul Hoffman.

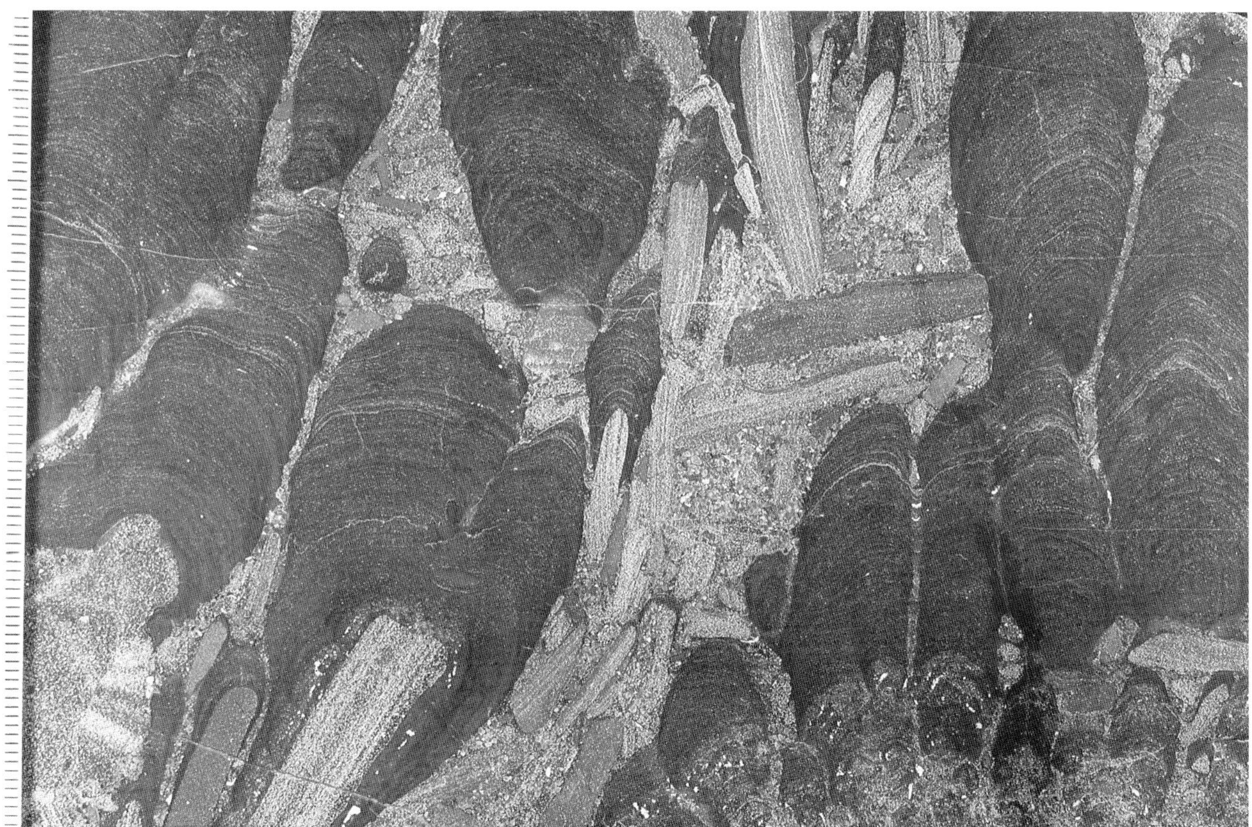

Fig. 83.-Stromatolites nucleated on flat intraclasts of planar and cross-stratified peloidal grainstone. Upper Cambrian Conococheague Limestone, western Maryland. Scale in mm.

Fig. 84.-Dolomitic stromatolitic bioherm from the Lower Proterozoic Taltheilei Formation, Pethei Group, Northwest Territories. Coin in left center is approximately 30 mm in diameter. Photograph courtesy of Paul Hoffman.

Fig. 86.-Exhumed bedding surface of elongate, columnar stromatolites from the tidal flat facies of the Lower Proterozoic Taltheilei Formation, Pethei Group, Northwest Territories. The elongation direction is normal to the ancient shoreline. Photograph courtesy of Paul Hoffman, hammer gives scale.

Fig. 85.-Margins of a large shelf-edge stromatolitic bioherm from the Lower Proterozoic Taltheilei Formation, Pethei Group, Northwest Territories. Hammer in center of photograph gives scale. Photograph courtesy of Paul Hoffman.

Fig. 87.-Bedding plane exposure of dolomitic columnar stromatolite from the Lower Proterozoic Taltheilei Formation, Pethei Group, Northwest Territories. Photograph courtesy of Paul Hoffman. Scale is in inches.

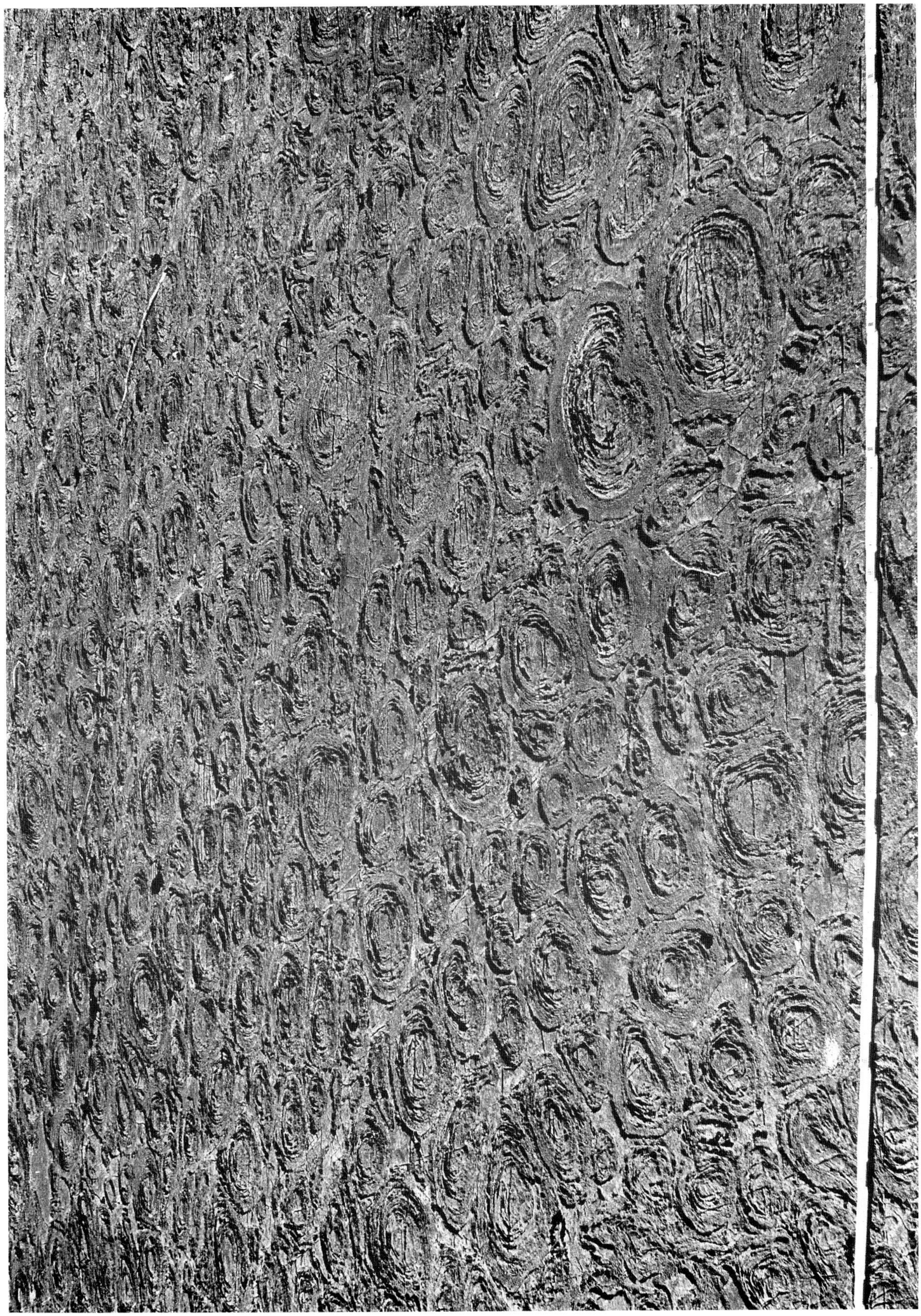

Fig. 88.-Bedding plane exposure of columnar stromatolites from the Lower Proterozoic Wildbread Formation, Pethei Group, Northwest Territories. Photograph courtesy of Paul Hoffman. Folding ruler is approximately 15 mm wide

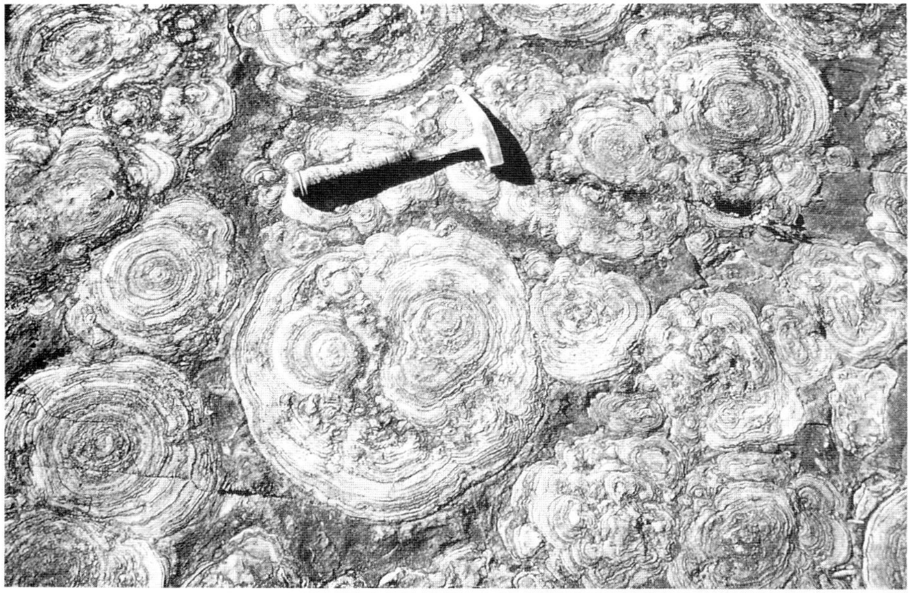

Fig. 89.-Bedding plane exposure of Upper Cambrian stromatolites originally described by James Hall (1847, 1883) as "*Cryptozoon*" from the Hoyt Limestone "Petrified Sea Gardens," Saratoga Springs, New York. Hammer for scale. Photograph courtesy of Jim Sorauf.

Fig. 90.-Conical stromatolites of the "form genus" *Conophyton* from the Middle Proterozoic Dismal Lakes Group, Northwest Territories. Hammer for scale. Photograph courtesy of John P. Grotzinger. See Walter (1972, 1976) and Semikhatov and others (1979) for an introduction to form genus terminology for stromatolites.

Fig. 91.-Conical stromatolites of the "form genus" *Conophyton* from the Proterozoic Top Camp Formation, Bangemall Basin, Western Australia. Hammer for scale.

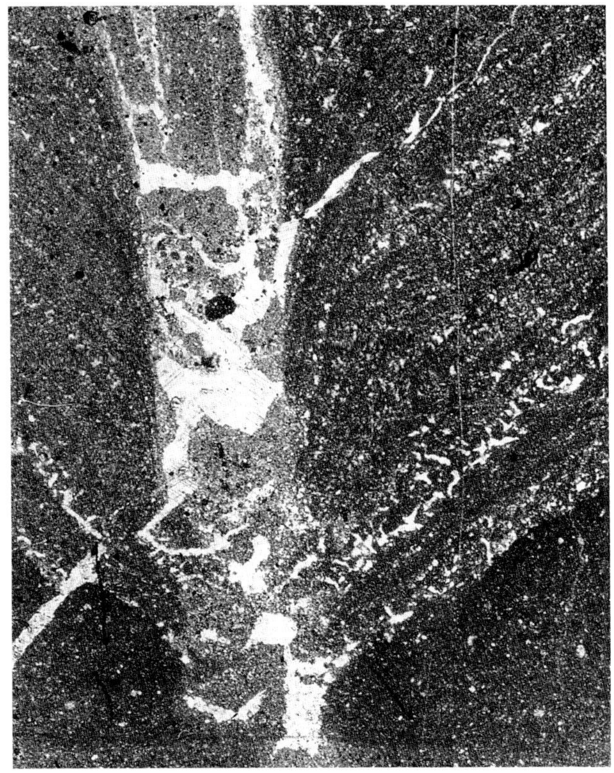

Fig. 92.-Thin-section photomicrograph of the stromatolites shown in Figure 83. Note that the laminae comprise couplets of fine sand- to coarse silt-sized peloidal grainstones alternating with clotted mudstones. Peloidal laminae commonly exceed the angle of repose, implying the former presence of a microbial mat. Compare with Figure 71C. Upper Cambrian Conococheague Limestone, western Maryland. Field of view approximately 5 mm.

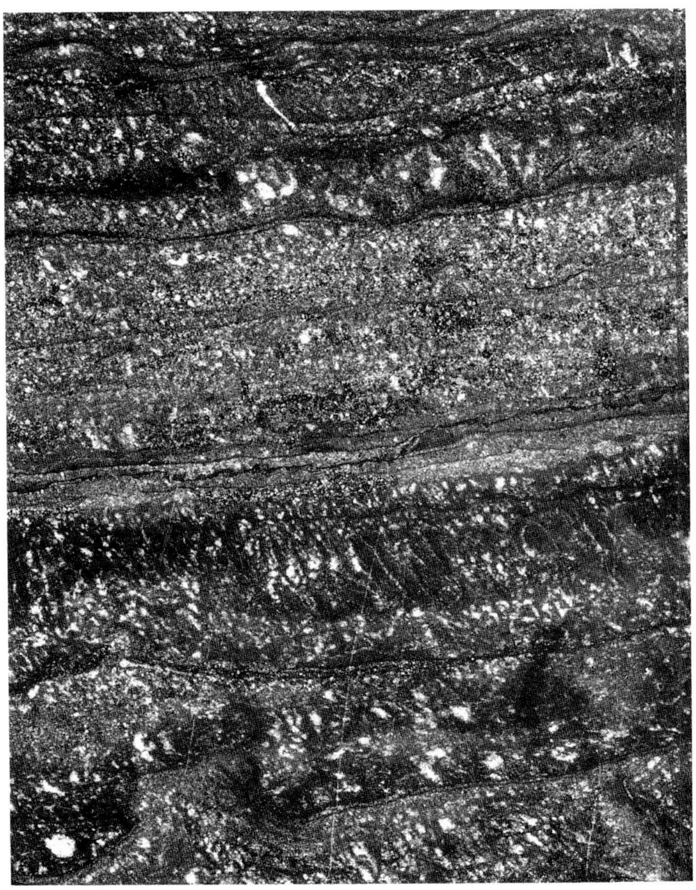

Fig. 93.-Thin-section photomicrograph of stromatolite from the Upper Cambrian Conococheague Limestone, western Maryland. Field of view approximately 6 mm. Note the variety of laminae textures including couplets of fine sand- to coarse silt-sized peloidal grainstones alternating with clotted mudstones (center of field), and vertical network of spar-filled tubes surrounded by dense micrite, interpreted as cyanobacterial filament molds (compare with Figure 134C). This latter fabric may be a chemical one, see section on cemented cyanobacterial crusts in Chemical Features. Finally, note the rare, thin, dense micrite laminae.

TABLE 3: A FLOWSHEET FOR INTERPRETING STROMATOLITES.

STEP 1 Based on internal fabric of the laminae, the stromatolite under consideration belongs in one of the four following categories:

1a. At least some of the laminae are composed of grainstone, the rest is micritic mudstone (**Figs. 71**, **83**, **92**). Grainstone seen as well-sorted, generally silt- to sand-sized peloids with point contacts between grains, may include quartz and skeletal grains; mudstone seen as either structureless micrite or clotted micrite with fused contacts between the clots. **GO TO STEP 2**

1b. Laminae are all mudstone (**Fig. 75**), either structureless micrite (**Fig 122**), or clotted micrite (**Fig 93**), or as a framework of tubules (**Figs. 93**, **134C**), or as tiny arborescent shrubs (**Fig. 123**), or any combination thereof. **GO TO STEP 3**

1c. Laminae are composed of radial crystals (**Figs. 76**, **121**)
THIS IS A **CHEMICAL STROMATOLITE.** **GO TO STEP 5**

1d. Laminae are neospar or crystalline dolomite. **GO TO STEP 4**

STEP 2

2a. Grainstones are in pockets (irregular to concave-up lower boundaries and flat tops: **Fig. 68**) or ripple-form lenses surrounded by mudstone laminae (**Fig. 69**). Nowhere do grainstone particles go over the angle of repose. This indicates *DETRITAL MATERIAL TRAPPED IN SURFACE DEPRESSIONS OR AS RIPPLES.* Microbial mats may or may not have been present. To proceed further with the interpretation it is necessary to resolve the nature of the mudstone laminae. **GO TO STEP 3**

2b. Grainstone in lenses or laminae either alternating with finely laminated mudstones or comprising entire structure (**Fig 71**, **92**). Grainstone laminae taper off down the sides of the stromatolite, grains locally exceed the angle of repose, laminae may be discontinuous over the top of the stromatolite (**Fig 71**, **83**). This indicates *AGGLUTINATION OF DETRITAL SEDIMENT ON TO A MICROBIAL MAT.* Mudstone interlayers are likely to also have been agglutinated by microbial mats as detrital mud (structureless micrite) or soft peloids (clotted micrite).
THIS IS A **MICROBIAL STROMATOLITE.** **GO TO STEP 6**

STEP 3

3a. Laminae are all structureless micritic mudstone (**Fig. 75**). **GO TO STEP 4**

3b. Laminae dominantly have clotted fabrics (**Figs. 93**, **134**, **135A**). This indicates *PRECIPITATION EITHER WITH OR WITHOUT MICROBES (COCCOID FORMS) INFLUENCING PRECIPITATION, OR AGGLUTINATION OF SOFT PELOIDS BY A MICROBIAL MAT.* **GO TO STEP 6**

3c. Laminae composed of a complex framework of carbonate tubules (**Figs. 93**, **134C**). This indicates *PRECIPITATION OF CARBONATE SHEATHS AROUND FILAMENTS OF MICROBES IN A MICROBIAL MAT.* **GO TO STEP 6**

3d. Laminae composed of arborescent shrubs (**Fig. 123**). This indicates *PRECIPITATION OF CARBONATE EITHER BY INORGANIC PROCESSES OR DUE TO MICROBIAL ACTIVITY OR BOTH.* **GO TO STEP 6**

3e. Laminae are layers of structureless micrite or clotted micrite alternating with layers composed of arborescent shrubs or carrying carbonate tubules. This indicates a *COMBINATION OF MECHANISMS WITH AT LEAST SOME PRECIPITATED FABRICS.* **GO TO STEP 6**

STEP 4

4a. Fine laminae tens to hundreds of microns thick, laminae notably discontinuous with internal unconformities, laminae taper out down the sides of the stromatolite, enhance surface irregularities. This indicates *EITHER AGGLUTINATION OF FINE-GRAINED DETRITAL PARTICLES OR PRECIPITATION OF MATERIAL BY A MICROBIAL MAT (PROBABLY PHOTOSYNTHETIC).*
THIS IS A **MICROBIAL STROMATOLITE.** **GO TO STEP 6**

4b. Fine laminae tens to hundreds of microns thick, at least some laminae notably continuous and isopachous, however, theylocally damp out irregularities (**Fig. 75**) and tend to maintain thicknesses where stacks of laminae bend (i.e. kink fold geometry: **Fig. 122**). This indicates *INORGANIC PRECIPITATION WITH LITTLE OR NO MICROBIAL EFFECTS.*
THIS IS A **CHEMICAL STROMATOLITE.** **GO TO STEP 5**

TABLE 3 - CONTINUED

STEP 5

5a. Gravity-influenced fabrics indicate *FORMATION IN A SUBAERIAL (VADOSE) SETTING*, as a caliche soil crust (**Figs. 125, 126, 128**), splash zone crust, speleothem, or spring chemical stromatolite. **GO TO STEP 6**

5b. Abundant isopachous laminae (especially those that envelop associated coated grains: **Fig. 122**) together with a notable absence of gravity-influenced fabrics indicate a *CHEMICAL STROMATOLITE FORMED UNDER SUBAQUEOUS CONDITIONS* in a marine lagoon, lake, spring-fed pool, or cave pool. **GO TO STEP 6**

STEP 6

6a. Stromatolite generally surrounded by sorted and imbricated intraclastic conglomerates or cross-stratified or planar stratified grainstones (especially ooid and skeletal). Eroded clasts of stromatolite in the surrounding grainstones or conglomerates. Evidence of exposure lacking.
IF MARINE, THIS IS A **SUBTIDAL STROMATOLITE**; IF NON-MARINE, THIS IS A **SUBLACUSTRINE STROMATOLITE**.

6b. Stromatolite associated with mud-cracked laminites, or flaser, lenticular or wavy bedded units. IF MARINE, THIS IS A **PERITIDAL STROMATOLITE**, IF NON-MARINE, THIS IS A **SUPRALITTORAL LACUSTRINE STROMATOLITE**.

6c. Stromatolite associated with micrite coated grains, micrite matrix, brecciation of stromatolites and grains with recoating of the fragments by micritic laminae, vadose cements, etc.
THIS IS A **CALICHE** (see elsewhere in this atlas).

6d. Stromatolite dominated by carbonate tubules, may be associated with arborescent crusts and clotted fabrics. THIS IS A **MICROBIAL TUFA** (see elsewhere in this atlas). The sedimentary structures, fossils, etc., in the host rocks will determine the environmental setting of this tufa, whether marine or terrestrial, supralittoral or subaqueous.

6e. Stromatolite with arborescent crusts and clotted fabrics, associated with laminites with radial crystalline fabric, and with coated grains. THIS IS A **SPELEOTHEM** OR A **SPRING TRAVERTINE** OR A **SPLASH ZONE CRUST** (nature of the associated rocks required for the final decision, see elsewhere in this atlas).

Below we have attempted, using the features of both modern and ancient stromatolites discussed above, to put together a means of approaching the problem of diagnosing the origin and significance of ancient stromatolites. Table 3 is an annotated flowsheet that can be used as a step-by-step guide to interpreting the origin and possible environmental setting of an ancient stromatolite using the fabric and geometry of the stacked laminae as the critical elements.

Cryptmicrobial laminites.-

Terminology.-Cryptmicrobial laminite replaces the older terms "algal laminite" or "cryptalgal laminite" (Aitken, 1967) and the terms "stratiform stromatolites," or "planar stromatolites" as explained above. A quick review of Figures 70, 72, and 73 shows the range of geometry of layers from modern deposits where the influence of a surface microbial mat is undisputed. In the descriptive terminology introduced in Table 1, these modern deposits would be sets and, in some cases, composite sets of planar fine laminae, wavy fine laminae, and crinkled fine laminae. The interpretation of sets of laminae with these geometries as "cryptmicrobial" or "microbial" is not straightforward. First, there are a number of different mechanisms not involving cyanobacterial mats which can produce laminated deposits in any carbonate environment, as illustrated in Table 1 and discussed under the section on Strata Without Internal Cross-Stratification. Therefore it is essential to look for independent evidence that a cyanobacterial mat covered the surface. Unfortunately, such evidence (e.g., "jelly roll" structures, cyanobacterial "tufas," filament molds and casts, and so on, see below) is normally not easy to find. The morphology of the laminae is simply not an unequivocal indicator of former cyanobacterial mats as is well exemplified by modern cryptmicrobial deposits which carry laminae that range from planar to highly crinkled. These modern cryptmicrobial laminites also show a considerable range in the texture of the sediments making up the laminae, which may consist of sands, muds, or mixed sediment types. Again, we would recommend use of a descriptive terminology for lamination in carbonate rocks along the lines of Table 1.

Distinguishing characteristics and origin.-

1. MODERN MICROBIAL LAMINITES. Most of the work on layering that involves microbial mats in modern carbonate environments has concentrated on high intertidal to supratidal subenvironments of low-energy tidal flats where there is an organic mat dominantly composed of cyanobacteria ("blue-green algae"). Modern examples of laminated sedimentary deposits that were directly influenced by a surface mat of microbes have been described from: (1) the supratidal islands of Florida Bay (Ginsburg and others, 1954); (2) the intertidal and supratidal mud flats and coastal marshes of Andros Island (Black, 1933; Monty, 1967, 1972, 1976; Monty and Hardie, 1976; Hardie and Ginsburg, 1977); (3) the intertidal and supratidal mud flats of the Trucial Coast of the Persian Gulf (Kendall and Skipwith, 1968; Kinsman and Park, 1976); (4) the intertidal sand and mud flats of Shark Bay in Western Australia (Davies, 1970; Logan and others, 1974a; Hoffman, 1976a); and (5) the siliciclastic mud flats accumulating behind barrier islands of the Delmarva Peninsula, Virginia (Harrison, 1971). Hardie and Ginsburg (1977) have described how complex in detail these types of deposits can be, and their study raises questions of how useful general descriptions of microbial lami-

nated deposits are and what criteria can be reliably used to identify their ancient counterparts (see the discussion of the origin of organo-sedimentary lamination above).

In modern microbially-laminated deposits, the type of mat appears to be directly reflected in the resultant layer structure. The type of mat is itself dependent on a complex set of factors, the most important of which seem to be amount of wetting and drying by seawater and rainwater (related in turn to elevation above sea level on the flats and the regional weather and climate), frequency of sedimentation, amount and type of sediment, and to a lesser extent, salinity. Two general cyanobacterial mat types are found in modern settings: (1) motile filamentous oscillitoracean-dominated mats that grow within and on the sediment surface, and (2) immotile fleshy mats of non-oscillatorean cyanobacteria that grow on the surface. The fleshy mats can be further subdivided into smooth fleshy mats or crinkled fleshy mats composed of various combinations of filamentous and coccoid forms. Sediments found below crinkled fleshy mats of the Persian Gulf and Shark Bay are generally described as only vaguely laminated, clotted, or massive (cinder mats of Kendall and Skipwith, 1968; tufted and convoluted mats of Davies, 1970; and blister, pustular and pinnacle mats of Kinsman and Park, 1976). However, the crinkled fleshy mats of Andros Island leave behind a number of distinctive fenestral laminae, thin beds, and calcified cyanobacterial tufas (Hardie and Ginsburg, 1977). The smooth fleshy mats of the Persian Gulf, Shark Bay and the Delmarva Peninsula (respectively, the "smooth mats" of Kinsman and Park, 1976, and Davies, 1970; the "polygon mat" of Kendall and Skipwith, 1968; and the "algal mat" of Harrison, 1971) appear to be the most widely distributed mat type. The sediments under smooth fleshy mats can be beautifully laminated (see Kinsman and Park, 1976, Fig. 7) and are characterized by laminae of fine sands or silts, varying from a millimeter up to 10 mm thick that alternate with thinner, muddy, organic-rich layers (the buried cyanobacterial mats). The sand to silt laminae pinch and swell and, as seen in the photographs, a few clearly pinch out into isolated lenses whereas the muddy laminae appear to be more continuous (Kinsman and Park, 1976, Fig. 7). Laminae under smooth fleshy mats are flat to wavy (terminology as discussed above). However, it is only local stromatolitic bumps and "jelly roll" structures (see below) that signal a cyanobacterial origin (Davies, 1970; Harrison, 1971; Demicco, 1985). Subaerial desiccation of thick fleshy surface mats results in disruption into polygonal cracks (e.g., Kinsman and Park, 1976); the upturned edges of which become preferential sites for growth of the succeeding generation of cyanobacteria to produce oversteepened "stromatolitic" layering (e.g., the type C algal heads of Black, 1933). Oscillatoracean mats have been described from the supratidal flats of Andros Island (Hardie and Ginsburg, 1977) and these also leave behind beautifully laminated fabrics (see Hardie and Ginsburg, 1977, Fig. 36). This very flat, millimeter-thick lamination is also characterized by alternating lenses of fine peloidal-skeletal sands and more continuous mud laminae (see descriptions in Hardie and Ginsburg, 1977).

2. ANCIENT CRYPTMICROBIAL LAMINITES. It is a common practice to interpret any laminated carbonate as evidence of deposition on cyanobacterial mat-covered tidal flats. Such an approach is clearly fraught with problems as illustrated in Table 1. Where crinkled laminae in part comprise detrital grains that exceed the angle of repose of sediment, microbial influence during sedimentation is a good possibility. In cases of crinkled laminae composed of mud or where laminae have a planar or wavy geometry, it is essential to look for independent evidence that a cyanobacterial mat covered the surface. Indicators of former microbial surface mats include: (1) "jelly roll" structures; (2) cyanobacterial "tufas"; and (3) filament molds and casts (see appropriate chapters elsewhere). Unfortunately, such evidence is normally not easy to find. Modern cryptmicrobial laminites also show a considerable range in the texture of the sediments making up the laminae, which may consist of sands, muds, or precipitated crystalline carbonate (Table 2). Examples of crinkled fine laminae interpreted to be microbial laminites are shown in Figures 21 and 22. Examples of planar to wavy laminites interpreted to be microbial in origin are illustrated in Figures 16, 18, and 19.

Occurrence and preservation.-Laminated carbonate mudstones and grainstones are one of the most common components of shallow-water carbonate platform deposits. Such laminites are widely interpreted as cryptmicrobial, particularly where the laminae are disrupted by mudcracks or fenestrae, but as noted below it is not a straightforward task to differentiate among laminites produced by different mechanisms.

The abundance of laminites in ancient shallow water carbonate deposits testifies to the preservability of the laminae structure. However, it should be noted that post-depositional compaction can have a serious modifying effect on reducing the thickness of the individual laminae, changing their morphology, and obliterating original internal fabrics such as fenestrae, as the insightful compaction experiments of Shinn and others (1977) and Shinn and Robbin (1983) have demonstrated (see section on Mechanical Compaction Structures under Physical Structures).

Significance.-Identifying ancient microbial laminites is difficult, particularly where the laminae are flat and smooth. Even wavy and crinkled laminae cannot easily be ascribed to microbial mats, because there are non-organic processes that produce such laminae. For example, wavy laminae are characteristic of deposition beneath in-phase ripples, and crinkled laminae can result from soft-sediment deformation. Only where there are accompanying stromatolitic features or "jelly-roll" structures (see below) can a reliable identification of ancient microbially-influenced sedimentation be made. Chemical precipitation fabrics (see above) are good evidence against sediment trapping, and where crystal size is many times microbe diameter or length, almost certainly excludes a microbially-induced precipitation origin. The reader should refer to Table 1 and the section on Strata Without Internal Cross-Stratification above where these problems are discussed.

Cryptmicrobial "jelly rolls."-

Terminology.-"Jelly roll" structure ("Swiss-roll structure" in Europe, M. Tucker, 1984, pers. commun.) is a term proposed here for structures otherwise known as "enrolling of mats" (Davies, 1970), "encapsulated" and "enrolled" structures (Eriksson and Truswell, 1974), "encapsulated roll-up structures" (Demicco, 1985), and "roll-up structures" (Simonson, 1988).

Distinguishing characteristics.-"Jelly roll" structures resemble rolled up carpets (Fig. 94) and appear in cross-section as spiral folds up to a few hundreds of millimeters long and a few tens of millimeters thick (Fig. 94C). They have been found on the modern tidal flats of Shark Bay, Western Australia (Davies, 1970), Virginia (Harrison, 1971), and Baja California. The modern examples shown here are from Baja California and are composed of laminated muds or interlaminated sands and muds. The unattached end of the spiral of the overfolded laminae is located in the fold core. An ancient example from Cambrian rocks of the central Appalachians is compared with a modern specimen from Baja California in Figure 95. In the ancient example, a thin sheet of alternating millimeter-thick peloidal sand laminae and several adjacent dolomite laminae

Fig. 94.-Cyanobacterial "jelly roll" and associated structures from the modern siliciclastic tidal flats of the Colorado River Delta, Baja California. (A) General view of cyanobacterial mat-covered supratidal storm flat. Note the incomplete mudcrack system which serves as a starting point for overturned sets of laminated mud and decollemont of centimeter-thick sets of laminated muds. The latter commonly end in anticlines as in the foreground. Individual "jelly rolls" approximately 10 cm wide. (B) and (C) show "jelly rolls" in more detail.

Fig. 95.-Modern and ancient "jelly roll" structures. (A) Epoxy impregnated "jelly roll" from modern Baja California tidal flat (approximately 80 mm in diameter). (B) Set of wavy laminae in the Upper Cambrian Conococheague Limestone of the central Appalachians. Note especially how the dark peloidal silt lamina in the core of the "jelly roll" is folded back on itself. Also note the associated intraclasts. Scale in mm.

each less than a millimeter thick are doubled over into a "jelly roll" about 12 cm long and 2 cm high in cross-section. Overlying laminae drape and bury the "jelly roll." "Jelly rolls" are distinguished from oncoids by their spiral geometry and three-dimensional, tube-like geometry.

Origin.-"Jelly roll" structures on modern tidal flats occur where smooth fleshy cyanobacterial mats (typically non-oscillatoriacean) cover the sediment surface like a carpet. Storm flood-currents tear the cyanobacterial mat and the underlying sediment organically bound to the mat, undermine it, and roll it down-current. In addition to making "jelly rolls," bed shear stress from storm currents can slide sheets of cyanobacterially-bound sediment up to a meter in diameter along thrust faults. Displacement of the sheets is taken up by anticlines. Care must be exercised in the diagnosis of "jelly rolls" because spiral folding is also reported from slump fold sheets (Allen, 1984, p. 383). In addition, Paterson (1986, cover photograph on Geology) figures a spiral fold about 2 m long of unknown origin developed in unconsolidated outwash sands and gravels. Slump-generated structures have thicknesses on the order of 1 m (Allen, 1984, p. 380), and are found in sheet-like disturbed zones of at least outcrop scale whereas "jelly rolls" tend to be much smaller, isolated structures that give way laterally to undeformed sediment. Breccias associated with gravity-induced deformation have interstitial sediment, generally fine-grained, that surrounded the fragments during transport. In contrast, intraclastic layers associated with "jelly-rolls" formed by current-generated shear stresses are sorted and show primary mechanical structures.

Occurrence and preservation.-These are fairly rare structures in shallow-water platform deposits.

Significance.-"Jelly-roll" structures that have sandy laminae within them indicate that the sediment surface was cohesive. In modern examples, this cohesion is affected by cyanobacterial mats. "Jelly-rolls" are one of the most reliable criteria for inferring the influence of ancient microbial mats on sedimentation. They are probably not a reliable depth indicator however, as cyanobacterial mats and other microbial mats (e.g., bacterial mats) are not strictly depth controlled. Simonson (1988) reports "jelly-roll" structures in early Proterozoic basinal deposits. Davies (1970) reports overfolding directed landward in the modern Shark Bay examples. However, we know of no reports of the use of these structures as a paleocurrent indicator.

Thrombolites

Terminology.-

Aitken (1967, p. 1164) proposed the term thrombolite: "(from the Greek *thrombos*, blood clot) ... for cryptalgal structures related to stromatolites, but lacking lamination and characterized by a macroscopic clotted fabric." Pratt and James (1982) suggested that the term thrombolite be used to refer only to the individual millimeter- to centimeter-diameter "clots" that characterize thrombolites. They interpret the individual clots as microbial in origin. However, Kennard and James (1986) abandoned this usage. Instead, they would restrict the term thrombolite to discrete structures that have an internal clotted fabric (see below). Other terms that are commonly used to refer to these and similar structures are "bioherm" and "biostrome" (e.g., thrombolitic bioherm or biostrome). The term bioherm was introduced by Cummings and Schrock (1928, p. 599) from Greek roots meaning organic reef. Cummings' (1932) definitions of these two terms are: (1) Bioherm: "... meaning a reef, bank or mound; for reef-like, mound-like, lens-like or otherwise circumscribed structure of strictly organic origin, embedded in rocks of a different lithology" (p. 333); (2) Biostrome: "... purely bedded structures, such as shell beds, crinoid beds, coral beds, etc., consisting of and built mainly by sedentary organisms, and not swelling into mound-like or lens-like form, ... a layer or bed" (p. 334). Wilson (1975, p. 23) suggested that the term bioherm should refer to a "build-up whose internal composition shows it to be largely derived from in situ production of organisms or as framework or encrusting growth as opposed to mainly mechanical (hydrodynamic) piling." One of the main nomenclatural problems of thrombolites is the long recognized fact that any given thrombolitic bioherm may contain: (1) laminated stromatolitic patches or discrete stromatolites (stromatoids in the terminology of Kennard and James, 1986); (2) millimeter- to centimeter-diameter "clots" (mesoclots in the terminology of Kennard and James, 1986); (3) variable amounts of the problematic fossils, *Renalcis*, *Epiphyton*, *Girvanella*, and *Nuia*, as well as masses of micrite that superficially resemble these forms; (4) irregular, patchy, and mottled micrite fabrics that lack any of the previous features ("cryptmicrobial" fabrics of Kennard and James, 1986, who suggest that these fabrics be clearly separated from the mesoclotted fabrics of thrombolites); and (5) framework-building skeletal metazoans (most notably archeocyathids, stromatoporoids, and bryozoans). Furthermore, these different fabrics may be randomly distributed within any given bioherm, or may show a distinct zonation within bioherms recognizable on the formation level (see below).

Distinguishing characteristics.-

Features characteristic of thrombolites occur at three different scales: outcrop-scale size and geometry; hand-sample fabrics; and thin section fabrics. Thrombolites range in size from centimeter-diameter mounds (Fig. 96) up to masses ten meters or so thick (Fig. 97). The three dimensional geometry cannot be resolved without adequate exposure. Figures 98 and 99 show some typical outcrop sizes and geometries in bed-perpendicular views for Upper Cambrian and Lower Ordovician thrombolites of the central Appalachians. Commonly, meter-scale individual thrombolites are coalesced into a biohermal complex meters to a few tens of meters thick and tens to hundreds of meters long (see also James and Kobluk, 1978). Coalesced bioherms typically enclose irregularly-shaped pockets of grainstone or flat pebble conglomerate. The term Gymnosolen was applied to bioherms where these granular pockets, as seen in two dimensional cross-sections, appear as vertically oriented pipes a few decimeters across and up to a few meters high. Commonly, individual thrombolites, clusters of thrombolites, and mounds of coalesced thrombolites are imbedded in, and interbedded with, cross-stratified skeletal, peloidal, or ooid grainstones. Thrombolites typically have a laminated outer micritic rind that may be up to a decimeter thick.

The internal structures of thrombolites at the hand-sample scale (mesoscale in the terminology of Kennard and James, 1986) are complicated. The following structures visible in sawn slabs and in those outcrops delicately etched by selective weathering are widely reported: (1) millimeter- to centimeter-scale micrite masses surrounded by skeletal wackestones (Fig. 100); (2) digitate stromatolitic columns ("fingers") surrounded by skeletal wackestones (Fig. 101); (3) in-place metazoan skeletons; (4) geopetal cavities filled with laminated internal sediment and sparry-cements (Fig. 102); and (5) massive fabrics (Fig. 103). Most commonly, the internal structure comprises millimeter- to centimeter-scale micrite masses ("clots" or

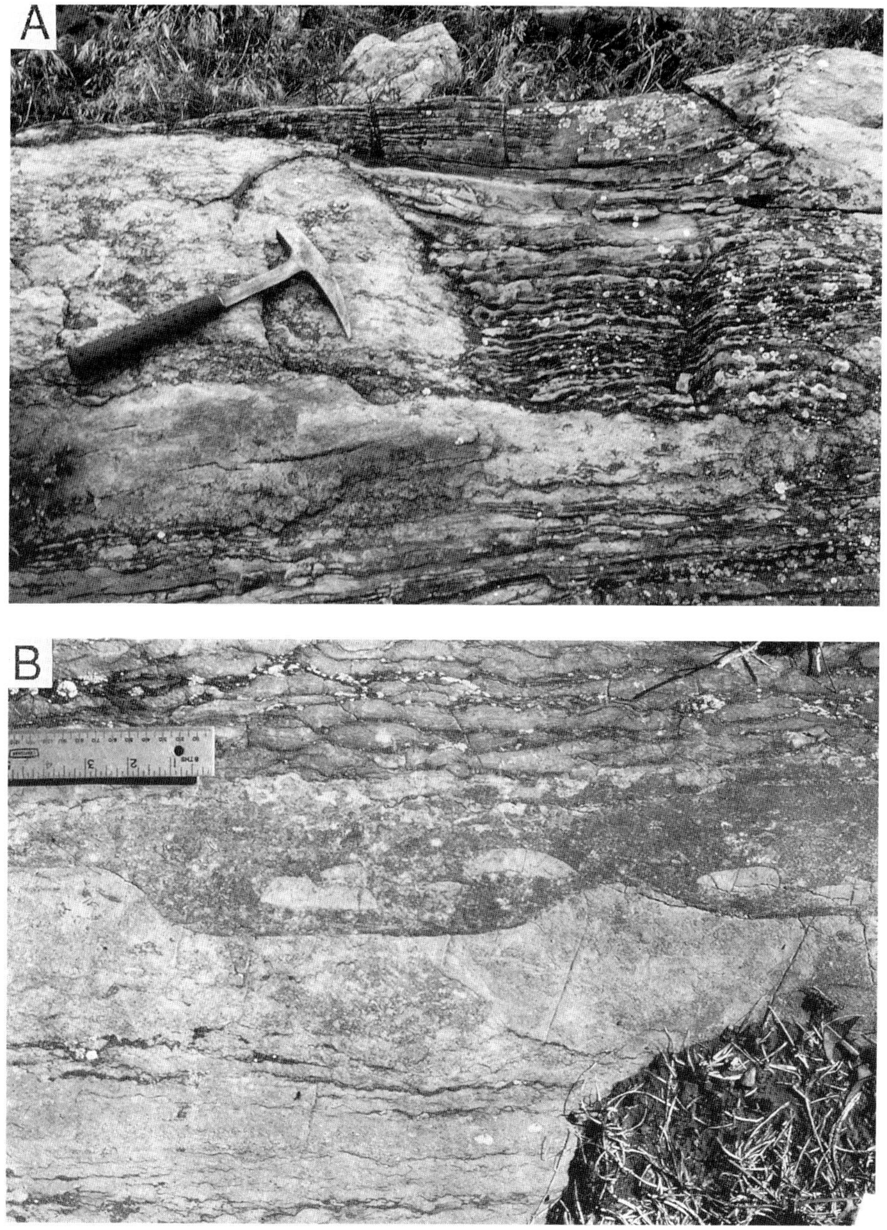

Fig. 96.-Field photographs of meter-scale thrombolitic bioherms from Upper Cambrian and Lower Ordovician carbonates of western Maryland. (A) Bioherms (under hammer and at upper right) separated by thin beds of graded dolomite. (B) Thrombolites and interthrombolite intraclast conglomerate truncated by a scalloped erosional surface. Note the clasts of thrombolitic material overlying the erosion surface. These relationships imply early cementation of the thrombolites (see text). (C) Irregularly-shaped thrombolites surrounded by coarse sand-sized, skeletal-peloidal grainstone. Penny for scale. (D) Columnar thrombolite with scalloped erosional upper surface. Thrombolite surrounded by coarse sand-sized, skeletal-peloidal grainstone. Scale in inches.

C

D

Fig. 97.-Large thrombolitic bioherms from the Upper Cambrian Conococheague Formation, western Maryland. In (A) four separate, nearly spherical thrombolites (arrows), each a few meters thick, are amalgamated into a large biohermal complex. Bioherms in (A) are separated by thin beds of graded dolomite. Solitary bioherm is shown in (B) (Following page) (also see Pettijohn and Potter, 1964, Plate 50B and 49B). Photograph courtesy of Francis J. Pettijohn. In both (A) and (B), the upper portions of the individual meter-scale thrombolites are composed of laterally-linked, decimeter-scale stromatolites in layers conformable to the upper surface of the mounds (see Plate 49A in Pettijohn and Potter, 1964).

B

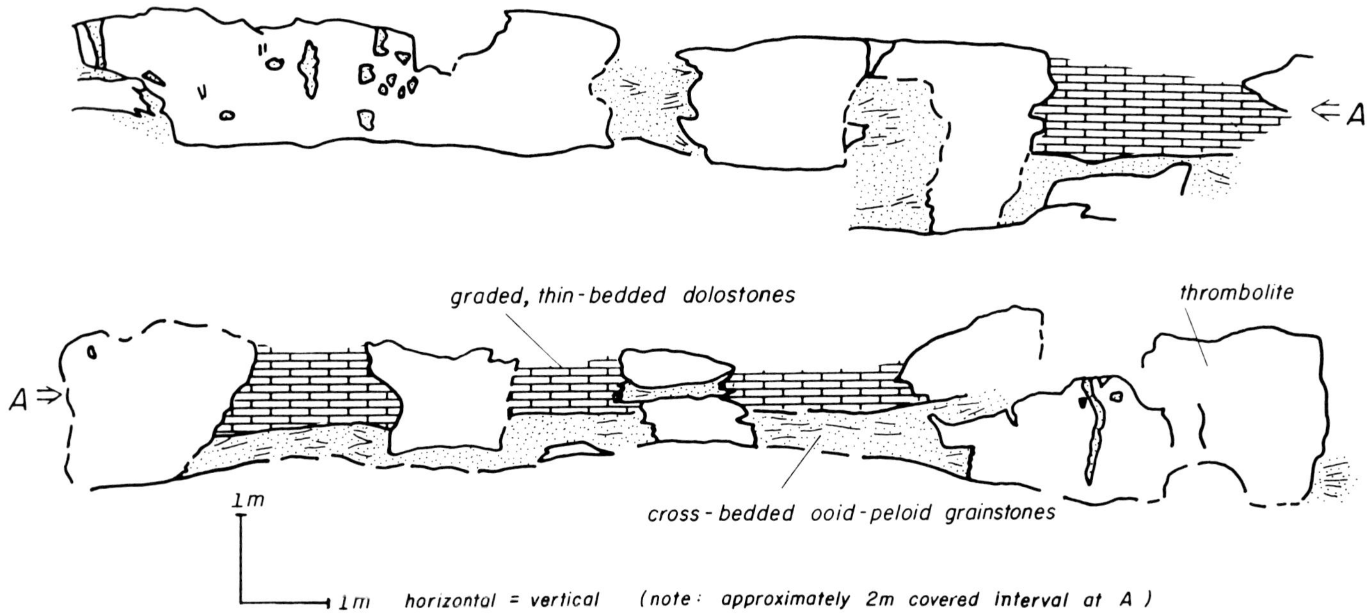

Fig. 98.-Two-dimensional cross-section geometry of a typical thrombolite bioherm layer from the Upper Cambrian Conococheague Limestone of western Maryland, prepared from a photomosaic. The irregularly-shaped mounds are separated by meter-scale channels filled with cross-stratified, medium to coarse sand-sized, ooid-peloidal grainstones overlain by graded, thin beds of dolostone. Note the smaller, irregularly-shaped patches of grainstone within the thrombolite mounds.

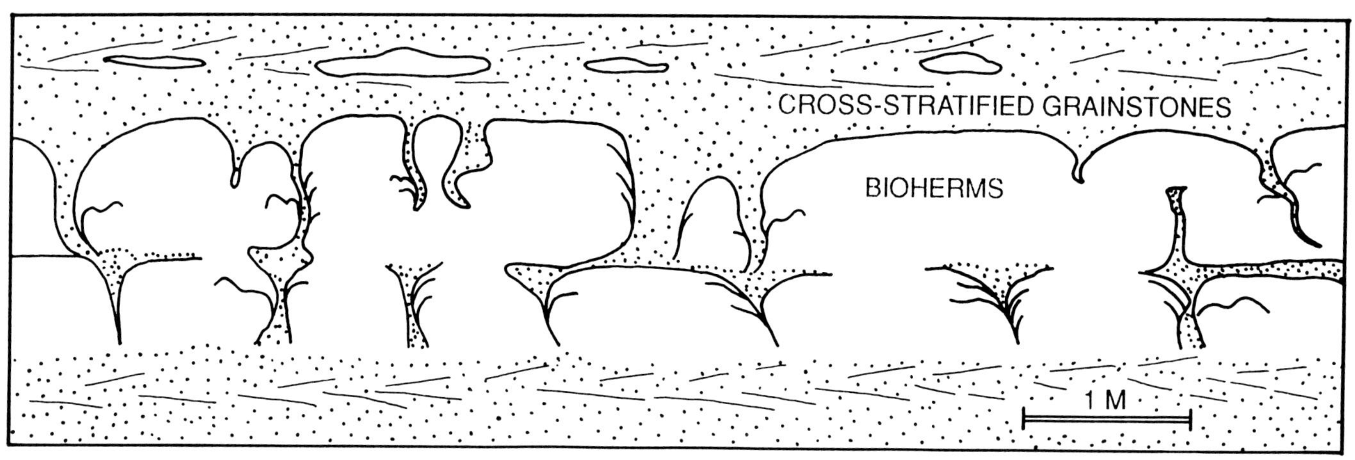

Fig. 99.-Schematic cross-sectional diagram of *Epiphyton-Girvanella* thrombolitic bioherm based on outcrop exposures of the Lower Ordovician Grove Limestone of the Frederick Valley, Maryland. These thrombolites are interpreted as shelf-edge bioherms. Barrel-like, coalesced mounds are surrounded by cross-stratified grainstones that also occur in intermound areas. Low profile thrombolites also occur above the coalesced bioherms. Courtesy of Chau Nguyen (see Nguyen, 1986).

Fig. 100.-Cross-sectional slabs of thrombolitic mounds. (A) Complicated structure of a thrombolitic layer (center of photograph) overlying disrupted thin beds of mudstone, and, in turn overlain by intraclastic conglomerate. Thrombolite layer comprises a number of irregularly-shaped "clots" of mudstone and an upper layer of stromatolites composed of grainstones that do not have well-developed internal laminae. Sample from the Conococheague Limestone of western Maryland. Scale bar along right side of specimen is 100 mm long. (B) Upward-directed irregularly-laminated micrite "fingers" (digitate stromatolites) surrounded by skeletal packstone and wackestones from a thrombolite in the Lower Devonian Manlius Formation of central New York. Specimen is approximately 20 cm high. Sample collected by Kathy Browne.

Fig. 101.-Examples of upward-directed, centimeter-scale micrite "fingers" in thrombolites from the Upper Cambrian Conococheague Limestone of western Maryland. Slabs taken from bioherm mapped in Figure 98, scale bar in each is 50 mm long. "Fingers" are surrounded by skeletal packstone to wackestone and are encrusted with *Renalcis* (small black clusters).

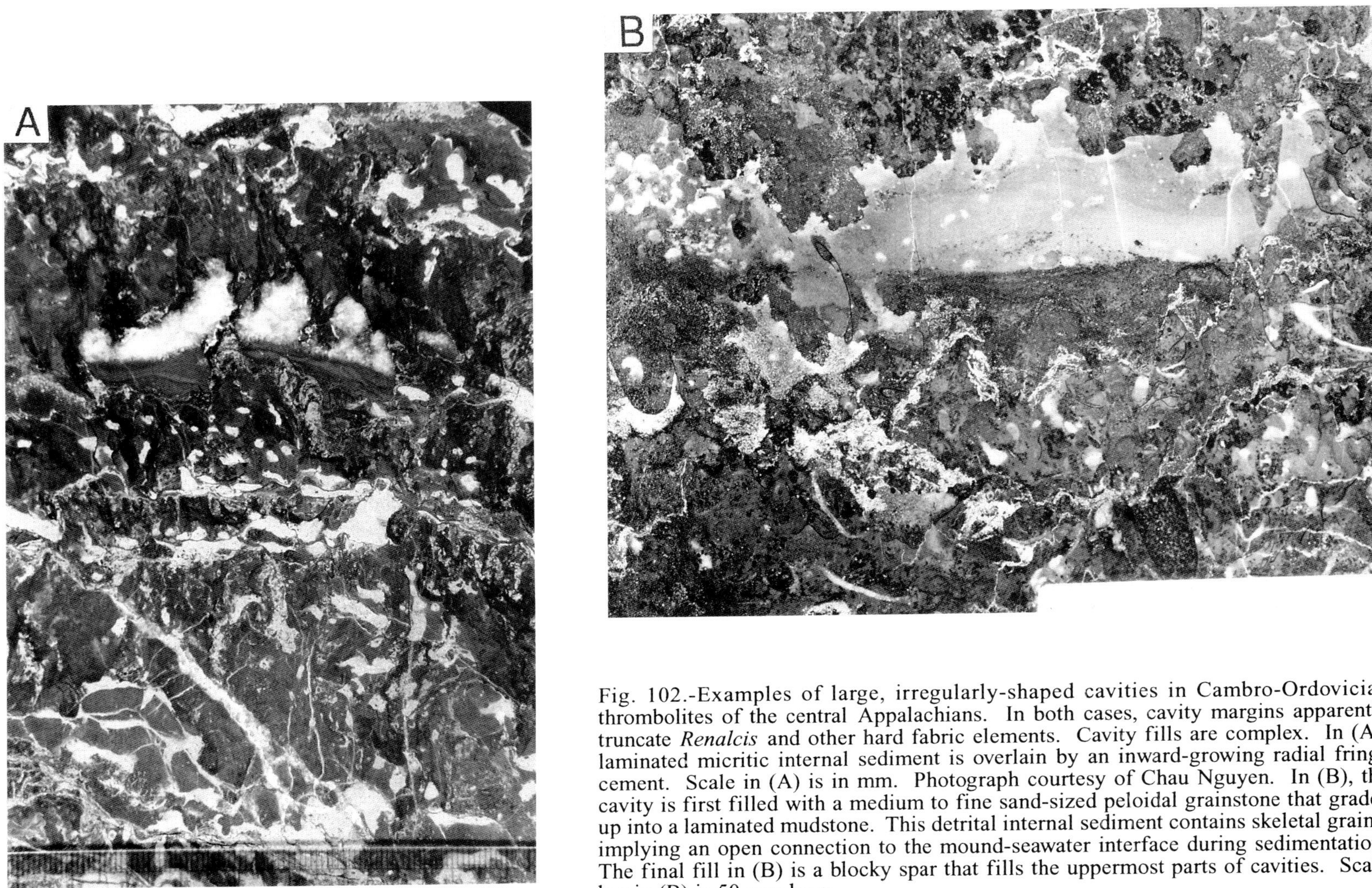

Fig. 102.-Examples of large, irregularly-shaped cavities in Cambro-Ordovician thrombolites of the central Appalachians. In both cases, cavity margins apparently truncate *Renalcis* and other hard fabric elements. Cavity fills are complex. In (A), laminated micritic internal sediment is overlain by an inward-growing radial fringe cement. Scale in (A) is in mm. Photograph courtesy of Chau Nguyen. In (B), the cavity is first filled with a medium to fine sand-sized peloidal grainstone that grades up into a laminated mudstone. This detrital internal sediment contains skeletal grains, implying an open connection to the mound-seawater interface during sedimentation. The final fill in (B) is a blocky spar that fills the uppermost parts of cavities. Scale bar in (B) is 50 mm long.

Fig. 103.-Indistinct, clotted fabric typical of most thrombolites. Scale bar is 50 mm long. Slab of a thrombolite from the Upper Cambrian Conococheague Limestone of western Maryland.

"mesoclots") surrounded by skeletal mudstones, skeletal wackestones, or skeletal packstones (Figs. 100, 101). Indeed, Kennard and James (1986) would restrict the use of the term thrombolite to bioherms in which this fabric predominates. The micrite masses typically make up 30 to 40% of the rock. The geometry of the micrite masses varies from highly irregular to vertically-oriented columns with circular cross-sections in bedding plane views. In hand-sample these micrite masses may contain variable amounts of fine, regular to irregular, stromatolitic laminae. A complete gradation apparently exists from micrite masses and columns with fine, regular stromatolitic laminae through to unlaminated massive micrite masses. Although both unlaminated and laminated micrite fabrics may occur side-by-side in a thrombolite, they are more commonly arranged with the unlaminated micrite clots in the lower portions of bioherms and the stromatolitic masses in the upper portions. Skeletal mudstones, wackestones and packstones fill the spaces between the mudstone masses. Skeletal remains include crinoid, brachiopod and trilobite fragments. Spar-filled and micrite-filled tubes up to a few millimeters in diameter are common in the fills between clots. These tubes apparently represent both burrows and borings because some tubes avoid skeletal fragments (burrows) and others cross-cut skeletal fragments and early cements (borings) (Nguyen, 1986). It is significant that some of these tubes also avoid the micrite masses whereas others cross-cut them, indicating that at least some of the clots were as hard as the skeletal fragments. The in-place framework-building skeletal metazoans include archeocyathids in the Lower Cambrian (James and Kobluk, 1978). In the Ordovician, corals, stromatoporoids, and bryozoans were important framework components of the bioherms. The spar-filled patches are commonly found beneath overhangs of the micrite-masses, and beneath metazoan skeletons, indicating shelter voids within the thrombolites (Fig. 102). Larger cavities are geopetally filled and contain laminated internal sediments beneath the sparry cements. In some cases, *Renalcis* hangs down from the roofs of the cavities. Some thrombolites appear massive in hand sample, with the distinctive clotted texture replaced by a color-mottled mudstone to wackestone that commonly contains a large number of stylolites (Fig. 103).

In thin section, the textures of thrombolites are also complicated. Individual micrite masses may be: (1) dense micrite, (2) clotted micrite with indistinct, micron-diameter tubes up to a few 100 microns long, or (3) *Renalcis* (and unchambered *Renalcis*-like forms; Fig. 104A, B), *Girvanella* (Fig. 104C), *Nuia*, and other problematic microfossils. *Renalcis* (Fig. 104A, B) may encrust the clotted and dense micrite masses. Certain thrombolites are composed of a framework of small bushes of *Epiphyton* that encrust plates of *Girvanella* (Fig. 105) (Nguyen, 1986; Read, 1980). Kennard and James (1986) suggest that thrombolites dominated by these types of problematic microfossils be distinguished by specifying the species involved. By far the most common thin-section fabrics are those where mottled micrite makes up the fingers or clots. The mottled character of these micrites is due to irregular patches of more or less dense micrite a few tens of microns in diameter surrounded by a mosaic of microspar. In addition, mottled micrites contain micron-diameter tubes and micron-diameter *Renalcis*-

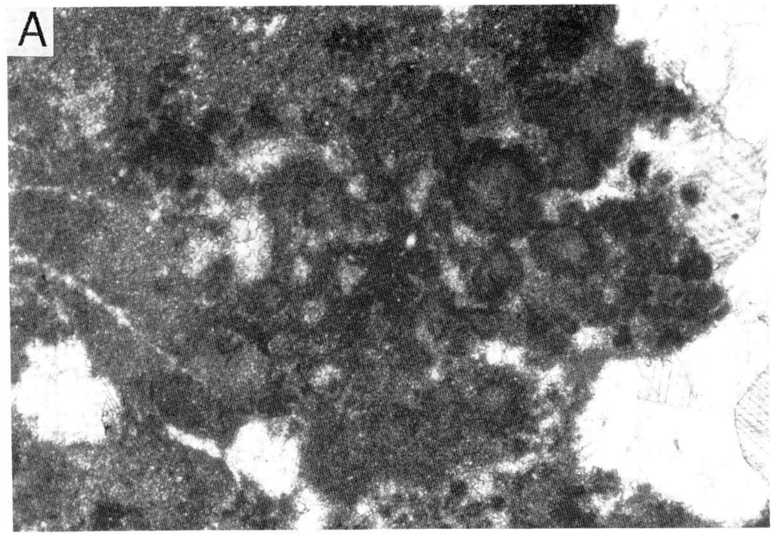

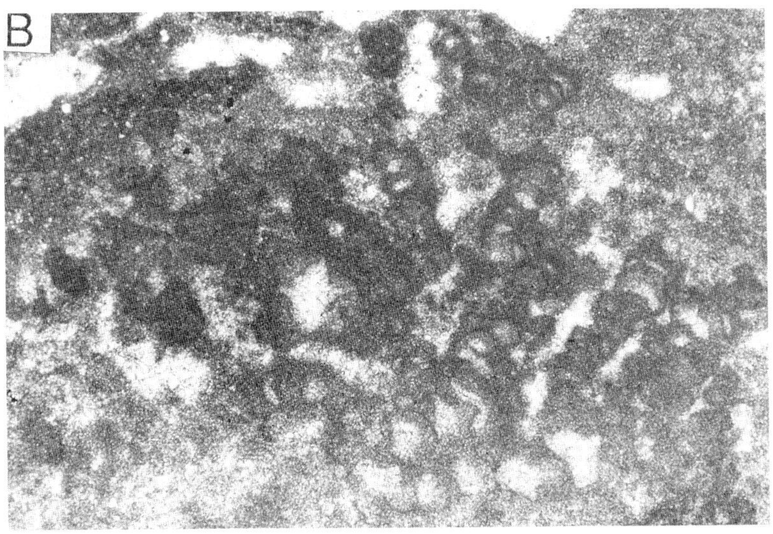

Fig. 104.-Problematic microfossils in thrombolites of the Lower Ordovician carbonates of western Maryland. (A) and (B) illustrate dark micritic clots with microspar cores typical of *Renalcis* clusters in Cambro-Ordovician thrombolites. (C) Parallel bundled filament molds of *Girvanella*. Width of view in all three illustrations is 2 mm. Photomicrographs courtesy of Chau Nguyen (see Ngyuen, 1986).

C

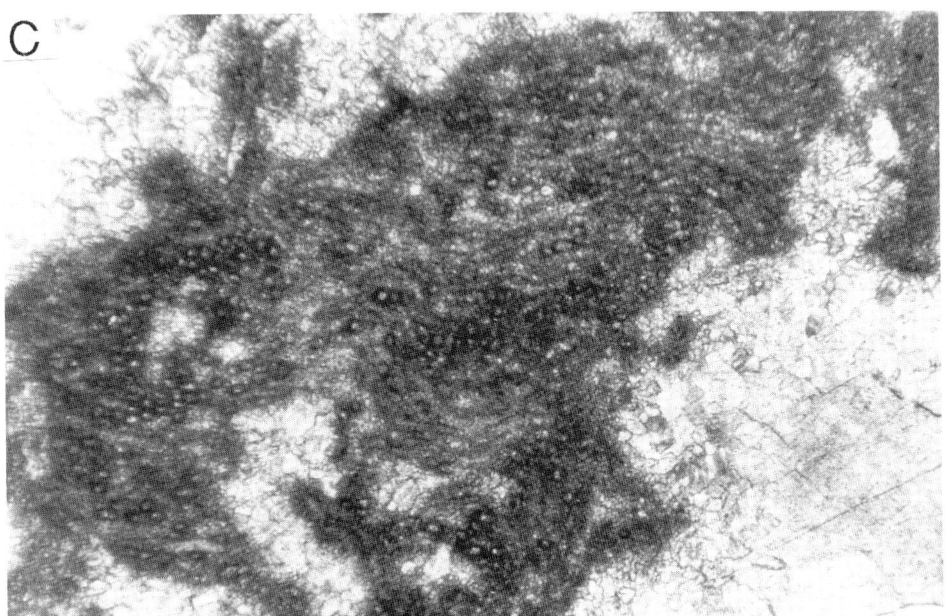

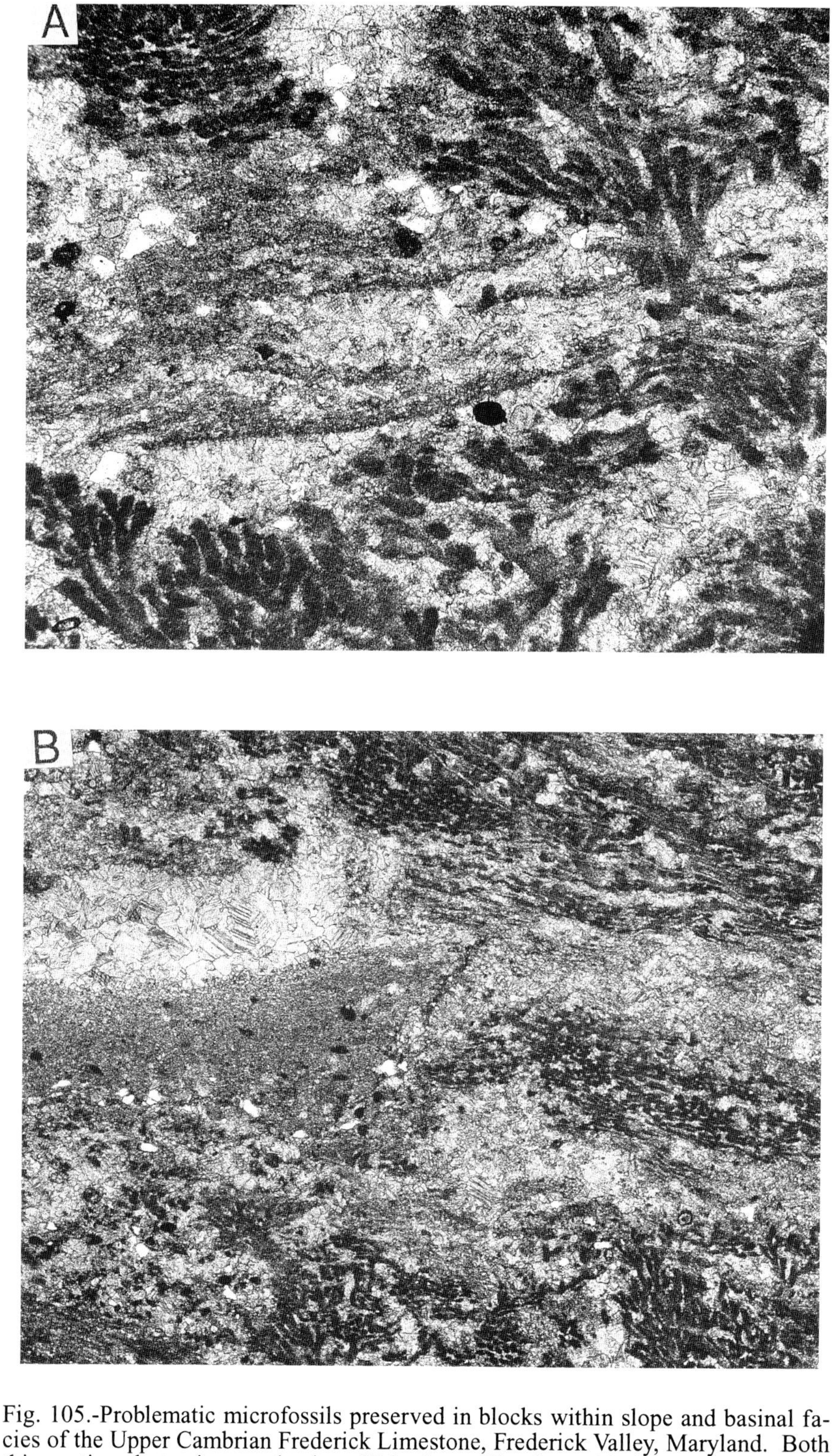

Fig. 105.-Problematic microfossils preserved in blocks within slope and basinal facies of the Upper Cambrian Frederick Limestone, Frederick Valley, Maryland. Both thin-section photomicrographs show (1) bushes of *Epiphyton* (dark branching micritic rods), (2) thin subparallel micritic plates (e.g., thin dark streaks roughly perpendicular to *Epiphyton* bushes in A) that show traces of poorly preserved filament bundles interpreted as molds of *Girvanella* , and (3) scattered quartz grains in a microspar mosaic (light-colored patches). Width of field in (A) is 2.5 mm. In (B), note the geopetal vug (upper left corner) in this *Epiphyton*-boundstone. Width of view in (B) is approximately 5 mm.

like clots. The packstones and wackestones surrounding the micrite clots and "fingers" contain skeletal fragments, burrows, and borings. As noted above, the burrows avoid the skeletal fragments, and significantly, the edges of micrite fingers, whereas the borings cross-cut skeletal fragments, micrite clots, and, in rare cases, cements.

In thin section, the margins of the geopetally-infilled vugs are outlined by a micritic rind a few microns thick that commonly truncates skeletal grains. The internal sediment beneath the final void-filling cements can contain skeletal fragments, suggesting that the cavities were open to the surrounding sea floor. The cements commonly comprise a radially-oriented, inclusion-rich fringe cement surrounding an internal blocky mosaic cement.

Occurrence and preservation.-
Thrombolites are most common in Cambrian and Lower Ordovician rocks and appear to be rare in later rocks. However, in a number of cases, much of what is referred to as "reef" between large, in-place, skeletal metazoans resembles thrombolites. This is true, for example, of the "Chazy" reefs of New York and the Devonian "reefs" of western Canada and the Canning Basin of northwestern Australia. Although there are scattered reports of Proterozoic thrombolites (Grotzinger, 1985), Kennard and James (1986) dismiss these accounts, and claim that thrombolites are restricted to the Lower Paleozoic. Shelf-edge bioherms preserved in place are rare around the North American platform in the Lower Paleozoic, however, an important example composed of *Epiphyton-Girvanella* bioherms is found in the Lower Ordovician Grove Limestone exposed in the Frederick Valley of western Maryland (Nguyen, 1986). Such "algal" reefs must have been quite widespread in Cambro-Ordovician times because blocks of *Epiphyton-Girvanella* boundstones are commonly found in breccias in off-platform facies (e.g., Ordovician of Virginia, Pfeil and Read, 1980; Cowhead Breccias of Newfoundland, James, 1981, and the Kicking Horse Rim of the Canadian Rockies, McIlreath, 1977). In western Maryland, thrombolites dominated by *Renalcis* and *Renalcis*-like forms are restricted to the interior of the wide Cambro-Ordovician carbonate platform and their morphology changes systematically across the platform (Nguyen, 1986). Thrombolites found in sections well away from the platform edge typically do not contain *Renalcis* or *Renalcis*-like forms.

Origin and significance.-
Thrombolites appear to represent in-place mounds that were hard, rigid bioherms with a biogenic framework when deposited. In other words, they were biological reefs (see Nguyen, 1986, for a detailed discussion of the petrographic evidence). The problematic microfossils making up the rigid biogenic framework are generally interpreted as calcified cyanobacterial filaments (*Girvanella* and perhaps also *Epiphyton*, Pratt, 1983), calcified coccoid cyanobacteria (*Renalcis*), green or red algae (*Epiphyton*), or foraminiferae (*Renalcis*). The enigmatic micrite clots that give thrombolites their name have also been interpreted to be the result of calcification of cyanobacterial mats and colonies. As with modern coral reefs, boring and perhaps dissolution of hard portions of the thrombolites apparently alternated with periods of internal sedimentation and cement growth, as judged by the vugs and their fillings.

Thrombolites are the dominant reef structures of the carbonate platforms and banks of the Lower Paleozoic and in the overall evolutionary trend of reefs they represent a transition from the cryptmicrobial stromatolite reefs of the Precambrian to the metazoan reefs of the Phanerozoic. In this context, the problematic framework builders of the Cambro-Ordovician thrombolite reefs, such as *Epiphyton* and *Renalcis*, can perhaps be regarded as colonial calcareous organisms transitional in development between Precambrian non-skeletal microbes and Phanerozoic skeletal algae.

Bioturbation and Bioerosion Structures

Burrows and other tubular structures.-
Terminology.-It is not possible here to adequately review the vast literature on burrows, which are the most common type of trace fossil and thus central to the subdiscipline of *ichnology*. The reader is referred to Bromley (1975), Basan (1978), Collinson and Thompson (1989), Ekdale and others (1984), Miller and others (1984), Curran (1985), Ekdale and Pollard (1991), and other SEPM atlases in this series. In addition, Shinn (1968b) and Garrett (1977) provide good studies of modern burrow systems in carbonate sediments.

Most burrows are tubular in shape with diameters in the millimeter-size range and thus they may be easily confused with other millimeter-sized tubular sedimentary structures, such as root casts and molds, migrating gas bubble trails, and the molds or casts of cyanobacteria or higher plant stems. Because existing criteria to distinguish among these various types of tubes are not without uncertainty, it would seem useful to include all such structures in a general category called *tubular fenestrae*. Although the term tubular fenestrae is commonly found in the literature as a descriptive expression, Shinn (1983b) would restrict the term "fenestrae" to those features of desiccation origin found in intertidal and supratidal settings. However, the difficulty with such a restricted usage is that desiccation is not the only mechanism by which fenestrae in peritidal sediments are formed (see under Mudcracks and Other Disruption Features in this atlas). We see the need to keep the term "fenestrae" as a nongenetic term and thus we recommend its continued usage to describe tubular structures where some uncertainty exists as to their origin.

Distinguishing characteristics and origin.-In general, *burrow* diameters range from less than a millimeter up to a few centimeters, but most have millimeter-scale diameters and are a few centimeters to several meters long (Figs. 106, 107). Burrows are commonly backfilled with pellets (Fig. 108), have internal spriete (or chevron structure), are lined, and where branched, maintain nearly constant diameter. "U-shaped" burrows are common whereas this geometry is rare in roots or bubble traces. Burrows also commonly carry distinctive ornamentation of their walls and many have a consistent and distinctive geometric pattern. These features are used to identify "ichno-genera" (see references cited above).
Root molds and casts described from subrecent caliche profiles (see below) and the roots and rhizomes of certain modern plants from tidal flat and shallow subtidal settings (Hardie, 1977a) could be mistaken for burrows, and vice versa. Most distinctive of roots is their downward branching pattern where there is marked and systematic decrease in tube diameter at junctions. In addition, roots may have a carbonaceous lining and may show branching out into clusters of root hairs. However, both Shinn (1968b) and Garrett (1977) made resin molds of large crustacean burrows that are cross-cut by smaller burrows of different organisms giving an apparent branching pattern. Also, the branches of some large crustacean burrows do vary in diameter, especially off of larger chambers and gallery systems. Burrows may take on any orientation, horizontal, vertical and random, the latter reflecting the meandering of the organism through the substrate. Roots are mainly vertically

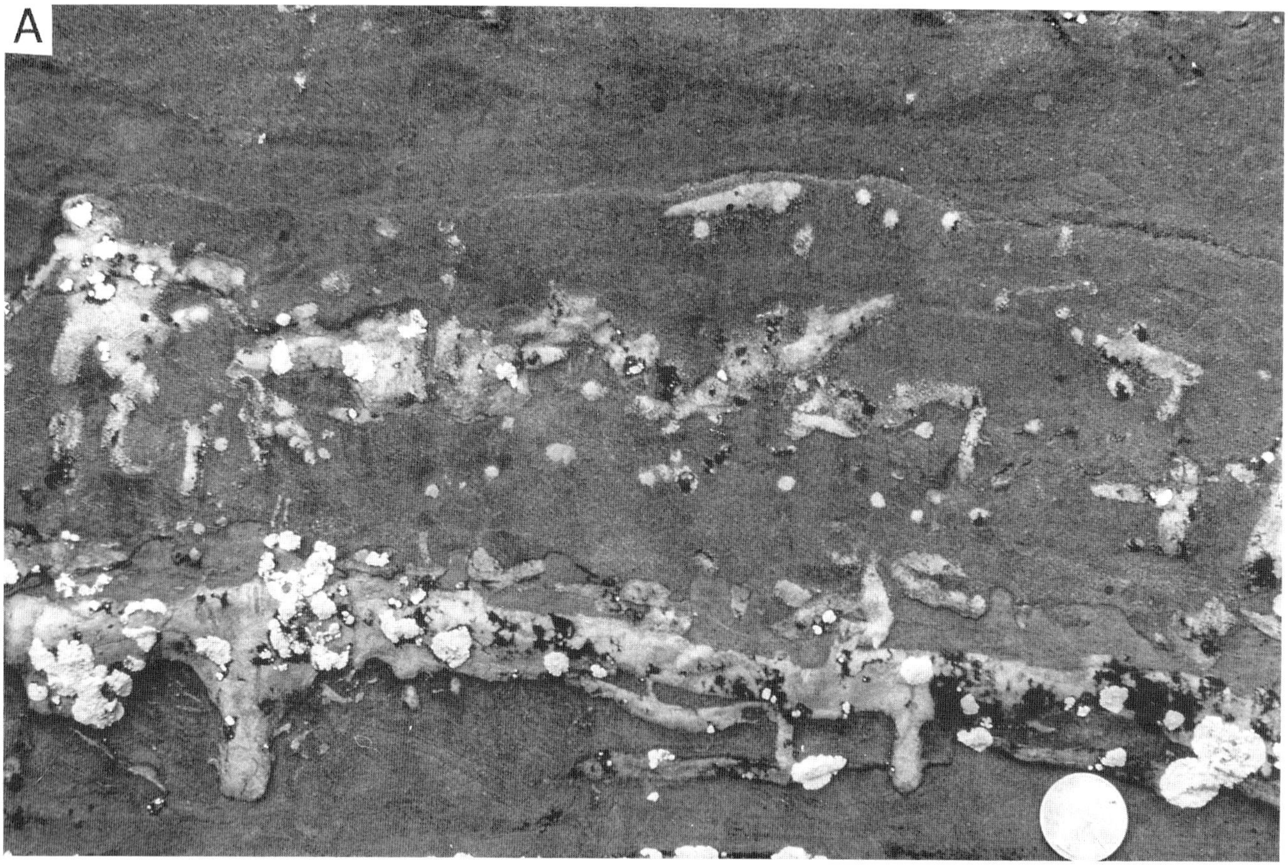

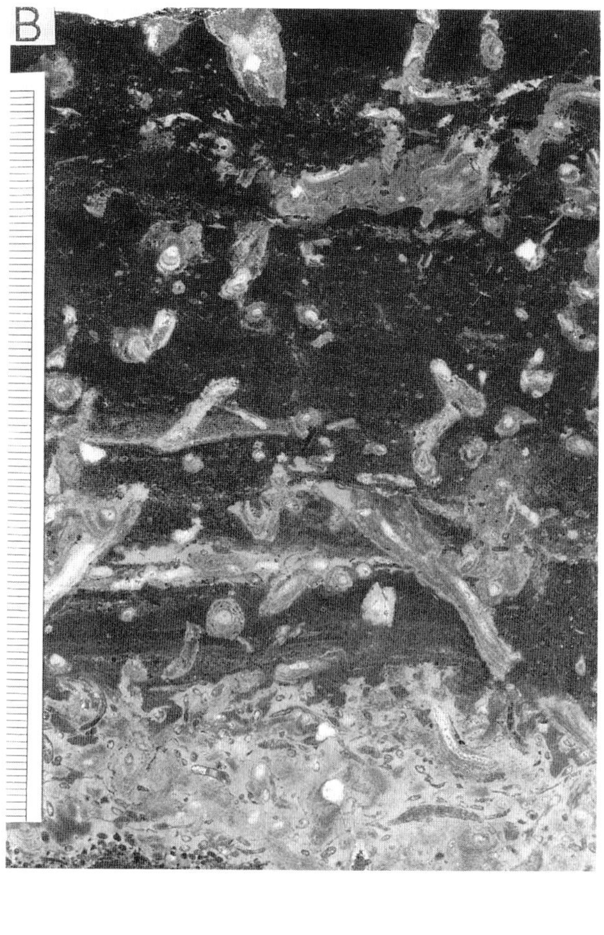

Fig. 106.-Cross-sectional views of burrow structures in shallow marine carbonates. (A) and (B) show dolomite-filled tubular burrows from Cambro-Ordovician carbonates of the central Appalachians. In (B), some of the burrows are only partly filled with internal sediment (now dolomite), the remainder of the burrow filled with a later calcite cement. Note the various circular to ovoid to cylindrical shapes where the tubular burrows have been randomly intersected by slab saw cut. Coin in lower right of (A) is approximately 25 mm in diameter; whereas scale bar in (B) is in mm.

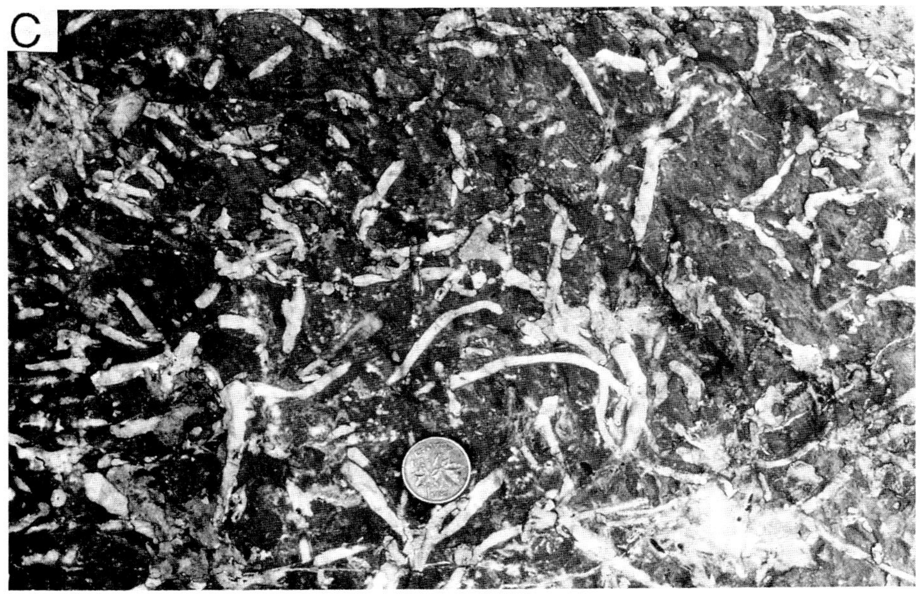

Fig. 107.-Burrow structures in shallow marine carbonates. (A) *Thallasinoides* traces from the Permian Winfield Formation of northern Oklahoma. Hand lens gives scale, photograph courtesy of Ray Mitchell. (B) Sediment casts of *Callianassa* burrows in Pleistocene Miami Formation, Florida Keys. Penknife for scale. (C) Bedding-plane views of dolomite-filled burrows in lime mudstone from the Middle Cambrian Eldon Formation of the Canadian Rocky Mountains in southwestern Alberta. Coin is approximately 15 mm in diameter. (D) (Following page) Bedding plane view of burrows disrupting Upper Cambrian dolomitic mudstone from the Gatesburg Formation of central Pennsylvania. Scale in cm, photograph courtesy of Glen Dolphin.

Fig. 108.-Bedding plane view of sediment casts of burrows of various sizes in the Middle Triassic (Ladinian) Knollenkalk member of the Livinalongo Formation, Dolomites of northeastern Italy. Note the pellet-stuffed burrow at left center (arrow). Coin is approximately 25 mm in diameter.

oriented. However, the rhizomes of modern sea grasses (*Thallasia*, *Zoastria*, etc.) are horizontal.

Differentiating *gas bubble trails* from burrows that do not have distinctive fill geometries or wall ornamentation, and so on, may also prove difficult. Indeed, no good criteria exist. Bubble trails can be nearly constant diameter or can have bulges of variable diameter, and thus look remarkably similar to burrow networks as seen in resin casts from modern sediments. Also, although bubble traces tend to be vertical, other orientations have also been observed. In modern subtidal and certain intertidal settings, bioturbation is intense enough to destroy all traces of physically-produced sedimentary structures, whereas primary physically-produced structures are apparently not totally destroyed by gas bubble migration. However, this criterion should only be used with extreme caution.

Cyanobacterial filaments (see section on stromatolites) are typically only microns in diameter and thus their molds are more likely to be confused with hair-root molds than with burrows. However, a number of genera of cyanobacteria with filament widths of several tens of microns (e.g., *Scytonema*, *Rivularia*) are common on modern carbonate tidal flats where the intertidal and subtidal lime muds carry tiny polychaete and nematode burrows of about the same size (Garrett, 1977).

Occurrence, preservation and significance.-Burrows are ubiquitous in Phanerozoic carbonate rocks and a substantial literature has evolved to exploit these features (Basan, 1978; Collinson and Thompson, 1989; Ekdale and others, 1984; Miller and others, 1984; Curran, 1985; Ekdale and Pollard, 1991; other SEPM atlases in this series). Many of these references also list criteria for the identification of burrows, roots, and gas bubble trails. Compaction can significantly alter burrow shapes. Shinn and Robbin (1983) in their compaction experiments on unconsolidated muds squashed out open burrows whereas pellet-stuffed burrows were flattened.

Burrow mottling (complete bioturbation).-

Organisms are capable of completely disrupting physical layering. Garrett (1977) showed that the organisms of the shallow subtidal and intertidal ponds of Andros Island, Bahamas, are capable of reworking the upper 5 cm of sediment many times within the course of a single year. It is of considerable significance that bioturbation is the dominant process in modern subtidal marine subenvironments where tidal and wave currents do not constantly rework the bottom sediments, so that the principal characteristic of such subtidal sediments is *lack of internal layering*. In point of fact, the subtidal facies of most ancient shallow platform carbonates of Phanerozoic age have been similarly homogenized by bioturbation (see, for example, Hardie and Shinn, 1986). Tedesco and Wanless (1991) documented the importance of a repetitive cycle of storm filling of *Callianassa* burrow networks that were subsequently re-excavated by the arthropods.

Partly bioturbated rocks are commonly referred to as mottled or burrow mottled (Fig. 109) (also see Pettijohn and Potter, 1964, Plate 14). Droser and Bottjer (1986, 1989) give a numerical index of degree of bioturbation.

Tracks and trails.-

A large literature exists on identification and interpretation of individual trace fossils (tracks, trails, etc.) and trace fossil assemblages. These are common in carbonates (see Droser and Bottjer, 1986). The reader is referred to Crimes and Harper (1970, 1977), Frey (1975), Basan (1978), Curran (1985), Bottjer (1987), Collinson and Thompson (1989), and Ekdale and Pollard (1991).

Scalloped and planar erosion surfaces.-

Terminology.-Read and Grover (1977) applied the terms scalloped and planar erosion surfaces to distinctive erosional layer contacts found in Ordovician rocks in southwestern Virginia.

Fig. 109.-Partly bioturbated carbonates (burrowing fabric is still recognizable, cf. "mottled bedding") and completely bioturbated carbonates. (A) is from the Middle Ordovician St. Paul Group of western Maryland. (B) (following page) is from the Rockdale Run Formation, Lower Ordovician Beekmantown Group of western Maryland (see also Pettijohn and Potter, 1964, Plate 14). Dark gray patches are limestone, light-colored areas are microdolomite. Note the tubular and circular cross-sections of burrows (now tan-colored dolomite) in the remnants of the gray limestone host. Photograph courtesy of Francis J. Pettijohn. (C) shows an epoxy impregnated core of intertidal pond deposits from the modern carbonate tidal flats of northwest Andros Island. Scale is in mm. This completely bioturbated texture was achieved by random burrowing of rather small annelids and gastropods (see Garrett, 1977, for further details). (D) The bioturbated "*Megalodont*" facies of the Middle Triassic Dachsteinkalk of Dachstein, Austria (compare with C). Lens cap is 50 mm in diameter. See Fischer (1964) and Important Associations section below for further details.

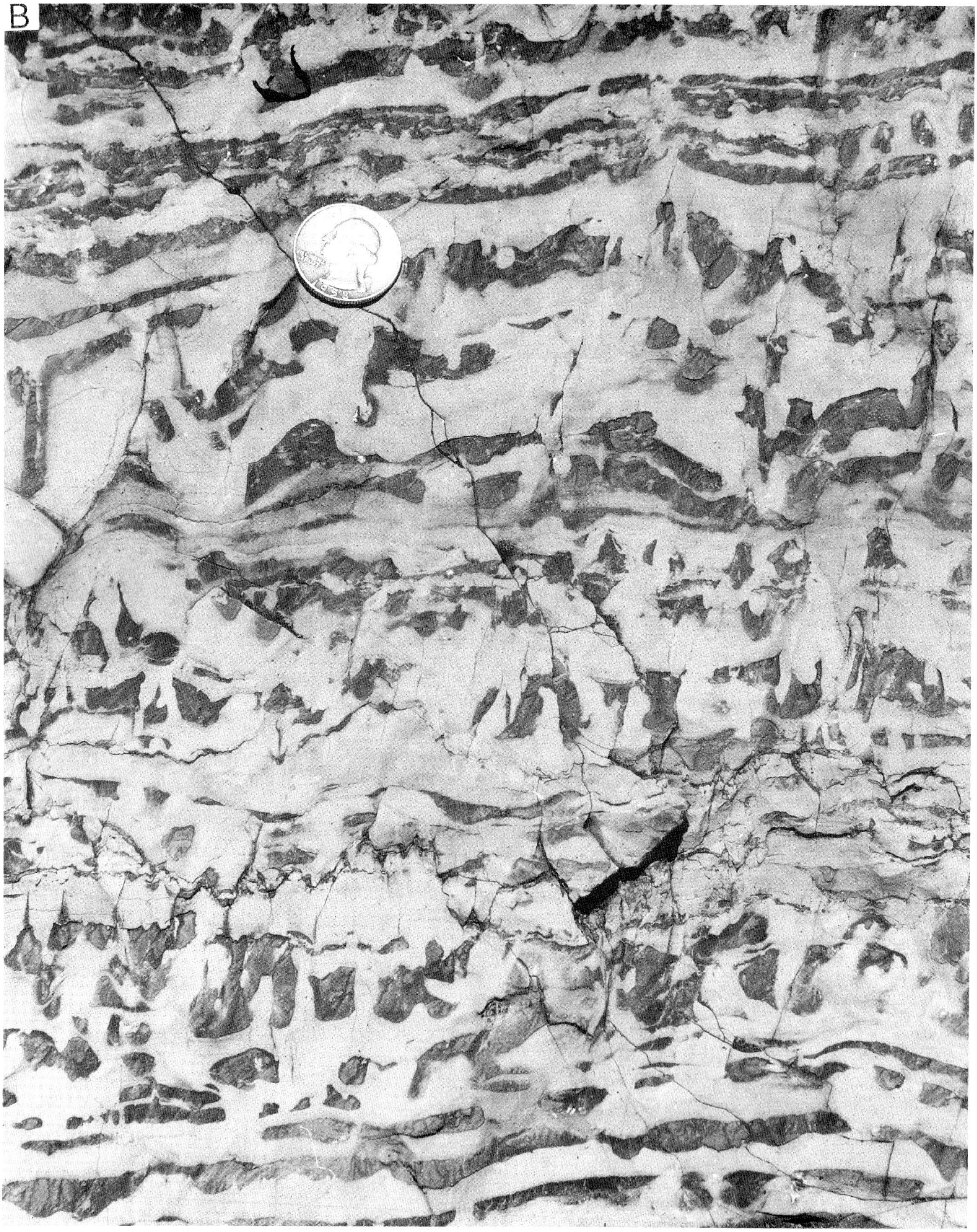

C

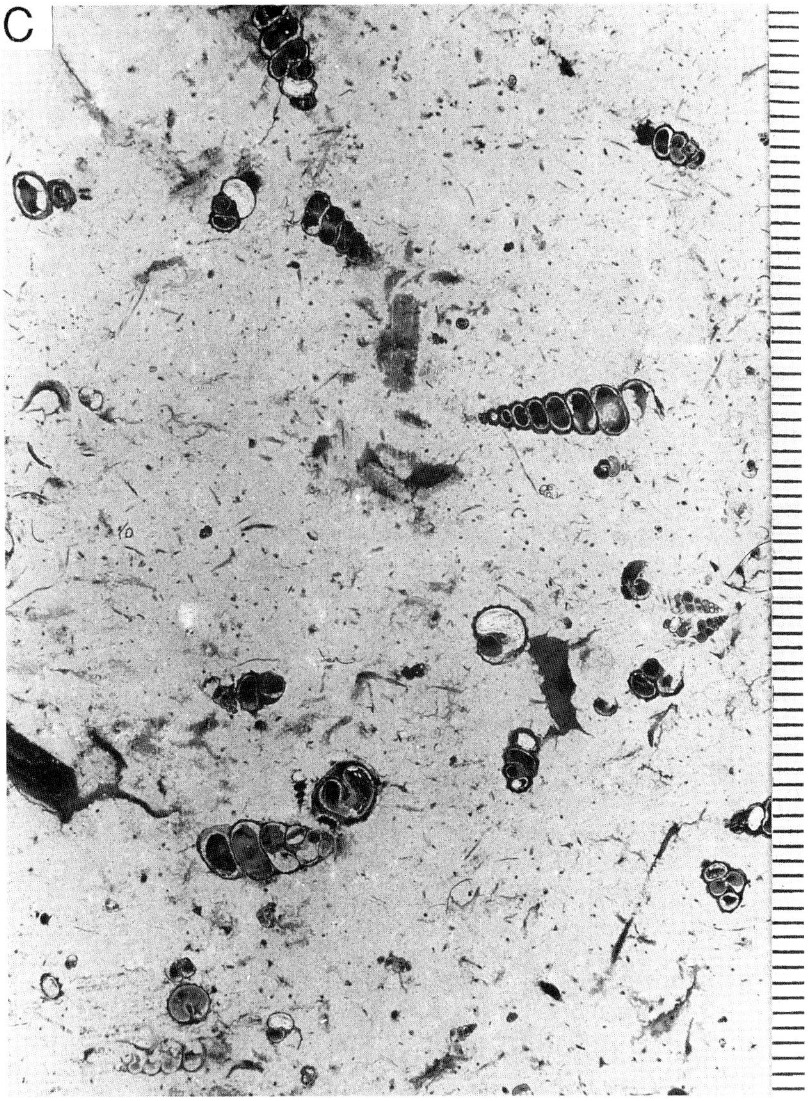

D

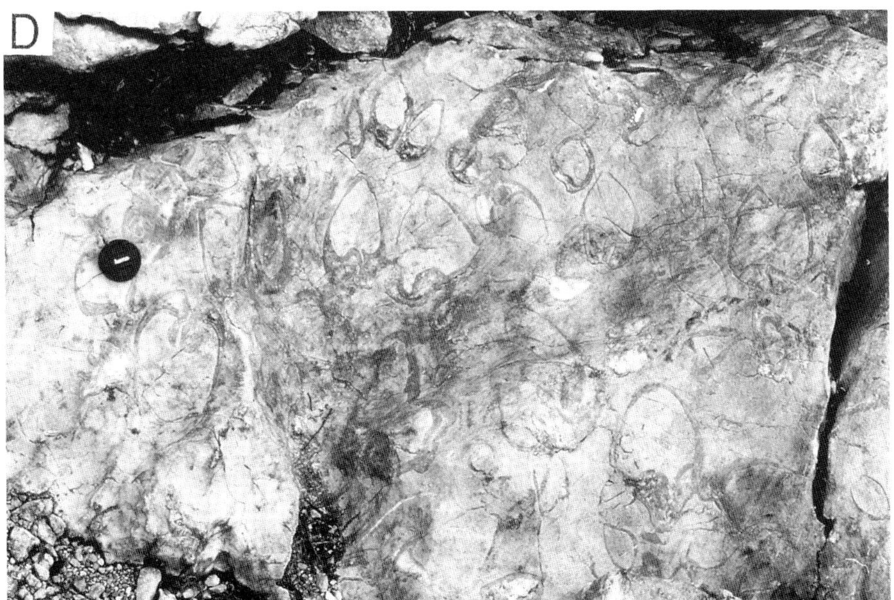

Distinguishing characteristics.-Scalloped and planar erosion surfaces as described by Read and Grover (1977) comprise contacts between fenestral mudstones and overlying skeletal-rich grainstones to packstones traceable for hundreds of meters in outcrops. In cross-section, the erosional contacts range from planar to smooth, highly curved to flat-bottomed depressions (100 mm to over 1 m wide) with smooth, curved, steeply-sloping walls separated by sharp, knife-edged pinnacles up to 0.3 m high. Pinnacle sides commonly have sharp overhanging lips (Read and Grover, 1977). Hard grains (such as skeletal fragments) and cements are truncated along the contact. In some cases the contact may be encrusted by skeletal algae and, in a few rare cases, molds of skeletal fragments and small vugs infilled by internal sediments and cements are found beneath the surfaces.

Origin and significance.-Read and Grover (1977) interpret the fenestral mudstones below the scalloped surfaces as intertidal deposits and the skeletal-rich overlying carbonates as subtidal deposits. Thus, they place these surfaces at the upper boundary of a peritidal cycle and suggest they are exposure features. They interpret the Ordovician scalloped surfaces to represent bowl-shaped depressions eroded into a pre-existing lithified intertidal rock platform similar to those developed in modern intertidal exposures of Pleistocene bedrock common on the Florida and Bahama Banks area. Read and Grover (1977) distinguish these Ordovician scalloped surfaces from submarine hardground surfaces (which may also have a scalloped geometry, see above) based on: (1) their position between inferred intertidal and subtidal rocks; (2) their lack of impregnation by opaque minerals; (3) and the differences between burrowed hardgrounds and the non-burrowed scalloped surfaces. It is not clear whether these characteristics are diagnostic enough to unequivocally distinguish between submarine hardgrounds and intertidal erosion surfaces.

Fig. 110.-Scalloped erosion surface developed on Pleistocene Limestones exposed in the present day rocky intertidal zone of Big Pine Key, Florida Keys. Bioerosion by chitons is responsible for much of the scallop morphology. Soft drink can at bottom of foreground pool gives scale.

In assessing the significance of scalloped erosion surfaces in shallow marine carbonates, it is important to recognize that in modern rocky intertidal areas the bedrock is Pleistocene or older and the highly scalloped surfaces are primarily if not entirely the result of bioerosion by chitons, gastropods, and other grazing and boring organisms (Fig. 110) (Multer, 1975). Similar scalloped bioerosion surfaces produced by sea urchins and other rock grazers are found in modern reefs, particularly on reef crests and back reef areas.

It is recommended that considerable caution be used in assessing the significance of ancient scalloped and planar erosion surfaces. This caution applies particularly with regard to whether such surfaces are or are not notable disconformities that record a significant period of subaerial exposure and lithification followed by bioerosion that occurs at the leading edge of a rising sea level (as has been the case for the Pleistocene-Holocene transition in Florida and the Bahamas). For example, in the Cambro-Ordovician carbonates of western Maryland, U. S.A., scalloped hardgrounds are exceedingly common in subtidal subfacies where they typically separate peloidal and ooid grainstones from overlying lithoclast breccias. Recently, Paul Zell of Pennsylvania State University (pers. commun. to LAH) has discovered abundant disarticulated valves of chitons (*Matthevia* or *Chelodes* species, cf. Runnegar and others, 1979, p. 1383) in lithoclast breccias overlying scalloped surfaces in the Upper Cambrian Conococheague Limestone and the Lower Ordovician Beekmantown Group, lending strong support to a bioerosion origin for the many scalloped and planar surfaces in these thick shallow water carbonate deposits. In these deposits, scalloped and planar surfaces may in some cases stack vertically one above the other less than a centimeter apart (Nguyen, 1986). Although this is not the typical spacing, it does suggest, along with the completely subtidal nature of the host rocks, that the scalloped surfaces are submarine hardgrounds (Nguyen, 1986) rather than subaerial disconformities (see below under Submarine Hardgrounds). It should be noted that modern chitons inhabit a wide range of marine environments from intertidal to the deep sea and they are capable of bioeroding any bacterial-cyanobacterial covered hard surfaces on the sea bottom at any water depth. They will produce scalloped surfaces in ancient fully lithified submarine outcrops with the same ease as they bioerode lightly cemented Holocene hardgrounds.

CHEMICAL FEATURES

Tufas, Travertines, and Speleothems

Terminology.-

A variety of non-marine carbonate laminites, stromatolites, and thrombolite-like mounds are known from modern lakes, marshes, springs, rivers, waterfalls, and caves. The terminology for these calcareous chemical structures has varied considerably with different workers in the past and there is little consensus today. *Travertine* is one of the oldest terms used to describe laminated carbonate spring and waterfall deposits (Lyell, 1842, v. 1, p. 397; Cohn, 1864; Weed, 1889). The name stems from the locality in Italy near Rome originally known as Tivertino in Roman times but now called Tivoli. The many quarries in the Campagna area between Rome and Tivoli have provided travertine to build, as Lyell (1842, v. 1, p. 407) has described it, "a large proportion of the most splendid edifices of ancient and modern Rome." The Tivoli travertine is found as facings on and within many buildings in the U. S. and other parts of the world. Deposits similar to travertine, but which are porous and spongy and found in a wide variety of modern terrestrial aqueous environments, have been called *tufa* or calcareous tufa (e.g., Lyell, 1842, v. 1, p. 397; Scholl and Taft, 1964; Irion and Muller, 1968), although other terms such as plant tufa, spring sinter, calc-sinter, algal bioherm, and so on, have been used (see Julia, 1983). *Flowstone* is a term that is used to describe smooth sheets of laminated carbonates that cover the walls and floors of caves and surface spring pools. Flowstones may carry the fabrics of both dense travertines and porous tufas in adjoining laminae. The entire suite of laminated flowstones, stromatolites (e.g., cave popcorn), ooids and pisoids, and other precipitated carbonate features of cave deposits are usually referred to collectively as *speleothems*.

Distinguishing characteristics.-

In view of the absence of any consensus about terminology for the whole group of chemical carbonates, we will in this atlas reserve the term *tufa* primarily for laminae, laminites, stromatolites, and discrete unlayered knobs and mounds that internally display a framework of calcareous tubules and pustules (Fig. 111). In modern tufas, the tubules are micron-thick calcareous sheaths that have precipitated around the filaments of cyanobacteria, bacteria, and fungi, and around the stems and fronds of such hogher plants as green algae and mosses (Fig. 111; see also Monty and Hardie, 1976, Figs. 4A, 9, 10, 11, 12). The pustules are sheaths around coccoid cyanobacteria and coccoid bacteria (Monty, 1976, Figs. 26, 27, 28). Perhaps also to be included under the term tufa, but much more loosely, are layers or mounds that consist of a framework of micrite clots and peloids of cryptocrystalline carbonate that may have been precipitated by bacterial colonies (Chafetz, 1986; Chafetz and Folk, 1984). However, these clotted tufas may be difficult to differentiate from non-biogenic clotted micrites.

In modern tufas found in lakes (Dean and Eggleston, 1975; Eggleston and Dean, 1976; Halley, 1976; Monty, 1972; Scholl and Taft, 1964), springs (Irion and Muller, 1968), rivers (Slack, 1967; Braithwaite, 1979), and coastal marshes (Gleason and Spackman, 1974; Monty and Hardie, 1976; Hardie, 1977a), the calcareous sheaths around filaments with diameters of a few tens of microns or more typically consist of a dark, very thin inner micrite layer (<10 microns) made of a mosaic of tiny micron-sized rhombs of calcite (magnesian calcite in some of the coastal marshes) and a thicker outer layer (several tens of microns wide) of calcite microspar with a crude to well-defined radial fabric and euhedral crystal terminations (see Monty and Hardie, 1976, Fig. 12). The calcified sheaths are loosely bound together with blocky microspar that partly fills the interstices between them and which may show a meniscus fabric in tufas formed in wet-and-dry settings such as springs, marshes and ephemeral lakes (e.g., Monty and Hardie, 1976, Figs. 12A, C). Moss tufa is composed of an open framework of large, anastomosing, ornamented tubes that are millimeters in diameter and centimeters in length (Fig. 112). Charophyte tufa looks very similar to moss tufa, but with a different internal ornamentation to the tubes (see, for example, Dean and Fouch, 1983, Fig. 16).

Tufas of special significance to students of shallow marine carbonates are those found in the coastal lakes and marshes bordering the modern carbonate tidal flats of Andros Island, Bahamas. Descriptions of these tufas and their depositional environments can be found in Monty (1967, 1972), Monty and Hardie (1976), and in several papers in Hardie (1977a). The Andros tidal flats are environmentally zoned with the main tidal channel/levee/tidal pond belt passing landward into a supratidal coastal marsh, which in turn passes into a belt of shallow freshwater lakes (Shinn and others, 1969). The surface of both the marsh and the lake subenvironments are covered by fleshy mats (tens of millimeters to a few centimeters thick) of the large filamentous (20-40 microns wide), non-motile cyano-bacterium *Scytonema*. In these landward subenvironments, flooding by seawater occurs only during hurricanes when laminae and thin beds of peloidal aragonite mud are deposited in the marsh (Hardie and Ginsburg, 1977). Normally, however, both the marsh and the lakes are either dry or under freshwater brought in by the frequent summer rains that characterize the climate of this wet tropical island setting. *Scytonema* is a freshwater cyanobacterium that can survive a limited submergence in seawater (it grows along the shores of the tidal ponds in the channelled belt, see Hardie, 1977a), but it thrives in shallow freshwater wetlands undisturbed by sedimentation for periods of years (Hardie, 1977b, p. 165-170). In these Andros freshwater marshes and lakes, the *Scytonema* calcifies to produce calcareous tufa biostromes and very small bioherms (Hardie and Ginsburg, 1977; Hardie, 1977b; Monty, 1972; Monty and Hardie, 1976), such as those illustrated in Figure 111. The tufa consists of a porous friable framework of thin microcrystalline calcite tubules (20 to 40 microns in inside diameter with walls 10 to 30 microns thick, and up to 3 mm long) which are lightly welded by calcite microspar. The tubules are calcified molds of *Scytonema* filaments, as can be demonstrated in those tufas where oxidation and bacterial degradation has not yet destroyed the host filaments. The calcite consists of blocky rhombs less than 4 microns across and varies in $MgCO_3$ content depending on the location of the tufa. At the seaward side of the marsh, where minor flooding by seawater during some onshore storms and unusually high spring tides occurs on occasion, the biostromal tufas typically consist of 16 mole% $MgCO_3$ magnesian calcite, whereas the small tufa heads are made of 12-13 mole% $MgCO_3$ magnesian calcite. Further inland, the calcite of the tufas gets progressively depleted in magnesium and in the freshwater lakes is low magnesian calcite with as little as 1.5 mole% $MgCO_3$ (Monty, 1972). The morphology of the tufas also depends on where the tufas were formed. Biostromal tufas form by calcification of carpet-like fleshy *Scytonema* mats that cover the bottoms of the inland lakes and large, very shallow depressions (kilometer-scale) in the marsh that are under a sheet of freshwater in the summer but dry in the winter. In these depressions on the marsh, the biostromal tufas are millimeter to centimeter thick layers of calcified *Scytonema* mats interbedded with peloidal laminae and thin beds deposited by hurricanes (Fig. 113). These biostromal microbial tufas are a most significant component of the Andros Island Holocene tidal flat deposit in that they make up between 50 and 60% of the 1- to 2-m thick sediments underlying the marsh subenvironment which covers a present day area of about 400 km^2. On the slightly elevated ridges bordering the depres-sions, particularly the ridges that separate the marsh from the main tidal channel belt, the tufas form small biohermal "heads" (centimeter scale, see Fig. 114) that are the calcified remains of small domes and mounds of patchy *Scytonema* mats that grew under slightly drier conditions than the thicker continuous mats of the depressions. These patchy mats are located preferentially over mudcracks (Fig. 114) up which groundwater seeps during the wet season. In the inland freshwater lakes, the lightly calcified sheaths around the *Scytonema* filaments are normally disaggregated by wind waves and the fragments dispersed across the bottom of the lakes as lime mud that becomes trapped by the mats to form unlithified stromatolites, which are especially well-developed along the lake shores (see above under Stromatolites; see also Monty and Hardie, 1976, Fig. 3). Freshwater calcitic mud is also being produced today in the same manner by calcification of *Scytonema* mats across the huge expanse of the Everglades in Florida (see Gleason and Spackman, 1974; Monty and Hardie, 1976). Further details, descriptions, and numerous illustrations of the *Scytonema* tufas of the inland algal marsh of Andros Island can be found in Hardie (1977a, b)

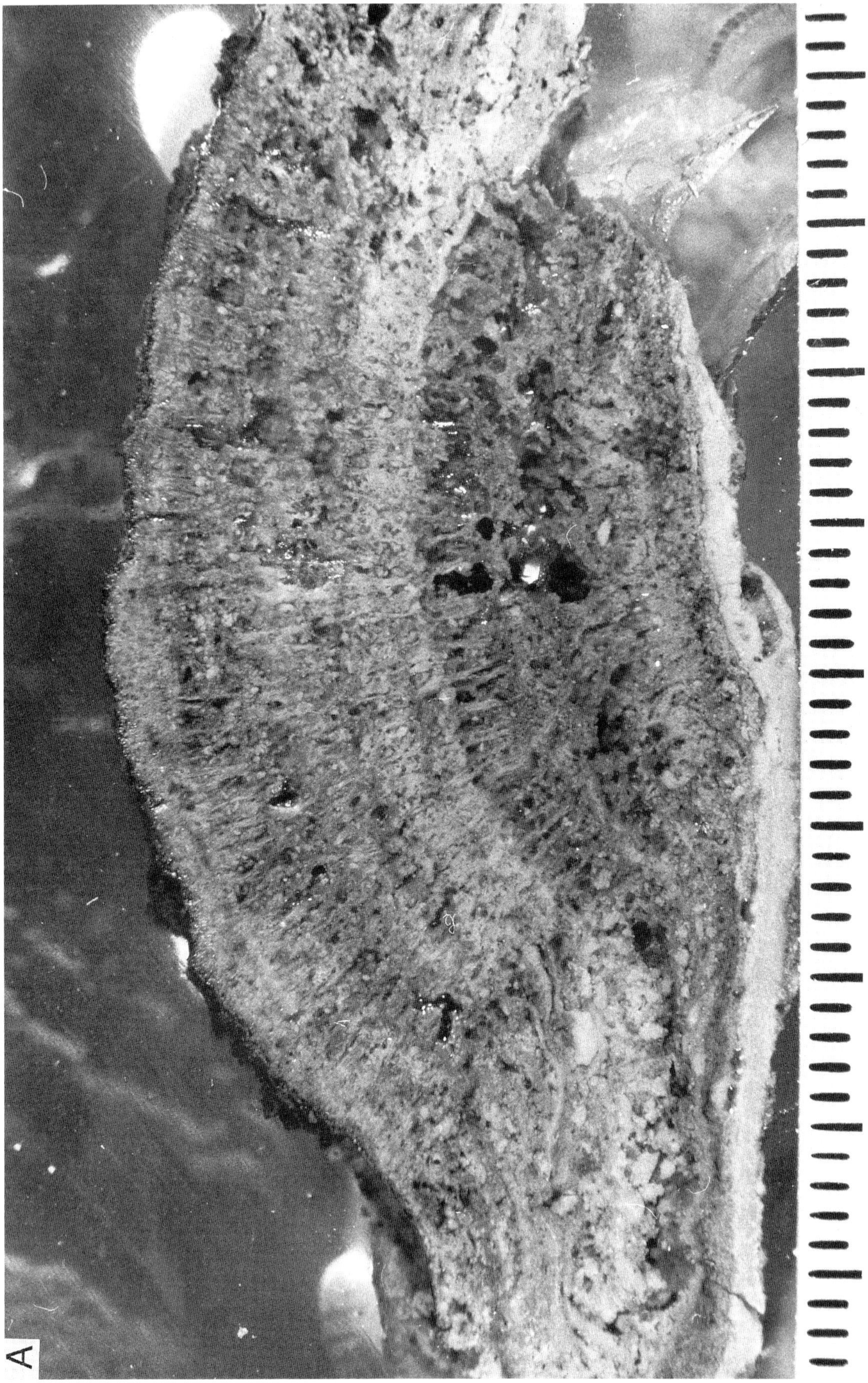

Fig. 111.-Epoxy impregnated slabs of modern cyanobacterial tufa from the supratidal cyanobacterial marsh subenvironment of the carbonate tidal flats of northwest Andros Island. The cyanobacterial marsh is landward of the channel belt. (A) The radial fabric elements are calcified cyanobacterial filament molds surrounded by unlithified peloids. Note the abrupt termination of the cyanobacterial tufa against intraclastic sand on the left. Scale in mm. (B) (Following page) This tufa is 30 mm across and is composed entirely of calcified cyanobacterial filaments, no detrital sediment is present. Note the fibrous "palisade" structure and the open fenestral pores.

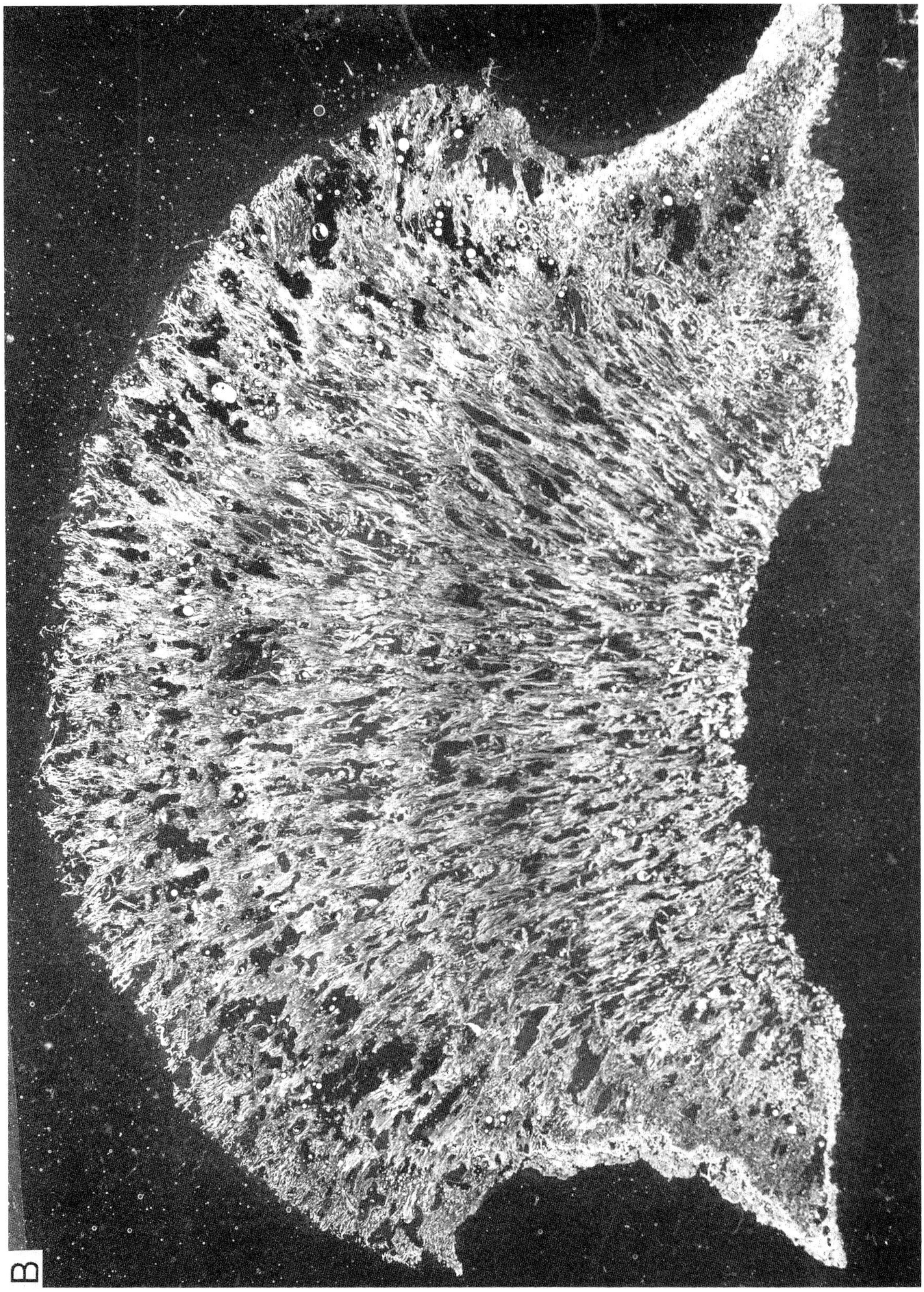
B

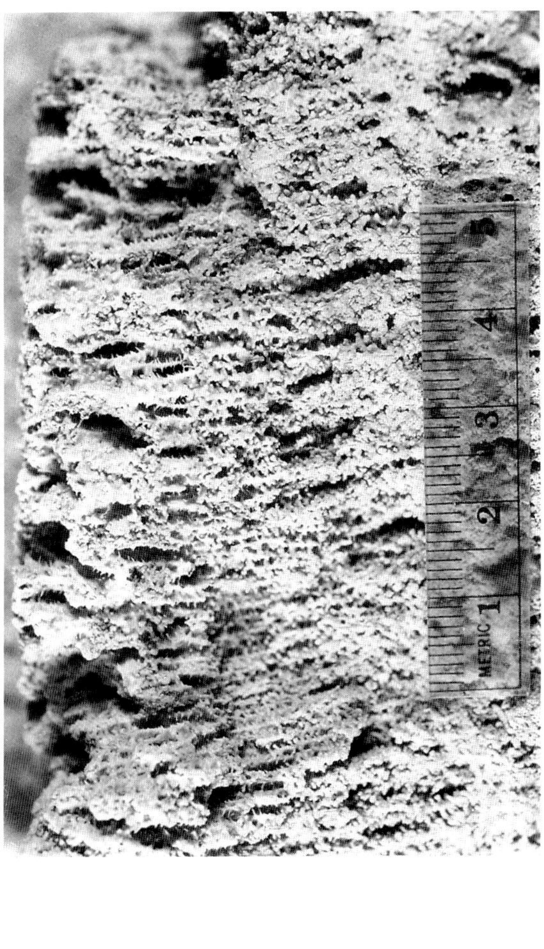

Fig. 112.-Modern moss-tufa from Sitting Bull Falls, New Mexico. Here, filaments of moss have been calcified. Note the larger carbonate crystals growing outward from the calcified moss filaments. Scale is in mm.

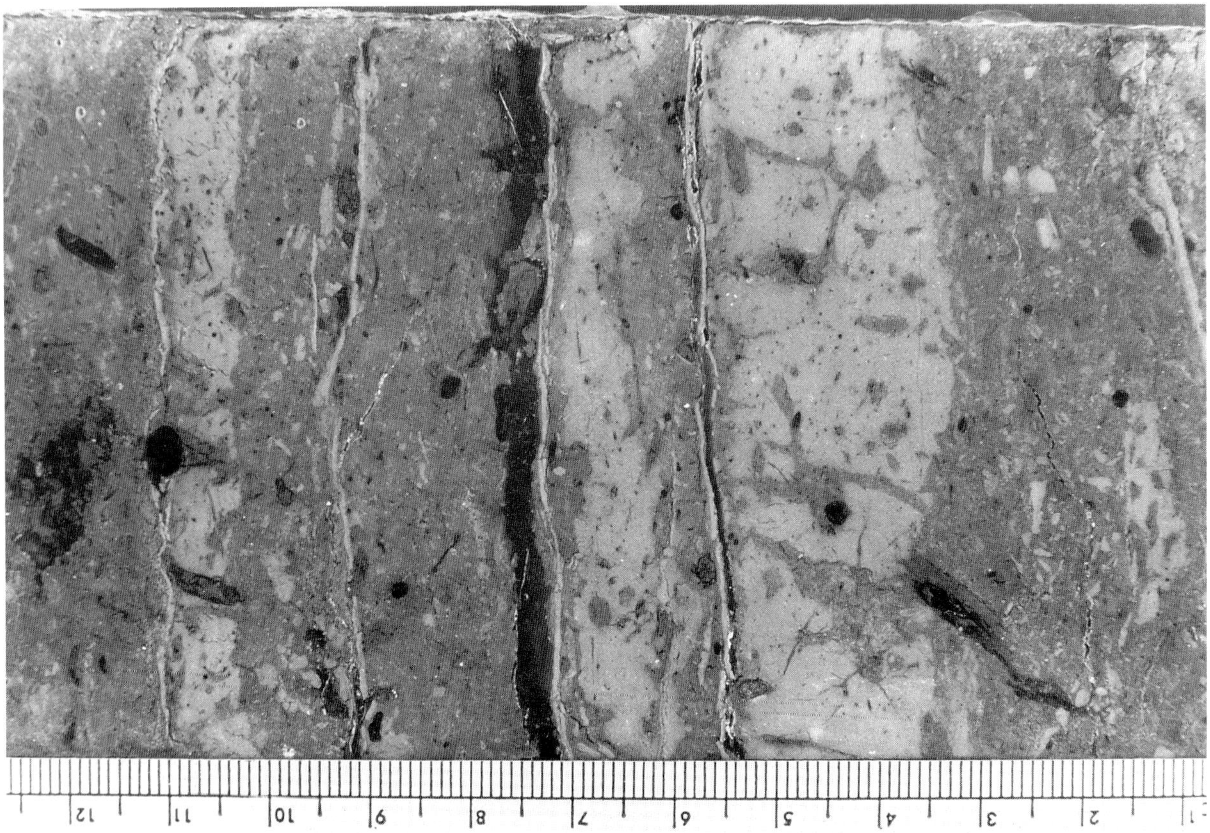

Fig. 113.-Epoxy impregnated slab of Holocene deposits beneath the modern supratidal cyanobacterial marsh landward of the channel belt of the tidal flats of northwest Andros Island. The layering here is cyclic on the centimeter scale and comprises centimeter-thick, dark gray layers of cyanobacterial tufa (Mg-calcite) overlain by centimeter- to millimeter-thick, white, hurricane-deposited peloidal layers (aragonite). The top of each peloidal layer is a thin white crust of aragonite-cemented peloidal sediment (aragonite). The very dark tubular to circular structures are black mangrove roots while the gray tubular to circular structures in the white hurricane layers are tufa-filled root holes. Rooting can completely disrupt hurricane layers leaving remnants in the form of scattered intraclasts in the tufa layers. See text for details (see also Hardie and Ginsburg, 1977, and Monty and Hardie, 1976). Small divisions of the scale on the left are mm.

Fig. 114.-Surface view of the elevated ridge that separates the cyanobacterial supratidal marsh from the main tidal channel belt of northwest Andros Island modern tidal flats. The dark patches are tufas that form small biohermal "heads" composed of calcified remains of small domes and mounds of patchy *Scytonema* mats. These patchy mats are located preferentially over mudcracks up which groundwater seeps during the wet season (compare with Fig. 77). Small circular tufas in the foreground are 20 mm across.

and Monty and Hardie (1976). Monty (1967, 1972) and Monty and Hardie (1976) have extensively described and pictured the tufas of the freshwater lakes of Andros Island.

Many of the huge carbonate mounds that formed in Holocene and pluvial Pleistocene lakes of the western U. S. (see for example Figs. 115 and 116) are called tufas or bioherms (e.g., Radbruch, 1957; Scholl, 1960) but not all of them display the tubular or clotted fabrics described above. Instead, some show radiating crystalline fabrics (such as thinolite fabric, see below under travertine) that undoubtedly are of a purely chemical origin. These properly belong in the travertine category.

In our treatment of chemical structures, we have chosen to use the term *travertine* for those laminae, laminites, and stromatolites that are characterized by a primary crystalline fabric which most typically takes the form of a radial array of needles or bladed crystals of calcite or aragonite with compromise boundaries (Fig. 117). With this definition, we have attempted to put into one category those macroscopic carbonate structures that have precipitated inorganically out of a supersaturated aqueous solution by nucleation and crystal growth of alkaline earth carbonate minerals onto a passive but stable inorganic substrate. The primary crystalline fabric is differentiated from neomorphic fabrics using the criteria for cements outlined so lucidly by Bathurst (1975, p. 417-425). The laminae of travertines are normally (but not exclusively) continuous and isopachous in flat laminites (e.g., flowstones) and low amplitude wavy forms with the morphological characteristics of laterally-linked, hemispherical stromatolites (see Logan and others, 1964). However, in domal travertines with morphologies like those of solitary, hemispherical stromatolites the laminae thicken over crests and thin down the sides of the mounds or knobs. Stalagmites and cave "popcorn" are good examples of non-isopachous non-biogenic stromatolitic travertines with radial crystalline fabric (see Thrailkill, 1976, Figs. 1, 7, 8, 9). Travertines may include: (1) laminae that are composed of homogeneous micrite (commonly "color banded" with alternating dark and light colored layers of equivocal origin, but probably chemical), (2) laminae composed of clotted micrite and peloids aggregated into arborescent (shrub-like) shapes (Fig. 118); these have been ascribed, correctly or incorrectly, by Chafetz and Folk (1984) to bacterial precipitation as noted under tufas above, and (3) laminae composed of filamentous tufa of cyanobacterial origin. Other notable features of many travertines, particularly spring travertines, are: (4) their dam and pool configuration, as is spectacularly exemplified by the Mammoth Hot Springs in Yellowstone National Park, Wyoming; and (5) their common association with ooids and, more particularly, pisoids (Fig. 119A), which in travertine pools may show a reverse grading (Fig. 119B); isolated ooids and pisoids are commonly enclosed within travertine flowstones where they are draped or encapsulated by the crystalline laminae of the flowstones.

Modern travertines are known from rivers, springs, lake shores, and caves. The latter, under the group heading of speleothems, will be briefly treated separately below.

A very particular type of travertine, known under the name of *thinolite*, is found abundantly in lake deposits formed during the Pleistocene ice age in the western U. S. (see, for example, Radbruch, 1957). It occurs as mounds up to tens of meters in height (Fig. 120A). In outcrop sections cut through the coalesced meter-scale domes that make up these massive travertine structures, the thinolite has an eye-catching fabric which consists of concentric layers of radiating prismatic skeletal crystals up to 10 cm long made of low-magnesium calcite (Figs. 120B-E). Recently, Shearman and Smith (1985) and Shearman and others (1989) have convincingly demonstrated that these thinolite crystalline fabrics are calcite pseudomorphs after *ikaite*, which is hydrated calcium carbonate ($CaCO_3 \cdot 6H_2O$) that precipitates metastably from alkaline waters at temperatures near freezing. Thus, these thinolite travertines are indicators of very cold water conditions (Shearman and Smith, 1985; Shearman and others, 1989).

Of special interest in studies of shallow marine carbonates are the intertidal and supratidal splash zone travertines ("coniatolites") that occur on the rocky shores of the Persian Gulf (Purser and Loreau, 1973; Scholle and Kinsman, 1974). We have briefly described these in the discussion of modern stromatolites (see under Cryptmicrobial Structures in section on Biogenic Structures).

Chemically precipitated carbonate deposits of caves have been termed *speleothems* (Thrailkill, 1976). Although speleothems are a type of travertine, they are referred to separately here because of the wide usage of the term speleothem in the carbonate sedimentology literature. A crystalline fabric is the essential character of speleothems (Fig. 121). Descriptions of the structures and fabrics of stalagmites, cave popcorn, moonmilk, and cave pearls and discussions of their origin as chemical precipitates are given by Thrailkill (1976).

Origin.-

The origin of travertines that carry well-defined primary crystalline textures such as radiating needles or blades that exhibit compromise boundaries (see above) is clearly one of direct chemical precipitation out of supersaturated waters. On the other hand, the unraveling of the origin of tufas is not so straightforward mainly because the details of the mechanisms of calcification of microbial mats, non-calcareous eukaryote algae, and higher plants to make tufas is not certain. The lime coatings are not external skeletons, nor are the crystals precipitated within the outer rows of living cells by the organism in order to stiffen the thalli, as, for example, occurs with some modern soft calcareous green algae such as *Penicillus*. Instead, where living cyanobacteria such as *Scytonema* are found in the early stages of calcification, the carbonate crystals are seen to be suspended in the mucilaginous sheaths around the filaments (see Monty and Hardie, 1976, Fig. 4G). What triggered the carbonate precipitation is unknown, but in those cases where calcification is confined exclusively to the filament mucilage and does not nucleate on associated inorganic particles, it seems most probable that the normal metabolic activities of the living organism itself (such as photosynthesis) have induced inorganic precipitation from the surrounding calcium bicarbonate-bearing waters. On the Andros supratidal marsh, daytime pH of the water within moist living *Scytonema* heads was measured at as high as 9.5, so it is clear that photosynthesis can significantly increase the $[Ca] \cdot [CO_3]$ ion activity product in the films of water around the filaments. In a set of experiments carried out at The Johns Hopkins University in 1986 by C. Nguyen (unpublished data, see also Nguyen, 1986), calcite was found to have precipitated around living cyanobacterial filaments grown under artificial day-night light cycles but did not nucleate on filament-size inorganic fibers under the same conditions of very slow evaporative concentration. Field evidence for this selective calcification of living cyanobacteria is found in the *Scytonema* tufa heads on the Andros marsh low ridges where there is preferential growth of magnesian calcite around the filaments while the peloidal sediment caught between the filament tufts remains uncemented (Monty and Hardie, 1976, Fig. 13; Hardie, 1977b, Fig. 49). However, it is clear from completely encrusted sheaths welded by meniscus cements that inorganic processes must have played a large part in the calcification. In this regard, it is significant that Pentecost (1978) calculated that the measured CO_2 uptake rates of the

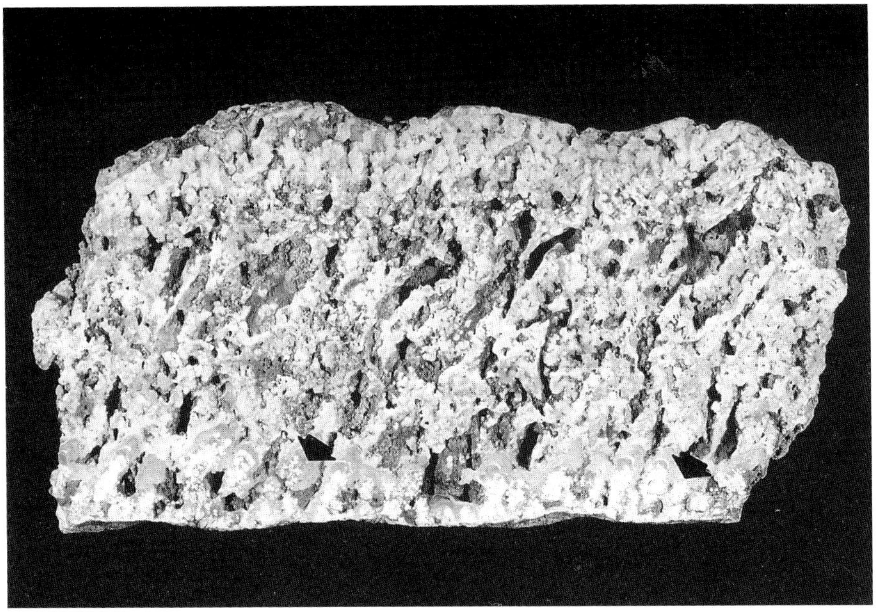

Fig. 115.-Slab of lacustrine tufa from Great Salt Lake. Note the thrombolitic fabric of the tufa mound. These centimeter-scale clots and fingers are composed in part of white masses that strongly resemble *Renalcis* (arrows). The clots are surrounded by peloids and ooids cemented by aragonite. Length of sample is 15 cm. Sample courtesy of Ron Spencer.

Fig. 116.-(A) Tufa mounds of Late Pleistocene to Recent age on the shores of Pyramid Lake, Nevada. Mainly composed of dendritic tufa and radiating thinolite crystals. Tallest tufa is 10 m high. (B) Large tufa mound of Late Pleistocene to Recent age in the Searles Lake valley, Nevada. (C) Subaqueous tufa mounds in modern Mono Lake, California, exposed by the recent lowering of lake level. Highest tufa approximately 4 m tall.

Fig. 117.-Travertine encrusting the Searles Lake tufa mound shown in Figure 116B. The travertine comprises carbonate paper-thin crusts that form the nucleation site for bladed carbonate crystals that impinge upon each other as they grow off the substrate. Scale in mm.

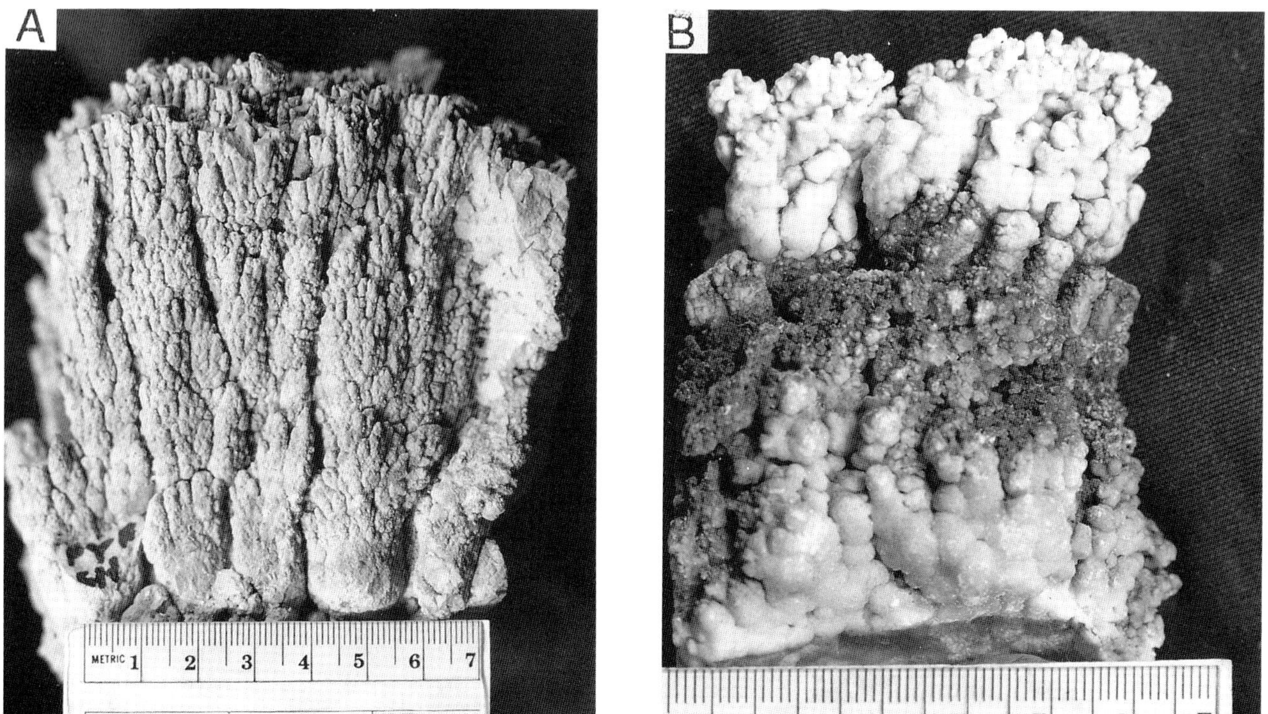

Fig. 118.-"Arborescent" or "dendritic" fabrics. (A) Hand sample of dendritic tufa from Pyramid Lake. Note the upward branching shrub-like shapes. (B) A more irregular example of arborescent tufa from hot spring pools of Pastos Grande, Bolivia (see Risacher and Eugster, 1979 for further details). Scale in mm. (C) (Following page) Cross-sectional slab of the famous Italian Tivoli travertine showing arborescent shrubs attributed by Chafetz and Folk (1984) to cyanobacteria and bacteria. Scale in mm. (D) (Following page) Dendritic fabric in halite developed on the inside walls of a glass container above a halite-saturated solution. A film of brine was continuously drawn up the walls of the container by a combination of capillary forces and evaporative pumping, and microcrystalline halite precipitated as the film of brine evaporated away. Note the close similarity in fabric to the Tivoli arborescent travertine shown in (C). Scale in mm.

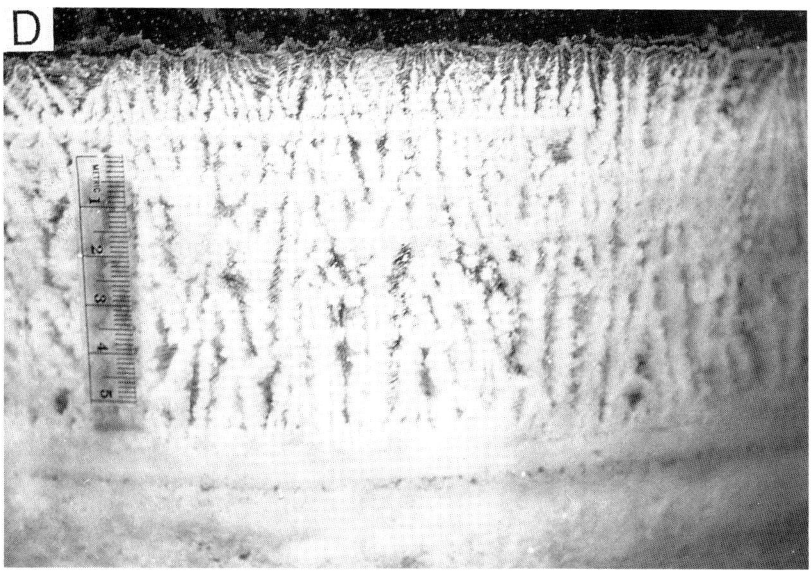

cyanobacteria in 27 deposits of freshwater tufas in the United Kingdom could only account for 1-2% of the calcite of these tufas (see also the study by Barnes, 1965). A possible explanation for selectively calcified cyanobacteria filaments in tufas that integrates the results of Nguyen's experiments and Pentecost's study is that the CO_2 uptake of the cyanobacteria selectively initiates nucleation of calcite crystallites on the filament sheaths and these crystallites then act as seeds for inorganic precipitation. In some cases, the calcareous coating on tufa-forming organisms may have been obtained quite accidently as the living or dead organisms, along with associated inorganic substrates, were coated by widespread inorganic precipitation out of supersaturated waters, such as occurs in evaporating alkaline lakes or in spring pools where carbon dioxide degasses from upwelling supersaturated groundwaters. In other cases, the calcification may have been the result of postmortem chemical changes brought about by bacterial degradation of the organism. This latter mode of origin may have been the principal cause of the calcification of the *Scytonema* mat layers interbedded with "hurricane layers" in the biostromal sediments that underlie most of the inland marsh of Andros Island (Fig. 113). A more complete discussion of this and other origins for the Andros tufas is given by Hardie (1977b, p. 170-175). Of significance is the observation that *Scytonema* mats and mounds on the shores of the seawater-filled tidal ponds of the channelled belt on Andros Island are not calicified. Thus, it appears as if the cyanobacterial calcification is an exclusively non-marine "freshwater" process on the Andros tidal flats. In this regard, the Andros tufas are like other occurrences of modern cyanobacterial and other microbial tufas, all of which formed in non-marine "freshwater" settings (see references given above).

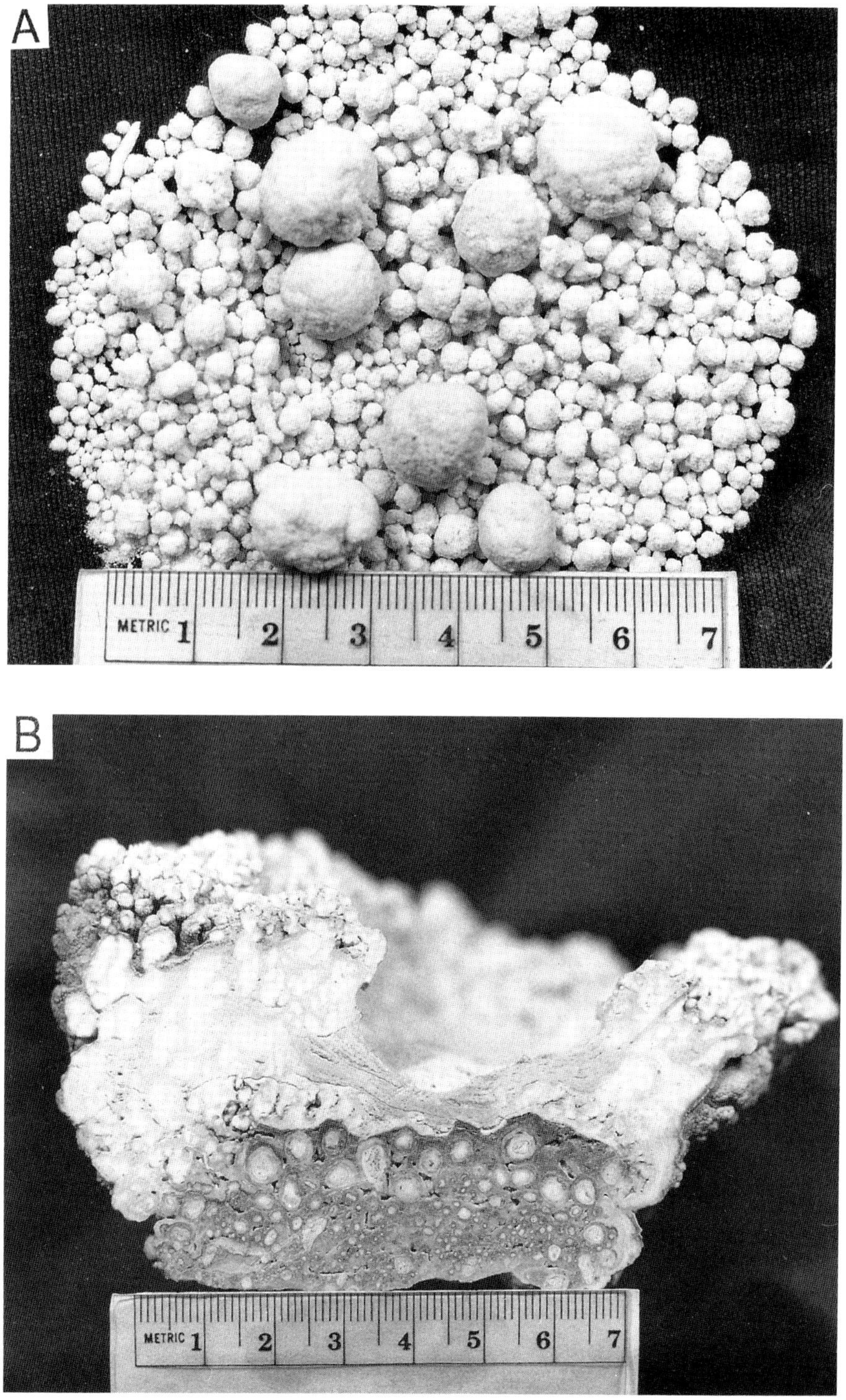

Fig. 119.-Ooids and pisoids from modern spring travertine pools. (A) Sample of loose ooids and pisoids (calcite) from Thermopolis Hot Springs, Wyoming. (B) Slab of reverse graded pisoids (calcite) from a hot spring pool and associated rim dam, Pastos Grandes, Bolivia (see Risacher and Eugster, 1979 for further details). Pisoids are cemented by calcite and overlain by laminated travertine crust. Scales in mm.

Fig. 120.-"Thinolite" travertine mounds of late Pleistocene age from Pyramid Lake, Nevada. (A) Clusters of large spheroidal structures with radial fabric typical of lacustrine thinolite travertine mounds. Spheroidal structure in center of photograph is approximately 5 m in diameter. (B) Detail of radiating thinolite crystals within a spheroidal structure of a thinolite travertine mound. Individual thinolite crystals can be over 30 cm in length. Lens cap in center of photograph is 50 mm in diameter. (C) and (D) (Following page) Hand samples showing the characteristic skeletal crystal morphology of thinolite crystal aggregates (originally ikaite, $CaCO_3 \cdot 6H_2O$, but now calcite). Oblique-basal view of the rectangular cross-sections of the prismatic crystals shown in C (paper clip for scale); side view of the elongate prism form shown in D (scale in cm and mm).

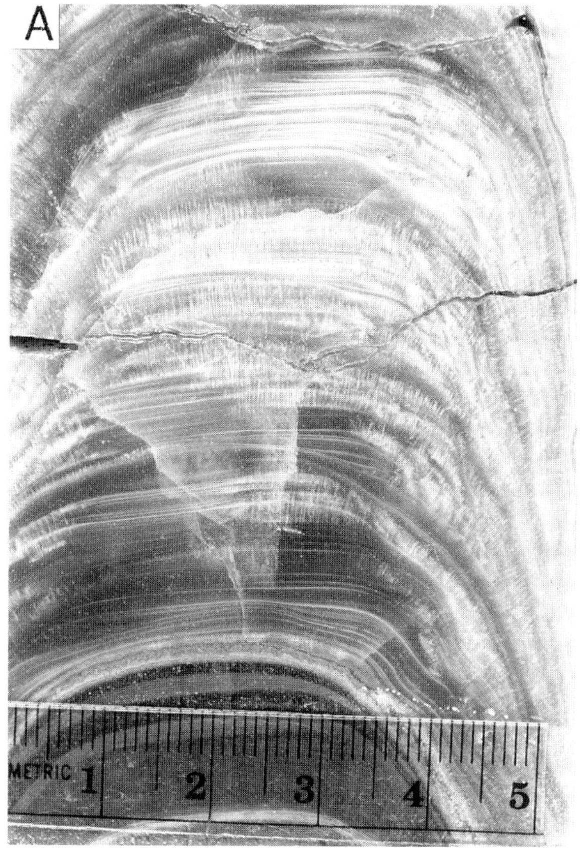

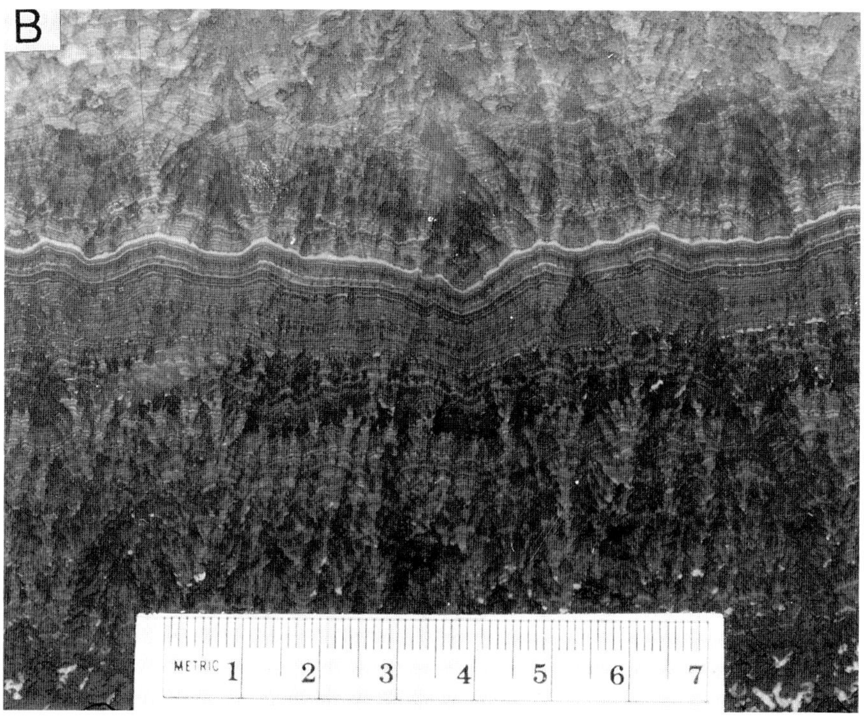

Fig. 121.-(A) Modern speleothem (stalagmite) from a cave in West Virginia showing concentric banding and radial crystalline fabric. Scale in mm. Sample courtesy of Ray Mitchell. (B) Lower Proterozoic travertine, from tidal flat carbonates of Pethei Group, Great Slave Lake area, Northwest Territories, Canada, showing concentric banding and radial crystal fabric similar to that of the modern speleothem in (A). Scale in mm. Sample courtesy of Paul Hoffman.

In mats composed of cyanobacteria with very small filaments (<5 microns in diameter, such as *Phormidium*), a mass of tiny acicular calcite crystals or stouter calcite rods may grow in the mucilaginous mass surrounding the filament network (Monty and Hardie, 1976, Figs. 12C, D, E) without producing a tubular moldic fabric. Precipitation of carbonate minerals that results from the biochemical activities of coccoid cyanobacteria and bacteria will also produce a non-filamentous layer or mass (Monty, 1976, Fig. 29). Knobs and mounds of such non-tubular tufa look very much like marine thrombolites (see earlier) in that they have an internal clotted fabric. The tufa mounds of the present day Great Salt Lake, Utah, which almost certainly are relicts of an earlier fresher water stage in the evolution of the lake (Halley, 1976), display in slabbed samples a thrombolitic fabric (Halley, 1976, Fig. 3). Both calcified filaments and irregular *Renalcis*-like clots make up a porous micritic framework in which intra-framework voids are partly filled with peloids and ooids cemented by aragonite (Fig. 115). Thus, these lacustrine tufas seem to be mainly precipitated subaqueously on both filamentous and coccoid cyanobacteria. Far more difficult to interpret are those non-tubular clotted "tufas" considered to be precipitates around the cells of tiny non-filamentous bacteria (smaller than coccoid cyanobacteria). The difficulty arises because of the lack of unequivocal cellular molds of the bacteria in the micrite clots. Chafetz and Folk (1984) have interpreted the arborescent clotted and peloidal fabrics of travertines as the products of biochemical precipitation of microcrystalline carbonate by non-filamentous bacteria. However, in the absence of identifiable bacterial nuclei to the clots, we are left with an ambiguous answer because such fabrics are also produced by inorganic precipitation. This is where the fabrics of microbial tufas and chemical travertines overlap (see below).

Occurrence and preservation.-

Holocene tufas and travertines are found in coastal marshes, lakes, springs, rivers, and waterfalls. Apart from the spectacular tufa and travertine mounds of the Pleistocene and subrecent glacial lakes mentioned above, ancient analogs of modern travertines have not been widely reported in the literature. Lacustrine stromatolites with tufa and travertine fabrics occur in the Eocene Green River Formation of Wyoming, Colorado, and Utah (Bradley, 1929b; Bradley and Eugster 1969; Surdam and Wray, 1976; Smoot, 1977, 1978), in the Upper Miocene Reis Crater deposits of southern Germany (Riding, 1979), and in the Precambrian Murky Formation (1,300-1,865 my BP) of the Athapuscow aulacogen in the N. W. T, Canada (Hoffman, 1976b). Perhaps some of the stromatolites reported from other ancient lacustrine deposits (see Dean and Fouch, 1983) may turn out on closer examination to be tufas or travertines. Reports of ancient spring and river tufas and travertines are rare (McGannon, 1975; Hay and others, 1986; Steinen and others, 1987).

Modern tufas and travertines, environmentally associated with marine carbonate tidal flat deposits, occur on Andros Island, the Everglades, and the Persian Gulf, as reported above. Recognition of ancient analogs of these deposits would be of considerable significance because of the special environmental conditions they record. Reinhardt and Hardie (1976) described a marine travertine horizon with encapsulated chemical pisoids from the Upper Cambrian Conococheague Group of western Maryland. Where this travertine is domal, it served as a template for the accretion of microbial stromatolites built by sediment trapping (Fig. 122; see also Reinhardt and Hardie, 1976, Fig. 7). Demicco (1981, 1985) correlated this singular travertine layer across a number of sections of the Conococheague Group in the central Appalachians. He interprets this travertine as a splash zone exposure crust analogous to that of the modern Persian Gulf "coniatolites." Hardie (1977b, p. 176-177) noted two possible ancient analogs of the Andros tufas, one in the Middle Ordovician St. Paul Group of western Maryland and the other in the Precambrian Pethei Formation of Canada. Mitchell (1981, 1985; see also Demicco and Mitchell, 1982), in a detailed study of the sedimentology of the St. Paul Group in the central Appalachians, has interpreted a stratigraphically restricted subfacies characterized by laterally-linked, hemispherical, stromatolites (see Logan and others, 1964) that occurs at the top of a major 3rd order regressive tongue (see Hardie and Shinn, 1986, Fig. 65) as deposited in coastal marshes and lakes analogous to those of modern day Andros Island. Hoffman (1975, 1976b) has interpreted digitate stromatolites with silicified filament molds in a subfacies that caps shallowing-upward platform cycles in the Precambrian Rocknest Formation (1,865-2,200 my BP) of the Wopmay orogen of Canada as tufas deposited in an Andros-type coastal marsh environment. Grotzinger (1985, 1986a) has an extended treatment of these Rocknest cycles and their tufas. In other Proterozoic carbonates of Canada, such as the Pethei Formation of Great Slave Lake (Hoffman, 1975), tufas with notably crystalline travertine fabrics indicative of chemical precipitation (Fig. 123) are found as part of shallowing upward cycles. The famous Upper Triassic Lofer facies of the Dachstein Limestone of the Northern Alps (Fischer, 1964, 1975) have shallowing upward tidal flat cycles (Goldhammer and others, 1990) analogous to the modern Andros Holocene vertical succession of burrowed subtidal lime muds capped by laminites (see Hardie and Ginsburg, 1977, Fig. 67). The comparison is made more complete by the presence of dolomite-rimmed filament molds (10-15 microns across), as seen under TEM, in the Lofer crinkled cryptmicrobial laminites (Fischer, 1975, Fig. 27-2B). These sheathed molds record the formation of microbial tufa around filaments of the same size as the filaments of *Scytonema* of the modern Andros tidal flats and marsh. Also, the very selective dolomitization of the Lofer filament sheaths suggests that the precursor mineral might have been magnesian calcite comparable to that of the Andros marsh *Scytonema* tufas.

Speleothems by their very nature as cave deposits have a low potential for preservability in that caves are formed under conditions of net loss of carbonate by dissolution. However, it is possible that some remnants of speleothems will be preserved along old karst surfaces, where their presence will be crucial to the identification of the surface as one of karstic dissolution. Esteban and Klappa (1983, Figs. 27, 28) show two outcrop photos of Cretaceous age speleothems filling paleo-karst cavities.

Significance.-

Biostromal tufas, particularly microbial tufas, associated with supratidal subfacies in ancient shallow marine carbonate deposits, would immediately suggest a coastal freshwater marsh or lake setting, if the few existing modern occurrences are typical of tufa environments throughout geologic time. Interpreted this way, such supratidal tufas would be important signals of a wet tropical climate (see Hardie, 1977a, p. 178-183). Thus, in the spectrum of climatically influenced marine carbonate deposits, supratidal tufa caps to ancient shallowing upward tidal flat deposits could be thought of as the rainy climate counterparts to the arid-climate, evaporite caps of sabkha cycles (Hardie and Shinn, 1986, Fig. 71).

Modern subaqueous tufa and travertine mounds are found only in lakes, where they have built edifices that rival the size of the huge Precambrian stromatolite mounds (Figs. 116, 120). And, in fact, it would seem worthwhile to examine the possibility that some of these Precambrian mounds might have been sub-lacustrine tufas or travertines rather than the marine peritidal

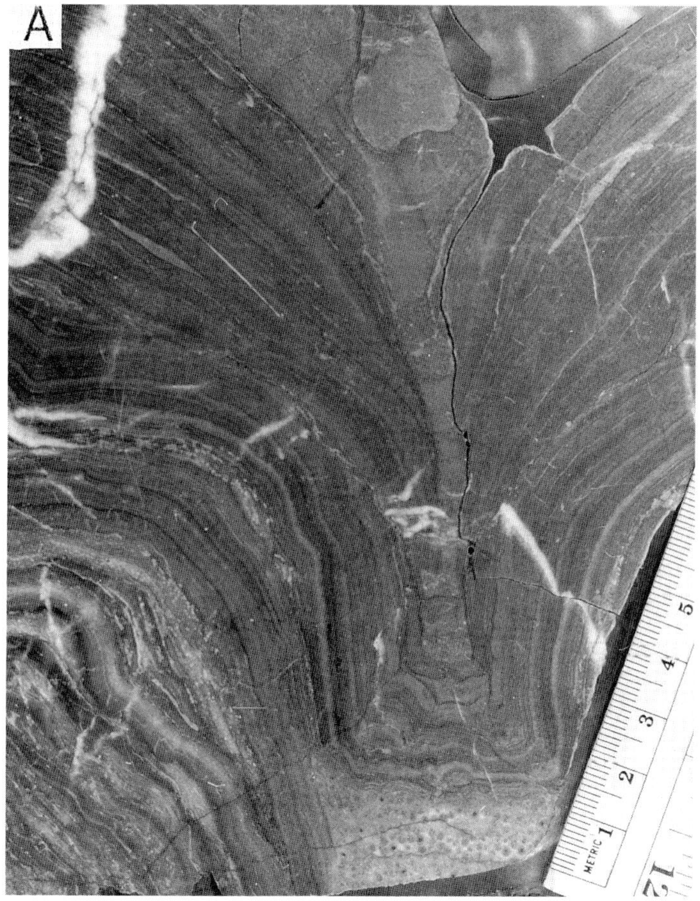

Fig. 122.-Banded travertine encapsulating ooids and pisoids developed in shelf carbonates of the Upper Cambrian Conococheague Limestone (see text for details). Note the fine, isopachous laminae in the center of (A) and in the lower portions of (B) composed of finely crystalline calcite. These laminac have morphological characteristics typical of modern chemically precipitated travertine, as, for example, the notable "kink-fold" geometry at sharp bends. The laminae in the upper half of the slab in (A) that pinch out down the sides of these stromatolitic heads are made of peloidal sediment that must have been trapped and bound by microbial mats (see text for details; see also Reinhardt and Hardie, 1976). Scale in (A) in mm, width of (B) is approximately 12 cm. Sample B courtesy of Paul Hoffman.

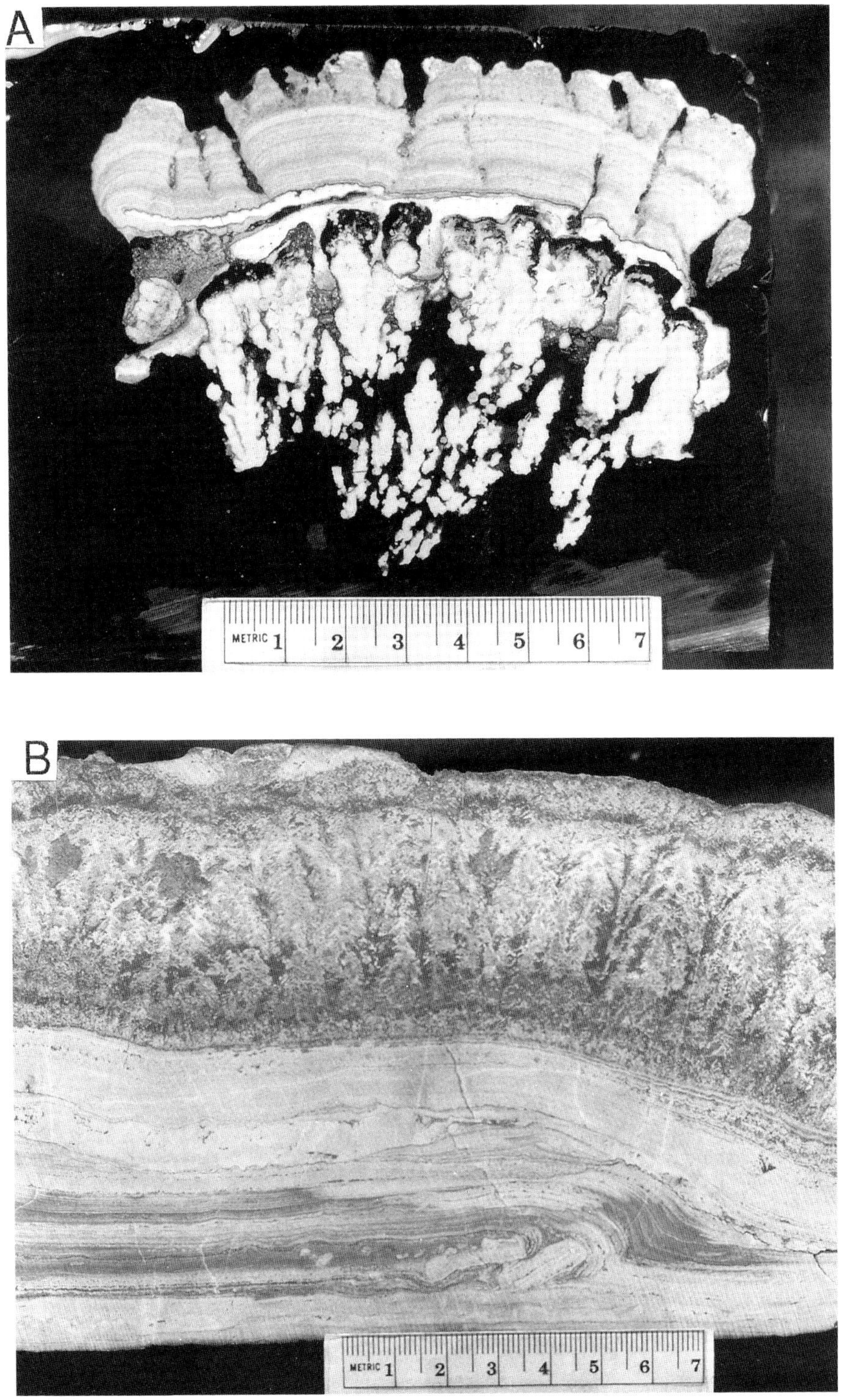

Fig. 123.-(A) Cross-sectional slab of epoxy-impregnated dendritic tufa overlain by wavy laminite (chemical stromatolite, i.e. travertine) of Late Pleistocene to Recent age from Pyramid Lake, Nevada. Scale in cm and mm. (B) Strikingly similar couplet, although reversed, from the Lower Proterozoic Pethei Group carbonates of the Northwest Territories, Canada. In this sample, wavy and crinkled laminites (chemical or microbial stromatolite?) are overlain by dendritic tufa. Scale in cm and mm.

"algal" stromatolites they are conventionally assumed to be by almost every student of stromatolites (see papers in Walter, 1976a, and the review of Precambrian carbonates by Grotzinger, 1989). Certainly the common fabrics of late Precambrian stromatolites such as the film, tussock, vermiform, and micrite microstructures described and illustrated by Bertrand-Sarfati (1976) would be perfectly consistent with a lacustrine tufa or travertine origin. The possibility is made all the more reasonable in light of the obvious difficulty of differentiating marine from lacustrine deposits in Precambrian rocks due to the absence of marine fossils.

Most modern filamentous cyanobacterial tufas, such as the calcified *Scytonema* of Andros Island and the Everglades, look remarkably like the problematic filamentous fossils known as *Girvanella*, *Ottonosia*, and *Samphospongia* (see descriptions in Johnson, 1961), a similarity pointed out by Hardie (1977b, p. 177). It would be worthwhile to consider the possibility that some forms of *Girvanella* and the like record the deposition of freshwater tufa.

Filamentous cyanobacterial tufas should not be confused with uncalcified filament molds so common in supratidal and intertidal carbonate sediments and rocks. The two features carry quite different messages.

Singular but widespread thin crusts of travertine within shallow marine deposits point to splash zone conditions or rocky shore pools in an arid climate. Very localized deposits of travertines with encapsulated pisoids, perhaps with associated tufa layers, are most likely to be records of spring deposits.

Traces of speleothems within a *terra rossa* soil separating shallow marine units are powerful evidence of a substantial relative sea-level fall and exposure of a carbonate platform top to karstic dissolution by a meteoric groundwater lens, as has occurred a number of times during the Pleistocene Epoch.

Caliche and Associated Soil Features

Terminology.-

Caliches, also known as calcrete, kankar, nari, and duricrust, among other terms, are carbonate deposits precipitated within or on a host sediment or sedimentary rock in the vadose zone principally under semi-arid or arid climatic conditions. Caliches range from isolated nodules of white chalky microcrystalline calcite that grew displacively just below the subaerially exposed land surface, to complex features many meters thick. The complex, meter-thick caliches are generally organized into well defined vertical successions of carbonate accumulation zones (Fig. 124), as follows: (1) a basal nodular chalky zone, which passes up into (2) a coalesced nodule zone that may have a platy semi-indurated structure, which is capped by (3) a hard laminated crust ("hardpan"), which in turn is covered by (4) a red-brown soil carrying pisoids (coated grains) (see Esteban and Klappa, 1983). Our initial understanding of the properties and genesis of caliches was based mainly on studies of Pleistocene examples in the southwestern U. S. (e.g., Bretz and Horberg, 1949; Brown, 1956; Swineford and others, 1958; Gile and others, 1966; Chafetz and Butler, 1980), but in the last decade or two Quaternary caliches have been examined in many other parts of the world (Goudie, 1973; Reeves, 1977; Watts, 1977, 1980; Esteban and Klappa, 1983; Wright, 1989). There is today a consensus that caliches are semi-arid to arid soil zone features and, in the terminology of soil scientists, they belong to the pedocal group in which the chalky calcium carbonate accumulation in the B-zone is designated the Cca-horizon and the indurated caliche crust is known as the K-horizon.

Fig. 124.-(A) Pleistocene caliche profile near Amarillo, west Texas. Hammer in upper left gives scale. Note the isolated nodules below, the coalesced nodules in the center, and the laminated hardpan at the top of the profile (at the level of the hammer). (B) Detail of branching tubular nodules from the lower isolated nodule zone; these may represent calcitic casts of plant roots.

Distinguishing characteristics.-

Of particular interest here as criteria for recognizing ancient caliches in carbonate deposits are the features of those caliches formed in and on limestone bedrock underlying modern carbonate platforms subaerially exposed by the succession of glacio-eustatic sea-level drops during the Pleistocene period (e.g., James, 1972; Read, 1974, 1976; Multer and Hoffmeister, 1975; Harrison, 1977; Perkins, 1977, Coniglio and Harrison, 1983). In the relatively thin (centimeter- to decimeter-thick) but well indurated caliche crusts that preserve and record the development of pedocal soils on exposed Pleistocene bedrock, the following features are noteworthy: (1) color-banded micrite laminae (millimeters to sub-millimeter in thickness) (Fig. 125); (2) stromatolites (Fig. 125), (3) micritic peloids, coated grains, and pisoids (collectively called glaebules by some workers, see Esteban and Klappa, 1983) (Fig. 126); (4) micritic matrix in which the glaebules are suspended; (5) irregular fenestrae and vugs, commonly lined with gravitational and meniscus cements and clay cutans that indicate vadose conditions; (6) tubular fenestrae which range from micron sized molds of filamentous microbes such as cyanobacteria and fungi, to branching networks of slightly larger tubes that are the traces of plant hair-roots, to vertically disposed anastomosing tubes of millimeter and centimeter diameter that are the molds of the roots of large plants; (7) skeletal remains of land snails and the calcified cases of insect pupae; (8) sub-millimeter petal-shaped clusters of calcite microspar known as Microcodium of uncertain origin but considered by Klappa (1978) to be a feature produced biogenically by soil fungi; (9) antiformal "tepee" structures (see Watts, 1977, and the discussion of tepees elsewhere in this atlas); and (10) fractures filled by micrite and microspar cements. The fractures filled by micrite and microspar cements may cut vertically across the caliche crusts as veins, or occur as horizontal sheet cracks, or they may be circular cracks within glaebules that follow the curvature of the grain surfaces making what have been termed "circumgranular cracks" (see Swineford and others, 1958). A final, and most important feature of soils developed on exposed Pleistocene bedrock, are *rhizoliths*. Rhizoliths are carbonate casts or linings of plant root-holes, the carbonate sheaths or clusters of concretions precipitated around roots as a result of the biochemical activities of the living plants or the postmortem decay of their roots, and the replacements of the organic matter of the roots that preserve the original cellular structure (Esteban and Klappa, 1983, Figs. 88, 89) (Fig. 127). Esteban (1974) coined the term *alveolar texture* for a network of millimeter to sub-millimeter sized walls made of micrite that appears to be coalesced rhizoliths. According to Esteban and Klappa (1983, p. 28) the boxworks of tangential needle fibers of calcite in caliches described by James (1972) are a form of alveolar texture. These caliche features are briefly described but well illustrated with color photographs in the review of exposure features in carbonate rocks by Esteban and Klappa (1983).

In the past, a number of laminated caliche crusts and their associated glaebules in ancient limestones have been mistakenly identified as peritidal stromatolites and oncolites (Wright, 1989). However, with the increasing accumulation in recent years of information on the characteristics of Quaternary caliches, such misidentification is much less likely.

Origin.-

Caliches are formed by the precipitation of calcite (or in some cases, magnesian calcite or calcian dolomite) in the vadose zone from films of pore water drawn up to the surface by a combination of capillary draw and evaporative pumping. In semi-arid or arid regions where the perennial groundwater table is deep, the source of the pore water is rainwater, either as direct rainfall or as storm sheetwash from nearby highlands, that percolates under gravity down into the aerated vadose zone. This water is then sucked back toward the surface by evaporative pumping after the rain shower or sheet flooding ceases. The small amounts of calcium and other metal ions in rainwater, which is in essence a carbonic acid solution of pH about 5.7, are increased significantly by rapid interaction (minutes to hours) with the soil, sediment, or bedrock through which the acidic rainwater percolates (see Garrels and Mackenzie, 1971, for a classic paper on acquisition of solutes by interaction of acidic rainwaters with rocks during chemical weathering; see also Eugster and Hardie, 1978). Rainwash from uplands will already have more or less undergone this chemical transformation by contact with the ground surface material and the wash load detritus transported by the sheet flow as bedload and in suspension. Under the driving force of evaporative pumping (Hsu and Siegenthaler, 1969), these meteoric waters are transported back up through the aerated vadose zone toward the surface as meniscus films that line the connected pores and undergo both carbon dioxide degassing and evaporation in the subsurface by exchange with the vapor phase (air) in the pores. This leads to supersaturation and eventually precipitation of calcite within the pore spaces of the host soil, sediment, or rock in the vadose zone. Unless this host is rich in ferromagnesian minerals such as olivines and pyroxenes, the calcium/magnesium ratio of these bicarbonate pore waters will be equal to or less than 1.0, and the first carbonate phase to precipitate out will be low magnesian calcite (see Hardie and Eugster, 1970; Füchtbauer and Hardie, 1976, 1980). If sufficient evaporative concentration of the pore-water films takes place as the calcite continues to precipitate, then the Mg/Ca ratio of the residual pore waters will rise and lead to the formation of magnesian calcites (Füchtbauer and Hardie, 1976, 1980; Eugster and Hardie, 1978; see also discussion in Watts, 1980, of the origin of the magnesian calcites and dolomite of the Kalahari Quaternary caliches). Rapid nucleation and crystal growth of calcite from thin films of meteoric water favors the development of micron-sized crystals, which may explain why most caliches consist of micrite and microspar. Development of thick caliche profiles (e.g., meter scale; note that the thickest Plio-Pleistocene caliches known are in the Kalahari and reach over 50 m, see Watts, 1980) under these conditions almost certainly would require long periods of time measured perhaps in tens to hundreds of thousands of years (see Goudie, 1973). However, where a perennial shallow groundwater reservoir exists under semi-arid climatic conditions, substantial caliche crusts should be able to develop at rates considerably faster than caliches formed entirely from surface infiltration. This is because a continuously available and, in essence, unlimited source of solutes can be easily drawn up through the vadose zone to the surface by evaporative pumping. Under such conditions, expansive cementation (Watts, 1978) flourishes and leads to extensive lithification of the host soil or sediment to make indurated crusts, and eventually to intracrustal growth of glaebules, growth of micritic matrix, brittle fracture, polygonal disruption, and buckling of the crust into tepees (see Watts, 1977; see also this atlas under Tepee Structures).

In hydrologically closed or restricted basins under arid climates where the perennial groundwater table is very close to the surface (within meters), more soluble minerals such as gypsum, halite, or trona will precipitate in the vadose zone instead of alkaline earth carbonate caliches (see this atlas under evaporite features).

The extent to which biochemical activities of biota are responsible for promoting the precipitation of the alkaline earth carbonates of caliches is uncertain. Where rhizolith concre-

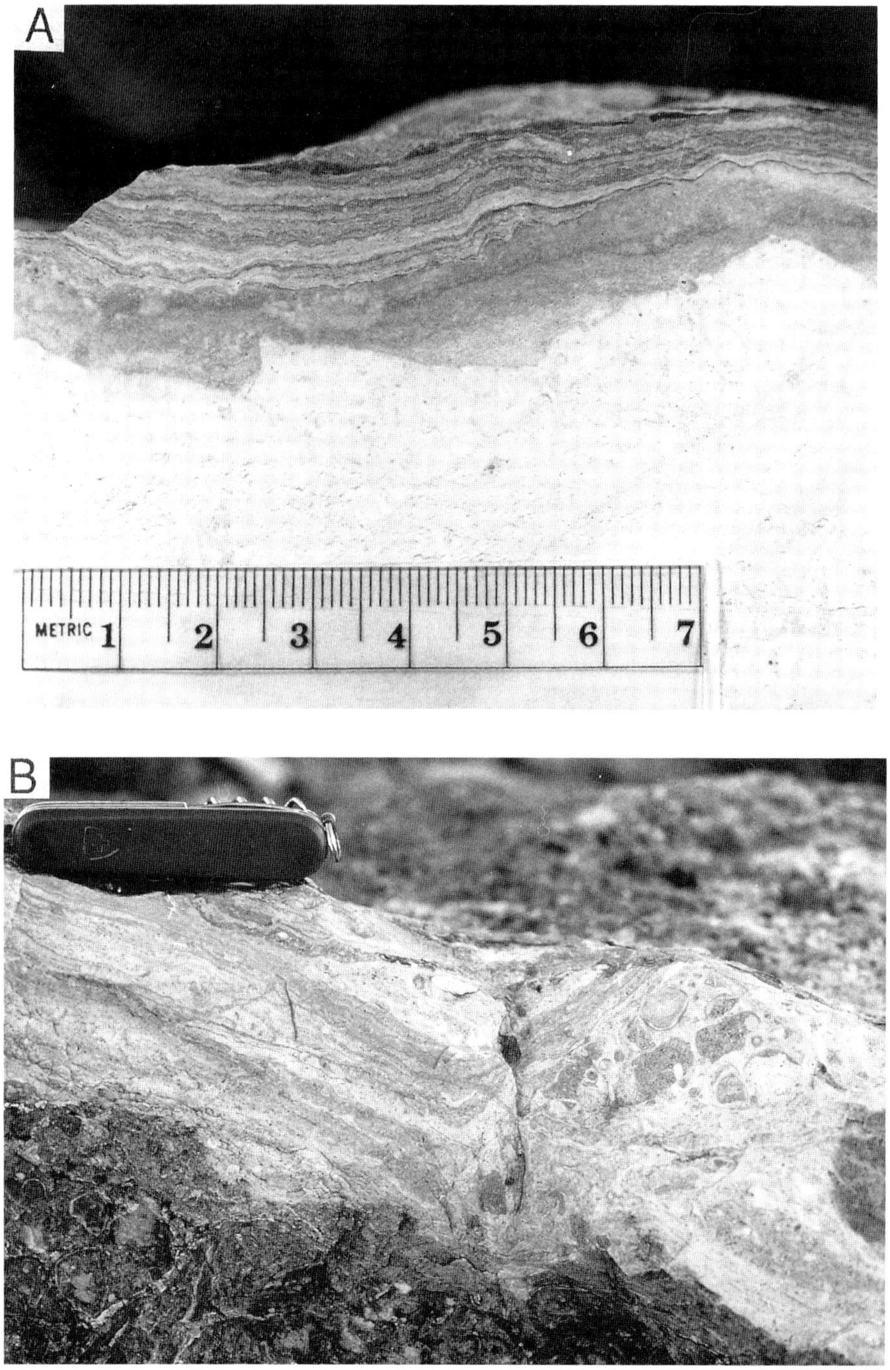

Fig. 125.-Laminated caliche crusts developed on top of Pleistocene carbonates. (A) is from the Florida Keys (scale in mm) and (B) is from the Carbla Oolite, Carbla Point, Shark Bay, Western Australia (pocket knife is 80 mm long).

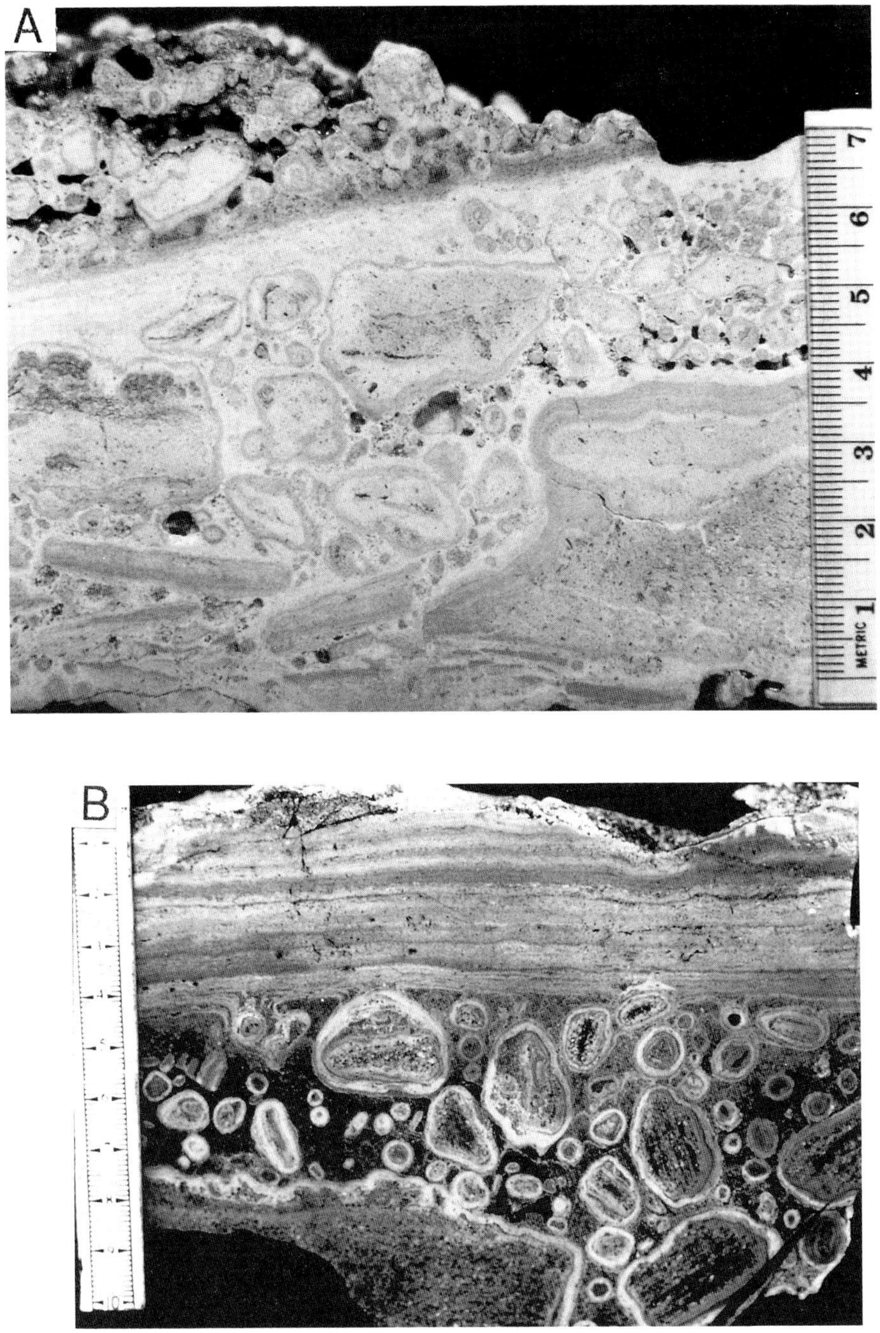

Fig. 126.-Pisolitic caliche crusts developed on top of Pleistocene carbonates. (A) Slab of coated grain zone of a caliche crust from near the Texas-New Mexico border (Route 176 between Andrews, TX and Eunice, NM). Scale in mm. Note the meniscus cements between grains in the upper portion of the photograph. Also note the pendant laminae beneath coated grains in the lower portion of the photograph. (B) Cross section of pisolitic lense in laminated caliche crust from the Carbla Oolite, Shark Bay, Western Australia. Scale in cm and mm. Slide courtesy of Fred Read.

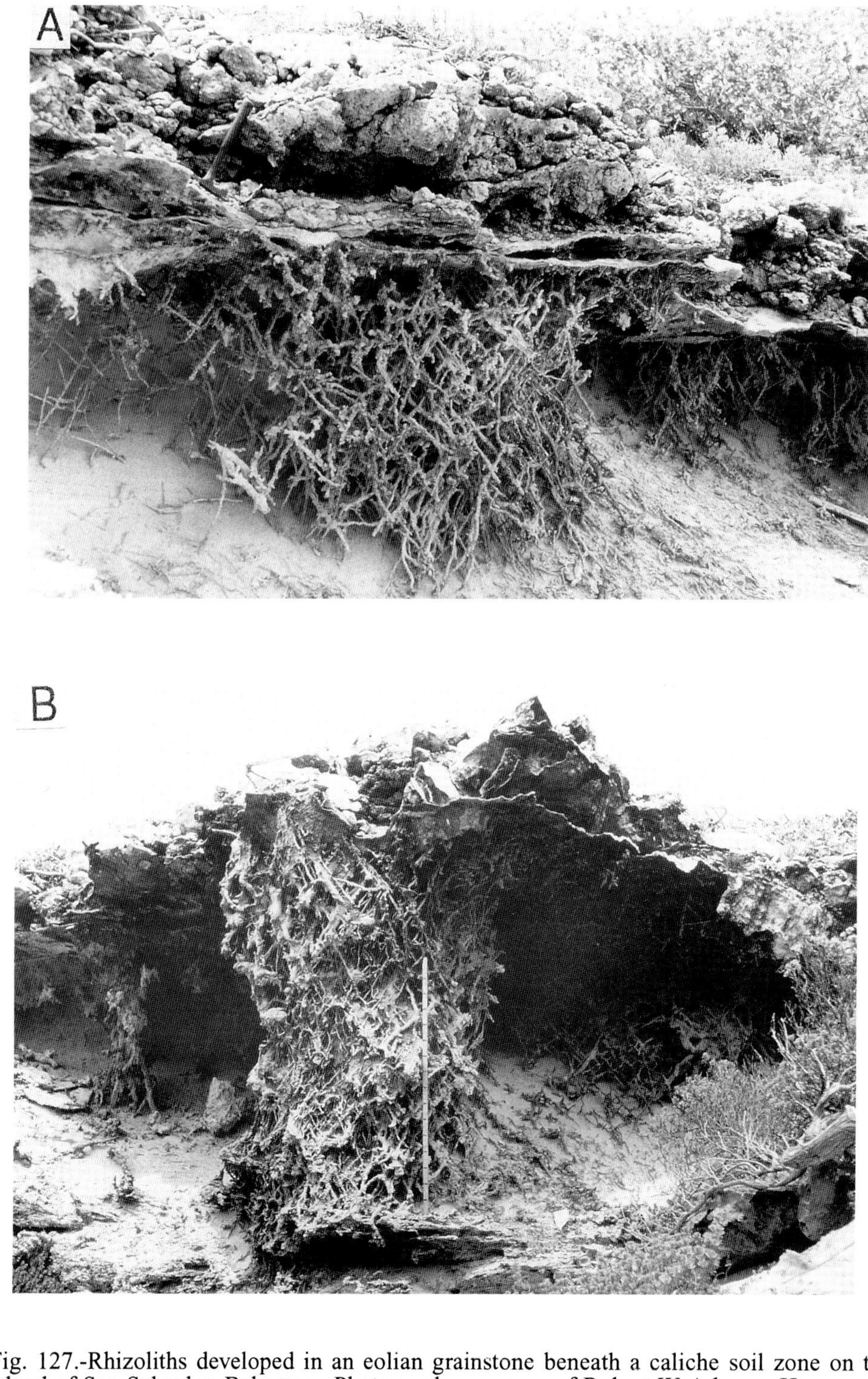

Fig. 127.-Rhizoliths developed in an eolian grainstone beneath a caliche soil zone on the island of San Salvador, Bahamas. Photographs courtesy of Robert W. Adams. Hammer in (A) provides scale. Staff in (B) is 1.5 m long.

tions around living roots or decaying roots form part of a caliche it is likely that calcification was mediated in some way by organic processes of the roots themselves or by the bacterial decomposition of the dead root material. Other rhizoliths may simply be inorganically precipitated fillings (casts) or linings of root molds. Klappa (1978) has called on the interaction between roots and fungi to explain the origin of the problematic calcite micro-crystal clusters in caliches known as Microcodium, and he has identified "lichen stromatolites" formed by calcification of laminated lichen "mats" that encrust caliche hardpan surfaces (Klappa, 1979).

In wet tropical climates, caliche-like soils can form at the subaerially exposed surface of limestone bedrock undergoing subsurface, karstic dissolution by flow of acidic meteoric groundwaters through fracture systems. Such *terra rossa* soils, red with the oxidized clayey residues of weathered iron-bearing detrital components of the limestone, may carry some of the features found in semi-arid and arid climate caliches, particularly laminated crusts, coated grains and pisoids, and the micritic matrix in which these accretionary grains are imbedded. The well known laminated crusts and the associated red-brown pisolitic breccias that fill solution pockets in the exposed Pleistocene coral reefs of the Key Largo Limestone of the Florida Keys (Multer and Hoffmeister, 1968) are good examples of such caliche-like features in karstic soils (Fig. 125).

Occurrence and preservation.-
Relict caliches are best known from Plio-Pleistocene deposits, both siliciclastic and carbonate, and most of our criteria for identifying caliche are based on these examples (see the comprehensive compilation of publications given by Esteban and Klappa, 1983, p. 49-51). Caliches analogous to those of the Pleistocene Epoch have been reported from siliciclastic and carbonate deposits throughout the geologic record, for example, in the Siluro-Devonian Old Red Sandstone of England (Allen, 1974), the Carboniferous of Kentucky (Walls and others, 1975; Harrison and Steinen, 1978), the Triassic of the Dolomites, Italy (Goldhammer and others, 1987), and the Triassic of Connecticut (Hubert, 1978).

Well lithified caliches that cap shallowing upward cycles in semi-arid climates have a good chance of being preserved, although, as is true for any subaerially deposited sediment, removal by erosion before deposition of the next sedimentary unit always remains a distinct possibility.

Significance.-
A caliche deposit in an ancient sedimentary succession is of particular significance because caliches, first and foremost, are indicators of subaerial exposure without detrital sedimentation for a significant period of geologic time. A caliche implies the existence of a subaerial "omission surface" that results from the complete shut down of sedimentation in what once was an active site of deposition. This shut down may be due to abandonment of a terrestrial subenvironment by river avulsion or abandonment of a supratidal marine subenvironment by shoreline progradation (see Hardie and Shinn, 1986, p. 45). The shut down may have been caused by a relative sea-level fall that, in turn, was driven by either eustasy or tectonic movement. Caliches are crucial stratigraphic markers of relative sea-level change. As a consequence, considerable care must be devoted to the interpretation of their origin. In cases where marine subtidal deposits are directly capped by caliches, such as is found with a number of Pleistocene carbonates (James, 1972; Multer and Hoffmeister, 1968) as well as certain Triassic carbonates of northern Italy (Fig. 128; Hardie and others, 1986; Goldhammer and others, 1987, 1990), subaerial exposure recorded by the caliche caps could not have been due to abandonment of supratidal flats by shoreline progradation but, instead, must have been the result of a relative fall in sea level. Where caliches cap marine supratidal deposits, the cause of the subaerial exposure could have been either abandonment by progradation without a sea-level change or progradation accompanied by a relative sea-level fall. In such cases, it may not be possible without additional clues to reach an unequivocal conclusion about the stratigraphic significance of the caliche.

The time required to develop a caliche deposit, thick or thin, mature or immature, is not well established because our models come from Plio-Pleistocene deposits that are too old to be directly dated by C-14 methods. However, we do have some approximate measures of their rate of formation by extrapolating back from the age of the host to the present. Such estimates from Pleistocene caliches on Barbados, formed on marine limestones deposited during the 83,000 and 104,000 yr sea-level high stands, indicate rates of net accumulation of about 0.006 m to 0.03 m/1000 yrs (from data in James, 1972, p. 820). This range of rates is supported by C-14 dates made by Robbin and Stipp (1974) on thin Holocene laminated crusts on the Florida Keys that indicate crust accretion of about 0.22 m in the last 6000 years. At such rates, caliches only tens of centimeters thick signal exposure of carbonate platforms lasting tens of thousands of years. Higher rates may certainly apply in some cases, but perhaps centimeters per millennium is the right order of magnitude. In support of relatively slow rates is the evidence from a Holocene example found by one of us (L. A. H.) in archeological excavations made in overbank sediments on the Nile Delta at Tell El-Retaba on the banks of Wadi Tumilat, Egypt. Here, scattered calcite nodules millimeters to a centimeter or so across have grown displacively in the mud bricks of buried Late Bronze Age buildings (occupied by Moses' Israelites about 1470 BC) that stand only a meter or two above the water table in this arid climate setting (Fig. 129). Clearly, at this rate of growth a mature caliche profile would require many tens of thousands of years to develop, even under the shallow water table setting and the arid climate of the Nile delta.

Supratidal Crusts

Cemented surface crusts are a striking feature of the modern tropical carbonate tidal flats of the Bahamas (Shinn and others, 1965, 1969; Hardie, 1977b; Hardie and Shinn, 1986), Florida (Shinn, 1968c; Carballo and others, 1987), Belize (Mazzullo and others, 1987), Caicos Island (Shinn, pers. commun.), Qatar in the Persian Gulf (Taylor and Illing, 1969; Shinn, 1969), Shark Bay, Western Australia (Logan, 1974), and Bonaire, Netherland Antilles (Deffeyes and others, 1965; Lucia, 1968). They are forming contemporaneously by cementation of the upper few centimeters of the surface sediments on subaerially exposed supratidal flats. The distinctive feature of these crusts is that cementation is the result of precipitation of aragonite, magnesian calcite, and calcian dolomite from seawater pore fluids in the upper vadose zone of the supratidal flats. These supratidal crusts attracted considerable attention when they were first discovered in the 1960's, because they marked one of the first known occurrences of Holocene dolomite formed in a marine environment. Since that time, these dolomitic crusts have been widely used, correctly or incorrectly, as models for the interpretation of the origin of dolomite in many ancient shallow marine deposits.

Distinguishing characteristics.-
The Holocene supratidal crusts are thin (less than 10 cm thick), firm to rock-hard sheets of cemented peloidal muds, as

Fig. 128.-Ancient caliches developed as exposure caps on subtidal shelf carbonates, Middle Triassic Latemar buildup, the Dolomites of northern Italy (see Goldhammer and others, 1987, 1990, and below under Important Associations). (A) and (B) Outcrop views of stacks of flat-lying cyclic platform carbonates of the Upper Cyclic Facies (A) and Lower Cyclic Facies (B) of the Latemar buildup. In both outcrops, the layering is defined by alterations of bioturbated limestone beds (gray protruding bands) and thin caliche caps (dark recessive ledges). Each alternating couplet of subtidal limestone and subaerial caliche cap constitutes a single depositional cycle. Scale bar in (A) is 2 m, but in (B) is 3 m high. (C) (Following page) Thin-section photomicrograph of a Latemar caliche cap showing: (1) vadose pendant cements; and (2) non-fabric selective, solution-enlarged pores with allochems truncated at pore margins. Scale bar is 3 mm long. (D) (Following page) Thin-section photomicrograph of a Latemar caliche cap showing composite pisoids (most of which have "pendant" coatings) and solution enlarged pores marked by truncated grain boundaries and floating relicts (designated by 2). Scale bar is 3 mm long. (E) (pg. 166) Thin-section photomicrograph of a Latemar caliche cap, which is a fenestral packstone consisting of pisoids, composite coated grains, and peloids. Note the poor sorting and the patchy distribution of interstitial micrite (a vadose microcement). Scale bar is 2 mm long. (F) (pg. 166) Thin-section photomicrograph of Latemar caliche cap characterized by fenestrae, peloids, micritized grains, coated grains and abundant interstitial micrite cement. Scale bar is 2 mm long. (G) (pg. 167) Thin-section photomicrograph of a Latemar caliche cap consisting mainly of composite pisoids containing altered skeletal and non-skeletal nuclei surrounded by concentric microspar laminae (note the coalescence of the outermost laminae). Scale bar is 2 mm long. (H) (pg. 167) Thin-section photomicrograph of a Latemar caliche cap showing bedding-parallel sheet cracks in peloidal wackestone matrix and large, vertical, solution-enlarged fenestrae. Scale bar is 2 mm long. All photographs courtesy of Robert K. Goldhammer.

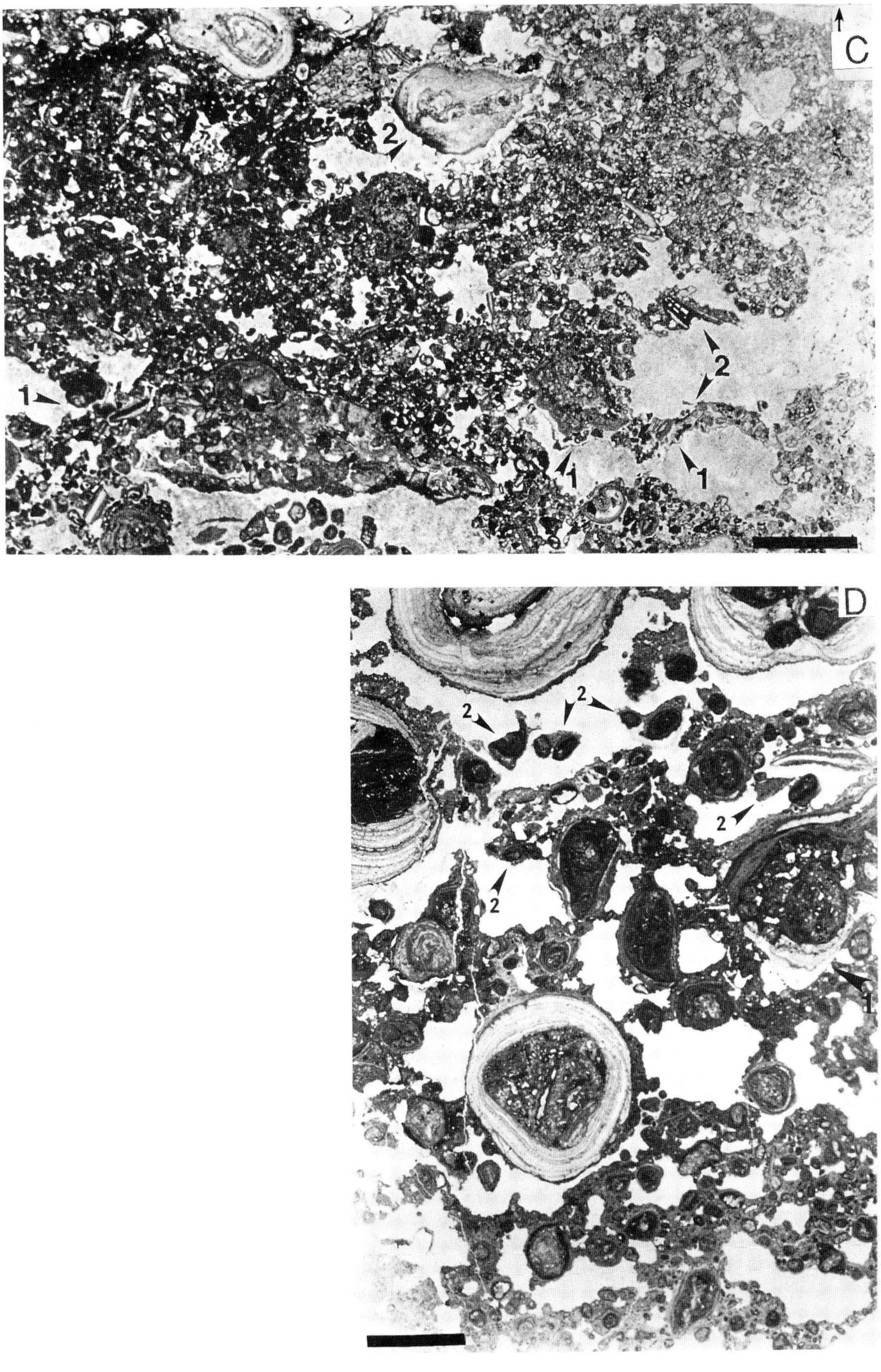
C
2
2
1
1
1
D
2
2
2
2
2
1

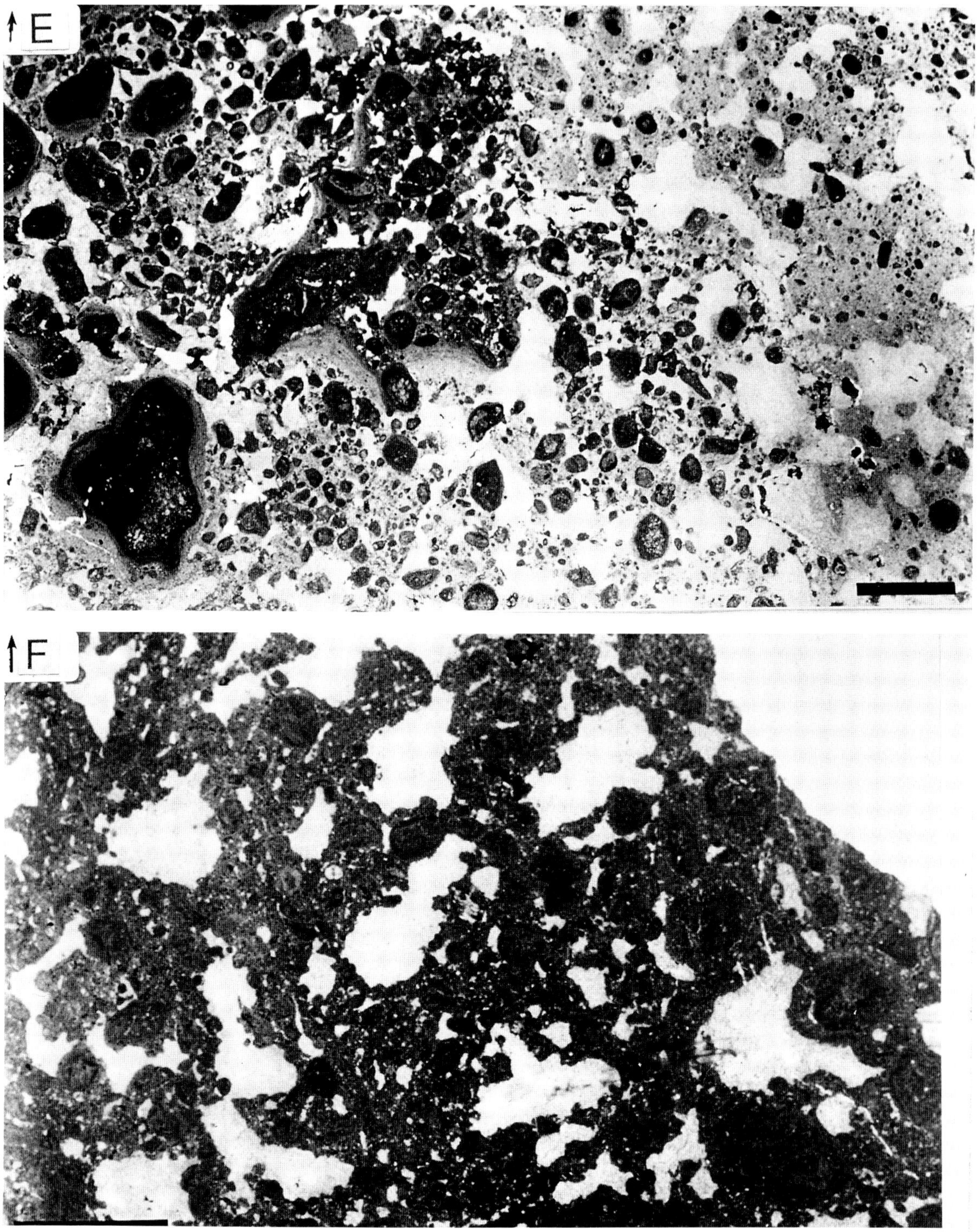
E
F

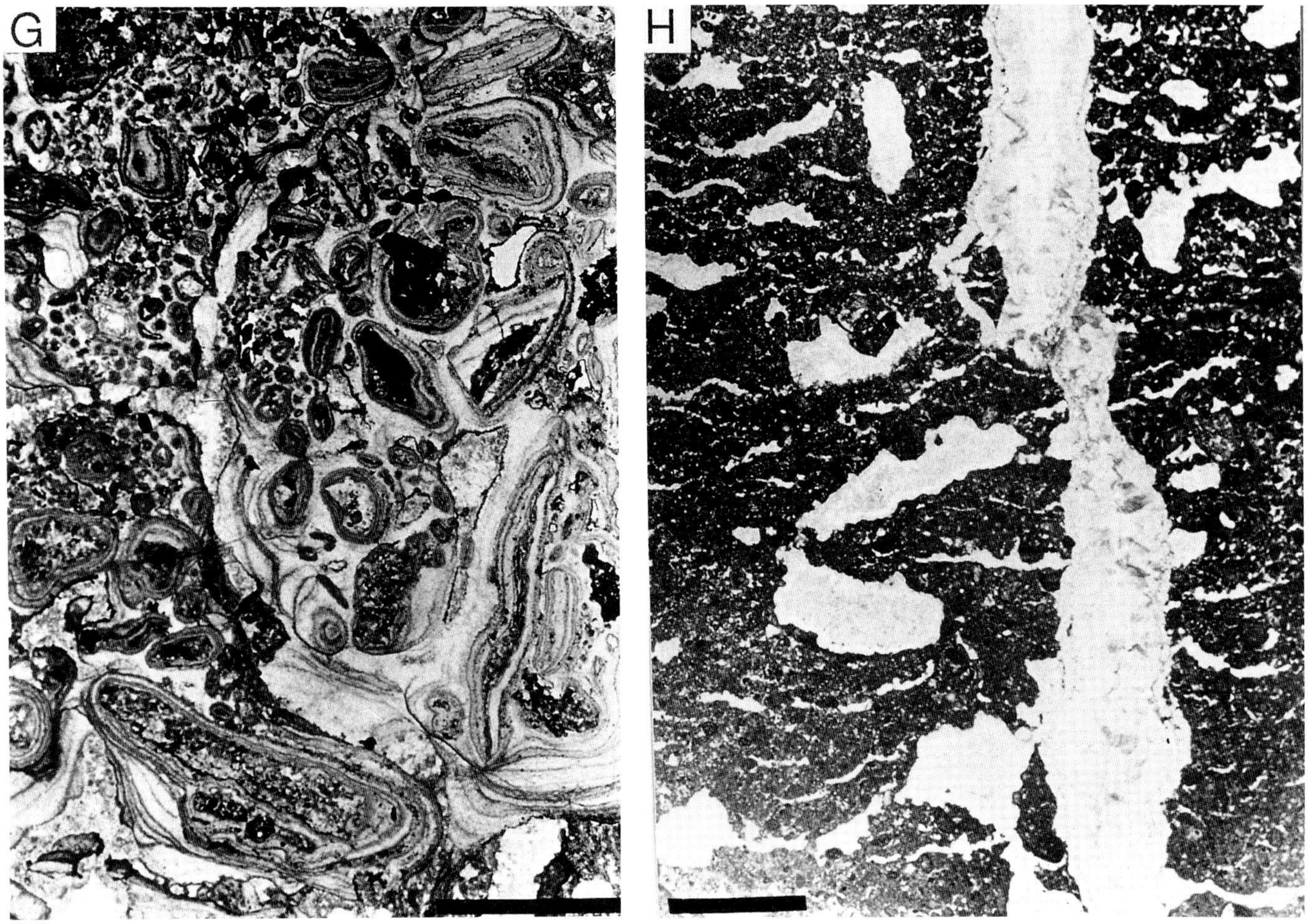

Fig. 129.-Holocene caliche found in archeological excavations in overbank sediments on the Nile Delta at Tell El-Retaba, on the banks of Wadi Tumilat, Egypt. Here, scattered calcite nodules millimeters to a centimeter or so across have grown displacively in the mud bricks of buried Late Bronze Age buildings (occupied by Moses' Israelites about 1470 BC) that stand only a meter or two above the water table in this arid climate setting. Pencil in upper right provides scale.

in the Bahamas and Florida, or cemented skeletal sands, as in Qatar and Shark Bay. The Andros Island crusts have characteristics common to many of the Holocene occurrences, and so we will use them as an example (Fig. 130).

In the channelled belt of the Andros Island tidal flats, the cemented surface crusts are restricted to the backslopes of channel levees and backslopes of the beach ridges along the shoreline that separates the tidal flats from the inner shelf-lagoon of the Great Bahama Bank. These surface crusts occur above the mean high water mark (exposure index 87-100; see Ginsburg and others, 1977, p. 7-11) within the "high *Scytonema* marsh" that rims the tidal ponds (Hardie, 1977b). They cover the surface as a dark, *Scytonema*-encrusted hardpan (Fig. 131) wherever there has been no measurable deposition of sediment since the last major hurricane to inundate the tidal flats (Hurricane Betsy of 1965). In places they are gently buckled into low antiformal ridges that trace out a rough polygonal pattern (tens of centimeters to about a meter across; Fig. 132). Aggressive growth of small mangroves on the levee backslopes has broken some of the crusts into loose plates that litter the surface (Fig. 133). These surface crusts pass beneath the levee crest sediments as continuous buried layers which can be easily traced by trenching and probing. Some of these buried crusts crop out on the tidal channel banks or along the open beach where they have been exhumed by erosional migration of the channels and the shoreline. The eroded fragments of these exhumed supratidal crusts are now accumulating as lithoclast flat pebble gravels (Hardie and Ginsburg, 1977).

In cross-sections of sawn slabs, the crusts are seen to range from massive burrowed peloidal mudstone to planar, wavy, or crinkled laminites (Fig. 134). These crusts preserve all the primary sedimentary structures, such as microbial lamination, cyanobacterial filament molds, mudcracks, fenestrae, burrows, intraclast lenses, and so on, found in their uncemented counterparts that make up the bulk of the lime sediments of supratidal-levee backslopes and pond shores (see Hardie and Ginsburg, 1977, for an extensive description). In thin section, it is clear that the crusts are simply the typical peloidal sediments deposited on the levees by overbank flooding during onshore storms (Hardie and Ginsburg, 1977) but which have been partly lithified by finely-crystalline intergranular cements (Fig. 134C). The degree of lithification is greatest near the surface of the crusts and progressively decreases downwards until the bottom of the crusts are so friable that they can be disaggregated by rubbing with the fingers. Vugs up to cm-scale have been produced by dissolutional enlargement of burrows, fenestrae, and sheetcracks as rainwater has percolated downward during the rainy summer season characteristic of the climate of the Bahamas (Fig. 135A). These vugs are commonly rimmed or partly rimmed with acicular and bladed aragonite microspar cements (Fig. 135B) that display floor-lining morphology (Fig. 135A) and meniscus and pendant fabrics typical of vadose zone cementation. The most finely crystalline of these vadose cements consist principally of microcrystalline high magnesian calcite (16-20 mole% $MgCO_3$) and microcrystalline calcian dolomite (35-43 mole% $MgCO_3$). Excellent SEM photographs of these cements are presented by Lasemi and others (1989). On northwest Andros Island, the calcian dolomite content of the crusts ranges from zero to near 60%.

Origin.-

The localization of the surface crusts to high supratidal areas, the internal sedimentary structures and textures indicative of supratidal deposition, and the meniscus and pendant fabrics of the cements unequivocally establish that these crusts originated by lithification of supratidal sediments at the subaerially exposed surface under vadose conditions. The mineralogy of the cements and the presence of subsurface seawater as a phreatic reservoir only a few tens of centimeters below the surface of the supratidal flats clearly points to seawater as the source of the cements, making this lithification a type of marine cementation process. That the source water for the cements was seawater was confirmed by hundreds of chemical analyses (unpublished data collected by O. P. Bricker from 1968 to 1971) of the pore fluids squeezed from the supratidal sediments at closely separated vertical intervals between the surface and the water table. These analyses also revealed that the salinity of the seawater in the pores of the sediments in the vadose zone increased significantly in the upper 5 to 10 cm to over 3 times normal seawater salinity. Clearly, evaporative pumping was not only operating to draw seawater up through the vadose zone to the surface but also promoted evaporative concentration. All these features, together with the increase in degree of cementation from the bottom to the top of the crusts, indicate that these surface crusts are in essence "marine caliches" in the early stages of development (note that Shinn and others, 1965, established that surface crusts on Andros Island near Williams Island have contemporary ages, less than 160 years BP). Carballo and others (1987) proposed that the supratidal crusts of Sugarloaf Key, Florida, are the result of cementation and dolomitization by seawater driven up through the tidal flat sediments by "tidal pumping" rather than evaporative pumping. This mechanism does not apply to the Andros crusts because: (1) there are crusts forming far back on the tidal flats where the tidal range is near zero; (2) in many pits dug on the tidal flats it was observed that vertical fluctuation in the subsurface seawater body was simply a subdued replica of the tides in the open channels and ponds, and the seawater table never reached the crusts except during abnormal high tides; and (3) the salinity increases observed in the upper vadose zone (see above) must be the result of evaporative pumping; in this connection it is worth noting that the oxygen isotope values measured by Carballo and others (1987) in the Sugarloaf Key crusts are heavier than normal seawater, also indicative of evaporative concentration.

The aragonite and high magnesian calcite cements of supratidal crusts are typical of early marine cements found in other marine environments such as shallow lagoons, sand shoals, and reefs (see below under submarine hardgrounds) except that the Mg content of the magnesian calcite cements of the supratidal crusts tends to be a little higher on the average than that of the submarine hardgrounds. Both aragonite and magnesian calcite are metastable phases that for kinetic reasons nucleate from seawater instead of the stable phase dolomite. Nonetheless, the Mg content of the metastable magnesian calcite cements in many modern marine carbonate sediments appears to be a predictable function of the Mg/Ca mole ratio and the temperature of the seawater from which they precipitated (Fig. 136; see also Füchtbauer and Hardie, 1976, 1980). The average $MgCO_3$ content of shallow submarine cements is about 12-13 mole%, whereas the Andros supratidal crusts have magnesian calcite cements with 12-16 mole% $MgCO_3$, indicating either a higher temperature, higher Mg/Ca ratio, or both for the seawater from which they precipitated.

The calcian dolomites in these modern supratidal crusts have been the subject of much study since they were first discovered in the 1960's (see papers in Pray and Murray, 1965, for example) because of their potential importance in helping to resolve the long-standing dolomite problem (see van Tuyl, 1916, for the most comprehensive review of the problem as it stood at the beginning of this century). These calcian dolomites are poorly ordered, non-stoichiometric phases with $MgCO_3$ contents of 38-43 mole%. Thus they are not crystallographically or thermochemically dolomites in the strictest sense, and so

Fig. 130.-Cemented surface crust (light areas) and patchy growths of *Scytonema* mats (dark areas) developed on the seaward-most edge of the inland supratidal marsh, modern carbonate tidal flats of northwest Andros Island. Stunted mangroves in background are 30-60 cm high and provide scale.

Fig. 131.-Cemented *Scytonema* mounds on the surface of the inland supratidal marsh, modern carbonate tidal flats of northwest Andros Island. Scale is 15 cm long.

for a long time have been called "protodolomites" (Graf and Goldsmith, 1956). The term calcian dolomite seems appropriate for those phases that show some ordering, but those dolomite-like phases with no ordering should be considered magnesian calcites. The nature and significance of non-stoichiometric dolomites (calcian dolomites) have been briefly reviewed by Hardie (1987). The thermodynamic properties of ordered and "disordered" dolomite have been reviewed by Carpenter (1980).

It has been widely assumed that Holocene calcian dolomites are the result of penecontemporaneous replacement of aragonite sediment particles, with the magnesium being supplied by seawater or its concentrated bitterns, as follows:

$$2\ CaCO_3\ \text{(aragonite)} + Mg^{2+}_{(aq)} = CaCO_3{\bullet}MgCO_3\text{(dolomite)} + Ca^{2+}_{(aq)} \quad (1)$$

or by a coupled reaction such as the Haidinger reaction (Lippmann, 1973, p. 150-151):

$$2\ CaCO_3\ \text{(arag)} + MgSO_{4(aq)} + 2H_2O_{(l)} = CaCO_3{\bullet}MgCO_3\ \text{(dol)} + CaSO_4{\bullet}2H_2O\ \text{(gyp)} \quad (2)$$

Initial petrographic evidence put forward in support of massive replacement of aragonite in Holocene carbonates via reactions (1) and (2) was based on thin-section observations at a scale inadequate to resolve the boundaries of the micron-sized calcian dolomite crystals. In this regard, more recent detailed SEM studies of both the Andros Island and Florida crusts by Lasemi and others (1989) demonstrate conclusively that *most, possibly all, of the calcian dolomite of these tidal flats is a cement* and not a replacement of aragonite. However, in the Florida crusts Carballo and others (1987), in addition to reporting abundant calcian dolomite cements, did identify in back-scatter electron images very small patches of calcian dolomite (1-2 microns) that appeared to be partially replacing Mg-calcite peneroplid foram shells (their Fig.7). These authors also suggested that at least some of the muddy matrix of these crusts has been replaced by calcian dolomite but the crystals were too small to allow microprobe analysis or to demonstrate unequivocally that aragonite crystals were being altered to calcian dolomite. For the Holocene Belize crusts, Mazzullo and others (1987), while noting abundant calcian dolomite cements, reported that replacement of original Mg-calcite sediment and foram tests by calcian dolomites was widespread, producing the thickest calcian dolomites in modern marine systems (up to 0.7 m). They attributed the unusual extent of such Holocene marine dolomitization to the fact that the host sediment was Mg-calcite rather than aragonite. But even in these Belize deposits, the small size of the calcian dolomite crystals (which average 70% of the sediment but can be as high as 97%) makes it difficult to demonstrate texturally beyond a shadow of doubt that the dolomite is a product of a solid-solid replacement reaction.

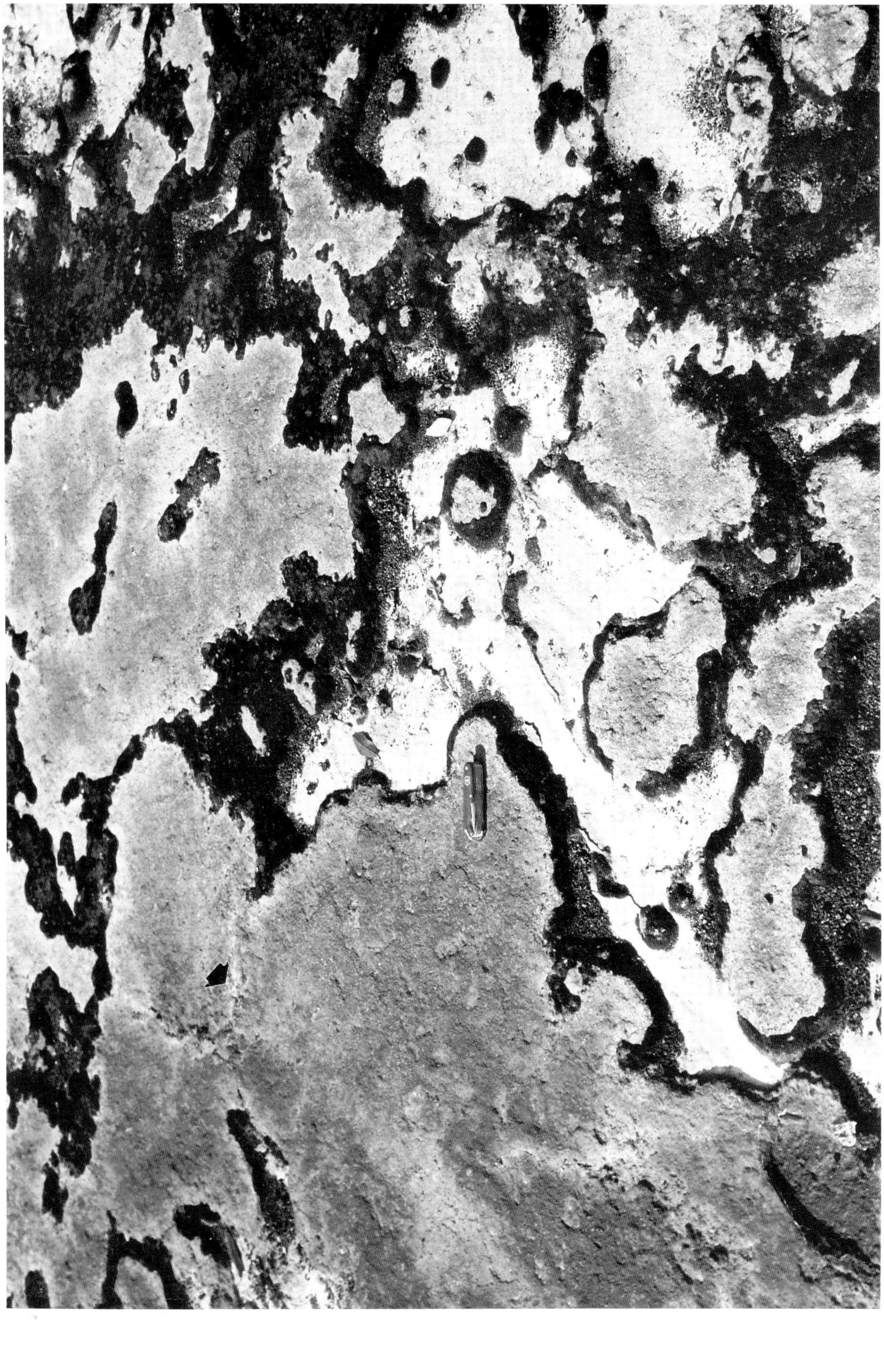

Fig. 132.-Surface crust of cemented *Scytonema* mats (gray areas) on the surface of the inland supratidal marsh, modern carbonate tidal flats of northwest Andros Island. Note the antiformal ridges in the upper left hand corner of the photograph where crusts have buckled (arrow). Living Scytonema mats (dark fringes) outline the edges of the crust. Pocket knife (left center) for scale.

Fig. 133.-Cemented surface crust overgrown by a thin *Scytonema* mat (dark-colored) on the backslope of a tidal channel levee on the modern tidal flats of northwest Andros Island. The pneumatophores of the mangroves have broken the crust into plates. The mangroves in the background are approximately 1 m high.

Fig. 134.-Cemented surface crusts from the backslopes of tidal channel levees of northwest Andros island modern tidal flats. (A) and (B) Two cross-sectional views of cemented cyanobacterial crusts with preserved *Scytonema* mat and filament bundle structures. Both photographs show flat and domal mat growth morphologies with an internal columnar "palisade" structure. White layered material is peloidal sediment; dark gray vertically oriented patches are aragonite-cemented sediment "palisades" (rough copies of spaces once occupied by thick tangled bundles of *Scytonema* filaments); vertical and horizontal black patches are open fenestral pores. In the upper photograph, note how adjoining columns in the upper domal part of the crust coalesce upward. In the lower photograph, note on the left how the fibrous structure has been partly destroyed by oligochaete burrowing. Pencil point for scale. (C) (Following page) Thin-section photomicrograph of the domal fibrous upper crust shown in (B). The radiating *Scytonema* filament molds are a notable feature of these crusts. The clotted peloidal mud matrix of this sample contains about 25% calcian dolomite cement (43 mole % MgCO3). Note the large irregular oligochaete burrows at top of photograph. Scale bar is 500 microns.

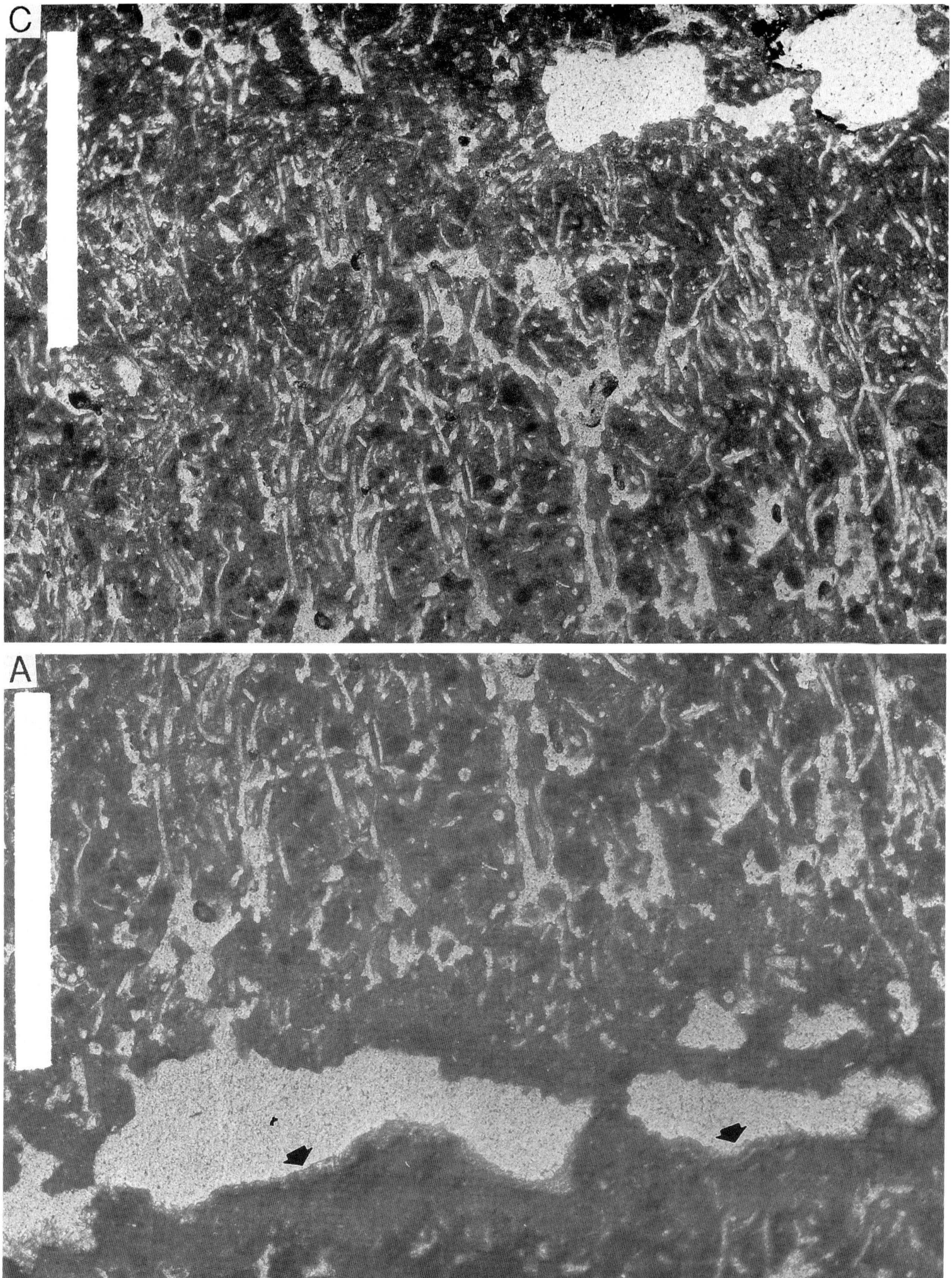
C
A

Fig. 135.-Cemented surface crusts from the backslopes of tidal channel levees of northwest Andros island modern tidal flats. (A) (Previous page) Thin-section photomicrograph of vuggy cemented cyanobacterial crust. This crust is notable for the rinds of aragonite cement that preferentially encrust only the bottom of the flat, elongate vugs (arrows). The cement is not found on the top of the cavities, indicating gravitational control on precipitation in the vadose zone (probably from downward percolating seawater that did not completely fill the connected pore system). The upper portion of the photomicrograph shows *Scytonema* filament molds similar to those in Figure 134C. (B) Scanning electron microphotograph of aragonite void-lining cements in (A). Note the acicular and bladed habit of the aragonite. 1500X magnification; the euhedral crystal in the upper left hand corner is about 30 long.

The problem of the origin of the very finely crystalline Holocene "dolomicrites" remains to be resolved. However, the possibility must be entertained of primary precipitation under ephemeral "lake" conditions (e.g., the Coorong or Deep Springs Lake) or precipitation within the pores of a host sediment in the vadose zone in the mode of Quaternary caliches where calcite micrite massively displaces the host sediment to become an almost 100% $CaCO_3$ deposit. Such a primary precipitation origin for calcian dolomites would imply that the operational chemical reaction is of the type:

$$(1+x)\ Ca^{2+}{}_{(aq)} + (1-x)\ Mg^{2+}{}_{(aq)} + 2\ CO_3{}^{2-}{}_{(aq)} = (CaCO_3)_{(1+x)} \cdot (MgCO_3)_{(1-x)} \quad (3)$$

This type of precipitation reaction is constrained by thermodynamic parameters, trace element and isotope distribution factors, and kinetic factors quite different from those that control replacement dolomitization reactions (such as reactions (1) and (2)). It follows that those Holocene calcian dolomites formed as free precipitates cannot be used as models for ancient calcian dolomites that show fabric-retentive textures resulting from coupled replacement reactions. Such primary calcian dolomite precipitates of modern supratidal crusts must be considered as unusual marine cements and the significance of ancient analogs of these dolomitic crusts must be judged with this in mind.

The rate of cementation of these crusts appears to be quite rapid, judging by the 0-160 years BP ^{14}C ages of the calcian dolomite in the surface crusts on Andros Island determined by Shinn and others (1965, Table 1). This rate is in keeping with the estimated accumulation period of 300 years for the supratidal laminites on the northwest Andros levees made by Hardie and Ginsburg (1977, p. 93) from short term sedimentation rate measurements.

Occurrence and preservation.-

Because supratidal crusts are early lithification features, they have a very high potential for preservation in ancient shallow marine carbonate deposits. In addition, they appear to be able to form easily and quickly on all types of carbonate tidal flats under a wide variety of climatic conditions, and we would expect them to be common features of ancient carbonate peritidal deposits. However, detailed studies that explicitly compare the characteristics of modern and ancient supratidal crusts appear to be rare. Where Holocene supratidal lithified crusts are used in the interpretation of ancient carbonate tidal

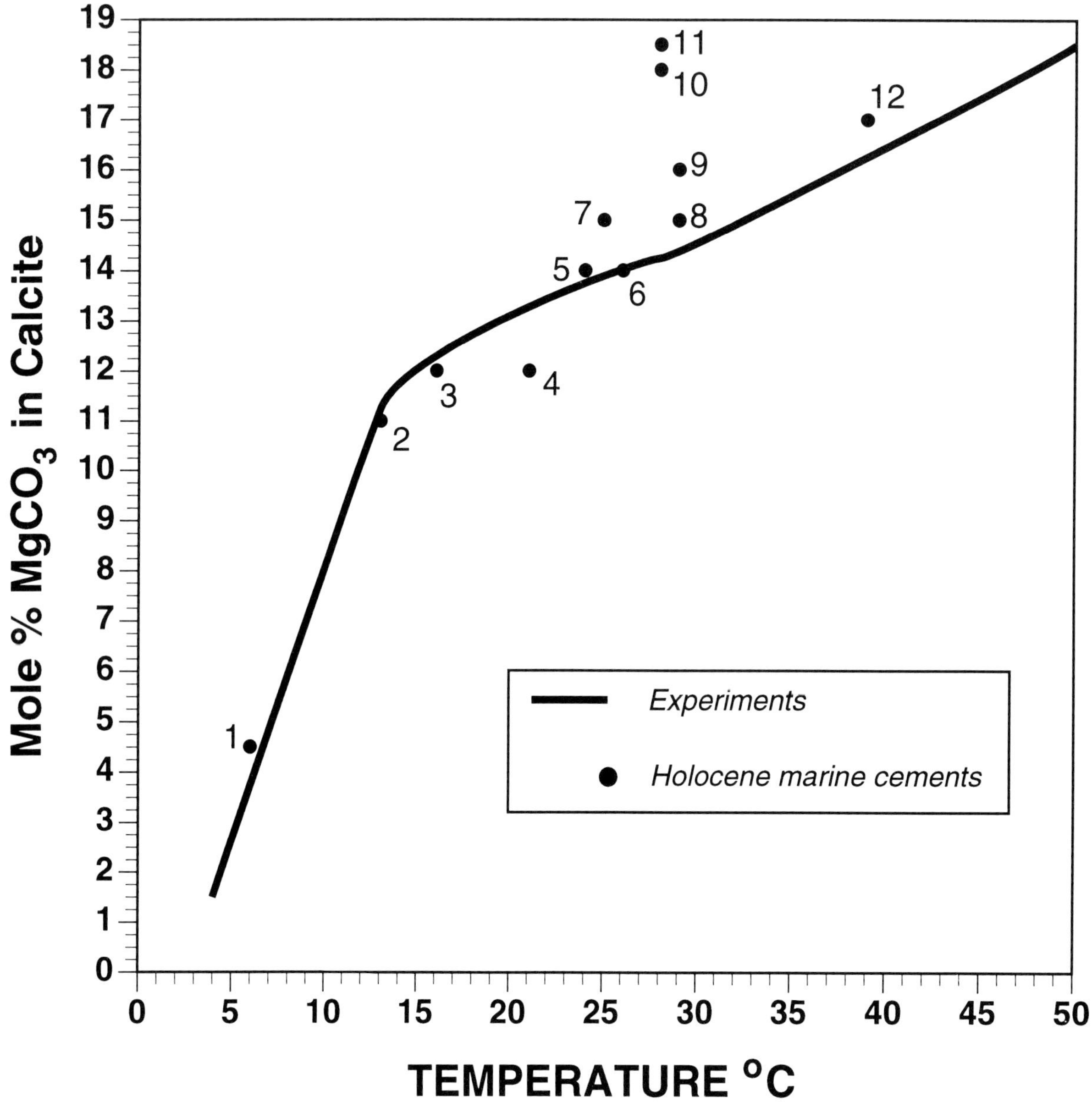

Fig. 136.-Mole % MgCO3 in calcite precipitated from modern seawater as a function of temperature. Solid line fit to experimental results obtained by Füchtbauer and Hardie (1976, 1980) using Ca-Mg-Na-Cl solutions with the same Mg/Ca mole ratio as that of modern seawater (5:1). Filled dots are natural Holocene marine cements from: (1) bottom of the Tongue of the Ocean, Bahamas (Schlager and James, 1978); (2) bottom of the eastern Mediterranean (Milliman and Muller, 1973); (3) Skagerrak, cement in red algae (Alexandersson, 1974); (4) bottom of the Red Sea (Milliman and Muller, 1973, and Milliman and others, 1969); (5) beachrock, Lechaion W. Korinth, Greece (Richter, pers. commun.); (6) Beachrock, eastern Mediterranean (Alexandersson, 1972); (7) beachrock, W. Neapolis, Peloponnes, Greece (Richter, pers. commun.); (8) cup reefs, Bermuda (Ginsburg and others, 1971a,b); (9) reefs, Bermuda (Ginsburg and Schroeder, 1973); (10) reefs, Belize (Purdy, 1968); (11) reefs, Jamaica (Land and Goreau, 1970); and (12) foreshore, Qatar (Taylor and Illing, 1969).

flat deposits, they are cited for their role as carriers of contemporaneous supratidal dolomite and not as sedimentary features in their own right. Goldhammer and others (1987) use the properties of the supratidal crusts described above to identify ancient analogs in a stack of almost 500 subtidal depositional cycles, each about 1 meter thick and capped by a centimeter-scale supratidal crust in the Middle Triassic Latemar buildup of the Dolomites of northern Italy (see discussion under IMPORTANT ASSOCIATIONS, below).

Significance.-
Supratidal crusts signify subaerial exposure of supratidal sediments without active sediment accretion or erosion for periods of time measured in decades to millennia. Local patches of thin crusts found in ancient deposits are valuable environmental indicators, but do not necessarily signal extended periods of exposure. In contrast, thick, laterally extensive crusts that may incorporate some of the expansive cementation features of soil caliches (pisoids, cement more abundant than original framework elements, etc., see above) should be considered as possible markers of important depositional hiatuses (diastems) caused either by abandonment of high supratidal flats during shoreline progradation, or by eustatic or tectonic sea-level falls (Hardie and Shinn, 1986, p. 59). For example, in the study of the Middle Triassic shallow marine carbonates of the Latemar buildup, Goldhammer and others (1987) used the evidence of meniscus and pendant early cements and caliche fabrics in the Latemar cycle caps (see Fig. 128) to infer that these caps were vadose crusts, which in turn registered stratigraphically significant periods of subaerial exposure (several thousands of years) between periods of active subtidal deposition. The remarkable stratigraphic repetition of subtidal subfacies with supratidal crusts in these rocks allowed Goldhammer and others (1987) to identify, using time series analysis, rhythmic cycles of high frequency sea-level oscillations in tune to Milankovitch astronomical cycles (see also Goldhammer and others, 1990).

It has been mentioned above that, up to now, the greatest application of the Holocene supratidal crusts by sedimentologists has been to the dolomite problem. Typically, this application has been a general one in the sense that they have been used as evidence that "dolomite" can form contemporaneously with deposition in supratidal settings, and the lithification aspect has been of only secondary concern at best. What may have been overlooked in this kind of application is that in most Holocene settings the supratidal "dolomite" occurs only as a supratidal "caliche crust" cement and not, for example, as a massive subsurface chemical mud (as is found in the modern Persian Gulf sabkhas, see McKenzie and others, 1980), or as widespread replacement of calcium carbonate sediment (as is typical of most ancient peritidal dolostones). The supratidal crust mode denotes subaerial exposure in a semi-arid subtropical climate or a tropical climate with a dry winter where limited caliche-type surface lithification can take place from slightly concentrated seawater. This is a vastly different "dolomite" environment from that of the Persian Gulf where the arid climate has induced significant quantities of micron-sized calcian dolomite crystals to precipitate from evaporating phreatic and vadose hypersaline brines in the soft lime muds beneath the sabkha surface (see Hardie, 1987). These "dolomitized" subsurface sediments of the modern Persian Gulf sabkhas are mainly older Holocene subtidal and intertidal muds rather than supratidal sediments. Instead, the more soluble saline minerals such as gypsum and anhydrite have precipitated near the surface in the supratidal lime and siliciclastic sediments (Kinsman, 1966; Butler, 1970; Shinn, 1983a). Thus, caution must be exercised in using modern supratidal "dolomite" to interpret the origin of ancient dolomitized peritidal carbonates. In this context, polygonally-fractured supratidal crusts are quite likely to be eroded and re-sedimented as lithoclasts in grainstones during major onshore storms (Hardie and Ginsburg, 1977) and such lithoclasts, if their vadose origin can be demonstrated, would obviously carry important evidence of the former presence of supratidal crusts in an ancient peritidal deposit. Lithoclasts of such dolomitic supratidal crusts are also resedimented into tidal channel lags by channel migration (Hardie and Ginsburg, 1977), and they too preserve indirect evidence of the former presence of supratidal crusts. There have been several reports of ancient peritidal carbonates with dolomite intraclasts in subtidal grainstone facies and these have been cited as evidence for massive early dolomitization of the underlying supratidal sediments (cf. Montanez and Read, 1992). If these dolomite intraclasts carry textural evidence that they were supratidal cemented crusts, then it remains entirely likely that before erosion and resedimentation, only the surface few centimeters of the underlying supratidal sediments were dolomitic, as is the case with the modern dolomitic supratidal crusts. Under such conditions, serious consideration must be given to the possibility that massive replacement dolomitization of the underlying supratidal limestones occurred later and selectively under burial conditions, as Nguyen and Hardie (see Hardie, 1989) have argued for the selective dolomitization of Cambro-Ordovician lime mudstones of the central Appalachians. It is in dolomitization problems of this kind that modern supratidal dolomitic crusts, with their very restricted vertical distribution, have their greatest significance in that they stand in marked contrast to ancient examples of massive dolomitization.

Submarine Hardgrounds

Terminology.-
"Hardground" is an informal term used mainly by carbonate sedimentologists to denote a submarine sediment surface that became lithified in the ambient depositional environment before the next sediment layer was deposited. No definition of hardground appears in the Dictionary of Geological Terms published by the American Geological Institute nor in Bates and Jackson's Glossary of Geology.

Widespread early cementation of modern carbonate sediments occurs both on shallow-water platforms and on deeper, off-platform slopes and basins (Fig. 3). On modern shallow-water carbonate platforms, contemporaneous cements are found in skeletal and ooid shoals along the bank margins (Purdy, 1963; Ball, 1967; Gebelein, 1974; Winland and Matthews, 1974; Harris, 1979; Dravis, 1979; Halley and others, 1983), in shelf-margin reefs (Ginsburg and Schroeder, 1973; James and others, 1976; MacIntyre, 1977; James, 1983), at or near the surface of the peloidal sands that cover the level-bottoms of shelf-lagoons (Shinn, 1969, 1983b; Taft and others, 1968; Logan and others, 1974b), in shelf-lagoon patch reefs (Ginsburg and others, 1971b; James, 1983), on beaches (see "beachrock" references in Bricker, 1971; Inden and Moore, 1983), and on tidal flats (Hardie, 1977b; Shinn, 1983a, 1986). In modern off-platform environments, cements are found in pelagic oozes, in mixtures of pelagic and periplatform-derived oozes, in turbidites on slopes and shallow basins, and in "lithoherms" (Milliman, 1966, 1974; Milliman and others, 1969; Fisher and Garrison, 1967; Bricker, 1971; Hsu and Jenkyns, 1974; Bathurst, 1975; Neumann and others, 1977; Schlager and James, 1978; Mullins and others, 1980; Cook and Mullins, 1983; James and Choquette, 1990a, b). The nodules of limestone produced by differential cementation of shallow slope (<500 m) deposits off the Bahama Banks may serve as modern analogs for differential cementation of ancient ramp deposits (Mullins and others, 1980).

In this section we will only treat the shallow subtidal hardgrounds. For more information on off-platform and reef cementation the reader should consult the appropriate references listed above. Supratidal crusts are described and discussed in a separate section (see earlier).

Distinguishing characteristics.-

Sands of modern carbonate shoals, the level bottoms of shelf-lagoons, and beaches are cemented by aragonite or high magnesian calcite (12 to 14 mole% $MgCO_3$) (Fig. 35). Within the shelf-lagoon environment, the degree of cementation decreases gradually from the marginal sand belts to the peloidal muds of the restricted inner shelf-lagoon (Purdy, 1963; Gebelein, 1974; James and Choquette, 1990b). This is closely paralleled by a decrease in intragranular cementation of peloids (Hardie and Shinn, 1986). Cementation within the skeletal-peloidal-ooid shoals of the shelf margin and peloid sands of the level bottom shelf-lagoon plain occurs in layers up to a decimeter thick. These cemented layers occur at the sediment-water interface as surface crusts (submarine hardgrounds *sensu stricto*) or within the near-surface sediment as layers parallel to the sediment-water interface. Cementation is most intense at the upper surface of the layer and decreases with depth (Shinn, 1969, 1983b, 1986). Multiple layers may be found at depth (e.g., the four layers found beneath the Persian Gulf, see Shinn, 1969, Fig. 7). The hardgrounds of the level-bottom peloidal sands of the Persian Gulf subtidal are buckled into tepees (Shinn, 1969; Assereto and Kendall, 1977). These submarine tepees occur as antiform and overthrust ridges that form polygons 10-40 m in diameter. Broken fragments of the hardground are found at the tepee ridges. Tepees are presumed by most workers to be the result of lateral growth of the hardground layer due to expansive cementation (see section on tepees elsewhere in this atlas). In the Bahamas, hardgrounds within the sand shoals of the shelf margins are broken into intraclasts by storms (Harris, 1979; Halley and others, 1983).

James and Choquette (1990b, p. 23) list many criteria for recognizing ancient submarine hardgrounds. The following criteria are based on that list, with some additions taken from the work of Bromley (1967), Bathurst (1975), and Nguyen (1986): (1) the suspected hardground has a sharp upper contact and a diffuse lower contact (Fig. 137), hard grains (skeletal fragments and ooids) and cement crystals may be truncated along the upper contact (Fig. 138); (2) the upper surface of the suspected hardground has a thin rind of micrite or is stained by iron, manganese or phosphate compounds (Fig. 137), in addition, tubes within the suspected hardground have micritic rinds and the tubes are fractured and sharply truncated along the upper surface (Fig. 139A); (3) the upper surface is encrusted by organisms that require a hard substrate; (4) borings (tubes which cross-cut hard grains and cements) are common along the upper surface; (5) intraclasts similar in composition to the suspected hardground occur above the upper surface, clasts have sharp edges that truncate hard grains and cements and are bored and encrusted by organisms that need a hard substrate; (6) sediments with marine fossils are interlayered with cements in cavities beneath the upper surface; (7) the suspected hardground is buckled into tepees; and (8) There may be multiple surfaces spaced millimeters to tens of millimeters apart (Fig. 139). As James and Choquette (1990b) point out, many of these features taken alone may be the result of subaerial exposure, so that it is necessary to use a combination of the criteria in order to identify a submarine hardground. In particular, difficulty arises in trying to separate submarine hardgrounds from beachrocks and from scalloped erosion surfaces cut into indurated limestones exposed within the intertidal zone (Read and Grover, 1977). Vadose cement fabrics (e.g., gravitational cements and meniscus cements, Dunham, 1969b; Bricker, 1971) are the only sure way to exclude the possibility of submarine lithification. However, the converse, that phreatic cements indicate submarine cementation, is not necessarily true. Keystone vugs (see elsewhere in this atlas) are common in beach deposits, but Shinn (1983b) reports large intergranular voids between lithoclasts from submarine hardgrounds on the Bahama Banks that could be confused with keystone vugs. Finally, it must be recognized that submarine hardgrounds could become exposed by relative sea-level changes and then become modified by subaerial diagenesis, so that the hardground will carry evidence of both submarine and subaerial lithification processes.

Nguyen (1986, p. 146, 157-159) describes surfaces that represent *firmgrounds* within interlayered grainstones and mudstones of the Lower Ordovician Beekmantown Group of western Maryland. Firmgrounds are sediments in the initial stages of hardground formation that can be easily broken up and transported as cohesive intraclasts (Jindrich, 1969; Hardie and Garrett, 1977). A typical Beekmantown firmground has an irregular erosional surface at the top of a small-scale cross-stratified or wave-ripple cross-stratified peloidal grainstone and this surface in turn is overlain by a flat pebble conglomerate (Fig. 140). The pebbles have rounded corners and commonly lie with their long dimension parallel to the erosive surface. Pebble boundaries are mainly traced out around individual sand grains that make up the lithoclast. Modern analogues of these firmground lithoclasts are found throughout the sand shoals of the Bahamas (Fig. 35) and are relatively friable in that sand grains and small clumps of aggregated sand grains can be rubbed off the surface with the fingers. It is very probable that the well-known grapestones of the Bahamas (Bathurst, 1975), particularly the sorted grapestone sands, are simply sand and granule sized lithoclast fragments derived by erosion of firmground crusts. The partly cemented walls of large tubular burrows are especially easily disinterred from firmgrounds by storm erosion and transported as tubular intraclasts recognizable in thin sections as grainstone "doughnuts" with a smooth sharp inner boundary and a rougher outer boundary. Such tubular fragments of *Callianassa* shrimp burrows are commonly found along with flat pebble lithoclasts at the base of lee slopes of the modern ooid shoals on the Bahama Banks (Fig. 141), and analogous tubular intraclasts have been identified in ancient shallow platform carbonates of the Lower Ordovician in the central Appalachians (Nguyen, 1986) and the Middle Triassic in northern Italy (Goldhammer, 1987).

Origin.-

There is a consensus that submarine hardgrounds are the result of early marine cementation by aragonite and magnesian calcite precipitated directly from seawater circulating through the uppermost few centimeters or tens of centimeters of porous sediments of a sandy sea bottom. Thus, the basic requirement for development of a hardground would be stability of the sediment-water interface for a long enough period of time to allow accretion of sufficient cement to firmly hold the sandy sediment together (Shinn, 1969). This necessitates a sea bottom that is undergoing neither net erosion nor net deposition during the cementation process. Judging by modern hardgrounds and firmgrounds, cements occupying as little as 5% of the intergranular void space can produce an indurated crust that can be broken into plates capable of withstanding erosional transportation. It is not certain how long it takes to reach and exceed this degree of cementation in any given submarine bottom environment. However, Shinn (1969) reports that early Grecian pottery is incorporated into hardgrounds in the Persian Gulf, indicating that these shallow submarine crusts can form in a matter of a few thousand years. Supratidal crusts have been

Fig. 137.-Scalloped submarine hardground surface developed on planar-stratified, medium to fine sand-sized peloidal grainstone and overlain by intraclast conglomerate carrying small thrombolites, Upper Cambrian Conococheague Limestone, western Maryland. Note the sharp, scalloped upper contact of the hardground. Thrombolites (light gray lobes) have nucleated on the upraised points with intraclastic conglomerate (slightly below coin) between the clotted fingers.

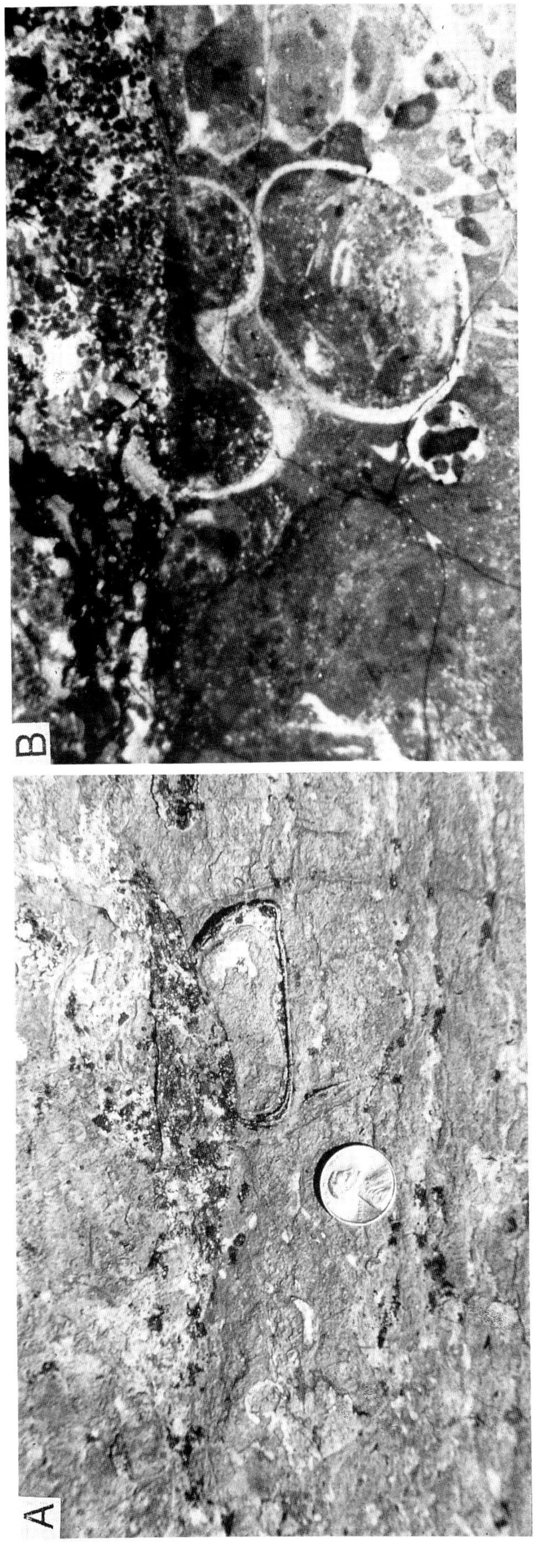

Fig. 138.-Outcrop photograph (A) and thin-section photomicrograph (B) of gastropod shells truncated along hardground surfaces. In (A) note that the packstone filling and surrounding the truncated *Maclurites magnus* shell is not preferentially scoured, indicating that the sediment was lithified prior to erosional truncation of the shell. Middle Ordovician St. Paul Group, western Maryland. Photograph courtesy of Ray Mitchell. (B) is from the Lower Ordovician Stonehenge Limestone, western Maryland. Width of field about 12 mm. Photomicrograph courtesy of Chau Nguyen.

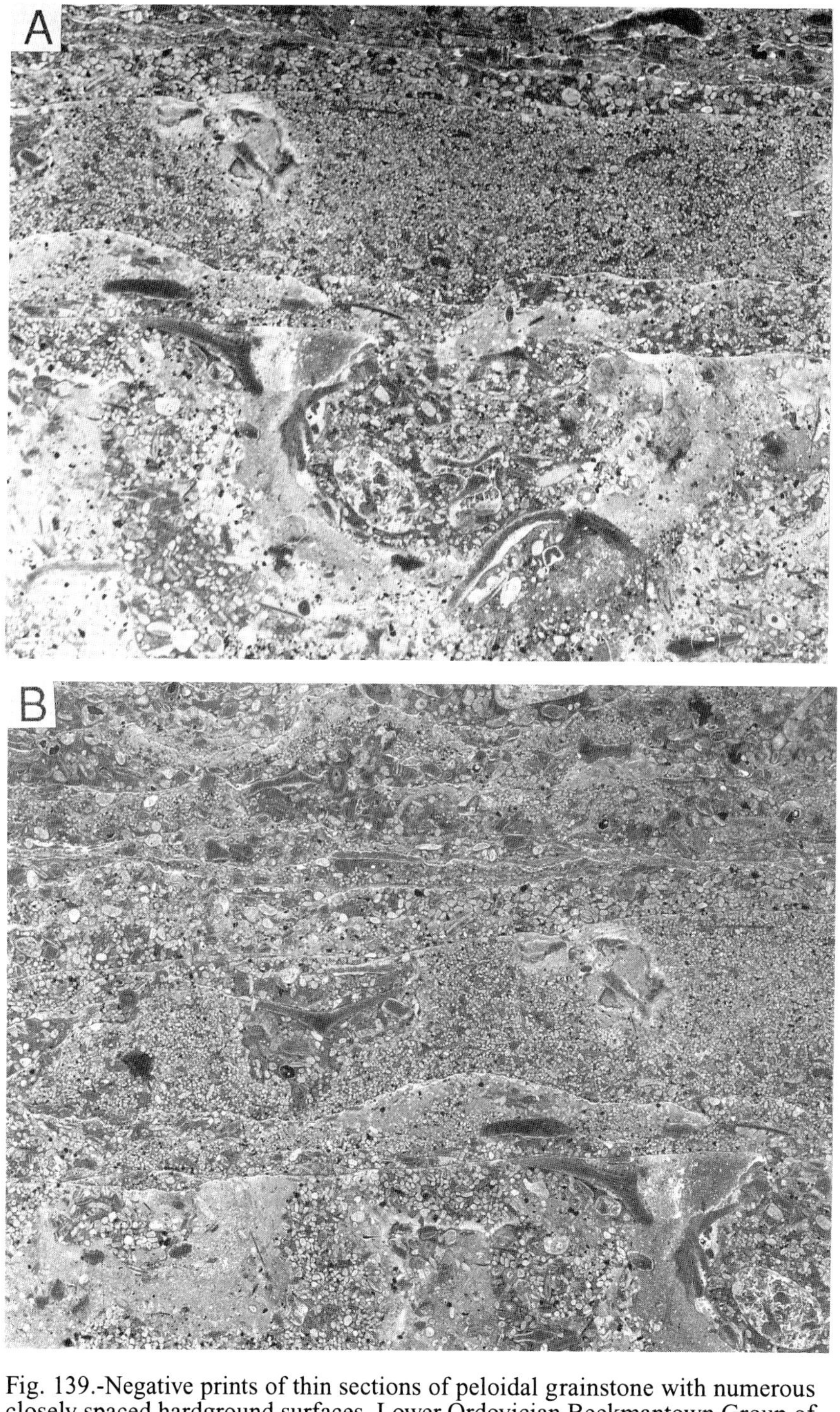

Fig. 139.-Negative prints of thin sections of peloidal grainstone with numerous closely spaced hardground surfaces, Lower Ordovician Beekmantown Group of the central Appalachians. (A) There are at least four, thin, light-gray micritic hardground surfaces in the field of view (approximately 11 mm across). The lowest hardground surface truncates a shell fragment (left center) and a dense micritic lining of a tubular burrow (right center). However, the sediment in the burrow itself is scoured suggesting that the burrow *lining* was as hard as the shell fragment whereas the sediment in the burrow was soft. (B) There are at least 10 thin, light-gray micritic hardground surfaces in the field of view. Width of field about 11 mm. Prints courtesy of Chau Nguyen.

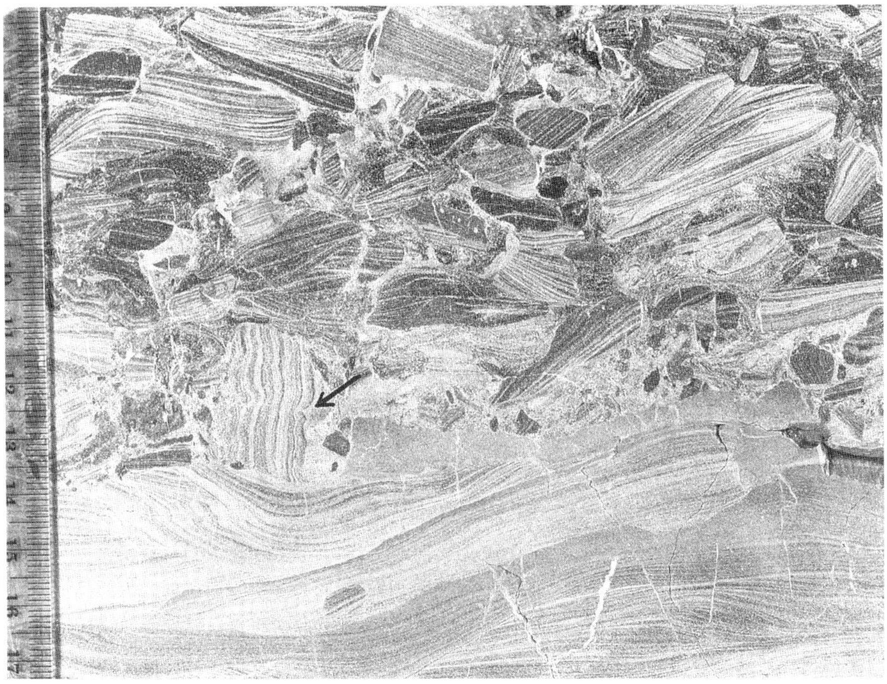

Fig. 140.-Slabbed sample of a flat pebble conglomerate overlying a small-scale cross-stratified, peloidal grainstone bed from the Lower Ordovician Stonehenge Limestone, western Maryland. Note the irregular erosional surface at the top of a small-scale cross-stratified grainstone. The pebbles, which are made of the same small-scale cross-stratified, peloidal grainstone as the underlying bed, have rounded corners and commonly lie with their long dimension parallel to the erosive surface. In thin section, the pebble boundaries are seen to be traced out around individual sand grains that make up the lithoclast. The most likely interpretation is that pebbles are intraclasts ripped up from a *firmground* made of lightly-cemented cross-stratified peloidal sand by storm-generated bottom currents. Scale is in mm, photograph courtesy of Chau Nguyen.

Fig. 141.-Tubular fragments of the lightly-cemented walls of burrows made by the shrimp *Callianassa* in ooid-peloid sands on the Great Bahama Banks. These fragments were ripped up along with other lightly-cemented ooid grainstone patches by storm-generated currents and deposited as lithoclasts at the base of lee slopes of modern active ooid shoals. Scale in mm.

dated at as little as a few hundred years old, while beach rocks have been found with incorporated fragments of middle to late 20th Century soda bottles. Thus, it is probable that submarine hardgrounds can develop over periods of tens to thousands of years.

Occurrence and preservation.-
Submarine hardgrounds have been reported from shallow subtidal carbonate rocks of all ages from the Holocene back at least to the Proterozoic. Such hardgrounds range from very thin seams (millimeters in thickness) that occur in abundance even on the thin section scale (e.g., Nguyen, 1986), through stacks of discrete early cemented storm layers (centimeters in thickness) (Purser, 1969; Aigner, 1985) to thoroughly bored surfaces, encrusted by skeletal organisms and impregnated with iron sulfides, manganese oxides, and phosphates (Bathurst, 1975). Potential for preservation of hardgrounds and their remnants in the form of lithoclast gravels is high because they are early lithification features.

Significance.-
As noted above, submarine hardgrounds require a stable sediment-water interface undisturbed by erosion or deposition for periods of tens to thousands of years. Thus, their main point of interest is that they are "omission surfaces" or "condensation surfaces" that mark hiatuses in submarine deposition. As such, the significance of each hardground varies in proportion to the duration of time it records. Hardgrounds that register extremely long hiatuses that span many thousands or tens of thousands of years clearly are of great stratigraphic importance because such extended periods of continuous nondeposition in a shallow carbonate environment must signify a major change in environmental conditions. Several possible changes might cause such drastic shut-down of the carbonate "factory": (1) shallowing to the point that seawater circulation becomes too restricted for significant carbonate production by biota, this shallowing could be a consequence of a relative sea-level fall (eustatic drop or tectonic uplift) or due simply to filling in of the shelf lagoon in front of prograding tidal flats during a relative stillstand of sea level; (2) deepening beyond the effective photic zone for prolific growth of calcareous algae; this deepening would be the result of eustatic sea-level rise or tectonic subsidence; and/or (3) significant increase in salinity due to the development of barriers to seawater circulation. On the other hand, hardgrounds that record only short duration hiatuses can be ascribed to local changes in sediment distribution patterns or random short term climatic fluctuations that affect the frequency and travel paths of storms on a decade or so scale. Where such short term hiatuses are a significant part of the ambient environmental conditions on shallow carbonate platforms, the subtidal sediments are likely to be dominated by lithoclasts derived from erosion and winnowing of short-lived firmgrounds. These lithoclast grainstones may, in fact, be the only record that periods between depositional events were of long enough duration for hardgrounds and firmgrounds to form. Clearly, if firmgrounds and hardgrounds, or their proxies in the form of lithoclast grainstones, that result from short duration changes occur stratigraphically in systematic arrays (e.g., become progressively more abundant toward a given subfacies boundary), then these collections of short duration changes are harbingers of larger, more systematic changes in the making, such as slow but progessive sea-level rise or fall. Unfortunately, the great difficulty with applying such relationships is that it is by no means certain how to accurately read the time value of these "omission surfaces." On the face of it, it might be expected that thickness of the hardground would be related to time, but on the great Bahama Banks just northwest of Andros Island, submarine hardground patches within a few kilometers of each other vary in thickness by an order of magnitude and also vary in the degree of encrustation by sessile invertebrates. Therefore, in ancient platform carbonates, only the laterally persistent hardgrounds that separate subfacies of different sedimentologic and faunal character should be regarded as stratigraphically significant "omission surfaces" until more is known about the importance of localized hardgrounds.

Tepee Structures and Associated Diagenetic Features

Terminology.-
Adams and Frenzel (1950, p. 308) introduced the term "tepee structure" to refer to antiform deformational structures found in the carbonates of the Permian Carlsbad Group in the Guadalupe Mountains of west Texas and New Mexico. Assereto and Kendall (1977) pointed out that similar antiform structures are found in modern peritidal deposits, in caliche profiles, and in submarine hardgrounds. They would reserve the term "peritidal tepee" for those tepee structures in ancient deposits inferred to have developed in peritidal settings. Along these lines we suggest that "tepee structure" be used as a nongenetic term (i.e., not depositional environment specific) with adjective modifiers where appropriate.

Distinguishing characteristics.-
Tepee structures, as seen in plan view, are a polygonal network of ridges arched along polygonal cracks in a cemented surface crust (as is spectacularly exemplified by tepee polygons in modern gypsum and halite crusts, see discussion below). The simplest cross-sectional forms of tepee ridge structures are antiform buckles that have rough, inverted-trochoidal geometry (Fig. 142). The limbs of each tepee consist of one or more beds sharply turned upwards. The opposing limbs meet along a fracture that marks the crest of the tepee. Beds from one side may be thrust over those from the other side along this fracture contact. Between these antiformal crests are broad, saucer-shaped zones where the arching dies out centripetally. Dimensions of simple tepee structures are variable. Distance between antiform crests ranges from a few tens of centimeters up to a few tens of meters. Vertical relief between crests and troughs ranges from a few centimeters up to a few meters. The thickness of beds being disturbed typically is proportional with the dimensions of the polygons.

Individual tepee zones are usually well defined stratigraphic units bounded both at the top and bottom by undeformed beds. Within each tepee zone, the tepee structures abruptly die out downward against a more or less flat bedding surface of the underlying undisturbed layer. The arched crests of tepee structures are commonly truncated by a flat, overlying bedding surface. Beneath this flat overlying bedding surface, the beds in the saucer-shaped intrabuckle depression typically onlap the tepees (Fig. 142). These relationships unequivocally demonstrate that the tepees are syndepositional features and not later tectonic deformation features.

The simplest forms of tepee structures generally occur in stacks where disturbed zones a few tens of centimeters or more thick are separated by undisturbed beds. From these relatively simple forms, tepee structures increase in complexity by horizontal and vertical fracturing until, in the extreme case, the deposit is a chaotic breccia.

Commonly associated with tepee structures are a suite of genetically related features that include sheet cracks, cross-cutting fractures, and shelter voids that are filled with a complex assortment of spectacular cements and internal sediments (Fig. 143). Pisoids are common features in the voids within tepees

Fig. 142.-Cross-section view of large tepees in the shallow marine platform facies of the Triassic Latemar Limestone, Dolomites, northern Italy. Note Paul Dunn, 2 m tall, at right center for scale.

Fig. 143.-(A) and (B) Cross-section views of tepees in the saw-cut walls of a quarry in the Middle Triassic (Ladinian) "Calcare Rosso", Lombardic Alps, northern Italy. Field of view in (A) is approximately 4 m wide. Paul Dunn provides scale in (B).

and in the depressions between tepees. In general, as the tepee structures become more structurally complex and brecciated, the number of different generations of fractures and cracks and their fills increases.

Sheet cracks are bed parallel fractures filled with early cements and internal sediments that are widest and most abundant in crestal zones and die out toward the relatively undeformed saucer-shaped intercrestal depressions (Fig. 144). Complex systems of cross-cutting vertical and oblique fractures are also common. In many cases, these have matching walls that can be fitted back together indicating simple tensional rupture without rotation. As the number of sheet cracks and cross-cutting fractures increase, the host rock becomes increasingly brecciated, with rotated blocks producing a complex network of shelter voids. The voids formed by the fracturing are filled with a variety of internal sediments, pisoids, and cements that display geopetal fabrics (Fig. 143). The vast and stunning diversity of these features cannot be fully illustrated here. The reader is referred to Dunham (1969b), Assereto and Kendall (1977), Assereto and Folk (1980), Esteban and Pray (1983), Kendall and Warren (1987), and Dunn (1991) for details.

The internal sediment may be red siliciclastic mud (*terra rossa*), which can contain volcaniclastic debris, carbonate mud (commonly peloidal), or microspar (crystal silt or vadose silt of Dunham, 1969a, b). The latter have been interpreted as broken crystals of penecontemporaneous cements (however, see Shinn, 1986). In a number of cases, the carbonate muds were clearly derived from beds overlying the tepee structures. Not uncommonly, the internal sediment contains fossils, including gastropods, nautiloids, and ostracods.

The cements can include a number of types indicating both phreatic and vadose conditions. The vadose cements are a critical element in identifying those tepees of supratidal origin. Among such vadose cements are those of gravitational (mainly micrite or microspar) and meniscus origin (Fig. 145A). Isopachous cements indicative of phreatic conditions include thin rinds of laminated micrite (that some workers interpret to be at least partly biogenic), and thick layers of radially oriented crystals (including radiaxial fibrous and fascicular optic calcite, the "coconut-meat" calcite of Assereto and Folk, 1980). Another common cement type consists of small hemispheres of radially oriented crystals (similar to the botryoidal cements described by Ginsburg and James, 1976, from the modern reefs and sediments of the Belize shelf). Most spectacular of all are the large, radially arranged arrays of calcite crystals up to decimeters in length (the "raggioni" of Assereto and Folk, 1980). These are interpreted from their square crystal terminations and high Sr contents as having originally been aragonite.

Pisoids occur in a number of tepee deposits (Fig. 145B), the most famous being in the Permian Capitan Group of west Texas and new Mexico (Fig. 146) (Dunham, 1969a; Esteban and Pray, 1983). These are commonly found in the labyrinthine cavity network of tepees as part of the internal sediment-cement complex, but also occur in the saucer-shaped depressions between polygonal buckles, where they likely formed on the bottom of open pools outlined by the tepee crests (in the manner of pisoids formed in spring pools, e.g., Fig. 119, see also Risacher and Eugster, 1979). In a number of cases, the outermost laminae of pisoids show downward, gravitationally influenced growth and fitted polygonal fabrics (Dunham, 1969 a, b) indicating growth under vadose conditions. Adjacent pisoids can merge, sharing a number of their outer laminae. Pisoids also commonly show multiple episodes of fracturing and recoating.

Origin.-

Tepee structures have been interpreted by most workers to have originated by the expansive growth of early cements within a host sediment. It is likely that, initially, minor buckling is the result of precipitation of fine grained cements within the intergranular voids of permeable sandy beds to produce a compressionally disrupted hardground. With the development of open fractures at the crests of the buckles, pore fluid flow patterns are radically changed so that the fractures become the principal conduits through which large volumes of water can be preferentially channelled. This results in the preferential growth of sparry cements within these fractures, which in turn leads to accelerated polygonal disruption, in particular arching and pushing apart of polygon edges as cementation is focussed beneath the ruptured antiform crests. In this way the initial lateral compressive deformation due to pervasive expansive cementation is replaced by localized upward wedging and push-apart deformation of the tepee ridges. Taken to the extreme, the multiple generations of cracks and fractures and their internal sediments and cements can almost completely brecciate the host sediment. Assereto and Folk (1980) describe "tepee-ized" rocks that are up to 75% diagenetic components.

In subtidal settings, where expansive cementation of submarine hardgrounds leads to the development of tepees, it follows that the pore water from which the cements precipitated was seawater. In peritidal and coastal salina settings, the pore water could be seawater, meteoric water, or both. For tepee formation in arid settings, the ambient groundwater (seawater, meteoric water, or mixtures of the two) is drawn up through the vadose zone by evaporative pumping (Hsu and Siegenthaler, 1969), although in some cases tidal pumping may also contribute some flow. The large volumes of cement precipitated within tepee structures demand flow through the sediments of a large volume of pore water supersaturated with respect to alkaline earth carbonates, and this in turn requires an extensive groundwater reservoir with a water table close to the surface to allow extensive pumping. Along these lines, Dunn (1991) has shown that the meter-high tepees in the shallow subtidal grainstones of the Triassic Latemar buildup of northern Italy were mainly cemented by early marine aragonite and magnesian calcite, under both phreatic and vadose conditions, indicating that seawater was the principal groundwater that flowed beneath exposed supralittoral lime sand flats of the Latemar buildup. However, several generations of these marine cements are separated by minor dissolution surfaces that testify to the occasional input of undersaturated meteoric waters.

Assereto and Kendall (1977) suggest that the polygonal pattern of tepee crests was inherited from an original desiccation crack network and that cycles of sediment wetting and drying, thermal expansion, and possible growth of saline minerals aided in initial fracture formation. However, the host rocks for most tepee structures are grainstones that cannot be mudcracked (see, for example, Goldhammer and others, 1990; Dunn, 1991). The alternative explanation we have offered above is based on observation of analogous polygonal rupture and tepee formation that occurs very rapidly in halite crusts during the desiccation stage of a salt pan cycle (Lowenstein and Hardie, 1985), as well as on the detailed observations of the characteristics of the spectacular tepee structures in the Middle Triassic platform carbonates of the Latemar buildup in the Dolomites of northern Italy (Dunn, 1991; see also Hardie and others, 1986; Goldhammer and others, 1990).

Fig. 144.-(A) Cross-section view of tepees developed in laminite subfacies of the Middle Triassic (Ladinian) Esino Formation of the Lombardic Alps, northern Italy. Paul Dunn provides scale. (B) Detail of the tepee in the lower right hand corner of (A). The dark gray layers are composed of plates of fenestral mudstones and wavy-crinkled fenestral laminites (see Figures 49, 51, and 53 for photographs of "untepee-ized" tidal flat cycles in this Triassic Esino Formation). Light gray material is vug-filling cement and internal sediment now mostly dolomitic in composition. Width of field of view is approximately 1.5 m.

Fig. 145.-Detail of tepee structures in the saw-cut walls of a quarry in the Middle Triassic (Ladinian) "Calcare Rosso" of northern Italy (see Fig. 143). (A) Light gray areas are plates of host-rock fenestral packstones and wackestones that are broken up and arched into tepee antiforms. The darker layers are mostly laminated radial cements that in some cases expand into large sprays ("raggioni") such as that to the right of the pocket knife (arrow). Intercalated with the laminated cements are pockets and layers of pisoids and intraclasts. Field of view is about 0.75 m. (B) Laminated cements encasing and overlying pockets of pisoids (lower portion of photograph) and intraclastic grainstones in the "cavities" between large broken plates of host rock. Field of view is approximately 0.5 m.

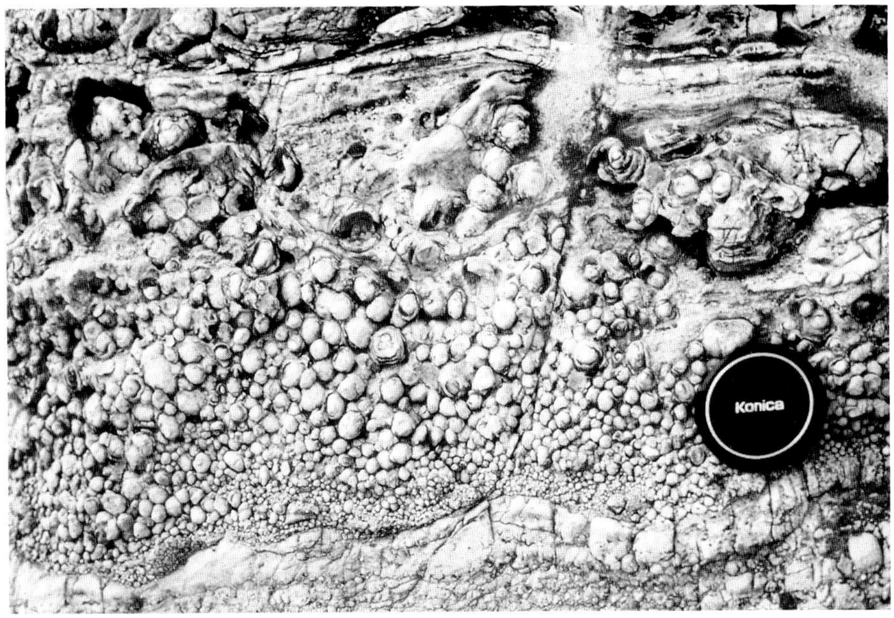

Fig. 146.-Inversely-graded pisoid bed from the tepee zone of the Permian Capitan "Reef Complex", west Texas. Compare with Figure 119B. Lens cap about 50 mm in diameter, photograph courtesy of Ray Mitchell.

Occurrence and preservation.-
Modern tepee structures are known from caliche profiles (pseudoanticlines, Assereto and Kendall, 1977), submarine hardgrounds in the Arabian Gulf (Shinn, 1969), and most strikingly, from coastal salinas and tidal flats of Australia (Davies, 1970; Warren, 1982a; Ferguson and others, 1982; Handford and others, 1984). They are fairly common in ancient successions, notably the Proterozoic Rocknest Formation of Canada (Grotzinger, 1985, 1986a, b), the Permian Carlsbad Group of the Guadalupe Mountains (Dunham, 1969b; Esteban and Pray, 1983) and the Triassic of the Alps (Bosellini and Rossi, 1974; Assereto and Kendall, 1977; Assereto and Folk, 1980; Hardie and others, 1986; Goldhammer and others, 1990; Dunn, 1991). Once formed, tepees have a very high potential for preservation because of their high degree of early lithification.

Significance.-
Tepee structures can develop in submarine hardgrounds and caliche profiles and the reader is referred to those sections for details on their fabrics and importance. Tepee structures with demonstrable vadose features but lacking caliche features clearly were formed at or near a subaerially exposed surface and, although in a sense they constitute a type of limestone "soil," they carry a message different from that of caliches and other limestone soils. In particular, subaerial tepee zones record conditions quite the opposite to those that promote the formation of karstic soil zones. Supralittoral tepee formation requires a shallow groundwater table and upward pumping of supersaturated groundwater in the vadose zone; whereas, karstic dissolution requires a deeper seated water table and downward flow in the vadose zone of undersaturated meteoric water. Evaporative pumping can be a most effective process for drawing groundwater up through a shallow vadose zone and so an arid or semiarid climate along a coastal plain should be most favorable to massive subaerial tepee development, as for example is the case for the modern Australian coastal plain environments where tepees are found in surface crusts (e.g., Handford and others, 1984; Warren, 1982a). Under intensely arid conditions, such as those of sabkhas and playas, evaporative pumping can be so effective that saline minerals such as gypsum and halite, far more soluble than alkaline earth carbonates, are precipitated *en masse* in the vadose zone (see below) and can lead to the disruption of gypsum and halite surface crusts into tepee structures analogous to carbonate tepees (see below; see Castens-Seidell, 1984; Lowenstein and Hardie, 1985; Hardie and Shinn, 1986).

In ancient carbonate deposits, there appear to be at least two important settings for development of extensive tepee zones, one exemplified by the Precambrian Rocknest and Permian Carlsbad examples and the other by the Alpine Triassic examples. The former seem to be associated with a particular depositional setting, that is, an island environment (an exposed barrier) located at the seaward margin of a carbonate platform. In such a setting, seawater seepage through the barrier, coupled with evaporation, could produce shallow brines beneath the island. When drawn to the surface by evaporative pumping, they serve as the source waters for expansive cementation of the surface sediments that ultimately leads to tepee formation. Surface seeps of these brines may also give rise to terrace pools in which travertine and layers of reverse graded pisoids could form (compare with descriptions of Risacher and Eugster, 1979, for spring pools of an inland playa system; see also Esteban and Pray, 1983). The Alpine Triassic examples carry a different message. In these shallow carbonate platform rocks, the tepees are caps to 4th order (10^5 year duration) megacycles and are important stratigraphic elements in an 800 meter thick stack of meter-scale 5th order (10^4 year duration) depositional cycles. Hardie and others (1986), Goldhammer and others (1990), and Dunn (1991) interpret the tepee caps to represent subaerially exposure of the platform for extended periods (tens of thousands of years at least) due to low amplitude 4th order (10^5 year duration) sea-level low stands, and in this way the tepee zones were used to aid in the reconstruction of the sea-level history during Triassic times (Goldhammer and others, 1990). The argument for extended exposure was based on modern and Pleistocene examples of the effects of subaerial exposure of carbonate platforms that resulted from the Quaternary glacio-eustatic sea-level changes. Caliche development on the subaerially exposed tops of modern carbonate banks and shelves since the 125,000 year BP highstand of sea level has proceeded at rates of about 0.006 to 0.035 cm/1000 yr (James, 1972) whereas cements in Holocene tepees have grown at rates of 0.02-0.04 cm/1000 yr (Handford and others, 1984). Clearly, at these rates even simple tepees will require many thousands of years to develop.

Evaporite Minerals, Casts, and Molds

Evaporite minerals and their pseudomorphs are common components of shallow marine carbonates (see Pettijohn, 1975). This section concentrates on gypsum, anhydrite, and halite because they are by far the most common evaporite minerals in modern carbonate sediments and ancient carbonate rocks. These minerals occur as crystalline layers, as isolated crystals, or as crystal clusters. In addition, common evidence of the former presence of these minerals includes molds and casts of crystals, and a variety of pseudomorphous and nodular forms. As yet, no satisfactory descriptive classification scheme exists for this diverse group of features. Therefore, our approach here is a genetic one based on the processes of evaporite formation and their resultant products in carbonate sediments and rocks, drawn mainly from work in modern settings (e.g., Shearman, 1966, 1970; Kinsman, 1969; Butler, 1970; Shinn, 1983a; Lowenstein and Hardie, 1985; Castens-Seidell, 1984; Hardie and others, 1985; Hardie and Shinn, 1986; Spencer and Lowenstein, 1990; Handford, 1991; Smoot and Lowenstein, 1991; Warren, 1991). This approach will allow us to emphasize criteria whereby the products of different processes can be distinguished.

Evaporite crystals, casts and molds on bedding surfaces.-

Distinguishing characteristics.-Crystals of evaporite minerals and their casts and molds as seen on bedding surfaces are characterized by rectilinear outlines (Fig. 147). Crystal casts may be constructed of detrital sediment or some crystalline pseudomorph. In cases where the original mineral, such as gypsum, is preserved, there is no difficulty making a proper identification of the mineralogy. However, care must be taken to establish that pseudomorphous replacement of the original mineral has not taken place, for example gypsum replaced by anhydrite, by checking crystal morphologies. Where only a cast or mold is preserved, the matter of identification of the original mineral that left the imprint can be exceedingly difficult, requiring accurate measurement of the interfacial angles preserved by the cast or mold. Such measurements are rarely done, instead, identifications are typically made on the basis of general crystal shapes, e.g., the "swallow tail" twin morphology typical of gypsum grown under subaqueous conditions. It must be emphasized that superficial resemblances not supported by interfacial angle measurements can lead to misidentification of the original mineral, which in turn can lead to a misidentification of the chemistry of the water from which the mineral precipitated.

Origin.-The origin of crystals and their casts and molds preserved on bedding surfaces can be reconstructed from observations made on the precipitation of gypsum and halite in modern evaporite environments. Extensive evaporation at the brine-air interface of an ephemeral or perennial brine pool will cause the nucleation of gypsum and halite crystals at that interface. Crystals nucleated in this manner can remain floating and grow at the interface, bouyed by surface tension. Gypsum crystals are commonly tiny polygonal plates (see Neev and Emery's, 1967, descriptions of gypsum crystallization in the Dead Sea) or needles (see Castens-Seidell, 1984), whereas halite crystals grow as square or rectangular plates, as stepped skeletal "hoppers" (Fig. 148), or as cubes with slightly indented sides (Fig. 149A). Halite crystals can also coalesce into wafer-thin "rafts" (Figs. 149A, B, C). Once a critical size is reached, surface tension can no longer hold the crystals or coalesced rafts at the surface and they sink to the bottom. Molds of the settled crystals are formed on the sediment surface when a later influx of undersaturated storm waters redissolves the settled crystals (this is especially common with very soluble minerals such as halite, which has a solubility of 36 g NaCl/100 g H_2O at 25°C). Later filling of the molds by sediment or by cements produces casts of the original gypsum and halite crystals.

Occurrence, preservation, and significance.-Bedding plane casts and molds of halite are commonly encountered in shallow-water carbonate platform deposits. Figure 148B shows examples

Fig. 147.-Halite cube casts on the underside of a bedding surface of wave-ripple cross-stratified peloidal dolomite from the Middle Cambrian Arctomys Formation of the Canadian Rocky Mountains of southwestern Alberta. Scale in mm.

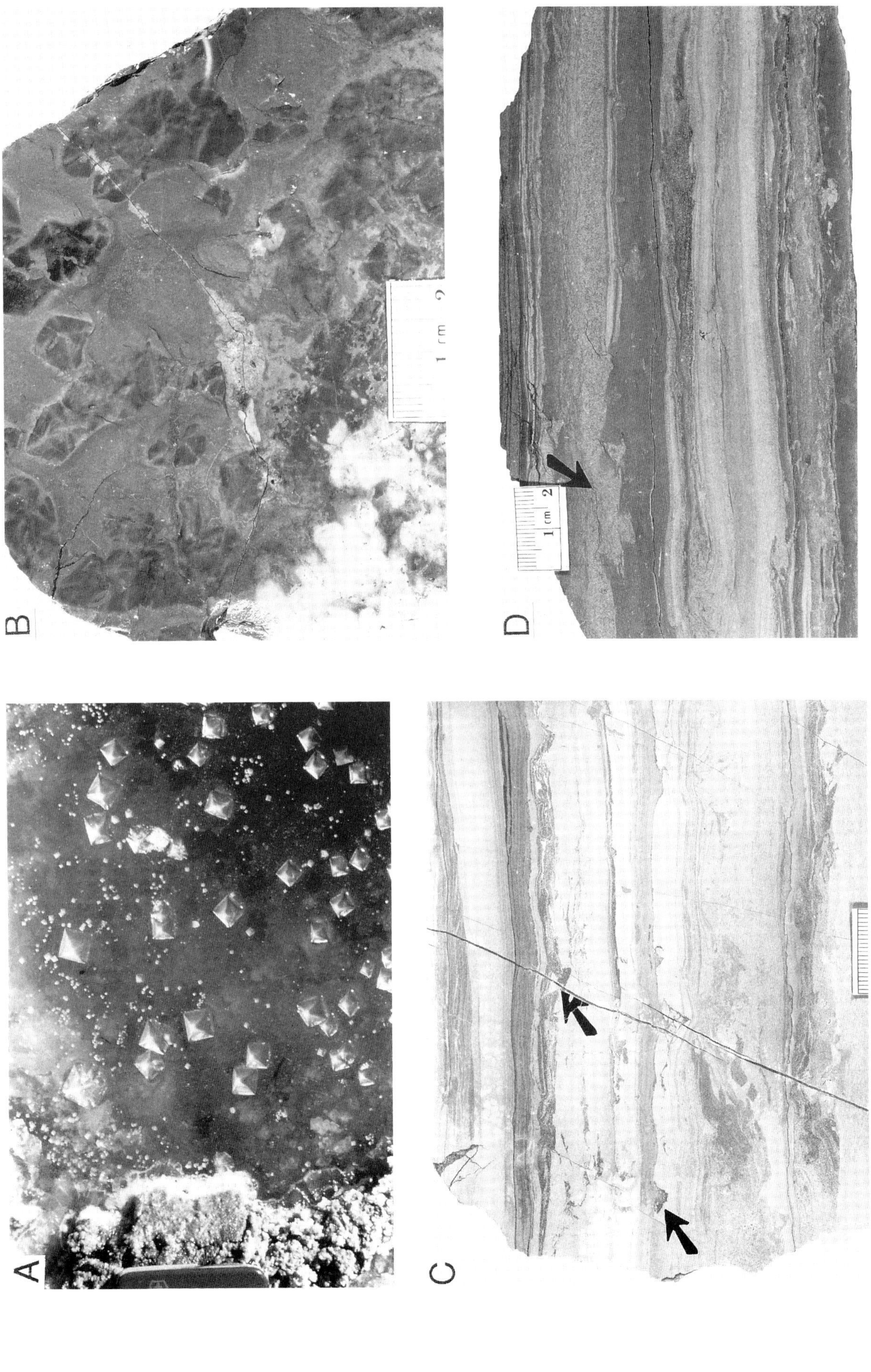

Fig. 148.-(A) Halite hopper crystals floating on the surface of a modern brine pool, Salina Omotepec, Baja California (see Lowenstein and Hardie, 1985, Fig. 10). (B) Halite hopper impressions preserved on a bedding plane of laminated dolomite from the Silurian Wills Creek Shale of the central Appalachians. Scale in mm. (C) and (D) Sawn slabs of planar to wavy laminated dolomite from the Silurian Wills Creek Shale showing halite casts (filled with peloidal grainstone) in cross section (arrows). Scales in mm.

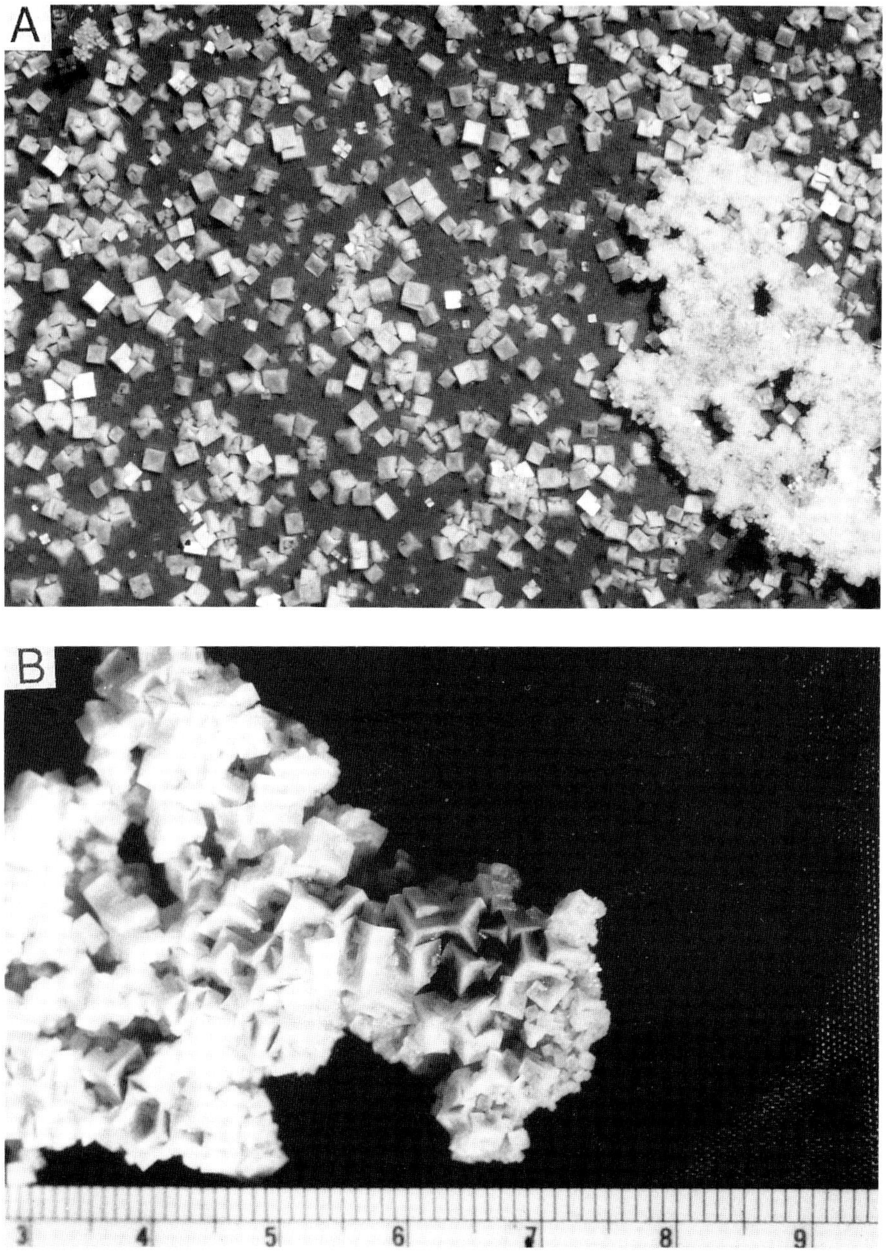

Fig. 149.-(A) Halite cubes, plates, and a raft of coalesced halite crystals (right) from a modern brine pool developed on a small playa near Carlsbad, New Mexico. Cubes and plates slightly out of focus are floating on the surface of the brine whereas sharply in focus cubes have fallen to the bottom and are partly embedded in the mud (dark background). Note the various angles at which the cubes have landed in the mud (compare with Fig. 147). Cubes vary from about 1 to 8 mm in diameter. (B) Shows detail of a cluster of bottom-grown hollow-faced cubes of halite. Small divisions of the scale are in mm. (C) (Following page) Permian halite. Thin-section photomicrograph of a laterally elongate sheet of connected halite crystals interpreted as a sunken hopper raft (arrow) surrounded by randomly oriented sunken halite cubes forming a crystal cumulate. The dark lamina at the base is microcrystalline polyhalite. Sample from the Permian Salado Formation of west Texas and New Mexico. Scale bar is 2 mm long. See Hardie and others (1983), Figure 3. Photomicrograph courtesy of Tim K. Lowenstein.

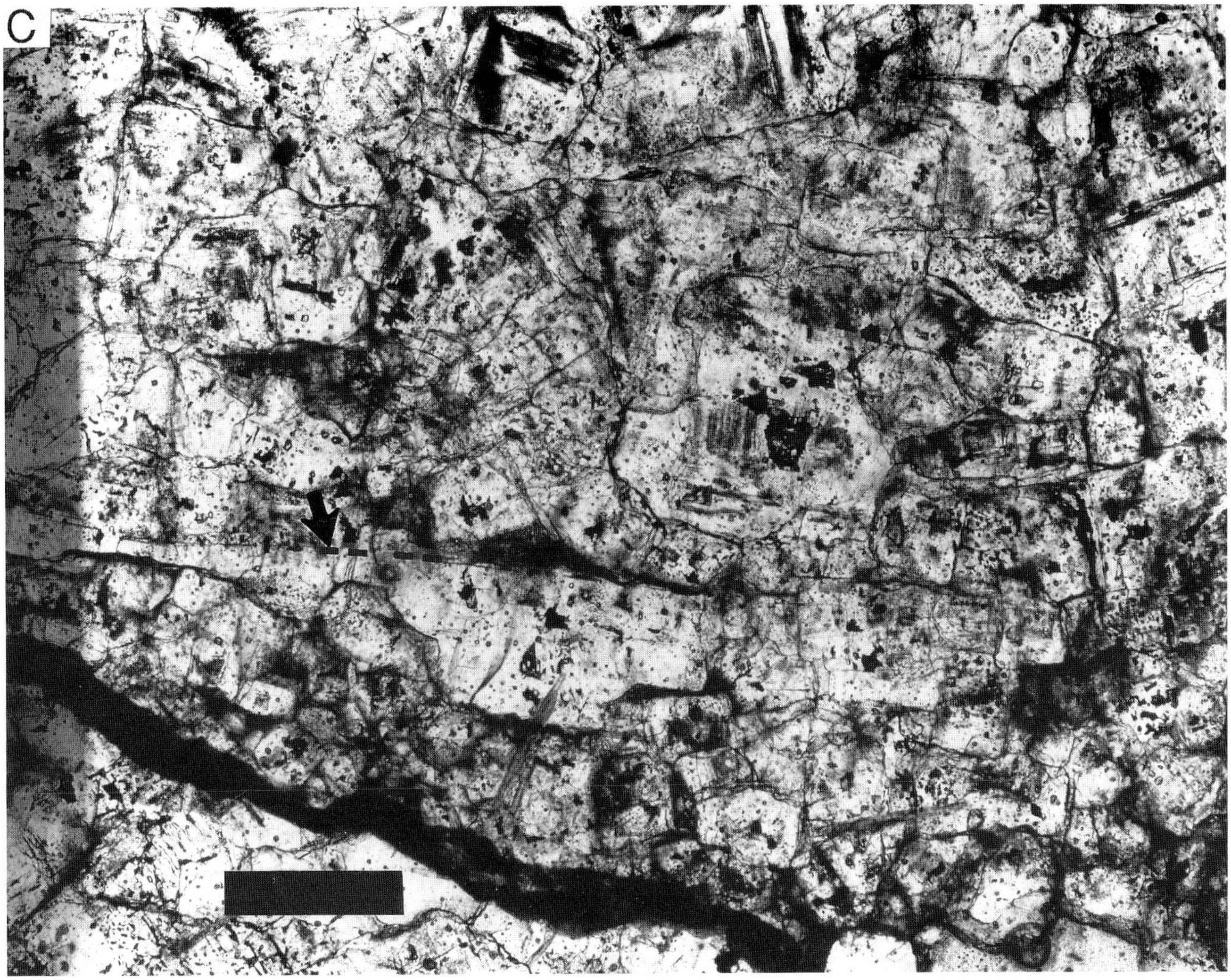

of halite "hopper" crystal casts from the Silurian Wills Creek Formation of the central Appalachians (see Tourek, 1970) and Figure 147 shows similar crystal impressions from the Cambrian Arctomys Formation of the Canadian Rockies. Note that clear impressions of triangular corners and edges of hopper-cubes are preserved along with square outlines, reflecting random orientations of crystals as they hit the bottom and became partly imbedded in the soft bottom sediment. Particularly significant are those Wills Creek examples where, as seen in cross-section, cross-laminated grainstones fill the halite impressions (Figs. 148C, D). This indicates that halite dissolution occurred while the partly imbedded cube was still at the sediment surface. Dissolution, probably by the next seawater flooding event, was followed by infilling of the crystal mold by sediment deposited from bedload. This surface detrital filling is a good criterion in establishing that the saline mineral formed in a brine pool at or above the sediment surface and was not a later intrasediment growth from saline pore fluids (see further discussion below). It may also be possible for gypsum and halite to nucleate on the bottom of extremely thin sheets of brine without seeds that originally formed at the brine-air interface. Certainly skeletal ice crystals can grow in films of water directly on the sediment surface in this way. Perhaps the spectacular filigree-like impressions of skeletal halite crystals such as those shown in Figure 150 grew in a similar way from a thin surface film of brine.

Evaporite crystals, casts, and molds within sediment layers.-

Distinguishing characteristics and origin.-Evaporite minerals commonly grow from groundwater brines within permeable sediments in the phreatic zone, the vadose zone, and at the subaerially exposed sediment surface of desiccated playas and sabkhas. Such intrasediment growth of evaporite minerals either incorporates the host sediment within the growing crystals (incorporative growth) or pushes the host sediment aside as the crystals grow (displacive growth). Incorporative growth of gypsum produces "sand crystals" and "sand rosettes" (Masson, 1955; Kinsman, 1966; Pettijohn, 1975) and poikilotopic cementation. Displacive intrasediment growth of gypsum (Fig. 151) and halite (Fig. 152) produces both isolated euhedra and clusters of intergrown crystals. Solitary euhedra to subhedra of gypsum (Masson, 1955; Kinsman, 1966; Cody, 1979; among many others) and halite (see Kendall, 1984, Fig. 33) are commonly reported growing within brine-logged marginal marine and marginal lacustrine sediments of modern evaporitic environments. The gypsum reported from these sites commonly has a "discoidal" or "lenticular" morphology with flattening normal to the c-axis (Fig. 151; see Masson, 1955; Cody, 1979). Crystals may also show ragged, corroded edges attributed to partial dissolution by contact with undersaturated groundwaters.

A

B

Fig. 150.-Halite skeletal crystal impressions from the Silurian Bertie Group of western New York. (A) Impressions of skeletal halite from a bedding plane in the Silurian Wills Creek Formation of the central Appalachians. Scale is in mm. (B) Halite skeletal crystal impression from within a cross-stratified peloidal dolomite. This large crystal apparently grew as an intrasediment skeletal crystal. Scale is in cm.

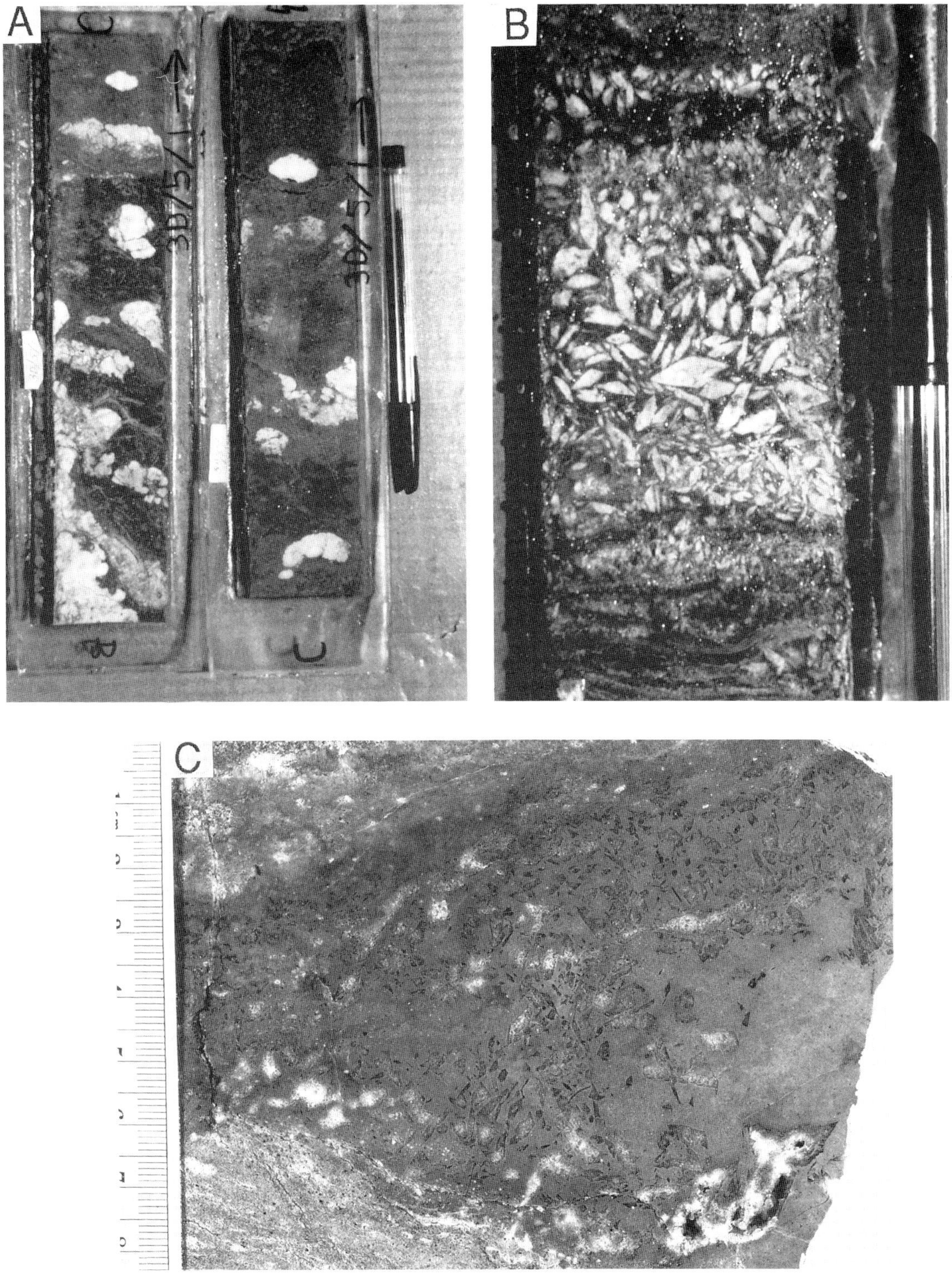

Fig. 151.-Displacive intrasediment growth of gypsum. (A) Plastic-impregnated core of anhydrite replacing clusters of intergrown gypsum crystals in supratidal mixed carbonate-siliciclastic sediment from Holocene coastal sabkha of Abu Dhabi, Persian Gulf. Pen for scale (B). Plastic-impregnated core of carbonate sediment crowded with displacively-grown gypsum crystals from the modern Persian Gulf sabkha at Abu Dhabi. Note the "discoidal" or "lenticular" morphology flattened normal to the c-axis and the sediment entrained within the crystals, two features characteristic of displacively-grown gypsum crystals in modern evaporitic environments both marine and non-marine. Pen for scale. (A) and (B) from Evaporite Core Workshop, Calgary, 1982. (C) Cement-filled molds interpreted to be pseudomorphous after gypsum, Mesozoic Morrison Formation, Colorado. Scale is in mm.

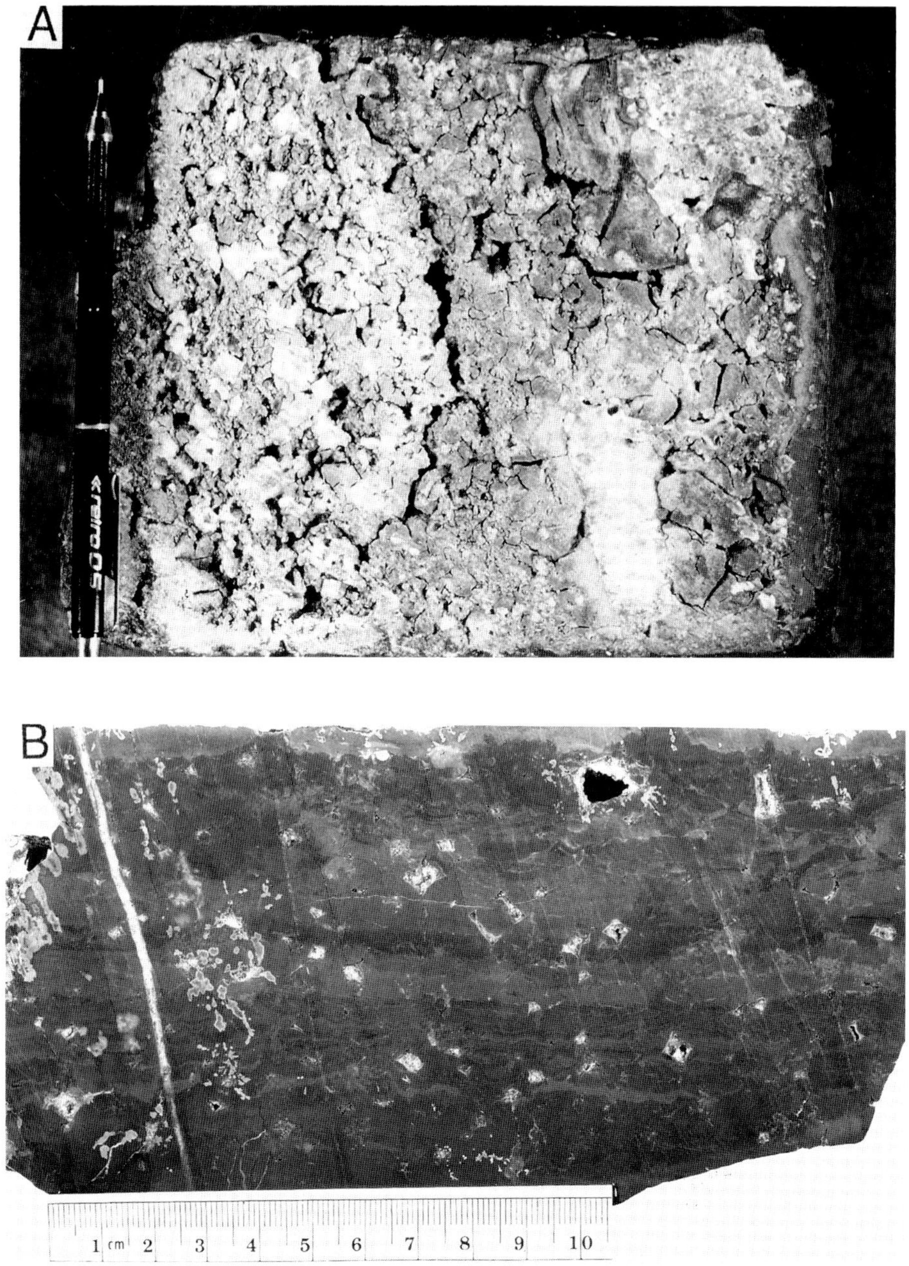

Fig. 152.-Displacive growth of halite crystals. (A) Epoxy impregnated box core of modern brine-soaked mudflat showing white, randomly-oriented halite crystals that grew in the gray mud. Cracks are from sample preparation. Pen for scale. Photograph courtesy of Tim K. Lowenstein. (B) Cement-filled crystal molds interpreted to be after displacive, intrasediment-grown halite. Upper Silurian Wills Creek Shale, central Appalachians. Scale is in mm. (C) (Following page) Thin-section photomicrograph of muddy halite rock from the Permian Salado Formation of Carlsbad, New Mexico (see Lowenstein, 1988, Figure 16). Clear halite crystals contain incorporated mud and few fluid inclusions. Dark areas between halite cubes are composed of zoned crystals of sylvite cement and minor mudstone. Scale bar is 5 mm long. Photomicrograph courtesy of Tim K. Lowenstein.

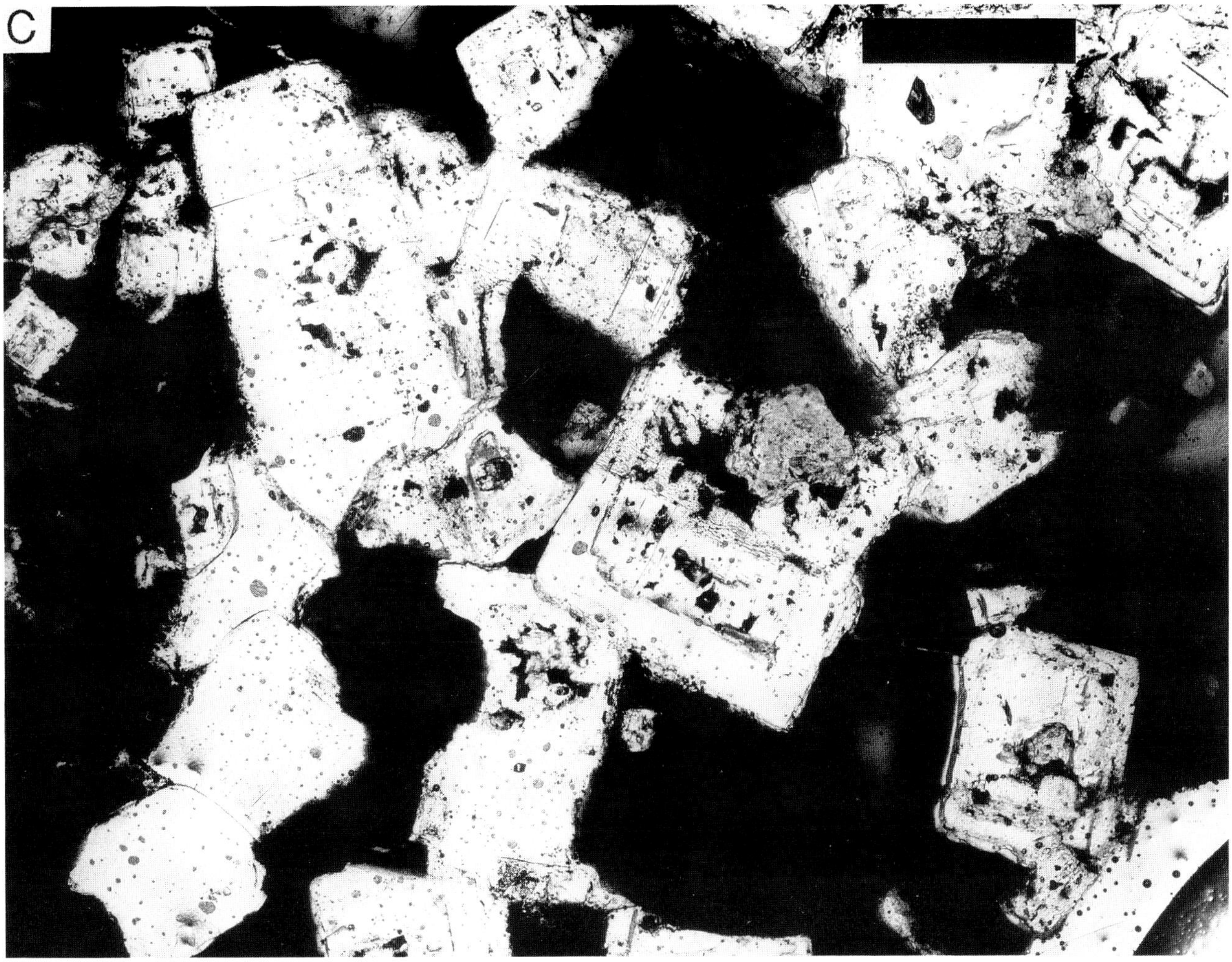

Also common are larger flattened bladed crystals, widely referred to as "selenite" (Kendall, 1984; Kinsman, 1966). These can be tens of centimeters in length. Both discoidal and bladed gypsum crystals typically carry variable amounts of inclusions of the host sediment (Fig. 151). These sediment inclusions may be traceable through larger crystals recording remnants of layering of the sediment in which the crystals grew (Hardie and Eugster, 1971; Warren, 1982b). Wind deflation commonly exposes crystals of discoidal and bladed gypsum that have grown within the sediment. The exposed crystals are then susceptible to erosion and transportation by wind and storm-floods. Large, vertically standing "selenite" crystals that grew within the sediment but then became partly exposed by erosion of the host sediment (as found at the surface of a number of modern sabkhas, see Masson, 1955; Kinsman, 1966) may be difficult to distinguish from bottom-growth gypsum crystals.

Halite crystals can also grow within a sediment mainly by displacing the host sediment, but typically they also incorporate at least some of this sediment along growth planes and between crystal aggregates. Figure 152C is a photomicrograph of muddy halite beds from the Permian Salado Formation of New Mexico (Lowenstein, 1988). These muds contain clear, randomly oriented cubes and subhedral to anhedral crystalline aggregates of halite. Detrital siliciclastic mud is incorporated within individual halite crystals where it follows crystal growth planes and occupies the interstices between aggregate crystals. Lowenstein (1988) interprets these halite crystals as intrasediment growths (also see Smith, 1971; Gornitz and Schreiber, 1981; Handford, 1981; Presley and McGillis, 1982).

Massive growth of gypsum and halite within the sediments in the phreatic and vadose zones beneath the dry surfaces of modern sabkhas, arid mudflats, and playas (the **saline mudflat** subenvironments of Hardie and others, 1978) quickly destroys primary sedimentary layering leading to complete "chemoturbation" of the sediments. This "chemoturbation" process within the brine-soaked sediments is accompanied by growth at the sediment surface of efflorescent crusts of halite (Fig. 153). These efflorescent crusts, which can be quite thick (tens of centimeters), are the finely-crystalline saline residues from complete desiccation of groundwater brines sucked up through the vadose zone to the surface by evaporative pumping. It is during the upward passage of these groundwater brines through the vadose zone that the intrasediment growth of the less soluble salts such as gypsum occurs. The rugged surfaces of these very soluble, halite-rich, efflorescent crusts, corroded by rain, act as traps for wind-blown sands. Later dissolution of these crusts by floodwaters may leave behind irregularly-shaped "sand patches" in the underlying muds that could easily be mistaken for adhesion ripples (Smoot and Castens-Seidell, 1982). Preferential dissolution of these efflorescent crusts by halite-undersaturated storm floodwaters is a significant process on both sabkhas and saline mudflats and has several important effects:

Fig. 153.-Saline pan halite crusts. (A) Polygonally ruptured halite crust deformed into polygonal pressure ridges (tepees), the "Devil's Golf Course", Death Valley, California. (B) Typical morphology of a surface halite crust in an ephemeral salt pan, Saline Valley, California. Note the polygonal pressure ridges that are approximately 1 m in diameter. Also note the puffy efflorescence growing in the gap at ridge crests in foreground. (C) Sawn slab of a modern saline pan halite crust (see Casas and Lowenstein, 1989, Fig. 1A). Photograph courtesy of Tim K. Lowenstein. Note the vertical dissolution pipes in the light-colored halite layers due to contact of the crust with undersaturated floodwaters. Dark layer in the middle is a detrital mud layer deposited during the initial stages of salt pan flooding. Pen at left for scale.

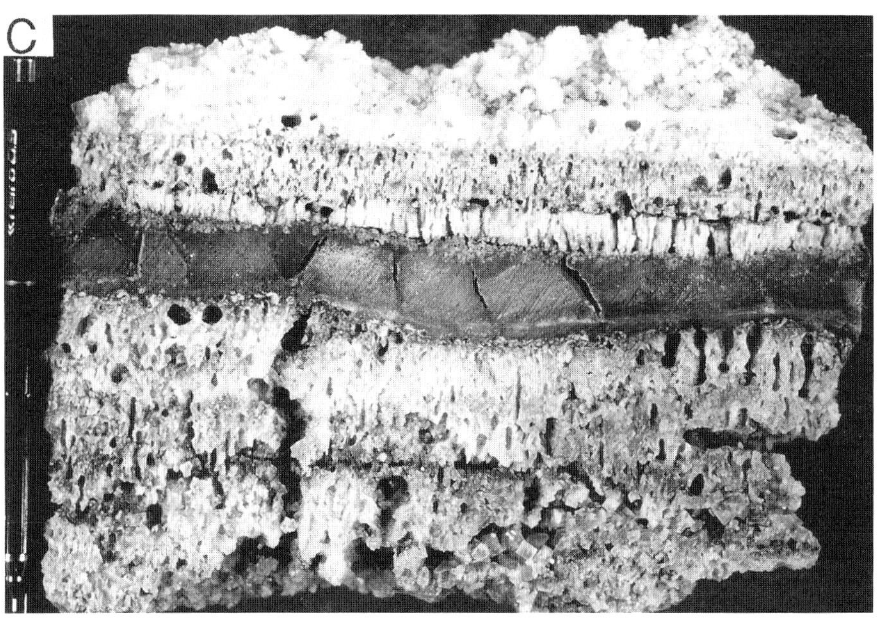

1. It charges the floodwaters with NaCl so that from the very beginning of the lake stage of the saline pan cycle (Lowenstein and Hardie, 1985), the lake waters are briny;

2. It significantly changes the chemical composition of the floodwaters so that their source is no longer easily recognized chemically (Hardie, 1984); this has major consequences on attempts to determine the composition of ancient seawater from the major ion or trace element chemistry of coastal sabkha evaporites;

3. The preferential dissolution of efflorescent crusts produces a variety of distinctive, vertically-oriented dissolutional features within the crust (Casas and Lowenstein, 1989); and

4. For coastal sabkhas, it effects separation of gypsum and anhydrite from halite. This separation occurs because major floods dissolve not only halite efflorescent crusts covering the sabkha, but also dissolve the layered halite of any halite pans on the sabkha. The bulk of these floodwaters flow back into the sea carrying most of the selectively dissolved NaCl, and leaving on the sabkha the less soluble gypsum to accumulate in the gypsum pans, as has been observed to occur on the modern sabkha of Baja California (Castens-Seidell, 1984; Hardie and Shinn, 1986, p. 48-49). This process accounts for the absence of layered halite from almost every ancient sabkha deposit capped by layered gypsum or anhydrite in the geological record (see below).

Occurrence and preservation.-Gypsum crystals (and pseudomorphs after gypsum) reported from ancient rocks commonly have euhedral crystal outlines including "swallow-tail" (or "axe head") twin morphology (Rubin and Friedman, 1977; Hardie and others, 1985; Spencer and Lowenstein, 1990). This euhedral shape is in contrast to the discoidal and flattened morphology of modern intrasediment displacive gypsum (however, see Schreiber and others, 1982; Cloyd and others, 1986, Fig. 9). Halite crystals are not typically preserved in outcrop. In cores, they can be common where they typically include some sediment along growth planes and between crystal aggregates (see Fig. 152; other examples are shown in Smith, 1971; Gornitz and Schreiber, 1981; Handford, 1981; Presley and McGillis, 1982). Molds and casts of intrasediment-grown halite that have been dissolved out and filled with internal sediment could be mistaken for hopper crystals that were once exposed at the sediment-water interface. For example, Figure 150B shows large step-faced hopper-like molds from Silurian dolostones of New York, but it remains a moot point as to whether these are molds of former intrasediment-grown or surface-grown halite crystals. As outlined above, surface efflorescent crusts are rarely if at all preserved in ancient carbonates.

Significance.-Intrasediment growth provides unequivocal evidence of post-depositional crystallization in an evaporitic environment. However, it is difficult to distinguish between minerals precipitated penecontemporaneously from shallow, near-surface brines and later diagenetic growth from deep formation brines during burial (see Hardie and others, 1985; Spencer and Lowenstein, 1990). Workers should take special care to confront this problem of the timing of precipitation when interpreting the significance of evaporite minerals (or their casts and molds) that have grown displacively within the host sediment.

Bedded evaporites.-

Distinguishing characteristics, origin, and significance.-Where evaporation at the brine-air interface of an ephemeral or perennial brine pool is not interrupted by influx of undersaturated waters and can continue unabated, crystals of gypsum and halite that nucleated at the air-brine interface (as described above) will accumulate and grow syntaxially on the bottom to make a gypsum or halite layer. Uninterrupted crystal accumulation by settling from the air-brine interface of both shallow and deep brine pools can produce gypsum and halite beds with cumulate textures (mechanical accumulation of crystals with grain-support textures, see Hardie and others, 1985, for further discussion). However, in shallow brine pools that would be typical of shallow marine carbonate settings such as sabkhas and saline lagoons, initial crystal settle-out is normally replaced by syntaxial overgrowth on crystals at the bottom of the brine pool to produce bedded gypsum and halite with very characteristic, vertically-oriented crystal fabrics (Figs. 154, 155, 156). In modern marine and nonmarine ephemeral halite pans and perennial saline lagoons and lakes (Shearman, 1970; Lowenstein and Hardie, 1985), halite grows at the bottom of the brine pool by syntaxial overgrowth on sunken "hoppers" and "rafts," and competing upward crystal growth rapidly buries those crystal growth faces with nonvertical orientations. This bottom growth at the sediment-brine interface produces vertically oriented "chevron-" and "cornet- shaped" crystals with internal growth bands defined by trains of cubic fluid inclusions (Fig. 155). Ephemeral salt pans go through well-defined cycles of flooding and desiccation that result in easily recognizable partial dissolution textures, vadose and phreatic halite cements, and tepee structures (Lowenstein and Hardie, 1985, have described in detail this salt pan cycle and the diagnostic features it imparts to the bedded halite). Likewise, in the supratidal gypsum pans of the sabkha in Baja California (Castens-Seidell, 1984; Hardie and Shinn, 1986), bottom syntaxial growth on foundered gypsum needles produces a hard crust of prismatic, blade-shaped, and swallow-tailed twinned crystals that are vertically-oriented (Fig. 154A). Like ephemeral halite pans, gypsum pans also go through flooding-desiccation cycles that modify the bottom-formed gypsum crust by partial dissolution, vadose and phreatic cementation by secondary gypsum, and disruption into tepee-structures (see below; Castens-Seidell, 1984, has described these features of gypsum pans in detail; see also Hardie and Shinn, 1986, p. 46-49). Figure 154C shows vertically grown selenite crystals from Holocene Marion Lake, Yorke Peninsula, South Australia. These crystals enclose layers of detrital material which overlie rounded crystal faces that were dissolved prior to the deposition of the detrital sediment. The rounded crystal faces were then repaired by syntaxial overgrowth of gypsum during the next evaporative stage. This type of internal fabric in gypsum crystals is subtle but strong evidence of accumulation under shallow, ephemeral, gypsum pan conditions. Because gypsum is relatively insoluble compared to halite and other saline minerals (gypsum solubility is about 0.2 g $CaSO_4$/100 g H_2O at 25°C), gypsum crystals and cleavage fragments can be transported by floodwaters and deposited as rippled sands and gravels on playas and sabkhas (Castens-Seidell, 1984). It is important to note that post-depositional dehydration of gypsum to anhydrite can severely modify the original textures of gypsum grown in a shallow brine pan. These special anhydritization features are discussed below.

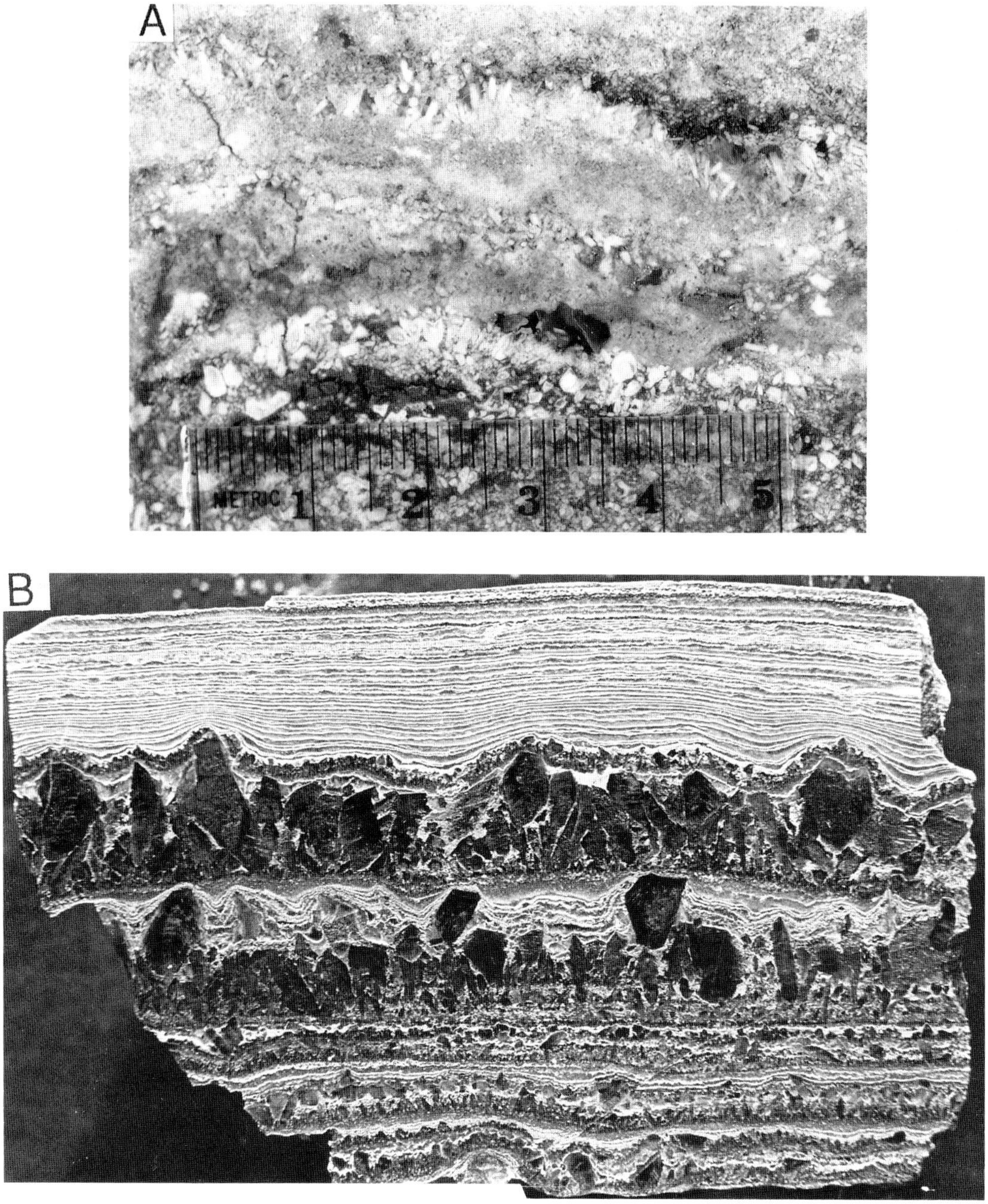

Fig. 154.-Vertically-oriented gypsum crystal layers. (A) Crusts of vertically-oriented gypsum crystals that grew on the bottom of shallow supratidal gypsum pans on the modern sabkha, western Colorado River delta, Baja California (Castens-Seidell, 1984; Hardie and Shinn, 1986, p. 46-49). The crusts are produced by bottom syntaxial growth on foundered gypsum needles which originally grew at the brine-air interface where they were buoyed by surface tension. Scale in mm. (B) An ancient example of layers of vertically-oriented gypsum crystals from the Upper Miocene Solfifera Series of Sicily. See Hardie and Eugster (1971, Fig. 21A). Sample is about 9 cm high. (C) (Following page) Selenite from Holocene Marion Lake, Yorke Peninsula, southern Australia. Layering is defined by aragonite layers that also fill pockets between selenite crystals. Note the prominent aragonite layers that pass through the selenite crystals which, in turn, remain in optical continuity across the aragonite layers. Note especially the rounded and rehealed pyramid apices along growth bands within the gypsum single crystals (best seen in center of photograph). Specimen is 10 cm wide. See Hardie and Eugster (1971, Fig. 21B).

C

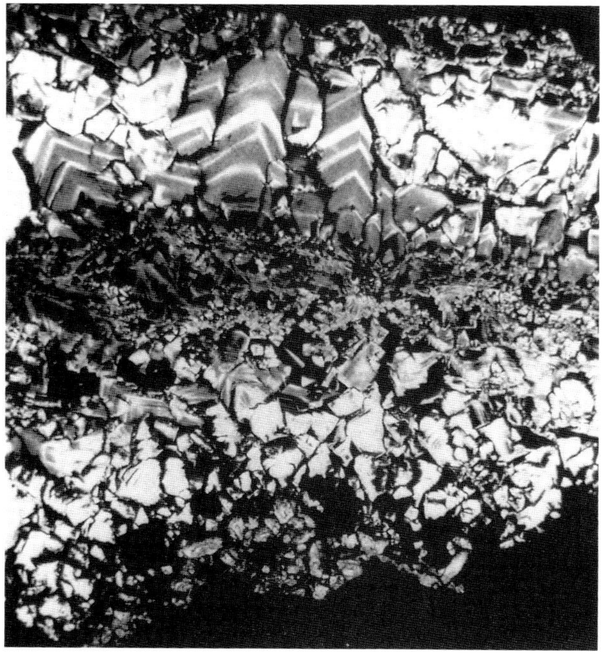

Fig. 155.-Thin-section photomicrograph of surface halite crust from Salina Omotepec, Baja California, Mexico, showing alternating layers of fine-grained cumulates of halite (sunken rafts, plates, and hopper cubes) and coarse-grained, vertically oriented chevrons and cornets that grew upwards from the saline lake bottom. Black areas are voids. In the chevron layers, bands of fluid inclusions can be traced stratigraphically across crystal boundaries from one crystal to the next. Width of field of view is approximately 100 mm. See Lowenstein and Hardie (1985, Fig. 12). Photomicrograph courtesy of Tim K. Lowenstein.

Fig. 156.-The fabric of ancient saline pan halite, Permian Salado Formation, Carlsbad, New Mexico. Thin-section photomicrograph of vertically oriented halite chevrons (upward directed edges and corners) and cornets (upward directed cube faces). Within single crystals, dark bands rich in fluid inclusions alternate with lighter bands containing few fluid inclusions. White areas are clear, inclusion-poor halite cement, usually present as rims on chevrons and cornets. Scale bar is 3 mm long. See Lowenstein (1988, Fig. 12). Photomicrograph courtesy of Tim K. Lowenstein.

Occurrence and preservation.-Because of their high solubility in dilute meteoric surface waters and shallow groundwaters, layers of halite are not commonly preserved in outcrops except in extremely arid climates. All that remains to testify to the former presence of halite layers are various collapse and dissolution breccias (see below). However, in the subsurface, drilling has established that a number of shallow marine carbonate deposits carry bedded halite. In contrast, gypsum and anhydrite are considerably less soluble in meteoric waters than halite and, as a result, it is not uncommon to encounter beds of gypsum and anhydrite in outcrops of shallow marine carbonates even in humid climates.

Nodular and "chicken wire" anhydrite.-

Terminology.-Discovery of solitary anhydrite nodules, small clumps of partially to completely coalesced anhydrite nodules, layers of coalesced anhydrite nodules, and folded layers of anhydrite in the Holocene sediments of the Persian Gulf (Curtis and others, 1963; Shearman, 1966) that were morphologically analogous to ancient nodular anhydrite (Fig. 157) kindled interest among sedimentologists in evaporites and led to the development of "carbonate-evaporite" or "sabkha" depositional models to explain some ancient shallow marine carbonate deposits (see James, 1984; Shinn, 1983a; Hardie and Shinn, 1986).

Imlay (1940, p. 35) introduced the term "chicken wire lattice structure," that, along with the terms mosaic anhydrite (Riley and Byrne, 1961) and flaser anhydrite (Jung, 1958), are commonly used to refer to coalesced anhydrite nodules. Bebout and Maiklem (1973) offer a descriptive terminology of analogous nodular anhydrites found in ancient rocks.

Distinguishing characteristics and origin.-In the sediments of the modern Persian Gulf sabkhas, isolated anhydrite nodules are spherical to ovoid in shape and commonly are tens of millimeters up to 0.25 m in diameter (Fig. 157). Small clumps of nodules are separated by the host sediments that were displaced during crystal growth. Under the microscope, these anhydrite nodules are seen as felted masses of anhydrite laths each approximately 100 microns long. Shearman (1966) and Kinsman (1966) regarded anhydrite as the primary nodule phase that grew in the sediment *de novo*. However, other workers considered that the anhydrite nodules formed from precursor gypsum crystals that, on dehydration, converted into easily-deformable slug-like masses made up of tiny anhydrite laths (Butler, 1970). This latter mechanism is strongly supported by the presence of partially dehydrated gypsum crystals in the sabkha sediments of Baja California (Hardie and Shinn, 1986, p. 49). In addition, West and others (1979) and Ali and West (1983) have described nodules composed of aggregates of small lenticular crystals of gypsum rather than anhydrite in modern sabkhas on the Mediterranean coast of Egypt, demonstrating that clusters of fine-grained gypsum can also form primary nodular structures. A continuous layer of densely-packed anhydrite nodules up to 0.25 m thick has been described from the Persian Gulf (e.g., Kinsman, 1966; Butler, 1970). This densely-packed layer of displacive nodules bears striking resemblance to ancient "chicken-wire" anhydrite. "Chicken-wire" anhydrite is a mosaic array of spherical to flattened nodules separated by thin films of sediment interpreted to be remnants of the sediment in which the nodules grew (Fig. 158A). In the Persian Gulf, the modern chicken-wire anhydrite beneath the subaerially exposed sabkha (Fig. 158A) can be traced laterally in a seaward direction into a gypsum crystal mush zone at the upper edge of the high intertidal-low supratidal zone along the lagoon shore. This gypsum crystal mush zone is formed by massive intrasediment growth of gypsum crystals at the tidal flat surface and in the vadose zone above the shallow seawater table. As the shoreline has prograded during Holocene time, the landward edge of this layer of gypsum crystal "mush" has been covered with supratidal storm sediments and windblown sand and undergone dehydration to "chicken-wire" anhydrite. Other settings than sabkhas have been suggested for the formation of "chicken-wire" anhydrite. For example, dehydration and deformation on burial of layered prismatic gypsum crystal mosaics originally deposited in a variety of subaqueous evaporitic environments such as lagoons and deep basins have been called on to account for nodular anhydrites with "chicken-wire" and other forms that resemble displacive growth textures (Dean and others, 1975; Schreiber and others, 1982; Arakel, 1980; Warren, 1982b; Warren and Kendall, 1985; Machel and Burton, 1991).

Occurrence and preservation.-In many outcrops and shallow borehole cores, nodular structures with all the characteristics of nodular anhydrite described above are composed of gypsum. This nodular gypsum could record primary gypsum precipitation in the form of nodules as West and others (1979) and Ali and West (1983) have demonstrated (see above). However, in our experience, most gypsum nodules display in thin section a fine-grained lath-like fabric that closely resembles the fabric of modern anhydrite nodules, which strongly suggests that they are replacements after anhydrite. The timing of such replacement, however, is difficult to determine. Murray (1964) pointed out that gypsum formed at the surface will dehydrate to anhydrite on burial at depths that depend on pressure, temperature, and groundwater salinity, and this secondary anhydrite in turn will be rehydrated to secondary gypsum on uplift and exposure to dilute meteoric groundwaters. Similarly, anhydrite formed by replacement of gypsum at the surface in a sabkha setting could be rehydrated *in situ* to gypsum if hydrologic conditions at the surface changed enough to bring in dilute meteoric waters. Or such sabkha anhydrite could pass through a burial-uplift cycle that finally results, millions of years later, in rehydration to gypsum. Judging from deeply buried deposits (e.g., Lowenstein, 1987) and deposits that have been buried and then uplifted (e.g., Bosellini and Hardie, 1973), many of the primary sedimentary structures of sabkha and lagoonal gypsum and anhydrite are still legible despite having undergone one or more cycles of dehydration and rehydration (Fig. 159).

Significance.-Since the initial reports and descriptions of nodular anhydrite from the Persian Gulf, there have been an astonishing number of papers published that have called on nodular anhydrite, or calcite and chert nodules presumed to be pseudomorphs after nodular anhydrite, as a criterion for syndepositional growth of evaporites in a sabkha environment. There are, however, a number of problems with interpreting the significance of these nodular features, especially the presumed pseudomorphs. The most serious problem is the fact that anhydrite nodules can form under a variety of conditions. For example, nodules of anhydrite have been discovered in cores of deep sea sediments taken from beneath the Red Sea hot brines in the Atlantis II Deep under 2000 m of water (see Degens and Ross, 1969, color plate between p. 366 and p. 367). Thus, modern nodular anhydrite is not unique to sabkha settings. But it also could be argued that burial dehydration of gypsum crystals or crystal clusters should produce anhydrite nodules in the same manner as occurs in sabkhas, because the nodular form simply reflects that a single crystal of gypsum is replaced by a cohesionless, and hence deformable, mass of small laths of anhydrite plus water. It is not the environment that is responsible for the nodular form but is rather the process of replacement by dehydration. As a consequence, determination of the timing of nodule growth becomes the central problem. Warping of host layers around a nodule is commonly cited as evi-

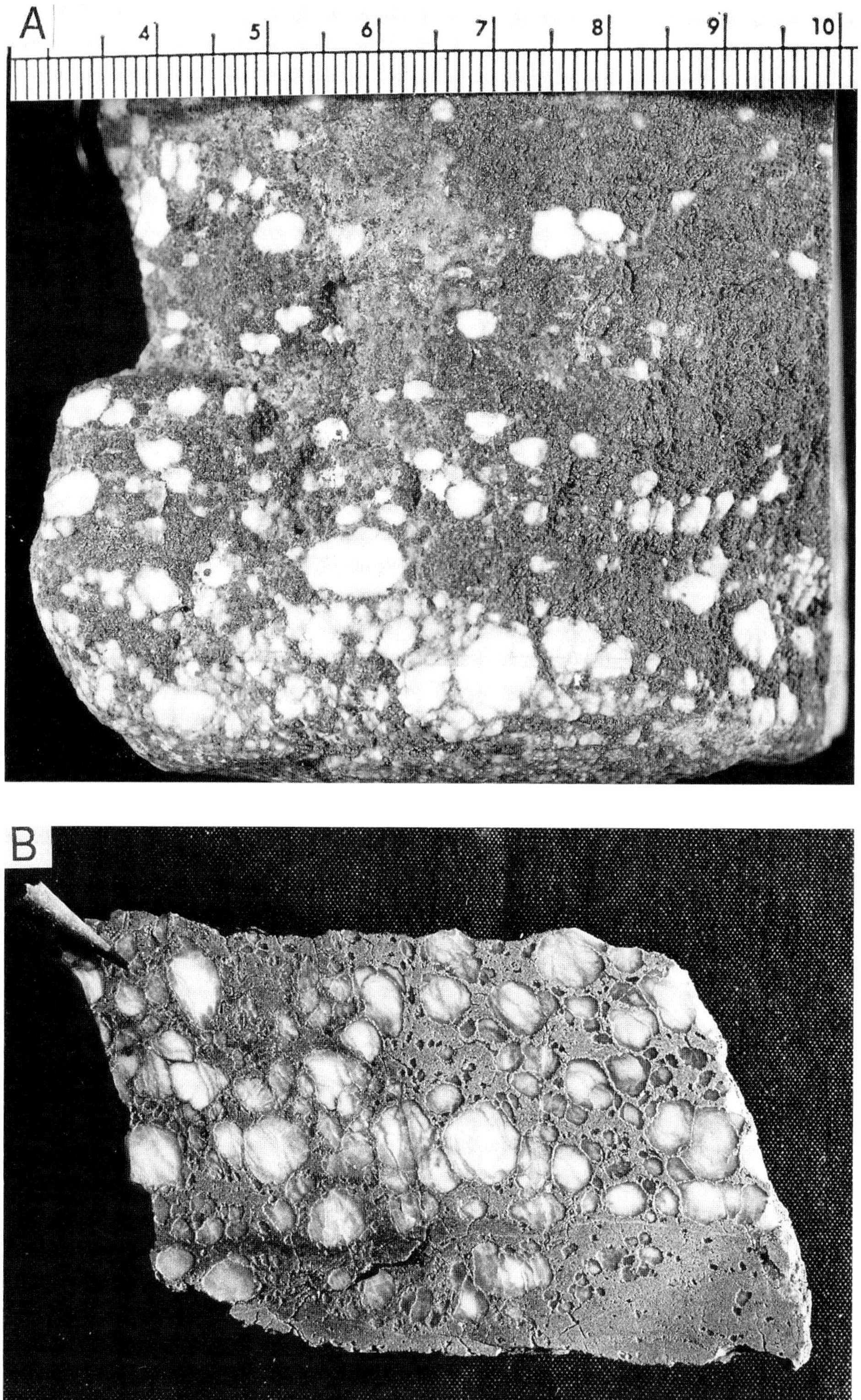

Fig. 157.-Anhydrite and gypsum nodules. Ancient gypsum nodules (originally anhydrite?) in fine peloidal-skeletal grainstones from the Permian Bellerophon Formation of northern Italy (A) and the Miocene Solfifera Series, Sicily (B). Scale in (A) is in mm; pencil for scale in (B). (C) and (D) Modern nodular anhydrite from beneath the sabkha, Abu Dhabi, Persian Gulf (photographs courtesy of Godfrey Butler). Scale bar in (C) is approximately 1 m. Machete in (D) provides scale. (Also see Bosellini and Hardie, 1973)

C

D

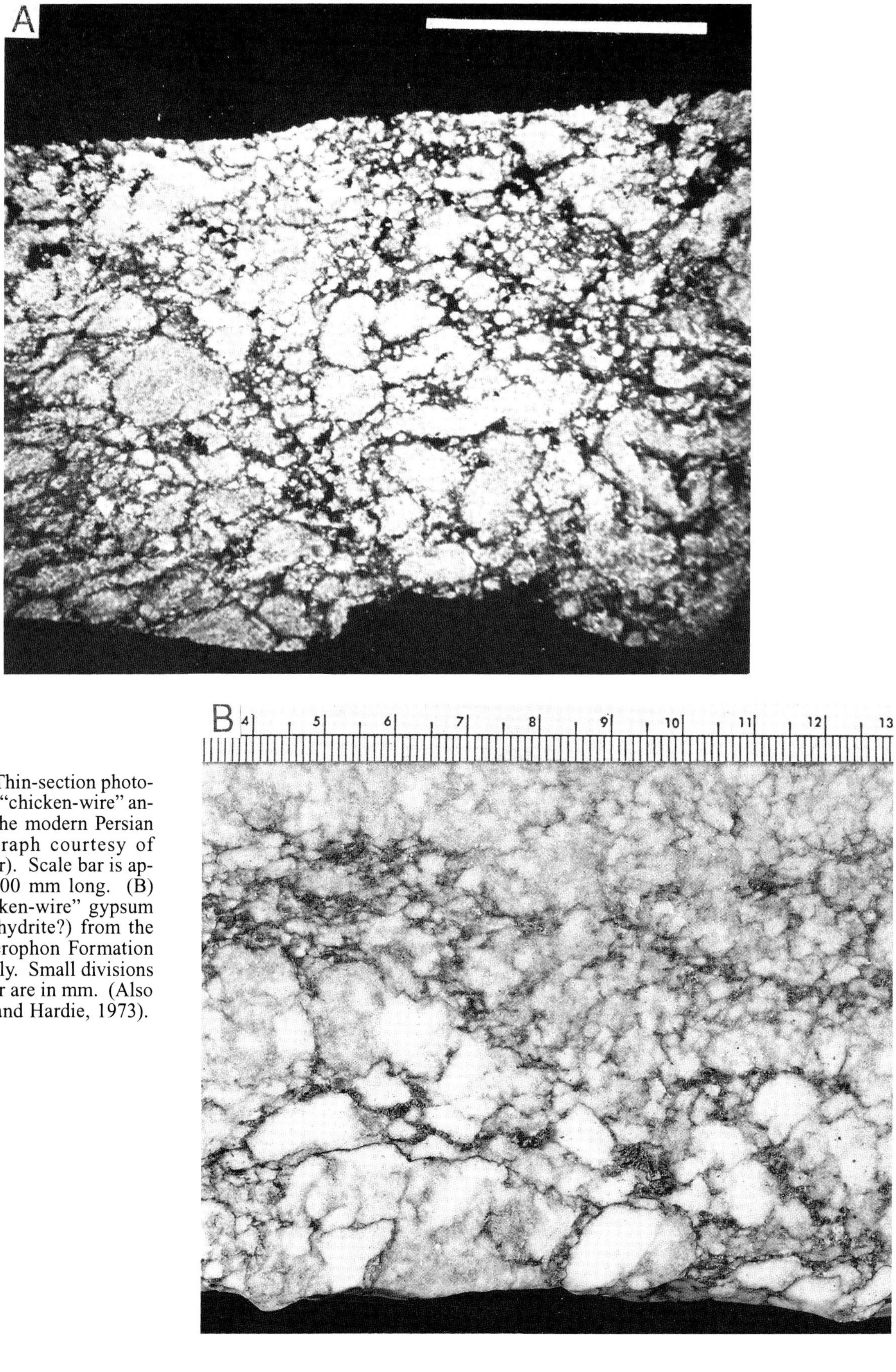

Fig. 158.-(A) Thin-section photomicrograph of "chicken-wire" anhydrite from the modern Persian Gulf (photograph courtesy of Godfrey Butler). Scale bar is approximately 100 mm long. (B) Ancient "chicken-wire" gypsum (originally anhydrite?) from the Permian Bellerophon Formation of northern Italy. Small divisions of the scale bar are in mm. (Also see Bosellini and Hardie, 1973).

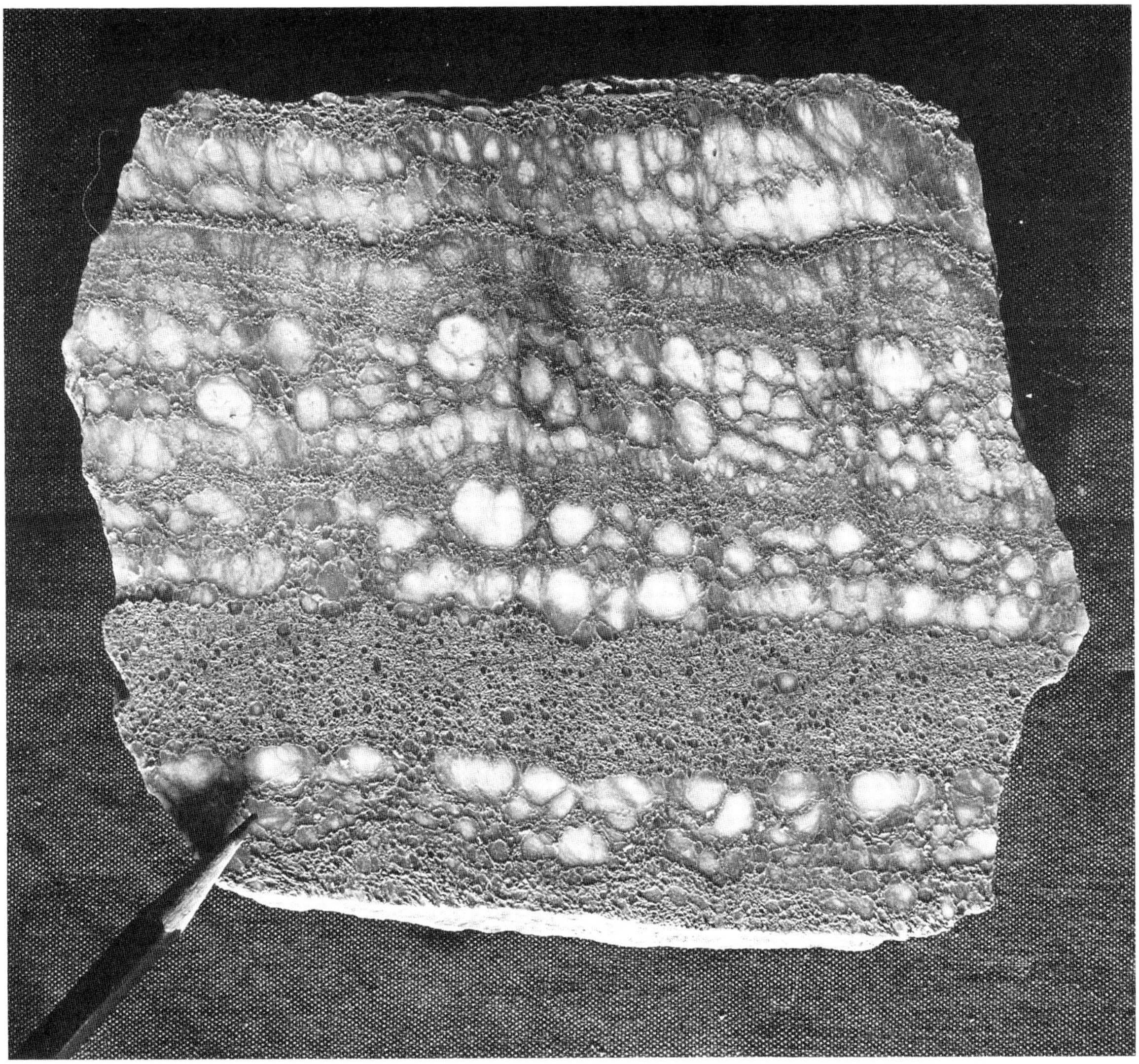

Fig. 159.-Nodular masses of microcrystalline gypsum (white patches) that range in shape from ovoid to rectilinear with outlines typical of vertically-oriented, bottom-grown gypsum single crystals (note the "swallow-tail" shapes of some of the rectilinear forms). Clearly the microcrystalline gypsum is pseudomorphous after large gypsum single crystals or crystal clusters It is possible that an early diagenetic stage of dehydration of bottom-grown gypsum crystals to microcrystalline anhydrite was followed by a final rehydration of the anhydrite to microcrystalline gypsum. Dark interlayers are composed of detrital gypsum grains of granule to sand size. Pencil for scale. Miocene Solfifera Series, Sicily (see Hardie and Eugster, 1971).

dence of penecontemporaneous growth of the nodule in a surface or near-surface sedimentary setting. However, this structure simply indicates that the growth of the nodule occurred prior to compaction. Without other independent evidence, nodules do not give unambiguous information about their time or environment of formation.

Nodules with morphologies similar to the anhydrite nodules from the Persian Gulf sabkha, but not composed of gypsum or anhydrite, are common in ancient carbonates. These typically consist of calcite, dolomite, celestite, megaquartz, or chert and are widely interpreted to be pseudomorphs after sabkha anhydrite nodules. However, little is known about the spectrum of possible origins of these non-evaporite nodules. Where small laths of included sulfates can be found in such nodules (Cloyd and others, 1986) it is indeed likely that they were originally gypsum or anhydrite nodules, but in the absence of some sort of corroborative evidence of this kind, such non-evaporite nodules should not be used as primary testimony of the former presence of anhydrite nodules. Folk and Pittman (1971) and Pittman and Folk (1971) suggested that length-slow chalcedony within a nodule was diagnostic of pseudomorphous replacement of sulfates. However, length-slow chalcedony is also found in DSDP cherts (Keene, 1983) that probably never contained evaporites (see also Jacka, 1974; Kastner, 1980; and Maliva, 1987 for further discussion of this problem).

In summary, although nodular gypsum and anhydrite and layers composed of densely-packed "chicken-wire anhydrites" are characteristic of deposition of evaporites under subaerial sabkha conditions, **they are not unequivocal criteria of such an environment.** For example, nodular gypsum and anhydrite can form by replacement of subaqueous gypsum crystals and crystalline layers (e.g., Loucks and Longman, 1982) or directly from hydrothermal fluids at depth (see above). Replacement could have taken place at the surface before or after burial, or it could have occurred on burial, either deep or shallow. And, indeed, no firm criteria exist to establish the time at which nodular anhydrite or gypsum grew. Nodules that resemble anhydrite nodules, but are composed of calcite, chert, or some other non-sulfate mineral, are even more difficult to interpret because there are no unambiguous criteria by which to decide whether such nodules were made of evaporite minerals in the first place.

Diapiric structures, tepees, and enterolithic folds in layered nodular anhydrite and gypsum.-

Distinguishing characteristics.-Complexly deformed layers of nodular anhydrite are found in ancient evaporite deposits as well as in the Holocene sabkha sediments of the Persian Gulf (Fig. 160). These layers may be up to tens of centimeters thick and are characterized by "enterolithic" folds and tepee-like antiform buckles that enclose diapir-like masses of nodular anhydrite and which are commonly planed off by an erosion surface and overlain by wind-blown sand (Butler, 1970; Shinn, 1983a; Figs. 38, 39, 40).

Origin.-These layers have been interpreted as the product of intrasediment growth of gypsum near the water table (e.g., Kinsman, 1966) with deformation resulting from the volume changes associated with conversion of gypsum to anhydrite. This mechanism for the production of bedded anhydrite in the Persian Gulf clearly presents a problem because it does not account for the bedded nature of the evaporites. The maximum of 10% increase in volume attending the dehydration of 1 mole of gypsum to 1 mole of anhydrite plus 2 moles of H_2O is not sufficient to account for the volume changes observed in some of the folds and thrusts.

An alternative mechanism for the production of layered gypsum-anhydrite was proposed by Castens-Seidell and Hardie (1983) (see also Castens-Seidell, 1984, and Hardie and Shinn, 1986, p. 46-49), based on study of a modern shallow gypsum pan on the siliciclastic sabkha of the Gulf of California (see Thompson, 1968; Weimer and others, 1982). Here, episodic flooding by meteoric water from inland storms and by seawater during onshore storms produces an ephemeral shallow lake in a large (100 km^2) linear depression that borders the huge alluvial fans at the landward edge of the sabkha (see Hardie and Shinn, 1986, Fig. 68). The flooding initially covers the bottom of the lake with a muddy storm layer, millimeters to centimeters thick, that soon after deposition becomes covered by a cyanobacterial mat. In the ensuing weeks, evaporation progressively raises the concentration of the lake water to a concentrated brine. Small gypsum crystals eventually precipitate at the surface of this shallow brine pool (< 2 m deep) and sink to the bottom, where they act as seeds for syntaxial growth of a "grass-like" bottom crust (millimeters to centimeters thick) of vertically-oriented gypsum crystals that blankets the muddy storm layer and its covering of cyanobacterial mats. This gypsum crust starts growing on the wrinkled cyanobacterial mats from which it inherits a somewhat rumpled morphology. But lateral pressure from expansive growth of the gypsum prisms contorts this subaqueous crust into tight enterolithic folds (centimeters to tens of centimeters in crest spacing), some of which may be isoclinal (see, for example, Brantley and others, 1984), and deforming at the same time the underlying muddy storm layer. In some cases, these folded subaqueous gypsum crusts may buckle along polygonal ridges (tens of centimeters to meters in crest spacing) as they grow. As the pan finally dries out, brine level drops just below the surface (typically centimeters to tens of centimeters) and intracrustal growth of gypsum cements takes place under vadose conditions as the groundwater brine is drawn to the surface by evaporative pumping. This diagenetic growth of gypsum cement within the existing voids of the crust leads to renewed compressive deformation of the already buckled and enterolithically folded gypsum crust which ruptures along a polygonal network of fractures with upturned edges making a polygonal pattern of antiformal ridges (Fig. 161). These structures are in essence tepee structures tens of centimeters high and with ridge spacings meters to tens of meters across (compare with carbonate tepees described elsewhere in this atlas). Continued expansive cementation elevates and further deforms the antiformal ridges, many of which develop overthrust anticlinal limbs. With extensive fracturing of the polygonal ridges, the flow paths of the groundwater brine are changed as it is drawn to the surface by evaporative pumping. Cracks and fractures along the crests of the antiformal ridges now become the dominant brine pathways because of their enhanced permeability and thus the antiform ridges become preferred sites for local concentration of gypsum crystal growth. The result is the development of diapir-like masses of gypsum crystals in the cracks and beneath the ridge crests which further uplifts and deforms the polygonal ridges and the underlying muddy sediments. Subsequently, both wind-deflation and storm floods corrode, erode, and plane off the upturned edges of the polygons. These erosional processes deposit the reworked gypsum in the dished depressions between the polygonal ridges as layers and lenses of gypsum sand.

In summary, the following features are characteristic of the layered gypsum as yet **unaltered to anhydrite** in Baja California: (1) thin-bedded character, (2) interlayering of bottom-grown crusts and detrital gypsum layers with layers of intraclastic sediment, and (3) deformational features such as enterolithic folds, tepee buckles with overthrusts, and diapir-like masses. These features are also characteristic of the lay-

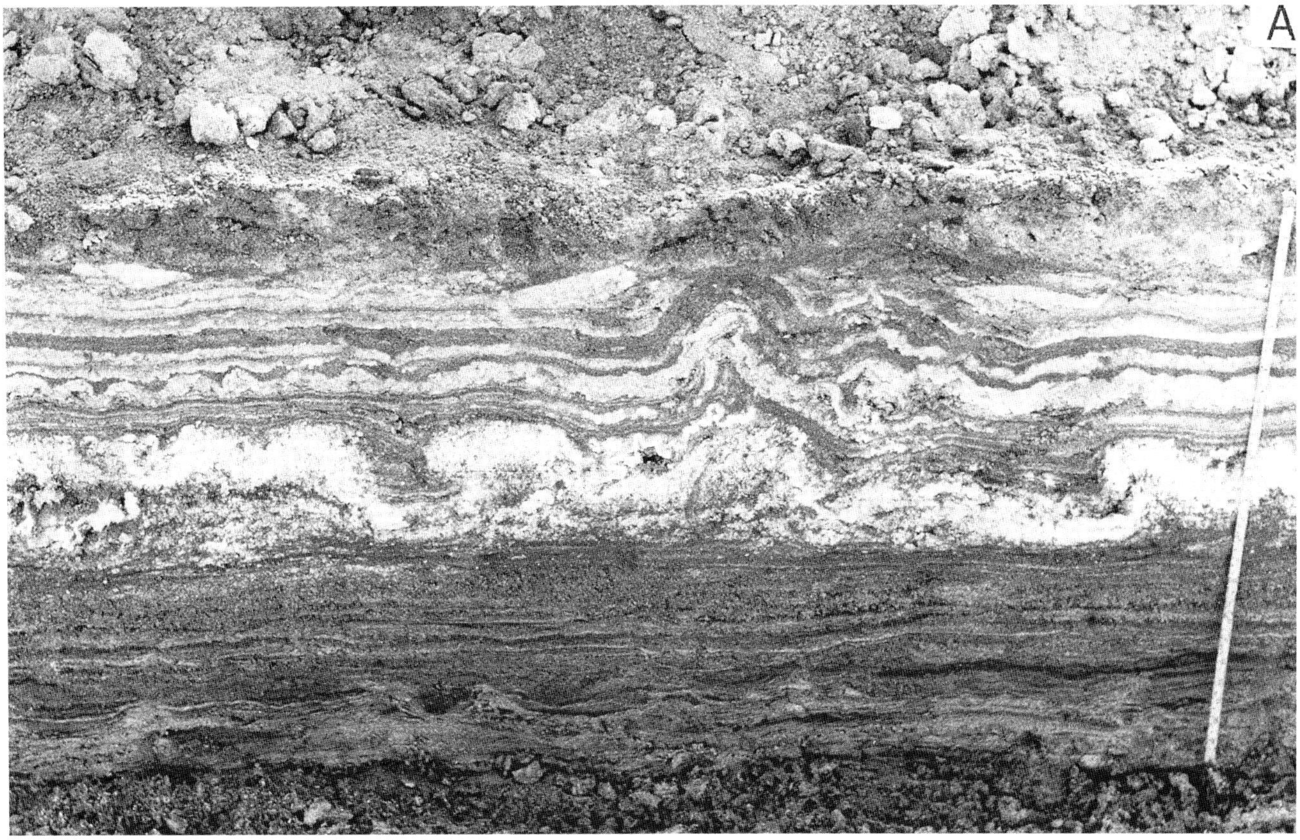
A

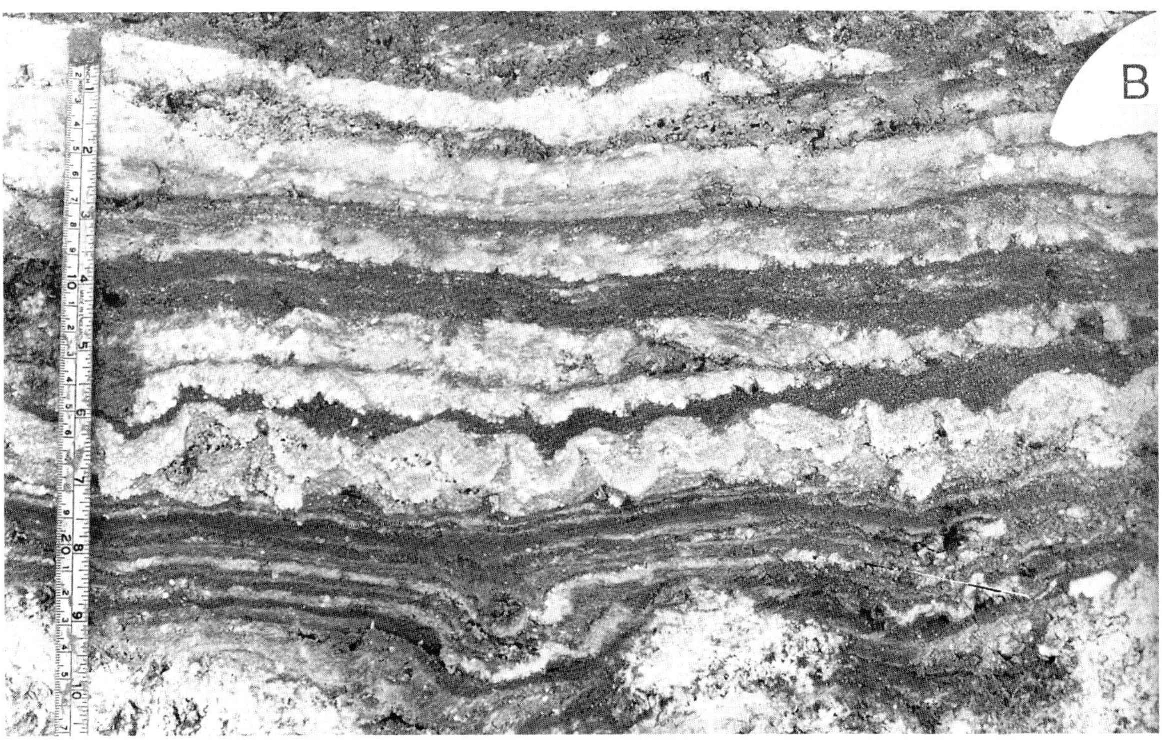
B

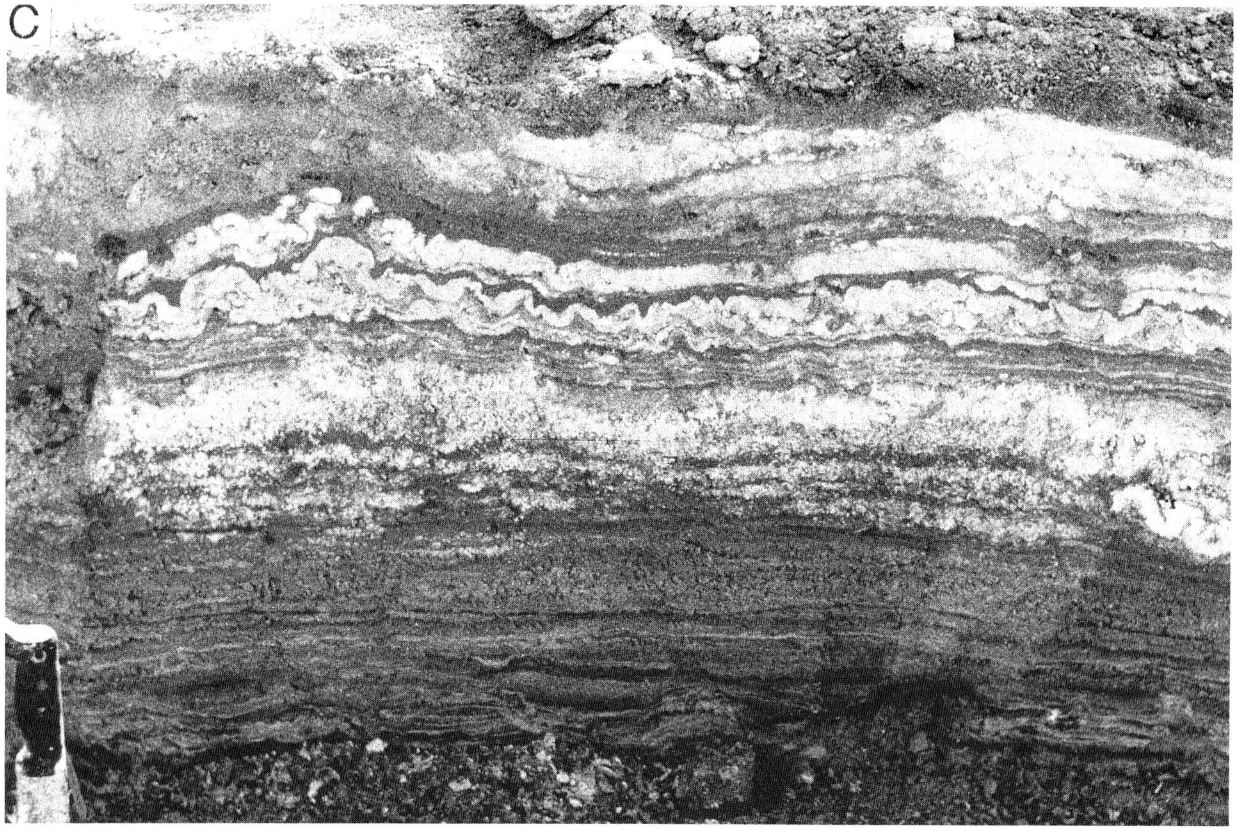

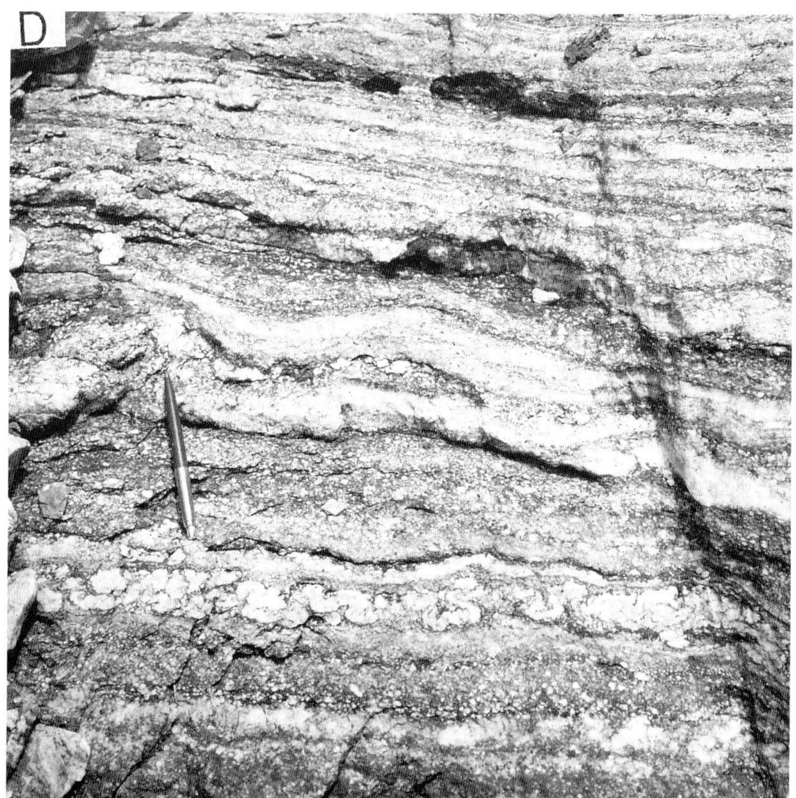

Fig. 160.-Layered nodular anhydrite. (A), (B) (Preceeding page) and (C) Complexly deformed layers of nodular anhydrite found in the Holocene sabkha sediments of the Persian Gulf. These layers are characterized by tepee-like antiform buckles (note the overthrust antiform in the right center of A), diapir-like masses of nodular anhydrite, and "enterolithic" folds. In (A) and (C) the upper layers have been planed off by erosion and overlain by wind-blown sand. Dark wavy layered units in the lower third of (A) and (C) are upper intertidal-lower supratidal cyanobacterial laminites. See Fig. 166 for stratigraphic relations of the sabkha evaporites to the cyanobacterial laminites and associated subfacies. Photographs courtesy of Godfrey Butler. (D) An ancient example of layered nodular anhydrite (now gypsum) with several layers displaying enterolithic folds (e.g., above and below the pen on the left). Permian Bellerophon Formation of northern Italy. Compare with modern layered anhydrite in (C). (Also see Bosellini and Hardie, 1973).

Fig. 161.-Polygonal antiform ridges (tepees) in a gypsum surface crust that floors an ephemeral gypsum saline pan on the modern siliciclastic sabkha of the Colorado River delta, Baja California, Mexico.

ered nodular anhydrite of the Persian Gulf sabkhas. We interpret this to mean that the deformation features in anhydrite are not the result of dehydration of gypsum to anhydrite, but are characteristic primary features of gypsum deposition under ephemeral saline pan conditions (for a full discussion of the saline pan cycle, see Lowenstein and Hardie, 1985). The Baja California model (Castens-Seidell, 1984) provides a rational explanation for the Persian Gulf layered anhydrite and analogous ancient layered anhydrites that cap shallowing-upward tidal flat cycles (Hardie and Shinn, 1986). We consider that the layered anhydrites of the Persian Gulf may represent "dead" gypsum pan systems that have been filled in and buried by aeolian sediments as the tidal flats have prograded seaward during the last 5000 years. Infilling of the pans has shut down active subaqueous deposition of layered gypsum and the buried gypsum pan deposits are now abandoned to subsurface vadose diagenesis by continental and mixed continental-marine brines. These brines converted much of the gypsum with all its original syndepositional deformation features to anhydrite, and also added calcian dolomite to the underlying intertidal and subtidal carbonates (see discussion in Hardie, 1987). In the Baja California gypsum pans, partial dehydration of gypsum to anhydrite is in the initial stages where the sand flats at the toes of the alluvial fans are beginning to encroach on the pan margins.

Occurrence, preservation, and significance.-The comments made in the section above on nodular anhydrite and gypsum for timing of replacement of gypsum by anhydrite and rehydration of anhydrite to gypsum, whether early at the surface, or during burial, or even later during uplift, apply equally well to the syndepositional deformation structures described in this section (see under Nodular Anhydrite and Gypsum).

Solution Features

Solution features arise where carbonates (or evaporites) are infiltrated by undersaturated waters. Despite the apparent simplicity of the dissolution process, solution features can be problematical. First, recognition of solution features is not straightforward and criteria whereby dissolution can be unequivocally recognized are lacking. Second, criteria that unequivocally establish the timing of the dissolution are lacking.

Solution can effect grains or crystals prior to deposition, during deposition, or anytime thereafter. Indeed, most carbonates are ultimately dissolved in weathering profiles. We are mainly concerned here with those solution features that carry information about the depositional environment. These post-depositional but pre-burial features are formed diagenetically soon after deposition by processes controlled by the existing depositional environment. They include some solution pores and some collapse breccias. However, solution pores and collapse breccias can also be post-burial features formed by late diagenetic or metamorphic-metasomatic processes controlled by the subsurface burial environment (Hardie and others, 1985). Choquette and Pray (1970) refer to the zone of early diagenesis as the eogenetic zone and the deeper zone as the mesogenetic zone. Choquette and Pray (1970, p. 220) also define a telogenetic zone, where "long-buried carbonate rocks are influenced significantly by processes associated with the formation of an unconformity," (i.e., weathering related to uplift and erosion). Karst and its ancient equivalent, paleokarst, are major results of telogenetic processes and include an array of features such as solution pores, collapse breccias, and depositional features (speleothems, internal sediments, etc.) as well. Karst and paleokarst are outcrop scale features and the reader is referred to the papers in James and Choquette (1988) for further details. One problem arises with any such terminology. Local relative sea-level changes (due to eustasy or local tectonism) can effect deposits from the surface down to burial depths of a hundred meters or more and can directly overprint earlier diagenetic features with a "telogenetic" signature. Familiar examples are the karstic features of south Florida, the Bahamas, and other parts of the Caribbean formed during Pleistocene sea-level fluctuations. Such boundaries are crucial to sequence stratigraphic analysis and this growing importance is evidenced by the papers in the symposium volume edited by James and Choquette (1988).

Here we shall consider pores (vugs, cavities, channels) of solution origin, solution collapse breccias, and stromatactis structures. Criteria that identify a dissolution origin for each of these features are discussed, as well as criteria for identifying timing. Stylolites are a very important solution feature of carbonate rocks (see Pettijohn, 1975; Scholle and Halley, 1985; Moore, 1989). Sutured stylolites form at burial depths of at least a few hundred meters. A number of geologists have postulated the existence of "pervasive" pressure solution without recognizable suturing (Logan and Semeniuk, 1976; Wanless, 1979; Bathurst, 1987). These types of features are discussed under the section on compaction above.

Solution pores, cavities and vugs.-

Terminology.-The most widely used classification scheme for porosity was proposed by Choquette and Pray (1970) and is essentially a genetic scheme. Pores and pore networks generally attributed to solution include moldic pores, vugs (equant pores), channels (elongate, inter-connected pores), and caverns ("person-sized" channels), and voids associated with solution collapse breccias. Choquette and Pray's classification includes a number of modifiers for pore descriptions based on the size of the pore.

Distinguishing characteristics and origin.-It has become standard practice to consider neomorphosed skeletal grains and ooids, or molds of these grains filled with cements, as evidence of a former aragonite or high magnesium calcite composition (Scoffin, 1987; Moore, 1989). Indeed, terms such as "diagenetic potential" have arisen to describe the aragonite and high magnesium calcite content of a sediment. Clearly, solution can enlarge a hole that started as a discrete grain (Fig. 162). However, at some point, the original shapes of the grains are obscured and it is impossible to recognize their former existence. The terms vug and channel refer to holes that are, in general, "not fabric selective" (Choquette and Pray, 1970). These are "cross-cutting" pore systems that have "indiscriminate position relative to fabric elements." It is important to note that cross-cutting relationships can exist at a number of scales: thin section, hand sample and outcrop scale, and that cross-cutting of some fabric elements alone is not sufficient to establish solution. For example, where the boundary of any hole comprises grain boundaries, solution is impossible to establish. Clearly this is a problem for muds where cohesion may allow open holes to be generated by fluid erosion (see discussion of Stromatactis). In short, the criteria to establish solution rest upon demonstrating the truncation of hard fabric elements, skeletal grains, ooids, cement crystals, etc. Even this may not be enough where pores have regular, tubular shapes and boring organisms may have been involved.

Occurrence and preservation.-Solution pores are common in carbonate rocks and contribute porosity to some of the largest oil fields on earth (Roehl and Choquette, 1985). However, from the considerations above, it is clear that establishing the timing of a solution feature may be difficult. Fossils in internal sediment within void fillings are the best way to establish the timing of the cavity fill. Certain cement-crystal morphologies are also commonly linked to post-depositional but pre burial conditions (Longman, 1980; Moore, 1989). However, given our current understanding of the controls on cement-crystal morphologies, the reliability of this approach is questionable.

Solution collapse breccias.-

Solution collapse breccias are an important feature of both karst and carbonate-evaporite successions encountered in outcrop. In both cases, the solution episode that generated the collapse may be classed as either of surface or subsurface (interstratal) origin (Wright, 1982). However, distinguishing between these two possibilities is difficult in the case of solution breccias in ancient deposits. Breccias formed by solution collapse are commonly clast support breccias and may be either the open work type or have the interstices between the clasts filled with a variety of internal sediments and cements. Solution of evaporites is commonly identified by lateral tracing of the layer into the subsurface where intact evaporites are found. Discussions of the features of paleokarst, the importance of paleokarstic collapse structures, and criteria for identifying these features are found in Middleton (1961), Clifton (1967), Blount and Moore (1969), Wright (1982), Esteban and Klappa (1983), James and Choquette (1988), Choquette and James (1990), and Loucks and Handford (1992).

Stromatactis.-

Terminology.-The name *Stromatactis* was first used by Dupont (1881) for irregularly-shaped, sparry patches he considered to be recrystallized stromatoporoids. These, and similar sparry patches have also been called: (1) reef tufa; (2) blue and black veins; (3) crystalline laminae; (4) bands of crystalline calcite; (5) sparry calcite; (6) patches of coarse crystalline calcite; and (7) *Stromatactis* and *Ptylostroma* (see Lees, 1964, p. 509 for references). Of these the British terms "reef tufa" or "tufa" (Pettijohn, 1975) are the most common synonyms.

Fig. 162.-Polished slab of an ancient dissolution surface, Middle Ordovician St. Paul Group, western Maryland. Chambers of gastropod tests are filled with internal sediment. The voids created by dissolution of the tests are filled with sediment from the overlying bed. Scale in mm. Sample courtesy of Ray Mitchell (see Mitchell, 1985, Fig. 5).

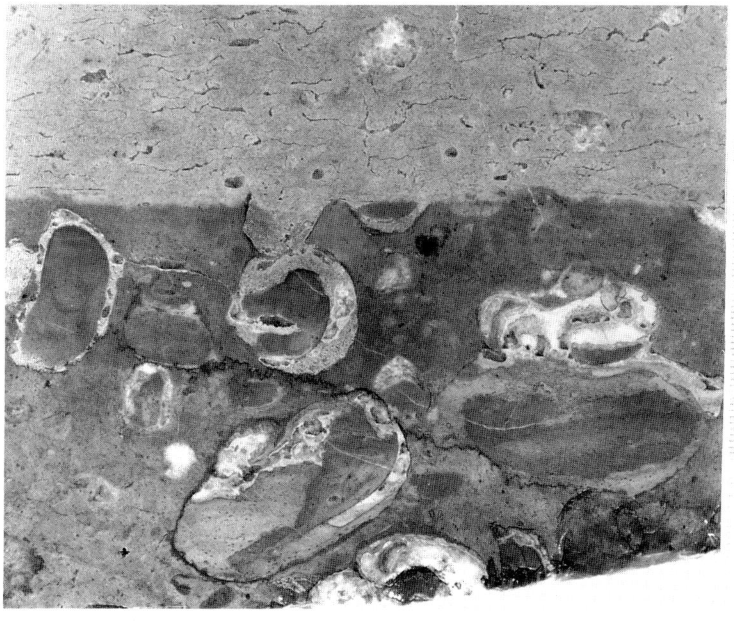

Heckel (1972) followed by Bathurst (1980b, 1982) used the term "stromatactoid structure" (uncapitalized and not in italics) based on their interpretations of stromatactis as an inorganic sedimentary structure. Tsien (1985) among others, suggested retention of *Stromatactis* (capitalized and in italics) based on his belief it was an organic sedimentary feature.

Distinguishing characteristics.-Bathurst (1982, p. 167) considers the following five criteria as essential to the diagnosis of stromatactis (Figs. 163, 164):

1. Stromatactis is an irregularly-shaped mass of sparry calcite (mainly radiaxial fibrous calcite) and internal sediment that ranges in shape from elongate to globose. Typically, the patches are 1-30 mm high and up to a few hundreds of millimeters long (although larger examples are known).
2. Stromatactis has a flat to undulose, smooth base.
3. The upper surface is irregularly digitate or dentate.
4. Stromatactis occurs in swarms with sub-parallel floors.
5. Individual patches within swarms are characteristically joined together by bridges of spar to form an interconnected, reticulate structure in three dimensions.

Bathurst emphasizes criteria 4 and 5 and does not consider individual examples to be diagnostic. Most other definitions usually include only criteria 1-3 above. In addition to the diagnostic criteria listed above, there are a number of less common but noteworthy features of stromatactis. Stromatactis structures are generally elongate and parallel the sides and tops of mud mounds (i.e. they are parallel to coarse bedding features in the mounds such as shale laminae or skeletal-fragment rich layers). At the edges of mounds, stromatactis can be inclined up to 50 degrees (Lees, 1964).

The internal sediment is commonly peloidal silt, although rare skeletal fragments (similar to those in the mudstone of the mounds) and broken crystal fragments (apparently cement crystals broken off wall linings) can be found. The internal sediment may be parallel laminated, show small-scale cross-stratification, or be a conical pile. In some cases, the internal sediment grades down into the host peloidal microspar (Wallace, 1987). Where the original floors of the stromatactis patch can be discerned, it is undulose to ragged, not digitate (Bathurst, 1982). In a number of cases, there is interleaving of successive layers of internal sediment and radiaxial fibrous calcite.

The spar is concentrated at the top of the structure (i.e. the structures are geopetals). The elongate crystals are oriented perpendicular to walls and increase in size away from host substrate into the sparry patch. The top of the cavity may be supported by a skeletal fragment in rare instances but such support is not typical. There may be a final, centrally located patch of blocky spar.

Stromatactis as defined above consists of equant to, more commonly, elongate forms. In many cases, the elongate forms grade laterally or downward into "zebra rocks," which are bedding parallel swarms of sheet cracks filled with radiaxial calcite spar (Fischer, 1964). The equant forms may be found grading into completely buckled, disrupted breccias (Bathurst, 1982; Logan and Semeniuk, 1976). The transition to brecciation is marked by en-echelon fractures that offset the cavity-supporting host mudstone. Small-scale buckled uplifts (tepees) are also reportedly associated with stromatactis.

Origin.-Stromatactis represents a cavity system that has been filled by internal sediment and cements. The timing of the internal sedimentation and cementation is commonly regarded as early based on the similarity of the internal sediment and its fossil content to the peloidal mudstones that comprise the bulk of the host mud mounds. Also, rare sedimentary breccias shed off of mounds contain stromatactis with orientations that demonstrate they have been transported as clasts. Radiaxial fibrous calcite is generally regarded as either a neomorphic replacement of an early magnesium calcite fibrous cement (Kendall and Tucker, 1973; Bathurst, 1959) or as a primary fibrous cement (Kendall, 1985). The interleaving of internal sediments and radiaxial fibrous calcite is also cited as evidence for an early syndepositional origin of the internal sediment of the cavity network.

The mechanism of cavity formation, however, is not yet settled. Current suggestions include: (1) decay of soft bodied organisms or influence of microbial encrustations; (2) burrow networks such as *Callianassa*; (3) dewatering; (4) downslope creep and slumping; and (5) differential cementation of host peloidal mudstone into successive, closely-spaced hardgrounds. In this latter mechanism, the hardgrounds were developed parallel to mound surfaces, and the cavity system represents unlithified intra-hardground areas through which large volumes of fluid has passed, depositing the internal sediment and cements (Bathurst, 1982; Wallace, 1987). Wilson (1975), Heckel (1972), Bathurst (1982), and Tsien (1985) provide reviews and literature citations. A central problem is the perceived inability of even cohesive mud to hold open networks of cavities. To do this requires early cementation as suggested by Bathurst (1980b, 1982). Wallace (1987) emphasizes the erosion, transportation and deposition of unlithified mound mud by pore fluids circulating through a system of interconnected cavities.

Occurrence and preservation.-Although mud mounds are reported to range from Paleozoic to Jurassic in age, the most spectacular examples occur in Paleozoic carbonates. Stromatactis is apparently restricted to these Paleozoic mud mounds (Wilson, 1975; Bathurst, 1982; James and MacIntyre, 1985). Stromatactis-riddled mud mounds are commonly referred to as "Waulsortian" mounds from Carboniferous examples in Belgium. Neuman and others (1977) report unfilled stromatactis-like cavities from modern cemented deep-sea lithoherms.

Details of the stromatactis structure are well preserved in outcrops, where their abundance and sparry nature make them quite spectacular visual displays.

Significance.-Clearly, until the origin of stromatactis structures is resolved, their significance is uncertain. Whatever their origin, they appear to be reliable geopetal indicators. Also, the evidence of early cementation of both the host and the cavities may help explain the apparent high slopes of some of these mounds. Finally, stromatactis indicates that large amounts of seawater passed through the system.

Fig. 163.-Stromatactis from the Devonian mud mounds of Belgium. Photograph courtesy of Wayne M. Ahr.

Fig. 164.-Stromatactis from the Devonian mud mounds of Belgium. Photograph courtesy of Wayne M. Ahr.

IMPORTANT ASSOCIATIONS

DIAGNOSIS OF ANCIENT CARBONATE DEPOSITIONAL ENVIRONMENTS

The Method of Comparative Sedimentology

The basic method of recognizing ancient shallow marine carbonate deposits in the geologic record is that of comparative sedimentology (Ginsburg, 1974), using as diagnostic elements the primary sedimentary structures and early diagenetic features described and illustrated in the body of this atlas, together with information derived from sediment textures, fossils, paleocurrents, and the geometries of individual depositional components (e.g., reefs, mud mounds, channel forms, etc.).

The method is most effective when used in a stratigraphic framework of careful measurement of vertical sections in outcrop or core. ***The primary objective is to subdivide a carbonate deposit into distinctive "microstratigraphic" units (vertical dimension on decimeter- to meter-scale) using the primary sedimentary structures, the early diagenetic features, the sedimentary textures, the fossils, the paleocurrents, and the geometries of individual depositional components.*** Such a subdivision into stratigraphic units could be a purely statistical one using "objective" criteria (that is, purely descriptive criteria) or it could be based on "genetic" criteria using particular associations of depositional and early diagenetic features derived from modern environments, flume experiments, theoretical arguments, and so on. We advocate the use of the "genetic" approach, one that divides the deposit into facies and their component subfacies defined respectively as the rock records of the ancient environments and their component subenvironments in which the sediments were deposited.

The following outline of the method has been taken, with some modifications, from Hardie and Shinn (1986, p. 37).

The fundamental unit of measurement is the **subfacies**, which is the rock record of a **subenvironment**. Modern sedimentary environments of all kinds consist of a mosaic of genetically linked subenvironments. For example, the carbonate tidal flat environment of Andros Island in the Bahamas is made up of a complex mosaic of interdependent channel, levee, and tidal pond subenvironments (Hardie, 1977a). Each subenvironment is a mappable part of the environment that has a distinctive physiography, distinctive surface features, and in which there operates a distinctive set of physical, biological, and chemical processes. As a result, the sedimentary record ofa subenvironment is characterized by a distinctive set of physical, biological, and chemical structures, sediment textures, early diagenetic features, biota, and, in certain subenvironments, sediment textures. It is these sets of primary sedimentary structures, sediment textures, early diagenetic features, and biota of each subenvironment that are the keys to the comparative sedimentology approach. Using such sets of distinctive features calibrated from modern environments as guides, stratigraphic sections are divided into subfacies. This approach carries a direct genetic link between the basic rock unit and its presumed depositional setting, that is, subfacies = deposit of a subenvironment. Based on the distribution of subfacies in the measured sections, genetically linked sets of subfacies are combined into larger stratigraphic units defined as **facies**. The definitions used here are as follows:

Environmental Unit		Rock Unit
subenvironment	=	subfacies
environment	=	facies

There are obvious dangers in using a rock terminology that is directly linked to the interpretation of rock origin, but the advantages are that it goes directly to the heart of the matter, the genesis of the deposit, and it forces us to ask critical questions about the significance of each observation. The definitions of facies and subfacies used here are significantly different from those employed for siliciclastic deposits (see Miall, 1978; Walker, 1984). A number of workers in siliciclastics essentially equate the record of an individual sedimentary structure, such as small-scale cross-stratification, with a facies. But a sedimentary structure is the product of a sedimentary processes (e.g. down-stream migration and vertical aggradation of ripples during bed-load transport of sand) that may occur in many different depositional environments. Therefore, such structures by themselves may not be environmentally diagnostic. Also, some vertical successions of sedimentary strucures may be deposited from the same flow (such as Bouma sequences and other waning current sequences), and so certainly do not signify the stratigraphic superposition of the deposits of different environments. Thus, the complexities introduced by considering individual sedimentary structures as facies can thoroughly obscure the basic picture of the paleoenvironment relations in space and time. It is only the whole set of interacting processes, operating in any distinctive subenvironment, that leaves a diagnostic record in the rocks. This is the rationale for the genetic approach to subfacies and facies outlined above and used in the treatment of ancient shallow marine carbonate deposits given below.

Importance of Sedimentary Structures and Early Diagenetic Features in the Diagnosis of Shallow Marine Carbonate Facies and Subfacies

Subfacies in carbonate rocks are most commonly defined on the basis of: (1) grain types, (2) grain textures, (3) fossil content, (4) sedimentary structures, and (5) early diagenetic features. Directional features and large-scale bedding geometry, although commonly used in defining subfacies in siliciclastic rocks, are only rarely used in studies of carbonates. Whereas all of these rock properties hold important information about the depositional environment, sedimentary structures comprise the single most useful set of parameters we can observe in sedimentary rocks. Unlike grain types, sedimentary structures were unambiguously made in place where they formed in direct response to physical, chemical or biological processes. Thus, sedimentary structures yield more basic quantitative and qualitative information about the depositional processes than do sediment types and textures. Grains can be transported from one subenvironment or another, from one environment to another. Although in modern marine environments, aragonitic ooids are formed in agitated settings at shelf margins they are commonly transported both into the shelf interior and off the margin into adjacent basins. Clearly, in such circumstances serious errors in interpretation would result if the presence of ooids were taken as primary evidence for a shallow shelf margin setting. The thick and extensive Mesozoic Vajont Formation of the Italian Dolomites stands as a worthy example of this cautionary tale. The Vajont Formation is comprised primarily of ooid grainstones which in the past have been taken as evidence of a shallow shelf depositional environment for the Vajont as a whole. However, later work using sedimentary structures, together with the identification of interbedded mudstones as radiolarian oozes, demonstrated that the ooids were transported by submarine slope processes into a deep basinal depositional environment (Bosellini and others, 1981; Zempolich, 1993, pers. commun.).

In many cases, grain types do not allow sufficient discrimination of ancient environmental settings. For example, from a petrologic point of view, 75% to 80% of the Cambro-Ordovician carbonates of the central Appalachians would be classified as peloidal mudstones, yet the array of sedimentary structures found in these peloidal mudstones is astonishing and allows sections to be differentiated into subfacies on the decimeter scale. Analyses of grain size, sorting, roundness, sphericity, shape, surface texture, and other textural parameters of sedimentary rocks likewise have failed to produce reliable paleoenvironmental discrimination, in spite of decades of research (Pettijohn, 1975). Indeed, in carbonate rocks these textural attributes are rarely, if ever, quantified (see Flugel, 1982). A number of these parameters are meaningless when applied to some carbonate textures. The sphericity of ooids and other coated grains is an example.

The depositional significance of *in situ* invertebrate fossils can only be ascertained by comparative morphological studies of recent forms and in this regard it must be recognized that the reliability of comparative morphological interpretations decreases back through the geologic record. Many, if not most, fossils found in carbonate rocks have been transported to one degree or another and any interpretations based on such transported fossil assemblages are less reliable than those based on *in situ* fossils.

Finally, sedimentary structures remain identifiable even after the host rock has suffered considerable recrystallization and mineral reconstitution during diagenesis or metamorphism. The same is not true of grain types, textures, and fossils which are commonly altered beyond recognition. For example, there are many published studies in which sedimentary structures have been successfully used to diagnose the paleoenvironments of ancient dolomites where the grain types and textures have been obliterated by dolomitization (cf. many of the papers in Grotzinger, 1989).

In light of these arguments, we have emphasized sedimentary structures and early diagenetic features in defining subfacies in our treatment below of ancient shallow marine carbonate deposits (see under Illustrative Examples).

Modern Analogs and the Building Block Approach

Comparison of a host of modern tidal flat systems reveals that no two Holocene tidal flat deposits are quite the same, a conclusion that seems to apply with equal certainty to ancient tidal flat deposits. It would seem a logical extension of this to conclude that it is unlikely that any ancient tidal flat deposits in the geologic column would be precisely analogous to any of the modern tidal flat deposits. Therefore, it clearly would be a mistake in using the method of comparative sedimentology to try to force ancient deposits to fit one of the few modern shallow marine carbonate environments that have been carefully studied, for example, those of the Bahamas, the Caribbean, the Persian Gulf, and Western Australia. Certainly first order comparisons of this sort are valuable guides to identifying the basic environmental settings in which ancient carbonates were deposited at the facies and even subfacies scales. However, as Hardie (1977c, p. 188-189) has pointed out, "if we are to go beyond broad comparisons and to exploit the wealth of environmental information stored in ancient tidal flat deposits then we must use an approach that stresses the individuality of each deposit. Our only recourse is to build 'thought models' by piecing together whatever we need from an array of basic 'building blocks.' These building blocks are the sedimentary structures, fabrics, textures, produced by the spectrum of processes that operate on tidal flats of all kinds, rainy or arid, high energy or low energy, carbonate or non-carbonate, tropical or temperate, land-locked or surrounded by the ocean. With a full storehouse of such building blocks, we should be able to construct any model we need, some pieces coming from one modern example, some from another."

We recommend that this basic "building block" approach be the cornerstone in any attempt to use the comparative sedimentology method to diagnose the paleoenvironments and paleosubenvironments of ancient carbonate deposits of all kinds. We have used this approach in the illustrative examples of genetic associations of sedimentary structures and early diagenetic features in ancient shallow marine carbonate deposits given at the end of this section.

One particular value of using the "building block" approach is that we should be able to go beyond general interpretations such as "supratidal," "intertidal," and "subtidal" and recognize different kinds of supratidal, intertidal, and subtidal subenvironments. More sharply-focussed interpretations of each subfacies could yield fundamental information on paleoclimate, paleotidal ranges, paleowave regimes, and so on (see INTRODUCTION, and Hardie and Shinn, 1986, p. 3). For example, in modern tidal flat systems:

1. the supratidal subfacies records with some clarity in sedimentary structures and early diagenetic features the prevailing climatic conditions (e.g., the arid coastal sabkhas of the Persian Gulf versus the rainy supratidal marshes of the Bahamas);
2. the intertidal subfacies not only houses the position of ancient mean sea level but it also preserves a legible record of the ancient mean tidal range (Klein, 1971; Terwindt, 1988); and
3. the shallow (<10 m depth) subtidal shelf-lagoon subfacies may record information on the ancient storm and tidal regimes (storm and tidal current velocities and directions; storm and fair-weather wave parameters) that can be read from wave, current, and hummocky cross-stratification.

Vertical Successions of Subfacies: Walther's Law and Progradation

A critical aspect in the diagnosis of the paleoenvironments of ancient shallow marine carbonates using the method of comparative sedimentology is the vertical succession of subfacies within facies. In the absence of unconformities, Walther's Law (Middleton, 1973) can be used to predict the order in which subfacies should appear in stratigraphic sections and thereby act as a rigorous test of an interpretation of a vertical succession of subfacies. Or, in cases of isolated subfacies of uncertain affinities, the vertical successions predicted by Walther's Law can be used as a diagnostic tool in the interpretation of the paleosubenvironment of the subfacies in question.

In modern shallow marine carbonate environments, the fundamental process that leads to a systematic vertical succession of subfacies in accord with Walther's Law is **progradation of tidal flats out over a shallow shelf-lagoon** (see Hardie and Shinn, 1986, p. 59-65). During the latest Holocene as the rate of sea-level rise slowed, the modern carbonate tidal flats of the Bahamas, the Persian Gulf, and Shark Bay (Western Australia) have grown seaward at net rates that range from 0.5 to about 5 km/1000 yr (Hardie and Shinn, 1986, p. 59). This shoreline progradation is the result of episodic sediment accretion as peloidal lime mud from the offshore "carbonate factory" has been added to the tidal flats by onshore storms (Hardie and Ginsburg, 1977) during the last 5000 years or so at rates faster than sea level has risen. **Such shoreline progradation produces a meter-scale, shallowing-upward, stratigraphic suc-**

cession of supratidal over intertidal over subtidal subfacies in classic Walther's Law fashion (Fig. 165). Progradation of carbonate tidal flats may be relatively continuous laterally, as in the Persian Gulf (Fig. 166), or it may be staggered, as in the Bahamas (Fig. 167).

Significantly, many ancient shallow marine carbonate buildups consist principally of vertical stacks of meter-scale, shallowing-upward successions (1-10 m thick) that are capped by supratidal subfacies, suggesting that progradation has been a fundamental element of shallow water carbonate depositional systems back at least into the Proterozoic (e.g., Demicco, 1983, 1985; Grotzinger, 1989; Hardie and Shinn, 1986; Koerschner and Read, 1989).

High Frequency Depositional Cyclicity

The vertical stacking of meter-scale, shallowing-upward successions that is so characteristic of many ancient shallow water carbonate deposits clearly involved repetition over and over again through the life of carbonate buildups of more or less the same cycle of conditions. It is the origin of this repetition of conditions and its resulting "cyclostratigraphy" that has been the principal concern of a number of recent studies of shallow water carbonates (see, for example, Grotzinger, 1986a, b; Hardie and Shinn, 1986; Goldhammer and others, 1987, 1990; Koerschner and Read, 1989).

The time scale of a single shallowing-upward depositional "cycle" is on the order of 10^4 years (5th order cycles) to 10^5 years (4th order cycles) based on data for total time/total number of cycles from ancient cyclic carbonates (e.g., Fischer, 1964; Hardie and Shinn, 1986, Table 2; Koerschner and Read, 1989). There are two main styles of 4th and 5th order high frequency shallowing-upward depositional cycle in ancient platform carbonates: (1) shallow subtidal subfacies capped by intertidal and supratidal subfacies that display a variety of primary sedimentary structures such as lamination, stromatolites, mudcracks, desiccation fenestrae, evaporite features, and so on, as described in the main body of this atlas, and (2) shallow subtidal subfacies directly capped by soils, caliches, tepee structures, and other features indicative of subaerial exposure without sediment accretion, as described in this atlas under Chemical Features. The critical differences between these two types of shallowing-upward depositional cycles lie in the different vertical successions of their component subfacies and in the nature of the boundaries between subfacies (Hardie and Shinn, 1986, p. 62).

In the first type, the vertical succession of supratidal over intertidal over subtidal subfacies with transitional boundaries between each subfacies unit follows a classic Walther's Law arrangement and clearly records a lengthy episode of seaward progradation of actively accreting tidal flats when sediment supply easily exceeded any relative sea-level changes (combined effects of eustasy, subsidence, tectonic movement). This type of Walther's Law "cycle" is typi-

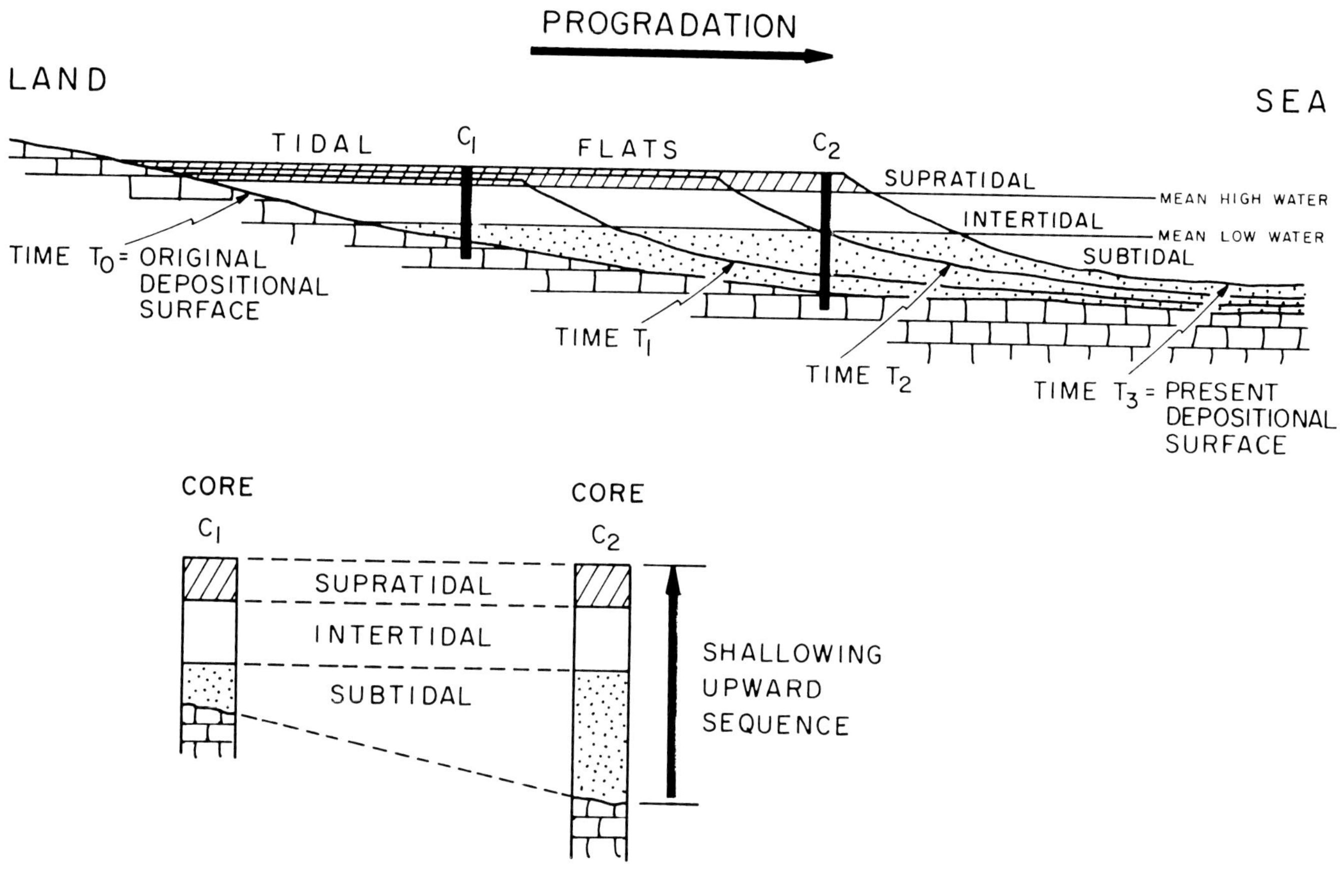

Fig. 165.-Generation of a shallowing-upward cycle by progradation of tidal flats (Hardie and Shinn, 1986, Fig. 72).

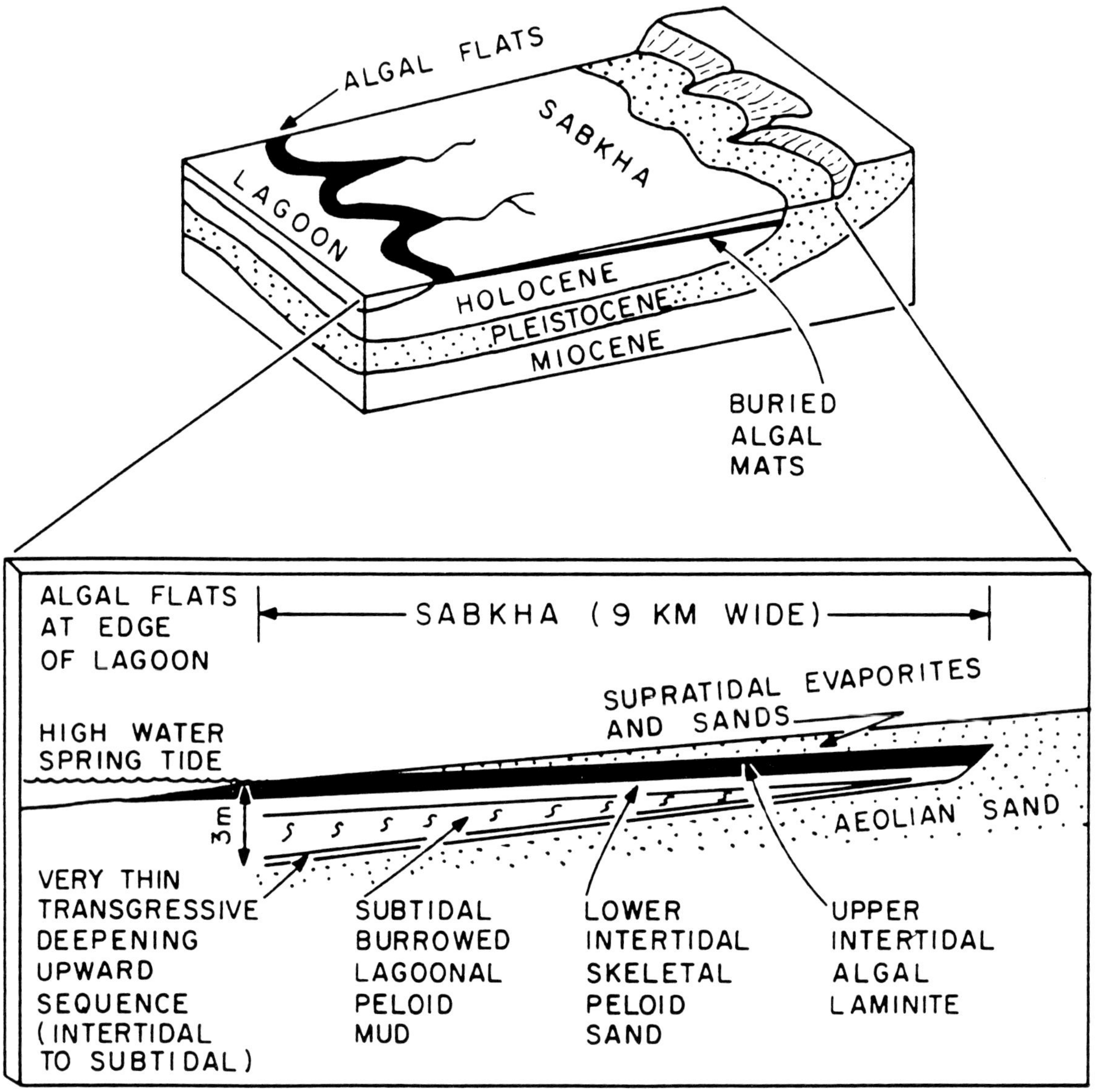

Fig. 166.-The Holocene deposits of the Abu Dhabi (Persian Gulf) arid tidal flats as an example of active accumulation of a shallowing-upward tidal flat cycle by progradation (Hardie and Shinn, 1986, Fig. 74, after Patterson and Kinsman, 1977).

cal, for example, of the shallowing-upward tidal flat cycles of the Upper Cambrian Conococheague Limestone of the central Appalachians. This and other similar cyclic shallow marine carbonates are described below under Illustrative Examples.

The second type of cycle, exemplified by the Pleistocene carbonates of south Florida (Perkins, 1977) and the Triassic of the Southern Alps (see under Illustrative Examples; see also Goldhammer and others, 1987, 1990), clearly records high frequency changes in relative sea level. In these systems, rapidly falling sea level led to subaerial exposure of subtidal sediments *without an intervening stage of tidal flat upbuilding and lateral progradation.* The period of emergence was extensive enough to superimpose disconformably a diagenetic caliche soil "cap" directly on the subtidal subfacies. Subsequent sea-level rise flooded the platform once again and allowed renewed accumulation of shallow subtidal deposits. The vertical stack of these soil-capped subtidal cycles are, in a sense, a strip-chart recording of the history of relative sea level that oscillated with a frequency on the order of 10^4 to 10^5 years.

The critical question of the cause of the repeated stacking of 4th and 5th order shallowing-upward cycles of both type 1 and type 2 in shallow marine carbonate platform deposits remains to be answered. The succession of these cycles in ancient shallow marine carbonates clearly records high frequency changes in relative sea level but whether such changes are the result of autocyclic (Ginsburg, 1971a) or allocyclic processes or a combination of the two has not been unequivocally resolved for most ancient shallow water deposits. For discussions of the current state of the problem the reader is referred to Hardie and Shinn (1986) and to the recent discussion/reply papers by Hardie and others (1991) and Read and others (1991).

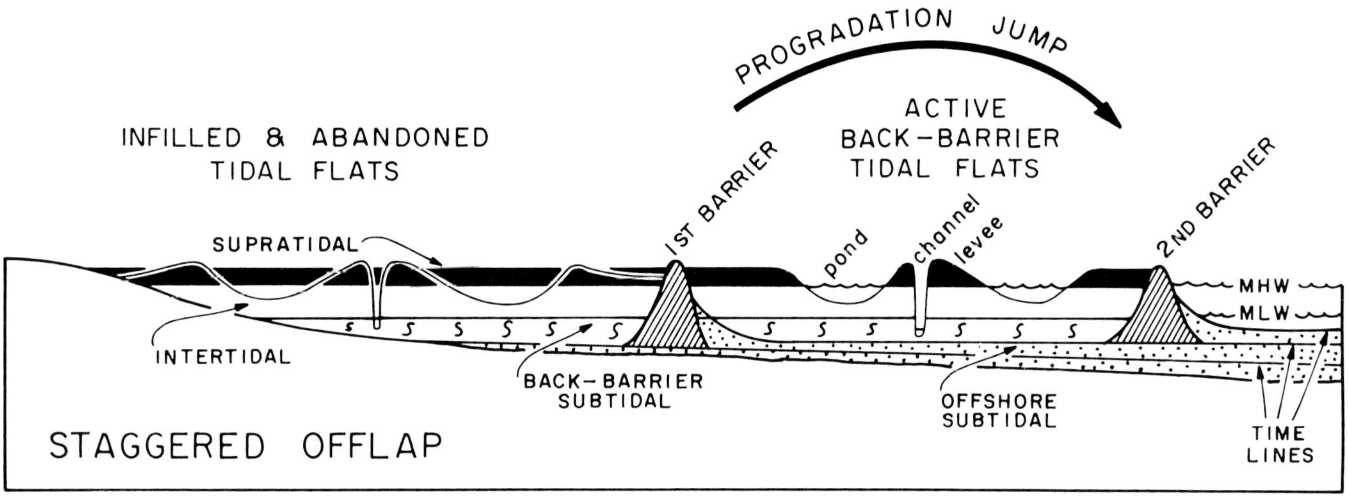

Fig. 167.-Progradation by "staggered offlap" as exemplified by the Holocene tidal flats accumulating on western Andros Island (Hardie and Shinn, 1986, Fig. 76)

AN INVENTORY OF COMMON SEDIMENTARY STRUCTURES AND EARLY DIAGENETIC FEATURES OF SHALLOW MARINE CARBONATE DEPOSITS

The shallow-water portions of modern carbonate buildups comprise the following succession of depositional environments from the buildup interior to its margin (Fig. 2):

1. shorelines (typically tidal flats),
2. very shallow inner shelf-lagoon,
3. deeper outer shelf-lagoon,
4. marginal sand shoals (typically, but not exclusively, oolitic), and
5. marginal reefs (present only where conditions were favorable for development of barrier reef-building organisms).

Ancient facies equivalents of deposits of these various modern environments (including karst and paleokarst overprints) are recognized from rocks as old as early Proterozoic (Wilson, 1975; Ginsburg, 1975; Hardie and Shinn, 1986; James and Choquette, 1988; Grotzinger, 1989; Walker and James, 1992).

Table 4 is a checklist of the various sedimentary structures and early diagenetic features found in these environments. Reef environments are not included in this table which is tabulated for Phanerozoic deposits. In Precambrian rocks, bioturbation structures so typical of subtidal rocks in the Phanerozoic are missing and in their place are found a wide variety of microbial structures, such as subtidal stromatolites.

ILLUSTRATIVE EXAMPLES

In the following sections, we present illustrative examples of the subfacies and facies "building block" approach in the interpretation of ancient carbonate shallow-water deposits. There are five examples of carbonate tidal flat and associated shoreline deposits, four examples of shelf-lagoon deposits, and one example of an ancient shelf-margin deposit. These examples are drawn mainly from our own work in the Cambro-Ordovician deposits of the central Appalachians and Alberta Rocky Mountains, and the Mesozoic carbonates of the Southern Alps of northern Italy and Northern Limestone Alps of Austria. Thus, they are not meant to be any kind of standard "facies models" but rather they represent specific individual examples of paleoenvironmental interpretations. For reviews of modern carbonate depositional environments, further discussions of paleoenviromental interpretation of ancient carbonate shallow water deposits, and references to many individual case studies of ancient deposits see Lucia (1972), Bathurst (1975), Wilson (1975), Ginsburg (1975), Scholle and others (1983), Hardie and Shinn (1986), Tucker and Wright (1990), and Chapters 14-19 in Walker and James (1992).

Tidal Flat Facies of the Upper Cambrian Conococheague Limestone, Central Appalachians

The Conococheague Limestone is part of a very thick succession of platform carbonates in the Appalachians that records continuous deposition at or near sea level from Middle Cambrian to Middle Ordovician time. The Cambro-Ordovician carbonate platform deposits of the Appalachians comprise a wedge over 3000 km long, more than 300 km wide, and up to 3.5 km thick (Colton, 1970). These carbonates accumulated on the ancient passive margin of eastern North America and constitute one of the largest accumulations of carbonates in the geologic record. In the central Appalachians, the Conococheague Limestone is 750 m thick and consists primarily of coarse-grained, noncyclic shelf lagoon facies (see later) interstratified with two separate sequences of cyclic, tidal flat facies (Demicco and Hardie, 1981; Demicco and Mitchell, 1982; Demicco, 1985).

The tidal flat facies are up to 100 m thick and consist of a succession of stacked, meter-scale, shallowing upward cycles capped by mudcracked, laminated mudstones (Root, 1964, 1968; Reinhardt and Hardie, 1976). Figure 168 is a bedding diagram and accompanying vertical logs of typical Conococheague Limestone cycles based on correlated measured sections (see Demicco, 1981, 1985 for details).

TABLE 4. Important Associations of Sedimentary Structures of Shallow Marine Carbonate Deposits

		Neptunian Dikes	Hummocky Cross-Stratification	Storm-Lag Lithoclast Gravels & Conglomerates	Channel-Lag Gravels & Conglomerates	Large-Scale Cross-Stratification	Submarine Hardgrounds	Thrombolites	Wave Ripple Cross-Stratification	Small-Scale Cross-Stratification	Bioturbation Structures [1]	Erosional Scour Marks	Scalloped & Planar Erosion Surfaces	Planar Laminae	Mechanical Compaction Structures
TIDAL FLATS	Rainy Supratidal Flats								●	●	R	●	?	●	●
	Evaporitic Supratidal Flats								●	●		●	?	●	●
	Low Tidal-Range (<1m) Intertidal Flats							R	●	●	●	●	●	●	●
	High Tidal-Range (>1m) Intertidal Flats							●	●	●	●	●	●	●	●
	Subtidal Ponds						?	?	●	●	●	●	?	R	●
	Tidal Channels				●	●	R	●	R	●	●	●	?	●	R
	Beach Ridges		?			●			●	●	●	●	?	●	R
	MUDDY INNER SHELF						?	●	●	●	●	●	●	●	●
	SANDY INNER SHELF		?	●		R	●	●	●	●	●	●	●	●	R
	OUTER SHELF	●	●	●		●	●	●	●	●	●	●	●	●	R
	MARGINAL SHOALS	●	●	●	●	●	●	●	●	●	●	●	●	●	R

		Fenestrae	Stromatolites [1]	Heterolithic Cross-Stratification	Planar Thick Beds	Wavy-Crinkled Thick Beds	Planar Thin Beds	Wavy Thin Beds	Wavy Laminae	Edgewise Gravels & Conglomerates	Imbricate, etc., Gravels & Conglomerates	Crinkled Laminae	Crinkled Thin Beds	Bioturbated Pseudobreccias [1]	Mudcracks
TIDAL FLATS	Rainy Supratidal Flats	●	●				●	●	●	●	●	●	●	●	●
	Evaporitic Supratidal Flats	●	●				●	●	●	●	●	●	●		●
	Low Tidal-Range (<1m) Intertidal Flats	●	●	●			●	●	●	●	●	●	●	●	●
	High Tidal-Range (>1m) Intertidal Flats	●	●	●			●	●	●	●	●	R	R	●	●
	Subtidal Ponds	●	●	?	●	●	●	●			●			?	R
	Tidal Channels	●	●	●			●	●			●				
	Beach Ridges	●					?	?	●	●	●				
	MUDDY INNER SHELF	●	●	?	●	●	●	●	●						
	SANDY INNER SHELF	●	●	●	●		●	●	●						
	OUTER SHELF	●	●	●			?	?							
	MARGINAL SHOALS	●	R												

		Prism Cracks	Sheet Cracks	Molar Tooth Structure	Cryptmicrobial Laminites [1]	Jellyroll Structures [1]	Tufas & Travertines [1]	Caliches & Associated Soil Features	Supratidal Crusts	Tepee Structures	Solution Pores, Cavities, etc.	Evaporite Minerals, Casts, & Molds	Nodular & "Chicken Wire" Sulfates	Bedded Evaporites	Deformed Sulfate Layers
TIDAL FLATS	Rainy Supratidal Flats	●	●	●	●	●	●	●	●	●	●				
	Evaporitic Supratidal Flats	●	●	●	●	●	●	●	●	●	●	●	●	●	●
	Low Tidal-Range (<1m) Intertidal Flats	●	●	●	●	●					R	●	●		
	High Tidal-Range (>1m) Intertidal Flats	●	●	●	●	●					R	●	●		
	Subtidal Ponds	●		●	?	?									
	Tidal Channels														
	Beach Ridges							●		●	?	?	?	?	
	MUDDY INNER SHELF														
	SANDY INNER SHELF									●					
	OUTER SHELF									●					
	MARGINAL SHOALS									●					

[1] Table applies to Phanerozoic. For Precambrian, no burrows and increased importance of microbially-influenced structures

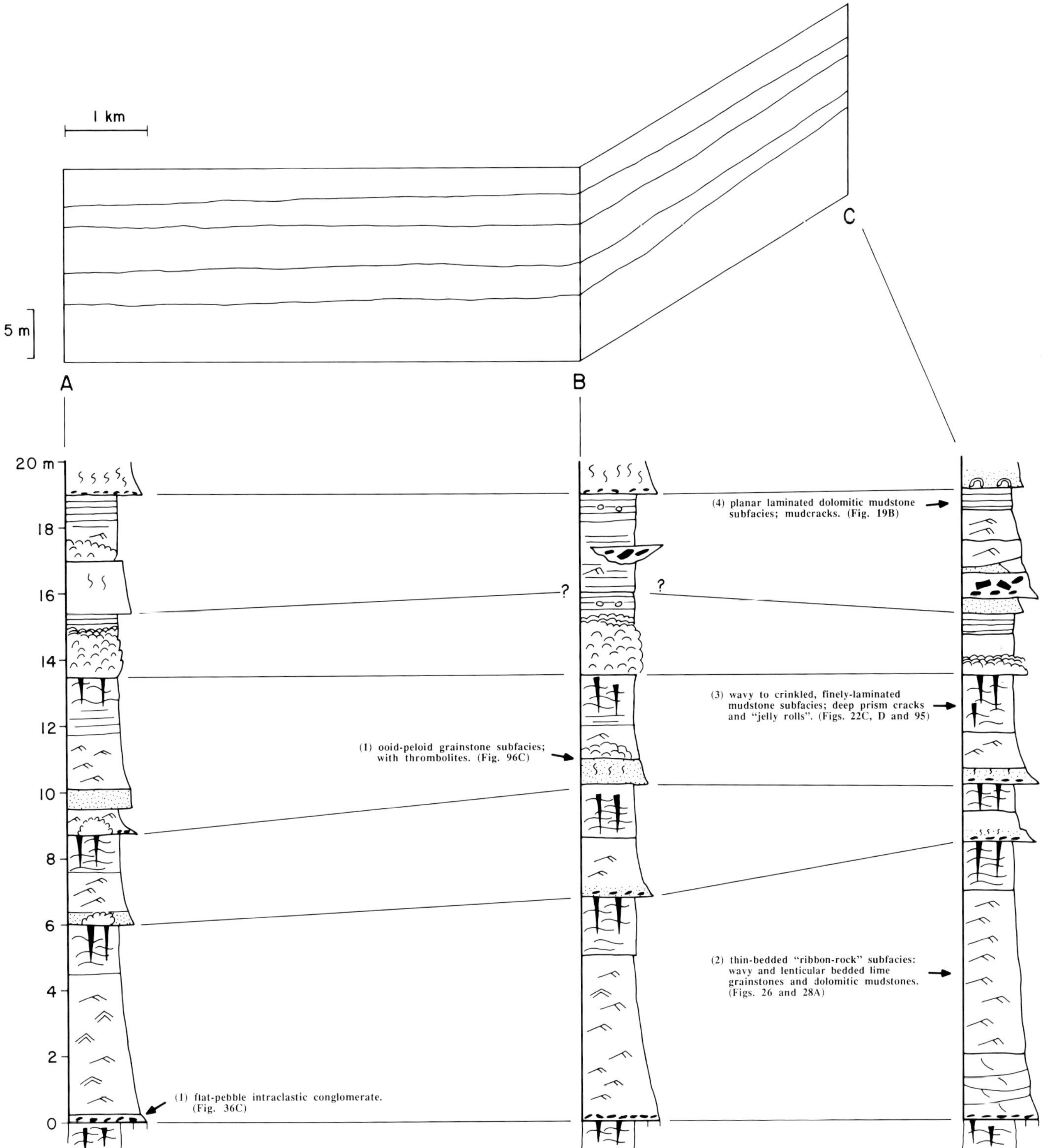

Fig. 168.-Conococheague cyclic facies. Upper portion of the figure is a *bedding diagram* that shows the inferred, layer-cake geometry of the cycles between cycle boundaries (represented by the black lines). The boundaries are based on exposures up to several hundred meters long, and correlated outcrops. The corners of the bedding diagram are the three stratigraphic logs below. The vertical scale of the logs is meters whereas the horizontal scale indicates relative grain size, coarsening to the right. Brief descriptions of the subfacies and numbers of text figures that show representative examples of the subfacies are given and arrows point to the symbol for that subfacies on the logs. *Please note that in this and succeeding figures representative photographs are from the same formation except where no good photograph is available and a close match from another formation has been substituted.*

Cycles are composed of four subfacies organized vertically as follows:

1. A basal subfacies, typically a few decimeters thick, of peloidal-ooid grainstones, flat pebble conglomerates, and small thrombolites. The grainstones are commonly burrow-mottled so that cross-stratification is not well preserved. The flat pebble conglomerates carry discoidal clasts tens of millimeters long composed mainly of peloid-ooid grainstones with some recycled conglomerate clasts. Thrombolites embedded in the grainstones are made up of clusters of small digitate stromatolites bound by encrusting *Renalcis*, *Girvanella*, and micritic rinds. Although in some outcrops there is up to 2 m of large-scale, planar cross-stratified, peloid-ooid grainstone at cycle bases, previous descriptions of Conococheague Limestone cycles (Demicco, 1981, 1983, 1985; Demicco and Mitchell, 1982) have overemphasized the importance of such thick cross-stratified grainstone in this subfacies. Recent work by one of us (RVD) indicates cycles with thin burrowed grainstones are more common than cycles with thick cross-stratified grainstones at their bases.
2. A thin-bedded "ribbon rock" subfacies comprised of wavy, lenticular and flaser beds in which peloidal grainstones alternate on a cm-scale with dolomitic mudstones. The grainstones are small-scale cross-stratified, wave-ripple cross-stratified, and in some cases burrowed, whereas the dolomitic mudstones clearly form drapes over the rippled upper surfaces of grainstone beds and may be mudcracked in the upper portions of some cycles. Ribbon rocks grade up section into the overlying subfacies 3.
3. Prism-cracked, undulatory laminated mudstone with rare "jelly-roll" structures.
4. Planar laminated dolomitic mudstones that can be quite extensively disrupted by small mudcracks. Isolated, displacive nodules of sparry calcite commonly occur in these mudstones.

Lower and upper boundaries of each cycle are sharp and easily located in the field but internal boundaries between subfacies are most commonly gradational, especially between (2) and (3), and between (3) and (4). Individual cycles can be correlated across depositional strike of the outcrops for a distance of at least 25 km (about 50 km on a palinspastic restoration, Demicco and Mitchell, 1982).

Each cycle is interpreted as a shallowing-upward succession produced by progradation of broad, low-slope tidal flats over sandy subtidal areas. The thin basal grainstones are here interpreted as winnowed subtidal lag deposits and are the commonest type of cycle bottom. The flat pebble conglomerates are storm-deposited, ripped-up hardgrounds that developed in both the sand and gravel lags. Thrombolites represent microbial patch reefs that nucleated on hardgrounds developed within the lags. Thicker cross-stratified grainstones may represent tidal channel deposits or shallow offshore "high energy" shoal areas. The wavy, lenticular and flaser bedded structures of the ribbon rocks are analogous to those in the sediments of the mixed sand-mud intertidal flats of the North Sea (Demicco, 1983). If the ribbon rock subfacies is interpreted as the deposits of intertidal flats, then the thickness of this subfacies in the cycles suggests a paleotidal range (Klein, 1971; Terwindt, 1988) on the order of 1-5 m. The prism-cracked laminites are interpreted as desiccated cryptmicrobial laminites of the upper-intertidal to lower-supratidal zone analogous to those in the modern Persian Gulf (Shinn, 1983a; Kinsman and Park, 1976). The isolated nodules in the uppermost mud-cracked dololaminite may have been anhydrite nodules formed in a mildly arid to semiarid supratidal zone.

A number of different cycle motifs are found in the tidal flat facies of the Conococheague. Cycles at the base of tidal flat facies are up to 12 m thick, and are commonly comprised of subfacies 1 - 2 - 3 - and rarely 4. There is a systematic decrease in cycle thickness upward. The last few cycles in a tidal flat facies commonly are only a meter or so thick and comprise subfacies 1 - 3 - 4, subfacies 2 being notably absent. Individual cycles within tidal flat facies are traceable laterally from outcrop to outcrop (Fig. 168). Outcrops A and B (Fig. 168) are separated by 5 km along strike whereas outcrop C is 5 km (restored) across strike in the direction of the inferred shelf margin. There is some lateral variability (notably the increase in thickness of subfacies 1 in the base of cycle 1 in column C, Fig. 168). However, on the whole, subfacies patterns are remarkably uniform among outcrops at this scale. Cycle bases and tops are sharp and well defined whereas within cycles the contacts between subfacies are notably gradational.

The laterally traceable cycles of the Conococheague Limestone imply fairly extensive progradation of low-slope tidal flats over adjacent shelf lagoons.

Tidal Flat and Coastal Marsh Facies of the Middle Ordovician St. Paul Group, Central Appalachians

The St. Paul Group of Chazyan age lies near the top of the Cambro-Ordovician carbonate succession in the central Appalachians and is 150-200 m thick in the central Appalachians. Like the Conococheague Limestone described above, the St. Paul Group also consists of shelf lagoon facies and tidal flat facies. However, both of these facies in the St. Paul Group are markedly different from those of the Conococheague Limestone.

The St. Paul Group has two types of shoreline facies interbedded with restricted shelf-lagoon facies and open shelf-lagoon facies (Mitchell, 1981, 1985; Demicco and Mitchell, 1982; Hardie and Shinn, 1986). The shelf-lagoon facies are described separately (see below). The shoreline facies of the St. Paul Group include a tidal flat facies and a coastal freshwater lake facies. Typical vertical logs of the tidal flat facies are shown in Figure 169A and a vertical log of the coastal lake facies is given in Figure 169B (taken from Demicco and Mitchell, 1982; Hardie and Shinn, 1986).

The tidal flat facies is composed of four subfacies:

1. Bioturbated mudstones with no internal layering. These occur as beds 0.1-1.5 m thick. They are characterized by up to 15% skeletal debris of low faunal diversity (mainly ostracodes) and innumerable mm-diameter tubular burrows with no preferred orientations.
2. Coarse packstone subfacies. This subfacies consists of beds up to 0.3 m thick of sand- to pebble-sized peloids and micrite intraclasts with a micrite matrix. Skeletal debris is rare and tubular burrows up to 20 mm in diameter disrupt what appears to have been a cross-stratified peloidal grainstone. Shallow erosional channels a few decimeters in depth with associated lateral accretion bedding defined by coarse intraclast grainstones (long axes of the mudstone chips lie parallel to the inclined accretion surfaces) can be found in some sections near the base of the St. Paul Group.
3. Laminite subfacies (0.2-1 m thick). This subfacies consists of a complicated interstratification of planar-, wavy-, and crinkly-laminated peloidal mudstones. Planar-laminated mudstones rarely have mudcracks. Wavy-laminated mudstones in some cases show a complex, interconnected network of mudcracks and sheetcracks producing an *in situ* breccia (pseudobreccia) wherein blocks have been moved

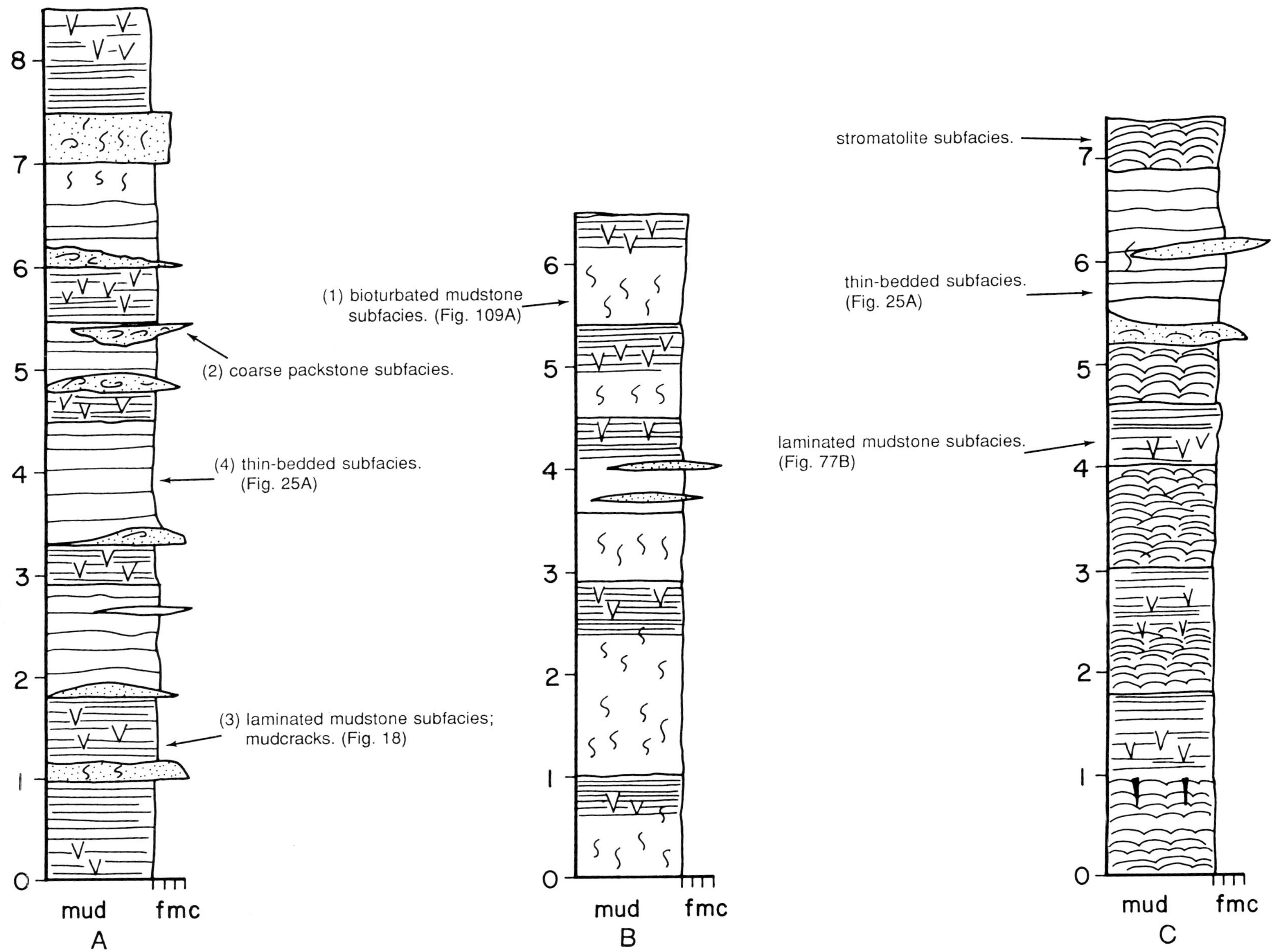

Fig. 169.-Typical vertical logs of the tidal flat facies (A) and (B) and a vertical log of the coastal lake facies (C) of the Middle Ordovician St. Paul Group of western Maryland (see text, and Mitchell, 1985, for further details). Vertical scale is in meters whereas the horizontal scale indicates relative grain size (m = mud, f = fine sand, m = medium sand, and c = coarse sand). The symbols for the subfacies that make up the cycles are indicated and keyed to text figures that show representative examples of the subfacies.

slightly relative to each other, but no wholesale transport has taken place as evidenced by the continuity of layering across the nearly in-place fragments. The crinkly-laminated mudstones carry mudcracks, very small stromatolites, and laminoid fenestrae.

4. Thin-bedded mudstone subfacies. In this subfacies, unfossiliferous mudstones containing molds of what appear to be cyanobacterial filament tubes (cyanobacterial tufas) are interbedded with laterally-linked hemispherical (Logan and others, 1964) stromatolitic mudstones and lenses of coarse intraclastic-skeletal packstones.

These subfacies are laterally continuous over about the 100- to 200-m length of individual outcrops and show few lateral changes in subfacies at this scale. However, cycles cannot be correlated laterally between outcrops spaced kilometers apart.

Mitchell (1981, 1985; Demicco and Mitchell, 1982), following the path of Matter (1967), made a detailed study of the St. Paul carbonates and concluded that the closest modern counterpart of the tidal flat facies is the Andros Island, Bahamas system. The abundance of lime mud, absence of ooid grainstones and algal patch-reef mounds, the restricted fauna, and the dominance of bioturbated mudstones and laminated subfacies make a convincing case. From a detailed comparison of the St. Paul rocks and the Andros Island sediments, Mitchell arrived at the following analogs.

St. Paul Subfacies		Andros Island Subenvironment (see Hardie, 1977)
1. Bioturbated mudstone	=	Intertidal-subtidal ponds
2. Coarse packstone	=	Tidal channels
3. Laminated mudstone	=	Supratidal levees
a. planar laminites	=	Levee crest
b. wavy laminites	=	Levee backslope
c. crinkled fenestral laminites	=	High "algal" marsh
4. Thin bedded mudstone	=	Inland freshwater "algal" (cyanobacterial) marsh

Mitchell (1981, 1985) noted that the subfacies of the St. Paul tidal flat facies are not arranged in a clear-cut succession of repetitive meter-scale depositional "cycles." However, it is possible to recognize within the complex succession of meter-scale subfacies individual shallowing upward sequences that record local progradation or vertical accretion such as pond-levee, pond-marsh-levee, marsh-levee (Fig. 169). In addition, there is a discernible stratigraphic pattern in that stacked units of bioturbated mudstone subfacies capped by mudcracked laminite subfacies (e.g., units of the type shown in Fig. 169, column B) dominate the lower part of the tidal flat facies and are progressively replaced upward by stacked units made up of thin-bedded subfacies capped by mudcracked laminite subfacies (e.g., units of the type shown in Fig. 169, column A) that become more abundant toward the contact with the overlying coastal freshwater lake facies.

The coastal freshwater lake facies (Fig. 169B) is characterized by a distinctive stromatolite subfacies which is cyclically interbedded with laminite subfacies and thin-bedded mudstone facies. The stromatolite subfacies forms meter-thick beds composed entirely of LLH-C (laterally linked hemispheres-concentric) stromatolites (cf. Logan and others, 1964). The laminae are composed of clotted micrite and peloids that outline numerous filament molds. Conspicuously absent are marine macrofossils or burrows of any kind. In composition, fabric, and morphology, the stromatolitic layers are strikingly similar to modern freshwater calcitic stromatolites of the coastal freshwater lakes landward of the tidal flats and algal marsh of Andros Island (Monty, 1972; Monty and Hardie, 1976). For this reason, plus the exclusive association with supratidal subfacies (laminated mudstones and thin bedded mudstones), Mitchell (1981, 1985, Demicco and Mitchell, 1982) interpreted this stromatolite-bearing facies of the St. Paul Group carbonates as a coastal freshwater lake deposit separated from the sea by a belt of rainy tidal flats of low wave and tide energy.

Features of the St. Paul Group worthy of note are:

1. The dominance of lime mud, the paucity of current- or wave-formed sedimentary structures, the thorough bioturbation, and the low diversity of fauna are all hallmarks of a low energy, low tide range tidal flat with restricted circulation.
2. Cyanobacterial tufa, stromatolites, and thin beds in the supratidal zone, and the absence of any traces of evaporites, are suggestive of a wet climate.
3. Mitchell (Demicco and Mitchell, 1982; Mitchell, 1985) noted that individual cycles could not be correlated among the different outcrops of either the tidal flat facies or the freshwater lake facies, although the facies themselves were correlatable. Instead, the complex arrangement of subfacies vertically and laterally signifies that accretion was not by simple progradation of the tidal flats across a flat-floored shelf lagoon, but, instead, carbonate mud filled in a complex mosaic of shallow tidal flat subenvironments separated from the open shelf lagoon by some form of barrier bars or beach ridges (perhaps in the manner of the "staggered offlap" progradation of Hardie and Shinn, 1986, p. 62).

Tidal Flat Facies (Lofer Facies) of the Norian Dachstein Limestone of the Northern Limestone Alps, Austria

The Norian-Rhaetian Dachstein Formation exposed in the Northern Limestone Alps of Austria is part of a carbonate buildup some 1.5 km thick. Around Salzburg, the buildup consists of a narrow belt of reef-margin facies and back-reef tidal flat facies a few tens of kilometers across. The tidal flat facies were the subject of Fischer's (1964) classic study. Fischer's descriptions and illustrations of the sedimentary structures (many of which he called attention to for the first time) remain unsurpassed. In addition, Fischer's original ideas on the origin of the stratigraphic repetition of individual meter-scale tidal flat units (the meter-scale cyclicity) in this formation called upon high frequency Milankovitch-forced sea-level oscillations.

The tidal flat facies of the Dachstein, known as the Lofer facies, consists of beds up to a few decimeters thick of laminated mudstones with abundant evidence of exposure (mudcracks, prism cracks, soils, sheet cracks) interlayered with massive, fossiliferous grainstones and packstones up to 20 m thick. Fischer originally interpreted the individual meter-scale tidal flat units as deepening-upward successions. Recently however, Goldhammer and others (1990) have re-examined the classic outcrops described by Fischer and have presented evidence to show that the majority of Dachstein tidal flat units are, in fact, shallowing-upward.

Figure 170 is a portion of a vertical log through the tidal flat facies at the Riemannhaus on Steinernes Meer taken from Goldhammer and others(1990). These authors recognized four subfacies:

1. Thick (up to 20 m), burrowed beds of grainstones to wackestones that contain hardgrounds and a diverse assemblage of in place and transported fossils dominated by megalodont molluscs but also including sponges, corals, gastropods, and echinoderms (Fischer, 1964, member C, "megalodont limestone");
2. Dolomitic, burrowed peloidal wackestones and packstones with a restricted fauna of thin-shelled gastropods and ostracods and rare prism cracks (Fischer, 1964, member B, "pellet and homogeneous loferites");

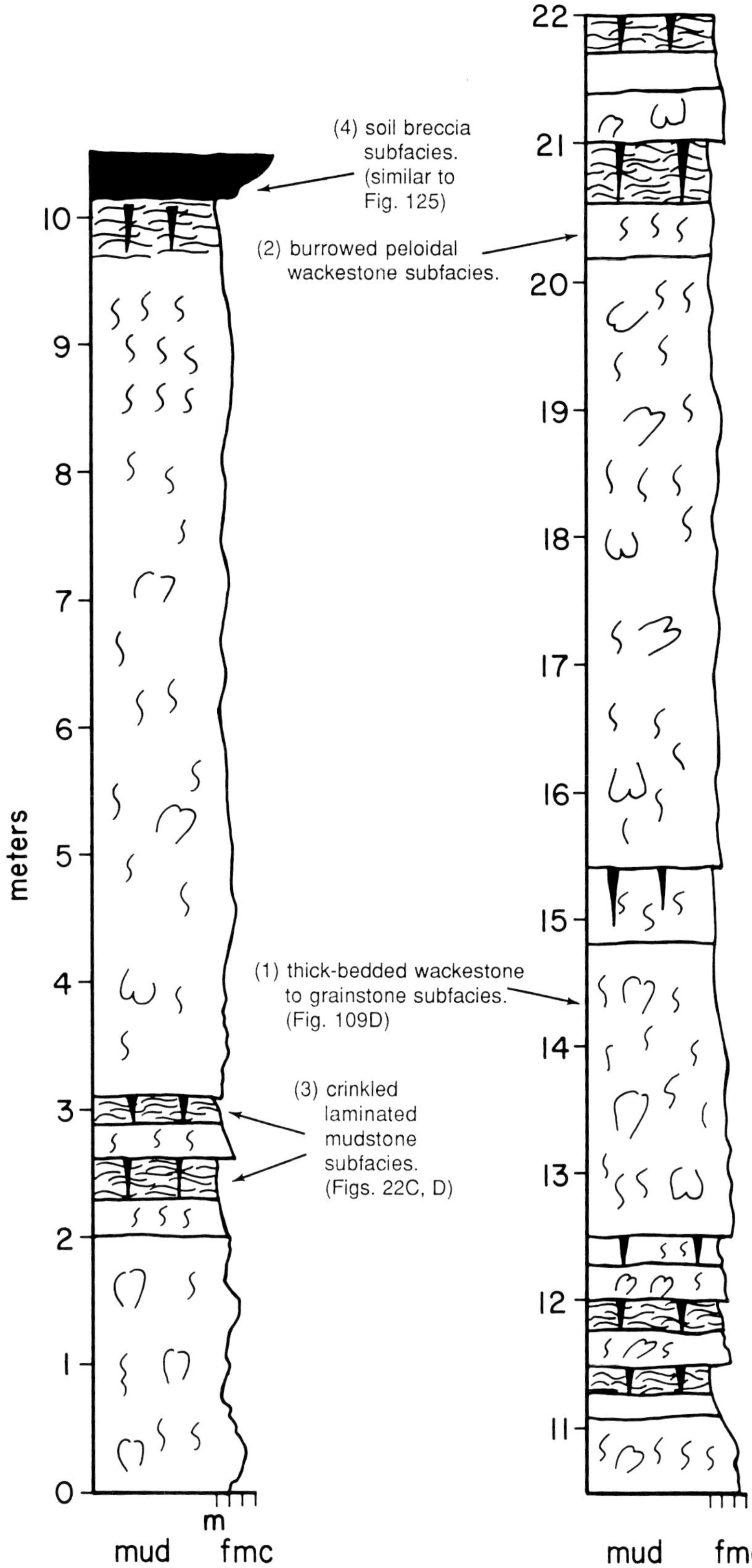

Fig. 170.-A portion of a vertical log through the tidal flat facies (Lofer Facies) in the Norian Dachstein Limestone at the Riemannhaus on Steinernes Meer, Northern Limestone Alps, Austria, taken from Goldhammer and others (1990). The vertical scale is in meters whereas the horizontal scale indicates relative grain size coarsening to the right. The symbols for the four subfacies that make up the cycles are indicated and keyed to text figures that show representative examples of the subfacies.

3. Dolomitic, crinkly-laminated mudstone with common prism cracks, mudcracks, sheet cracks, and fenestral pores (Fischer, 1964, member B, "algal mat loferite"); and
4. Intraformational conglomerate comprised of clasts of the other subfacies in a green, brown or red matrix carrying argillaceous material (Fischer, 1964, member A).

Fischer (1964) interpreted subfacies 1 as a burrowed subtidal mud deposit, subfacies 2 and 3 as intertidal deposits which were influenced by cyanobacterial mats, and subfacies 4 as a soil related to subaerial exposure. Goldhammer and others (1990) have further refined these interpretations by suggesting that subfacies 2 represents the deposits of intertidal ponds whereas subfacies 3 records supratidal deposition. Critical to the interpretation of these meter-scale depositional "cycles" is the position of the soils which presumably record rather lengthy exposure. Fischer (1964) originally thought that the soils capped subfacies 1 and that subfacies 2 and 3 overlie the exposure surface and grade up into facies 1, completing a transgressive cycle. Goldhammer and others (1990, Fig. 16), however, have shown that in the 190-m thick Riemannhaus section where there are 61 cycles exposed, the soils exclusively overlie subfacies 2 and 3 (see Fig. 170). In the Lofer section on the Dachstein mountain range described by Fischer (1964), Dunn, Hardie and Demicco (unpub. data) found that laminites which appeared in one part of the outcrop to be resting directly on soils could be traced laterally into massive soil zones that had incorporated large blocks of the laminated subfacies. Clearly, the laminite, rather than being deposited on the soil, had been partly destroyed by the development of the soil. The majority of cycles in the Lofer tidal flat cycles are shallowing-upward and the "modal" subfacies succession is 1 - 2 - 3 - 4.

The Lofer tidal flat facies of the Dachstein Limestone resembles the St. Paul tidal flat facies in the style of the sedimentary features in the intertidal and supratidal deposits, although freshwater coastal lake deposits are not found in the Lofer of the Steinernes Meer and Dachstein sections. In this regard, it is interesting to note that in the shoreward equivalent of the Dachstein Limestone, the Hauptdolomit, organic rich black shales are found. One such black shale contains Holostean fish remains. Paleozoic Holostean fish are most likely freshwater and the same is likely true for these Mesozoic examples (Vogeltanz, 1969; W. Schlager, 1990, pers. commun.). As with the St. Paul shoreline carbonates, the Lofer tidal flat facies does not carry evaporites. We infer that during Lofer time the local climate of the tidal flats was a rainy climate, like that of St. Paul tidal flats in the central Appalachians. Another possible line of evidence for a rainy climate rather than an arid climate is the presence of soil caps instead of sabkha evaporites (Hardie and Shinn, 1986). The presence of soils caps on some cycles and not on others (Goldhammer and others 1990, Fig. 19) would seem to be more related to the time intervals between depositional cycles than to climate variations from cycle to cycle.

Finally, the thoroughly burrowed subtidal deposits of the Lofer tidal flat facies resemble those of the St. Paul except that they have a higher faunal diversity. Notably absent are cross-stratification structures. These observations suggest that the ambient subtidal setting was one of low wave and tidal current energies where burrowers were able to easily rework any primary layering between storms, as occurs today on the rainy Bahama Banks (Hardie, 1977).

Evaporitic Tidal Flat (Sabkha) Facies in the Jurassic Buckner Formation, Alabama

The Buckner Formation is an Upper Jurassic evaporitic succession, up to 800 m thick, that was deposited along the northern margins of the ancestral Gulf of Mexico (Lowenstein, 1987, p. 108). The Buckner Formation (and evaporite-bearing equivalents) occurs in the subsurface over a large area beneath Texas, Arkansas, Louisiana, Mississippi, Alabama, and Florida where it commonly serves as an impermeable seal for hydrocarbons in the underlying Smackover Formation.

The sedimentary features of the Buckner Formation have been described by Dickinson (1968), Harris and Dodman (1982), and Lowenstein (1987). Lowenstein (1987) analyzed the sedimentary features of 34 m of evaporitic, tidal flat facies found in a core taken from the lower portions of the Buckner Formation in the Conoco #2 Mary S. Higgins well in southwestern Alabama. Figure 171 reproduces the lower 14.5 m of this interval. Although this core comes from a depth of 5.7 km, there is remarkable preservation of primary sedimentary structures similar to those found in less deeply buried evaporites (Lowenstein 1987). Lowenstein (1987) recognizes three subfacies:

1. Structureless dolomitic mudstones up to 1.5 m thick that contain tubular burrows, rare pelecypod fragments, and a clotted peloid texture in thin section.
2. Siliciclastic and carbonate mudstones up to 2.5 m thick that contain anhydrite nodules up to tens of millimeters in diameter. Nodules are commonly flattened and range from isolated nodules surrounded by mudstone to a tightly packed "chicken wire" mosaic.
3. Anhydrite beds up to 0.3 m thick that consist of one or more beds of vertically-oriented gypsum crystals pseudomorphed by anhydrite. The vertically oriented pseudomorphs are up to 15 cm long, are outlined by mud, and have internal mud films that trace out original gypsum growth planes.

Lowenstein (1987) interpreted the patterns in the vertical succession of individual subfacies in the Buckner Formation as recording shallowing-upward depositional cycles each of which represents the progradation of a sabkha over lagoonal sediments in the manner in which the Holocene sabkhas of the Persian Gulf have prograded out over shallow subtidal lagoons (Kinsman, 1966; Evans and others, 1969; Butler, 1969, 1970).

Subfacies 1 is interpreted to be burrowed pelleted mud deposited in a restricted lagoon analogous to the burrowed peloidal muds formed in the subtidal lagoons along the modern Trucial Coast. Subfacies 2 is interpreted to have formed by displacive growth of gypsum and anhydrite in the shallow, subsurface vadose and upper phreatic zones of supratidal mudflats. The interpretation of subfacies 3 is problematic. It records competitive subaqueous growth of gypsum up off the bottom of a shallow, hypersaline brine body. Two subenvironmental interpretations are possible. Gypsum growth may have taken place in shallow, supratidal brine pans that formed on the surface of the sabkha due to storm flooding. A similar modern supratidal gypsum pan on the supratidal flats of Baja California has been described by Castens-Seidell (1984) (see Hardie and Shinn, 1986). An alternative is that subfacies 3 represents times when the nearshore lagoon was so restricted that the waters became supersaturated with gypsum. Such a situation might occur in the lagoons of the Trucial Coast if they were to become further restricted.

The similarity of the burrowed, subtidal lagoonal subfacies at the bases of tidal flat cycles of the Buckner Formation to the basal lime mud subtidal subfacies of both the St. Paul Group and Lofer Facies of the Dachstein Formation tidal flat cycles is remarkable considering the notable differences in the supratidal subfacies at the tops of cycles of each of these deposits. The similarities in the subtidal subfacies but differences in the supratidal subfacies of these three deposits provide a striking example of the selective sensitivity of the sedimentary record of different subenvironments of carbonate tidal flats to environmental conditions (Hardie, 1977b, p. 188-189; Hardie and

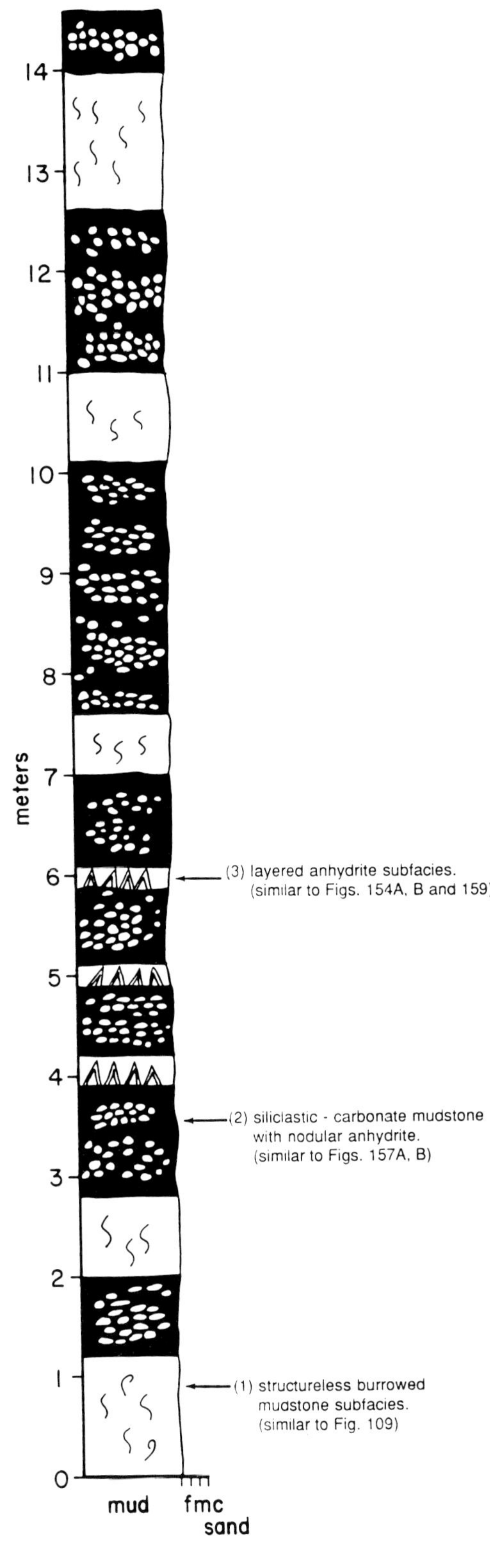

Fig. 171.-Stratigraphic log representing 14.5 m of tidal flat facies from the evaporitic Buckner Formation in the Conoco-2 Mary S. Higgins well in southwestern Alabama (see Lowenstein, 1987). Vertical scale is in meters whereas the horizontal scale indicates relative grain size (f = fine sand, m = medium sand, and c = coarse sand). The symbols for thc subfacies that make up the cycles are indicated and keyed to text figures that show representative examples of the subfacies.

Shinn, 1986). In particular, shallow subtidal marine subenvironments that undergo tidal exchange of seawater with the open sea are insensitive to climate and record primarily information on the tide and wave regimes and the seawater salinity. In contrast, supratidal subenvironments that remain subaerially exposed for more than 90% of the time are insensitive to the marine conditions and record mainly the prevailing climate and catastrophic weather conditions. These relationships are clearly demonstrated by the similarities and differences in the nature of the subtidal, intertidal and supratidal subfacies of the Buckner, St. Paul and Lofer tidal flat cycles. Despite the fact that the intertidal and supratidal subfacies of these three ancient shallow marine carbonates indicate a wide range of prevailing climatcs from rainy (supratidal freshwater lake deposits of the St. Paul) to arid (hypersaline sabkha deposits of the Buckner), the subtidal subfacies of all three deposits are remarkably similar to each other, recording shallow somewhat restricted bodies of seawater of near normal salinity (perhaps slightly elevated at most) uninfluenced by the climatic conditions. There is a direct analogy here with the modern carbonate tidal flat deposits in the arid Persian Gulf and in the rainy Andros Island (Bahamas). In both the Persian Gulf and Andros Island, the subtidal sediments consist of burrowed peloidal muds with low diversity faunas (the same species in many instances) while the supratidal sediments are radically different with evaporites characterizing the arid sabkhas of the Persian Gulf and freshwater coastal marsh and lake deposits characterizing the rainy Andros Island (see Hardie and Shinn 1986, Fig. 71, p. 54).

Barrier Beach and Associated Tidal Flat Facies from the Middle to Upper Cambrian Waterfowl Formation, Albertan Rocky Mountains, Canada

The Cambrian and Ordovician rocks exposed in the Rocky Mountains in southwestern Alberta, Canada form a westward thickening wedge up to 2.5 km thick composed of shales and carbonates deposited in a variety of tidal flat and shelf lagoon subfacies (Aitken, 1966, 1971, 1978). In the vicinity of the Alberta-British Columbia border, platform facies are replaced by a thick package of time-equivalent deeper-water, off-platform shales and minor carbonates to the west. The exact nature of the transition is not yet fully resolved.

The Middle Cambrian to Upper Cambrian Waterfowl Formation is part of the shallow-water platform deposits. The Waterfowl Formation is 165 m thick at the shelf margin and thins eastward (landward) until it is 35 m thick in the Front Ranges of the Rockies, a restored distance of 150 km. Large (up to 5 km long), three-dimensional outcrops of Waterfowl tidal flat facies make it possible in the field to walk out subfacies relations laterally among measured sections on the same thrust sheet and has allowed us to work out a number of important aspects of the geometry of the subfacies from these outcrops (Waters, 1986; Waters and others, 1989; Cloyd and others, 1990).

Figure 172A is a stratigraphic log through the complete 32-m thickness of the Waterfowl Formation exposed continuously along a 5-km long outcrop belt of the Waterfowl Formation. The lower 21 m of the Waterfowl Formation here comprises tidal flat facies that consist of fourteen laterally-traceable, shallowing-upward cycles each of which is composed of five subfacies, as follows:

1. Graded beds approximately 0.1 m thick. The base of each bed is a planar-stratified peloidal grainstone that grades up into dolomitic mudstone. Lenses of flat pebble conglomerates and wave-rippled cross-stratified grainstones are common as interbeds.

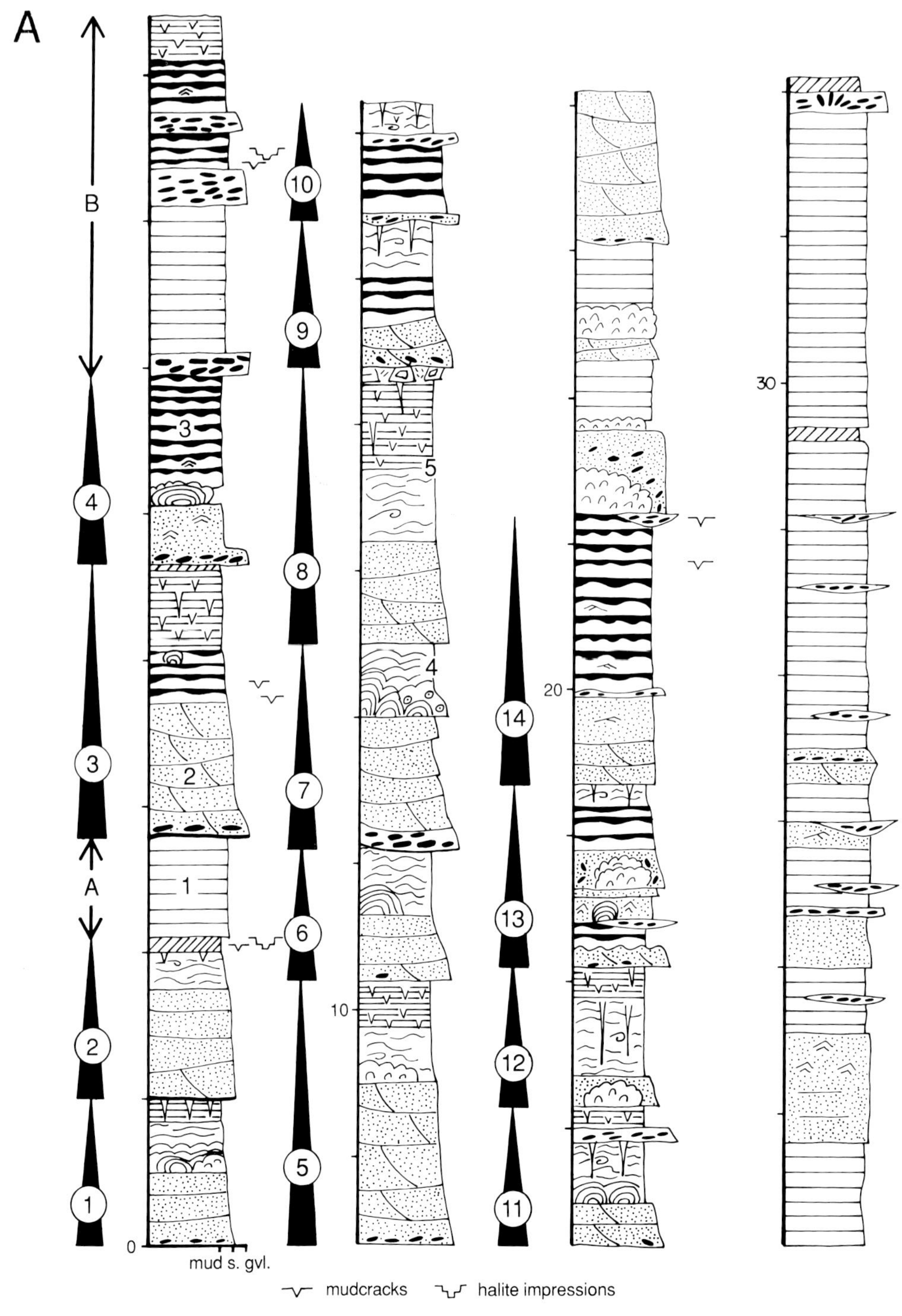

Fig. 172.-Tidal flat facies of the Middle-Upper Cambrian Waterfowl Formation of the southern Canadian Rocky Mountains in southwestern Alberta (see Waters and others, 1989 for details). (A) Stratigraphic log through the complete 32 m thickness of the Waterfowl Formation exposed continuously along a 5 km long outcrop. Vertical scale is in meters whereas the horizontal scale indicates relative grain size (s = sand, gvl = gravel). The symbols for the subfacies that make up the cycles are indicated by the numbers in the columns: (1) graded beds of planar stratified peloidal grainstone that grade up into dolomitic mudstone; (2) large-scale trough and planar cross-stratified peloidal grainstones; (3) wavy beds comprised of alternating layers of peloidal grainstones and dolomitic mudstones similar to Conococheague "ribbon rocks"; (4) thrombolites with overlying stromatolites embedded in very fine-grained peloidal grainstones; and (5) mudcracked planar, wavy- and crinkly-laminites with "jelly-roll" structures. Individual cycles are denoted by the numbered arrows to the left of the column (intervals A and B are not parts of the cycles).

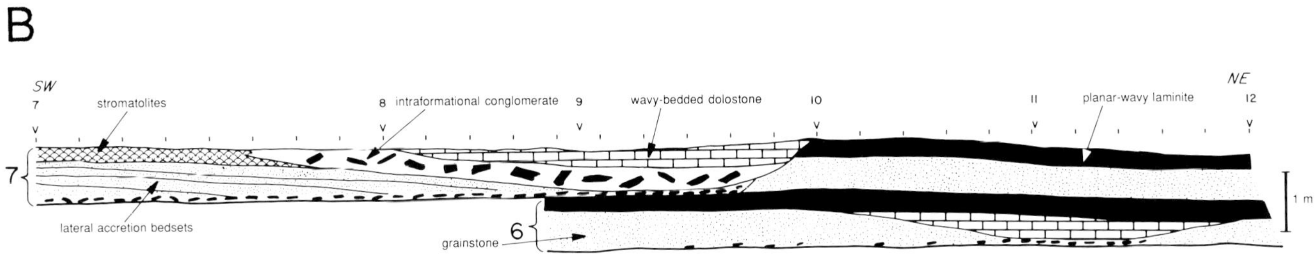

Fig. 172.-(Continued) (B) Lateral variations in cycles 6 and 7 documented by walking out outcrop and by analysis of photomosaics. Vertical scale bar at right = 1 m, tick marks across top = 1 m. Location of measured sections given by numbers across the top (see Waters and others, 1989). Subfacies labelled. Note the abandoned channel-fills and the lateral accretion bedding. (C) Lateral variations in cycles 3 and 4 and interval B documented by walking out outcrop and by analysis of photomosaics. Scale bar at left; vertical exaggeration 25X. Location of measured sections given by numbers across the bottom. Sections 6 and 7 are separated by approximately 500 meters (see Waters and others, 1989). Subfacies as labelled. Note the grainstone mound (interpreted as a beach ridge), and the two stromatolitic mounds (interpreted as barriers). Shoreward (to the right) of these mounds are restricted pond subfacies with halite hopper molds. Silhouettes below compare true scale profiles of grainstone ridge in cycle 3 (lower) and beach ridge from Andros Island (upper, from Hardie, 1977).

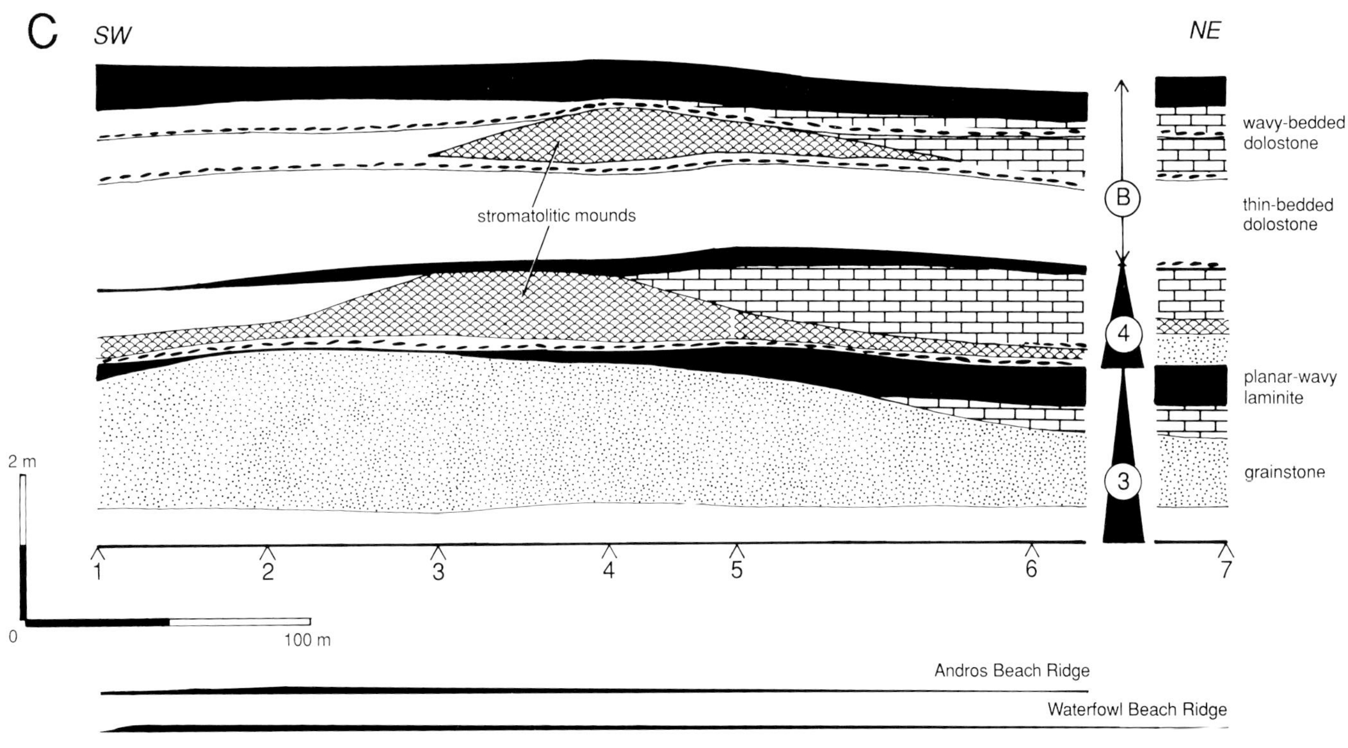

2. Meter-thick cosets of large-scale trough and planar cross-stratified peloidal grainstones. The cross-stratification is, in some cases, of the herring-bone type and contains mud-drapes and reactivation surfaces.
3. Cryptmicrobial bioherms (commonly thrombolites with overlying stromatolites) embedded in very fine-grained peloidal grainstones.
4. Wavy beds comprised of alternating layers of peloidal grainstones and dolomitic mudstones similar to Conococheague "ribbon rocks" (see above).
5. Mudcracked planar, wavy- and crinkly-laminites with "jelly-roll" structures.

Where traced along strike approximately half of the cycles have lateral accretion bedding and associated coarse- and fine-grained channel fills in their basal grainstone subfacies (Fig. 172B). These surfaces dip up to 14 degrees relative to the basal scour surfaces and are transverse to local paleocurrent directions. Channel fills are trough-shaped, 10 m wide, 1 m deep, and are filled with mudstones or rudstone breccia of fragmented stromatolites.

In four of the cycles there are notable mound-like thickenings of grainstones or stromatolitic bioherms along strike (Fig. 172C). The grainstone mounds are well-sorted, medium to coarse-grained coated-peloid and ooid grainstone in cosets of swash cross-stratification that dip at low angles in the inferred seaward direction. Landward of these mounds, the wavy bedded subfacies is characterized by large, spectacular halite crystal impressions on bedding surfaces.

The grainstone subfacies at the bases of the cycles are interpreted as the deposits of meandering channels. The reactivation surfaces, herring-bone cross-stratification, and mud drapes indicate that these were tidal channels. These channels presumably fed intertidal and supratidal flats. Intertidal flats were mixed sand and mud flats whereas the high intertidal to supratidal flats were covered by cyanobacterial mats. The grainstone mounds are interpreted as beach ridges that developed over shallow subtidal storm-deposits (the graded beds). These beach ridges may have formed offshore of the tidal flats by *in situ* buildup of sand in the zone of breaking waves (the "staggered off-lap" model of Hardie and Shinn, 1986). Alternatively, these barriers may represent beach ridge spits that prograded alongshore and built into the area from updrift sources (the "chenier plain" model of Shinn, 1986). In either case, the inshore deposits that initially accumulated were intertidal pond deposits. The abundant salt casts in these pond deposits not only indicate an arid climate, but also suggest that the barrier beaches restricted circulation into the ponds, which in turn promoted elevated salinities. The grainstone beach ridges are striking in that they are nearly identical in dimensions to the beach ridges in front of the tidal flats on Andros Island (Fig. 172C).

Current-Dominated Shelf Lagoon Facies of the Upper Cambrian Conococheague Limestone, Central Appalachians

A typical vertical measured section of the shelf lagoon facies of the Conococheague Limestone is shown in Figure 173 along with a two-dimensional representation of the logged outcrop based upon photomosaics and lateral tracing of beds in the field. Although there are only three subfacies within the shelf-lagoon facies of the Conococheague, their vertical and lateral relationships are complicated (Fig. 173). These three subfacies are as follows:

1. Ooid-peloid-skeletal grainstones characterized by large-scale trough and tabular cross-stratification with lesser interbeds of planar stratified grainstones, intraclastic conglomerates, and dolomitic mudstones. Skeletal grains include trilobite and echinoderm fragments, and the intraclastic conglomerates consist of pebbles made up of cross-stratified grainstones with early marine cement fabrics. The grainstones contain hardground surfaces which commonly serve as nucleation surfaces for thrombolites (see below).
2. Thrombolitic bioherms with *Renalcis* common as an internal framework and encrusted on the micritic "clots" of the thrombolites. Internal sediment composed of echinoderm- and trilobite-wackestone fills spaces among the clots. The Conococheague Limestone has well-defined stratigraphic intervals in which these bioherms are up to 10 m thick.
3. Ribbon rock subfacies characterized by wavy bedding that consists of wave-rippled and current-ripple cross-stratified peloidal grainstones alternating with dolomitic mudstones. The dolomitic mudstones thin over crests and thicken in troughs in the wavy bedded grainstones. Mudcracks are singularly absent from these muddy interbeds (compare with tidal flat ribbon rocks described above). Intraclast lenses up to tens of meters long characterized by "edgewise" flat pebble fabrics are common.

Demicco (1985; see also Demicco and Mitchell, 1982) interpreted these three subfacies to represent ooid-skeletal sand shoals, subtidal patch reefs, and intrashoal level-bottom deposits, respectively. The dominance of cross-stratification over burrowing suggests daily current working. The presence of hardgrounds indicates that penecontemporaneous cementation was a normal part of the sedimentary regime of the Conococheague shelf lagoon just as it is today on the shallow Bahama Banks and offshore Persian Gulf. The intraclastic conglomerates (which comprise broken fragments of hardgrounds) testify to episodic storms large enough to erode the hardgrounds and rework the broken fragments into intraclast gravel lenses, again much as happens today on the Bahama Banks. These interpretations are made from the basic characteristics of the subfacies, but further insights can be gained by looking at the vertical and lateral relationships of subfacies (Fig. 173). These are briefly described below.

The grainstones are organized into storeys which commonly have an intraclastic conglomerate overlying an erosion surface at their base. Within a storey, accretion surfaces with low dips of a few degrees separate bedsets of grainstones some tens of centimeters thick. Bedsets commonly fine upward both through their thickness and laterally along their lengths. At the tops of some bedsets, current ripples and wave ripples are preserved beneath a dolomitic mudstone drape. Paleocurrents within an individual bedset are typically variable but 180° herringbone cross-stratification occurs in some cases. Hardgrounds that serve as the nucleation surfaces for the thrombolites in the outcrop are developed at the burrowed top of a bedset. The heterolithic ribbon rock subfacies typically occurs in thin packages and, in a number of cases, appears to be gradational with the underlying grainstone bedsets.

These observations support the interpretation that grainstone deposition took place in shoals, probably in response to storm-generated unidirectional currents and tidal currents. The exact nature of these shoals is not yet understood. The grainstone storeys with internal low-angle accretion bedding probably represent laterally migrating channels. Mud drapes within storeys and rare reversals of paleocurrents within these deposits suggest tidal currents. Variable paleocurrents may result from ebb-flood segregation of flows within the channels, a phenomenon commonly observed in modern tidal channels (Elliot, 1986). Alternatively, wave-generated unidirectional flows may account for the variable paleocurrent patterns. The bar surfaces of the channels had low dips but no cutbanks. The grainstones with thrombolites may represent abandoned channels fills but this is quite speculative because there are no rec-

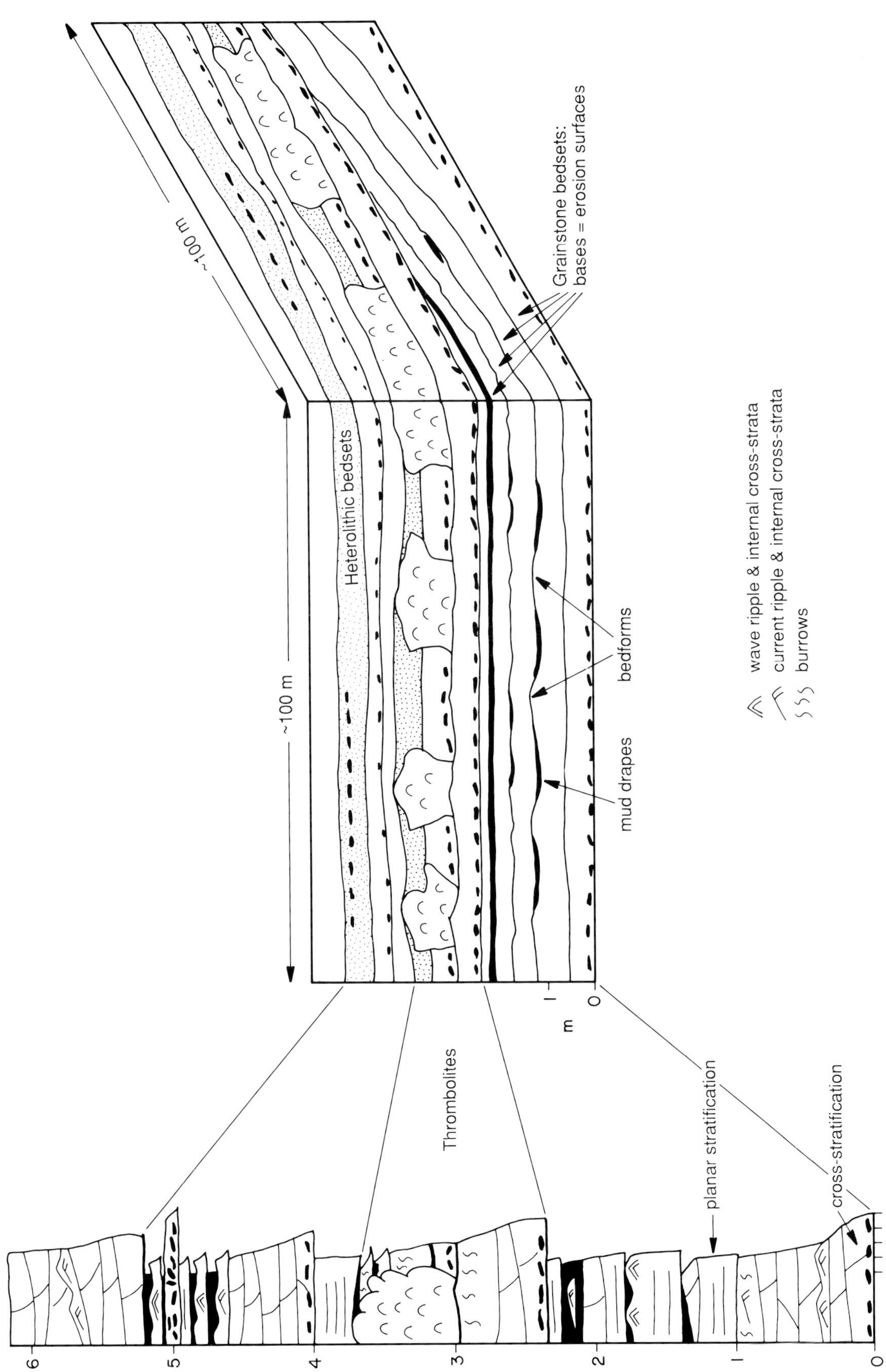

Fig. 173.-Vertical measured section (left) and two-dimensional bedding diagram (right) of the shelf lagoon facies of the Upper Cambrian Conococheague Limestone, Maryland. The bedding diagram is based upon photomosaics and lateral tracing of beds in the field. Scales are in meters. The horizontal scale of the stratigraphic log indicates grain size (coarsening to the right). The symbols for the component subfacies are indicated and keyed to text figures that show representative examples of the subfacies. Note the low angle, inclined bedset surfaces in the grainstones and the mud drapes.

ognizable channel forms at this outcrop scale. Alternatively, those grainstone sheets which do not appear to have lateral accretion bedding may represent storm-related spill-over lobes and sheets outside of the channels. These inactive areas became burrowed and cemented. Whatever was the exact depositional setting of the subtidal lagoon facies of the Conococheague Limestone, it is clear that it was a setting dominated by unidirectional currents of either tide or wave origin. Tidal currents are suggested by details of the stratification patterns and this is consistent with the meso- to macrotidal range inferred from the tidal flat deposits. Common storm events are also suggested by the intraclast conglomerates. Some of the thicker dolomitic mudstone interbeds may be analogous to rare muddy storm layers deposited on the modern Bahamian ooid shoals by hurricanes (E. Shinn, pers. commun., 1992, observed a peloidal aragonitic mud layer several cm thick in the Joulter's Cay shoals after Hurricane Andrew).

A striking modern analog for the Conococheague grainstone facies exists in shoal areas of the Exuma Islands area of the Bahamas, where broad channels cut through shoals between islands. Here, hardgrounds developed on old channel bottoms serve as nucleation sites for "stromatolites." Lateral migration of the shoals periodically buries then excavates the stromatolites. Internally, these stromatolites bear a striking resemblance to many ancient thrombolites (Dill and others, 1986, 1989).

Storm-Dominated Shelf Lagoon Facies of the Lower Ordovician Beekmantown Group, Central Appalachians

The Lower Ordovician Beekmantown Group is approximately 1.2 km thick in the central Appalachians and is sandwiched between the Upper Cambrian Conococheague Limestone and the Middle Ordovician St. Paul Group. The Beekmantown Group in the Great Valley is similar to the Conococheague Limestone in that it is composed of shelf-lagoon facies and cyclic tidal flat facies. However, in the easternmost outcrops in the Frederick Valley, Maryland, the Beekmantown contains a well-defined shelf margin facies (see further descriptions below), a facies not preserved in Conococheague outcrops. Nguyen (1986) made a detailed study of the shelf-lagoon facies of the lower formation of the Beekmantown Group, a formation known as the Stonehenge Limestone. He identified in the Stonehenge Limestone both an inner and outer lagoon facies dominated by storm-produced features. The outer shelf lagoon facies is similar to the shelf lagoon facies of the Conococheague Limestone in that it is composed of thrombolitic bioherms encased in cross-stratified grainstones. Figure 174 shows a typical vertical log of inner shelf lagoon facies along with a two-dimensional representation of the logged outcrop based upon photomosaics and lateral walking-out of beds. Two subfacies can be distinguished, as follows:

1. Ribbon-rock subfacies with heterolithic beds that are similar to Conococheague ribbon-rocks in that they consist of thin-bed scale alternations of limestone grainstones and dolomitic mudstones. The dolomitic mudstones are not mudcracked, and, in a number of places, siliciclastic shales take the place of the dolomitic mudstones. The grainstone beds are wave-ripple cross-stratified, planar-stratified, and hummocky cross-stratified. Burrows are common. Patchy early cementation followed by compaction has in many cases produced a sedimentary boudinage-like structure. Beds and lenses of intraclastic conglomerates with erosional bases are common. These are comprised of clasts of cross-stratified grainstones similar to the grainstone beds of this subfacies.
2. Hummocky cross-stratified peloidal-skeletal grainstones with thrombolite mounds, and common hardgrounds.

These subfacies are arranged in a notable layer-cake arrangement in outcrop (Fig. 174). The subfacies and their vertical and lateral distribution are virtually identical to those well-known from siliciclastic storm-dominated shelf deposits (e.g., Craft and Bridge, 1987). The grainstones, of both ribbon rocks and thicker grainstone beds, and the intraformational conglomerates are most likely storm deposits. The numerous hardgrounds with encrusted bioherms attest to periods of quiescence during which cementation took place. During these quiescent periods, the muds were bioturbated. These latter features together with the ribbon rocks indicate that the Beekmantown shelf-lagoon was normally a low energy shelf interrupted by periods of storm wave activity.

Bioturbation-Dominated Inner Shelf Lagoon Facies of the Middle Ordovician St. Paul Group, Central Appalachians

The shelf lagoon facies of the St. Paul Group (Mitchell, 1981, 1985) stand in sharp contrast to those of the Conococheague Limestone and Beekmantown Group in that they are dominated by bioturbated peloidal mudstones to wackestones rather than grainstones. Where packstones and grainstones do occur in the St. Paul Group, they are also bioturbated. St. Paul Group shelf lagoon facies comprise two subfacies organized into thick bedsets:

1. Burrowed-disrupted wackestones that contain a fauna with a very low to moderate diversity.
2. Burrow-disrupted packstone to grainstone (cross-stratification rarely preserved) with a diverse fauna including bryozoans and tabulate corals.

In both subfacies, which are interpreted as the deposits of restricted inner shelf lagoons and more open outer shelf lagoon environments respectively, there are numerous submarine hardgrounds. These can be separated vertically by millimeters. These facies are also notable for their "thick bedded" aspect, common to many ancient muddy shelf deposits. However, whereas many of the bedding planes in these shelf lagoon facies of the St. Paul are apparently defined by hardgrounds, other layering surfaces are followed by stylolites (see discussion under compaction above and Bathurst, 1987). Leached gastropod shells associated with some of the hardgrounds may be indicative of freshwater exposure (Mitchell, 1981, 1985). This is a further suggestion of very shallow-water, restricted conditions.

Exposure-Dominated Shelf Lagoon Facies of the Middle Triassic Latemar Buildup, Dolomites Region, Italian Alps

The Dolomites region of the Southern Alps in northern Italy is characterized by well-exposed Middle Triassic carbonate platforms surrounded by basinal sediments. The Latemar buildup is a subcircular buildup of Anisian-Ladinian age about 5 km in diameter that consists of a core of flat lying shelf-lagoon facies about 700 m thick surrounded by a narrow shelf-margin facies that grades outward into a wide apron of steeply-dipping foreslope deposits (Bosellini, 1984; Goldhammer, 1987; Goldhammer and others, 1987; Harris, 1988; Goldhammer and Harris, 1989). Goldhammer and others (1990) divide the Anisian-Ladinian platform limestones that core the Latemar buildup into four facies: (1) Lower Platform Facies (LPF, 250 m); (2) Lower Cyclic Facies (LCF, 90 m); (3) Tepee Facies (TF, 120 m); and (4) Upper Cyclic Facies (UCF, 210 m) (Fig. 175 and see Fig. 128).

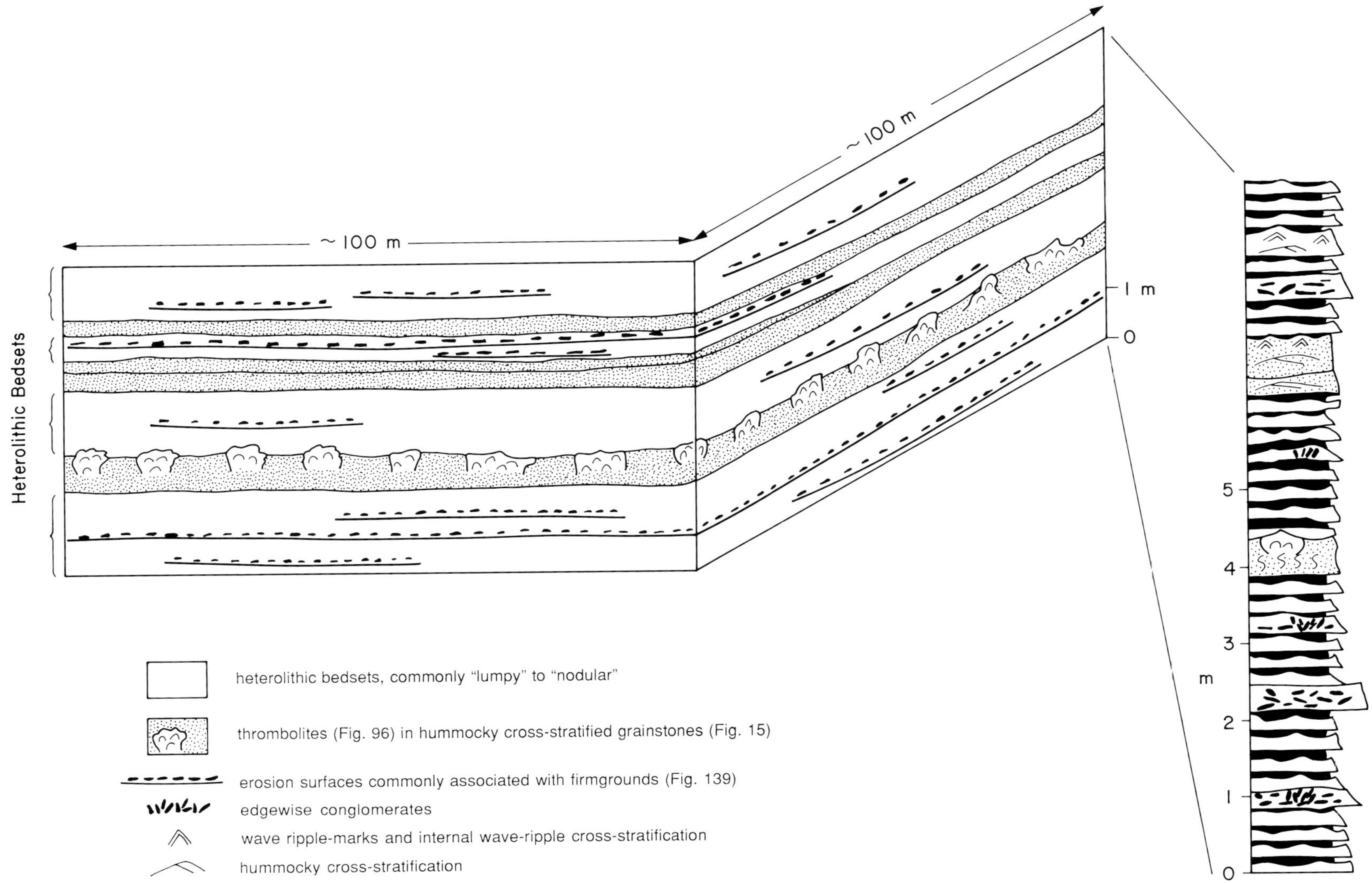

Fig. 174.-Vertical measured section (right) and bedding diagram (left) of the shelf lagoon facies of the Lower Ordovician Stonehenge Limestone, western Maryland. This bedding diagram is based on photomosaics and lateral tracing of beds in the field. Scales are in meters. The horizontal scale of the stratigraphic log indicates grain size (coarsening to the right). The symbols for the component subfacies are indicated and keyed to text figures that show representative examples of the subfacies.

One basic depositional cycle is characteristic of all four facies. This characteristic depositional cycle is made up of four subfacies (Fig. 175B):

1. A basal limestone subfacies of bioturbated, fine sand-sized, skeletal-intraclast grainstone to packstone. Skeletal grains include gastropods, crinoids, dasycladacean algae, and foraminifera.
2. Medium to very-coarse, sand-sized skeletal grainstone to packstone. Skeletal grains are commonly micritized.
3. Coarse-sand to gravel-sized oncolitic, lithoclast grainstone.
4. An upper dolomitic subfacies 0.05-0.15 m thick consisting of the skeletal-intraclast grainstone to packstone overprinted by vadose caliche fabrics. Diagenetic features such as meniscus and pendant cements and solution vugs are common in the vadose caliche caps and thicker examples of this subfacies contain composite pisoids, sheet cracks and micritic matrix (see Fig. 128).

Thicknesses of the various subfacies varies among the facies. In the LPF, thick limestone subfacies (1 to 3) are punctuated by caliche caps approximately one every 10 m. Submarine hardgrounds and sheetcracks, with attendant early marine cements, are common in the limestones in the LPF but are not found in the other Latemar buildup shelf-lagoon facies. The LCF consists of 73 laterally traceable, meter-scale depositional cycles whereas the UCF consists of 230 similar cycles. In the LCF and UCF, depositional cycles are generally organized into five-part "megacycles" (four and six part megacycles are common) characterized by systematic decrease in cycle thickness upsection (Figs. 128A, B and Fig. 176A). Finally, the TF facies of the Latemar Limestone is made up of 295 cycles averaging about 0.5 m thick. These cycles, although thinner, are also grouped into thinning-upwards megacycles. Here, however, the tops of megacycles are tepee zones up to 2 m thick that disrupt a number of the upper cycles in a megacycle.

Goldhammer and others (1987, 1990) interpret this remarkable shelf sequence as a record of repeated subaerial exposure of the Latemar buildup on a number of time scales. These cycles are unlike the peritidal cycles associated with tidal flat deposition in that intertidal and supratidal subfacies recording the aggradation and progradation of tidal flats are not found. Instead, sandy subtidal deposits accumulating on the Latemar buildup were repeatedly exposed directly to subaerial diagenesis. Further insight into the controls of this exposure can be found in the astonishing rhythmic stacking of cycles in the LCF and UCF of the Latemar Limestone (Figs. 176B, 177A). Fischer plots (see Fischer, 1964; Goldhammer and others 1987, 1990) graphically display the organization of the basic depositional cycles into five-part megacycles (up to 5 m thick) characterized by progressive upward thinning of successive cycles within each megacycle. Furthermore, autocorrelative time-series analysis, and high resolution spectral analysis (Fig. 176B) confirm this megacyclic organization of cycles within the cyclic facies of the Latemar Limestone. Goldhammer and others (1987, 1990) interpreted the megacycles of the Latemar Limestone to represent two superimposed sea-level oscillations in tune to Milankovitch astronomical rhythms. This interpretation is based on: (1) comparison of the Latemar megacylic sequences with the Pleistocene sea-level record, and (2) computer modeling of megacylic sequences using Milankovitch type sea-level oscillations (Fig. 177B). The average time duration of each limestone-dolostone couplet is on the order of 10^4 years (actual range 8,000 to 20,000 due to uncertainties in stage boundary dating). The composite megacycles are probably on the order of 10^5 years. Pleistocene dated coral reef terraces (from areas presumed to be undergoing constant uplift) and the oxygen isotope record from deep sea cores have been used to demonstrate that Pleistocene glacio-eustatic sea-level changes consisted of two superimposed sea-level oscillations: (1) asymmetric "sawtooth" oscillations with amplitudes of approximately 100 m and a duration of approximately 100,000 years, and (2) sinusoidal oscillations of about 20,000 year duration and some tens of meters amplitude. Time series analysis of the oxygen isotope data of deep sea sediments has established the case for Milankovitch forcing of these Pleistocene sea-level oscillations (Hays and others, 1976; Imbrie and others, 1984). Goldhammer and others (1987, 1990) were able to accurately simulate the five part megacycles of the Latemar Limestone using a one-dimensional computer model. Sedimentation rates, subsidence rates and soil formation rates input to the model were constrained by values from modern carbonate depositional systems. Their modeling employed a composite sea-level curve based on the composite Pleistocene record, namely, the durations and shapes of the superimposed curves were similar to the Pleistocene record, but with meter-scale amplitudes instead of the tens and hundreds of meters amplitudes of the Pleistocene curves. Thus, Goldhammer and others (1987) interpreted the Lower and Upper Cyclic Facies of the Latemar Limestone as recording Milankovitch-driven eustatic sea-level oscillations.

Goldhammer and others (1990) were able to extend their interpretation of the composite sea-level changes to take into account the Anisian Lower Platform Facies (LPF) beneath the Lower Cyclic Facies (LCF) and the Tepee Facies (TF) between the Lower and Upper Cyclic Facies (UCF). Qualitatively, it is apparent that the bulk of the LPF represents subtidal deposition with many hiatuses (the submarine hardgrounds) and only rare exposure. In the Tepee Facies on the other hand, the thick tepee zones argue for prolonged subaerial exposure with infrequent submergence. The vertical succession LPF -> LCF -> TF -> UCF preserved in the interior of the Latemar platform records a low frequency depositional cycle as follows: subtidal conditions with frequent hiatuses of deposition and infrequent subaerial exposure (LPF) -> roughly subequal amounts of subtidal deposition and subaerial exposure (LCF) -> long periods of subaerial exposure punctuated by rare submergence (TF) -> subequal amounts of subtidal deposition and subaerial exposure (UCF). This long-term shoaling upward trend followed by the beginnings of a deepening trend occurred over a period of approximately 9 my (Goldhammer and others, 1990). Goldhammer and others (1990) were able to model the entire succession by superimposing their 10^4 and 10^5 year duration composite sea-level oscillation curves onto a roughly sinusoidal sea-level oscillation with 10^6 year duration and amplitude on the order of a few tens of meters (Fig. 178). On the sea-level rise portions of the low frequency (10^6 year) sea-level oscillation, sea-level falls of the composite 10^4 and 10^5 sea-level oscillations only rarely "touched down" and exposed the platform surface. Instead, the variable water depth on the platform due to the higher frequency composite sea-level oscillations are recorded as submarine hardgrounds, etc. These conditions model deposition of the Lower Platform Facies. The two cyclic facies match conditions on the crest (LCF) and in the trough (UCF) of the low frequency sea-level oscillation. Under these conditions, the rate of change of the low frequency component is essentially zero and therefore nearly every sea-level fall of the composite 10^4 and 10^5 year duration sea-level oscillations exposes the platform. Goldhammer and others (1990) referred to these as "Goldilocks" cycles because everything was "just right" to preserve a sedimentary record of the 5:1 ratio of the composite high frequency sea-level oscillations. During the fall portion of the low frequency sea-level oscillation, many of the sea-level rises of the composite 10^4 and 10^5 year duration sea-level oscillations were not able to flood the platform. These "missed beats" are recorded by the thick tepee zones caused by the prolonged subaerial exposure. In summary, the subaerial exposure features of the Anisian-Ladinian shelf facies exposed in the center of the Latemar buildup record, with uncanny clarity, the superposition of three orders of sea level oscillations.

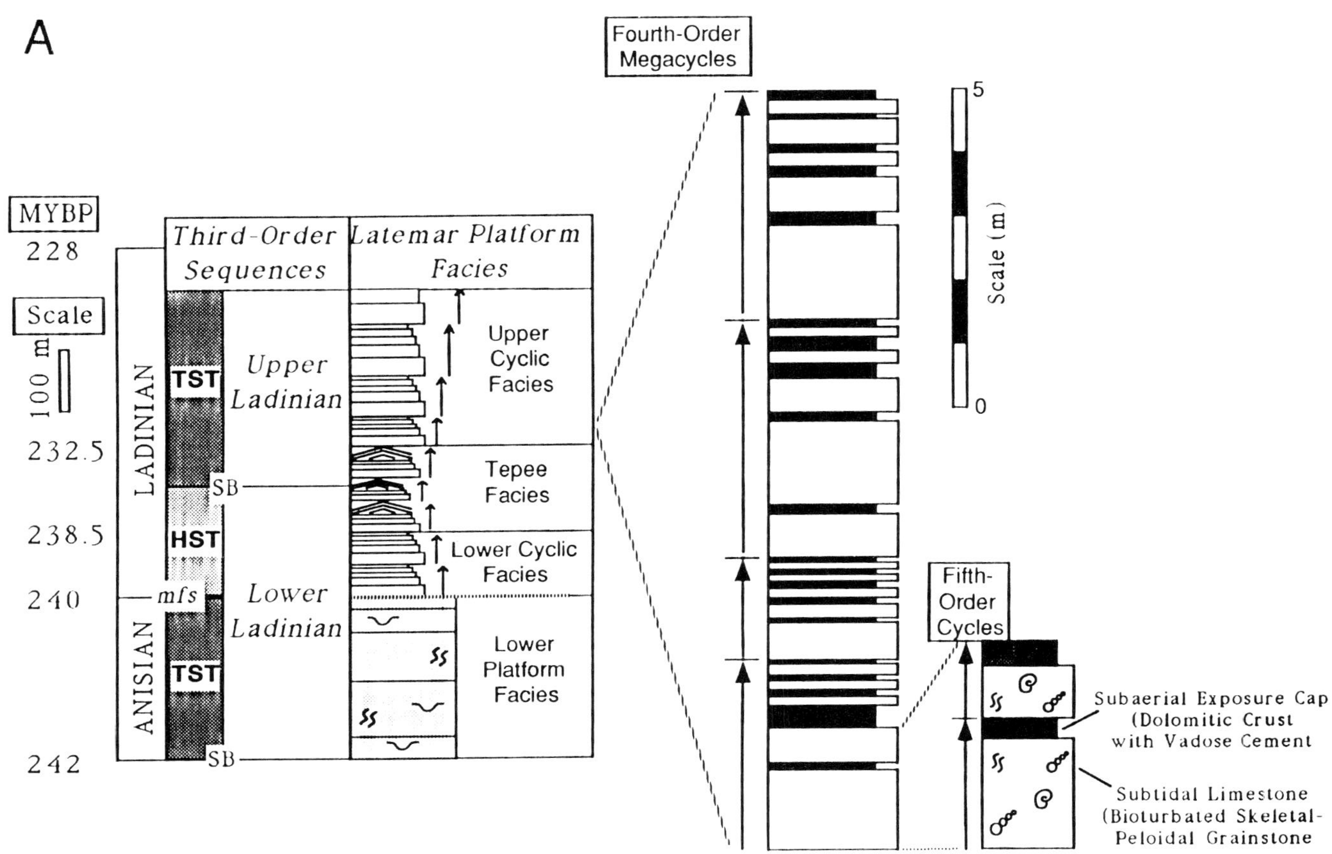

Fig. 175.- (A) Cyclic platform carbonates from the Middle Triassic Latemar buildup, the Dolomites, northern Italy (see text, and Goldhammer and others, 1987, 1990; Hardie and others, 1991, for details). Organization of the platform facies of the Latemar buildup into 3rd order depositional sequences is shown on the left side. TST = transgressive systems tract; HST = high stand systems tract; SB = sequence boundary; mfs = maximum flooding surface. On the right-hand side, part of a measured section is shown. This section illustrates the organization of 5th order depositional cycles into a 4th order "megacycle". See text for details.

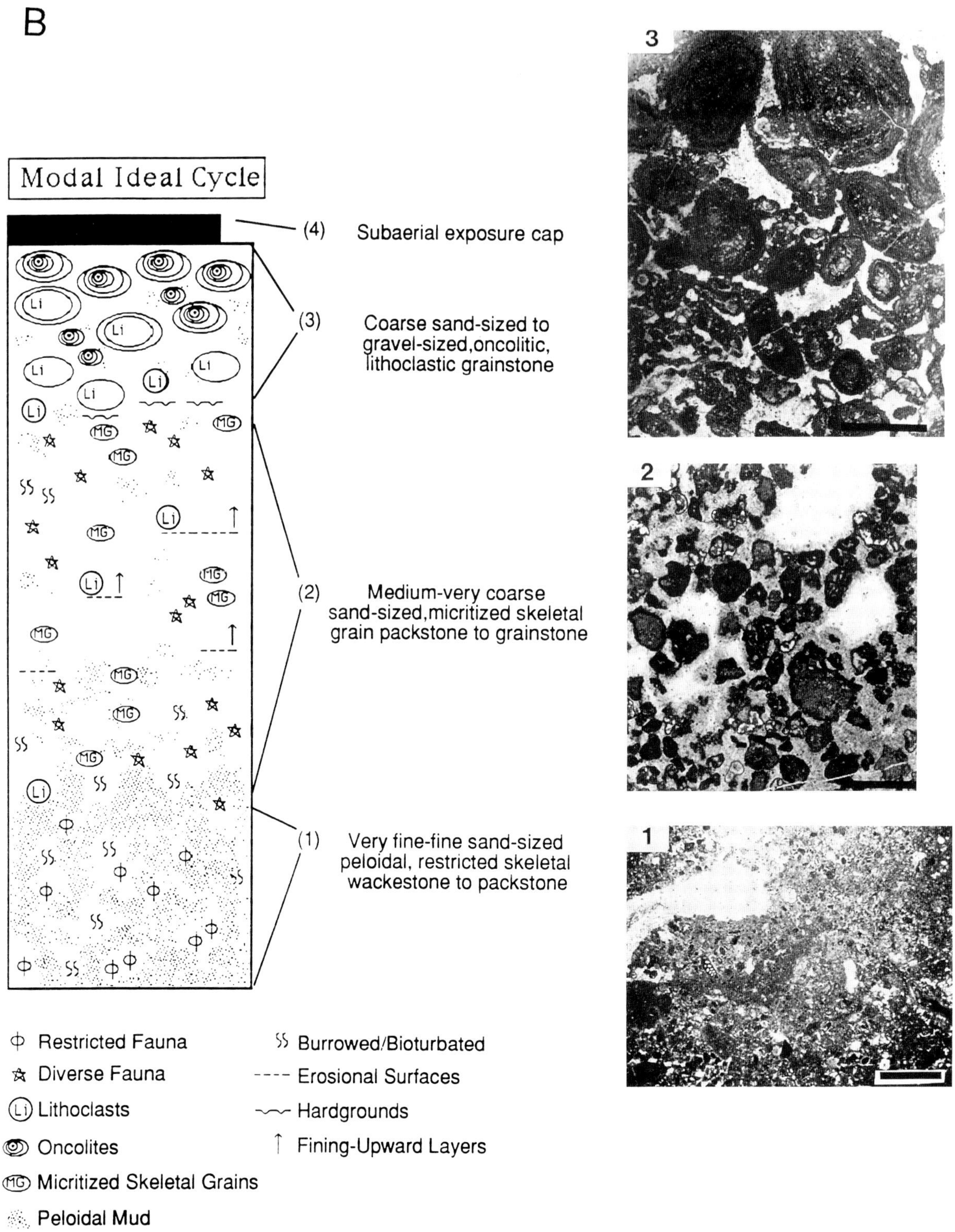

Fig. 175.- (Continued) (B) Vertical textural trends typical of the bioturbated subtidal member of the thicker 5th order depositional cycles of the Latemar cyclic facies. Note that grain size, grain support fabric, faunal diversity, and abundance of lithoclasts and oncoids increase upwards. Representative thin section photomicrographs of the three parts of this type of subtidal member are shown on the right. Scale bars are 1 mm for part 1, 2 mm for part 2, and 6 mm for part 3.

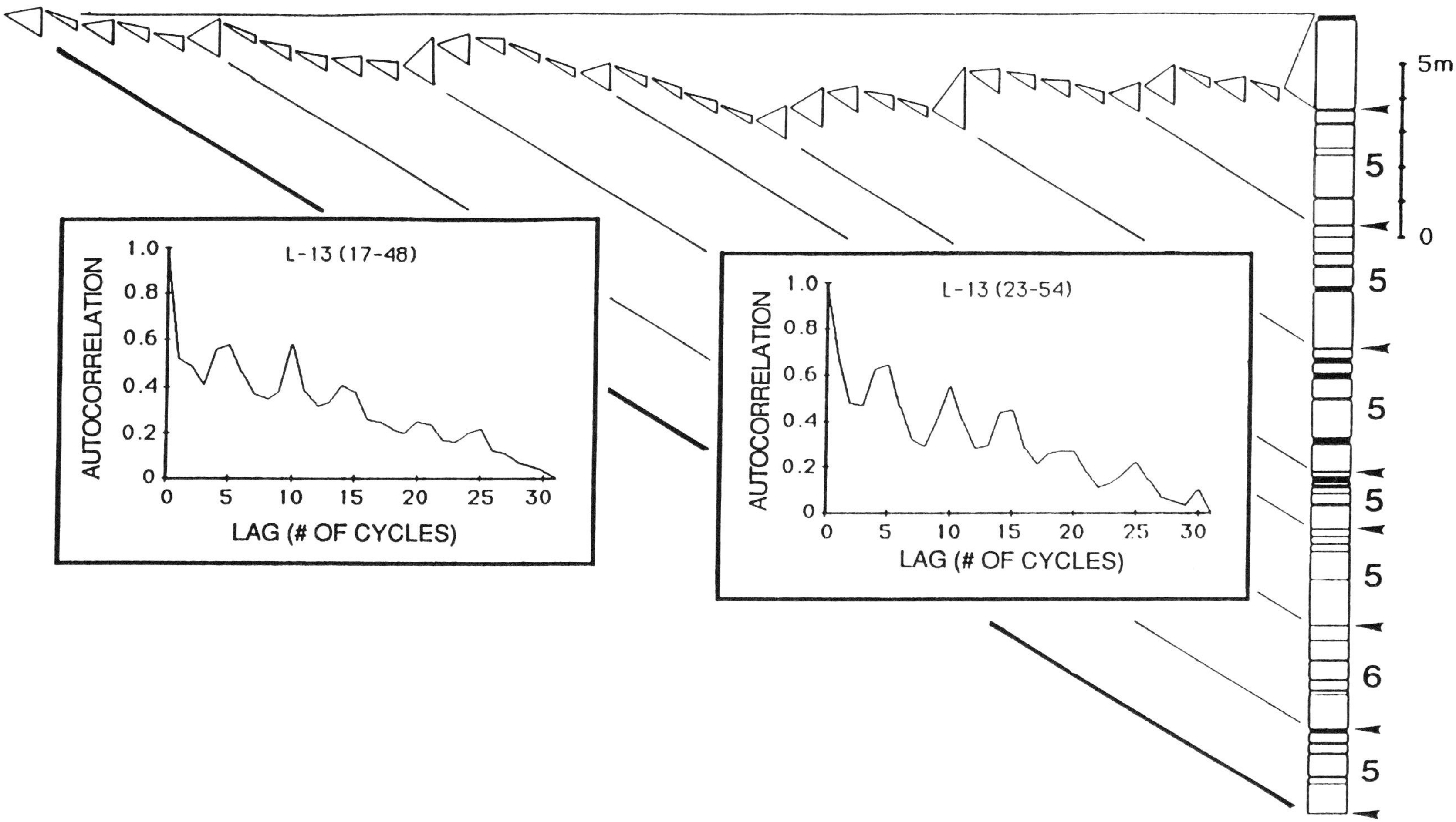

Fig. 176.-Cyclic platform carbonates from the Middle Triassic Latemar buildup, the Dolomites, northern Italy (see text, and Goldhammer and others, 1987, 1990; Hardie and others, 1991, for details). Fischer diagram of 37 consecutive meter-scale cycles from the upper cyclic facies of the Latemar buildup with time on horizontal axis ("younging" to the left) and stratigraphic thickness as vertical axis. This diagram reveals the organization of cycles into 5-part megacycles that appear as a temporal succession of waves representing systematic deviations from mean subsidence and hence systematic variations in accumulation space. The autocorrelation graphs (insets), which refer to two different but overlapping sets of 32 consecutive cycles, confirm the 5-part organization of cycles into megacycles.

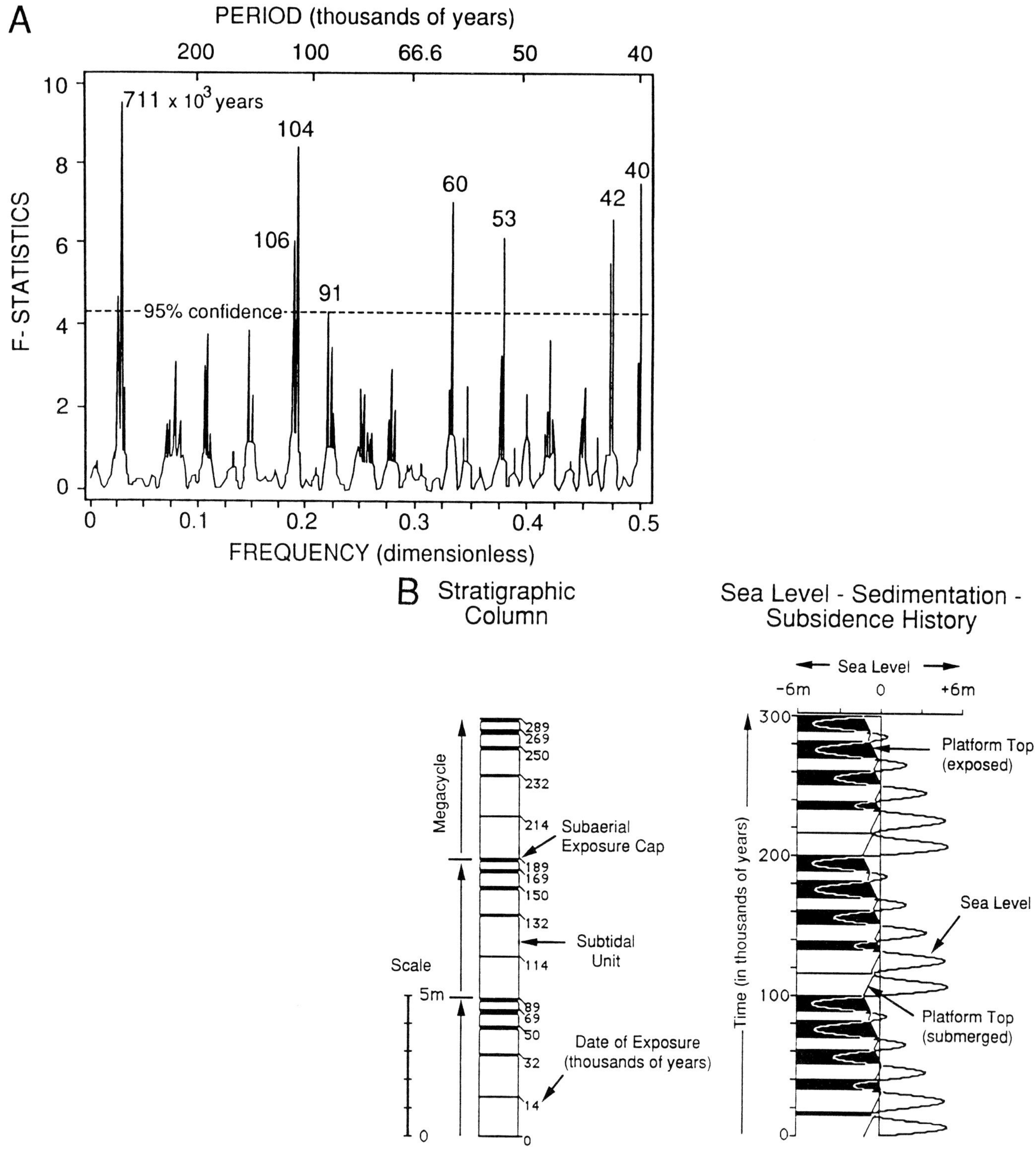

Fig. 177.-Cyclic platform carbonates from the Middle Triassic Latemar buildup, the Dolomites, northern Italy (see text, and Goldhammer and others, 1987, 1990; Hardie and others, 1991, for details) (A) High resolution spectral analysis of 140 consecutive depositional cycles from the upper cyclic facies of the Latemar Limestone of the Latemar buildup. Shown here are the results of the analysis of variance plotted against dimensionless frequency. If a 20,000 year sampling rate is used, then the frequency scale shown at the top applies. Those peaks exceeding the 95% significance level (F-statistics) are identified by their periodicity in thousands of years based on the 20 ky sampling rate (from Hinnov and Goldhammer, 1988). (B) Computer simulation of Latemar platform megacycles. Stratigraphic column (left) shows three complete 4th order megacycles consisting of five meter-scale 5th order cycles composed of subtidal limestone (white) and dolomitic exposure cap (black). Chart on right illustrates the interaction of composite sea level (asymmetric 100 ky sawtooth wave and superimposed 20 ky sinusoidal oscillations) with platform sedimentation and subsidence during the run.

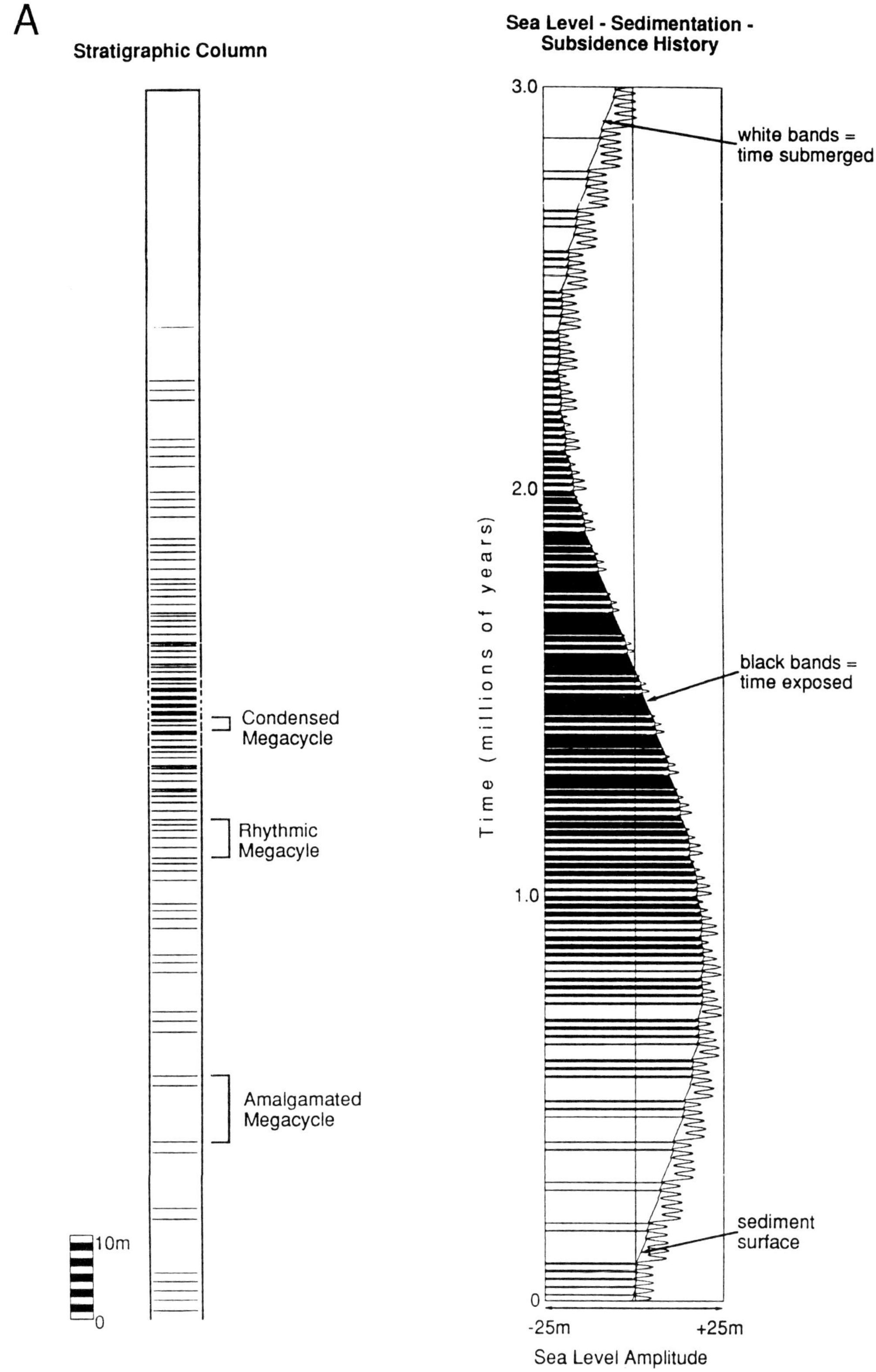

Fig. 178.-Cyclic platform carbonates from the Middle Triassic Latemar buildup, the Dolomites, northern Italy (see text, and Goldhammer and others, 1987, 1990; Hardie and others, 1991, for details). (A) Simulation using the "Mr. Sediment" computer program of a 3 my submergence-emergence record on a carbonate platform subjected to composite high frequency sea-level oscillations (of 20 ky and 100 ky) superimposed on a 3rd order (3 my) sea-level change. The amplitude of the 20 ky oscillation is greater than the amplitude of the 100 ky oscillation. Resulting cyclostratigraphy is shown on the left. Input data: sedimentation rate = 0.1 m/1000 yr; subsidence rate = 0.05 m/1000 yr; sinusoidal 20 ky wave with 3 m amplitude; asymmetrical 100 ky wave with 2 m amplitude; 3rd order wave with 40 m amplitude and 3 my duration.

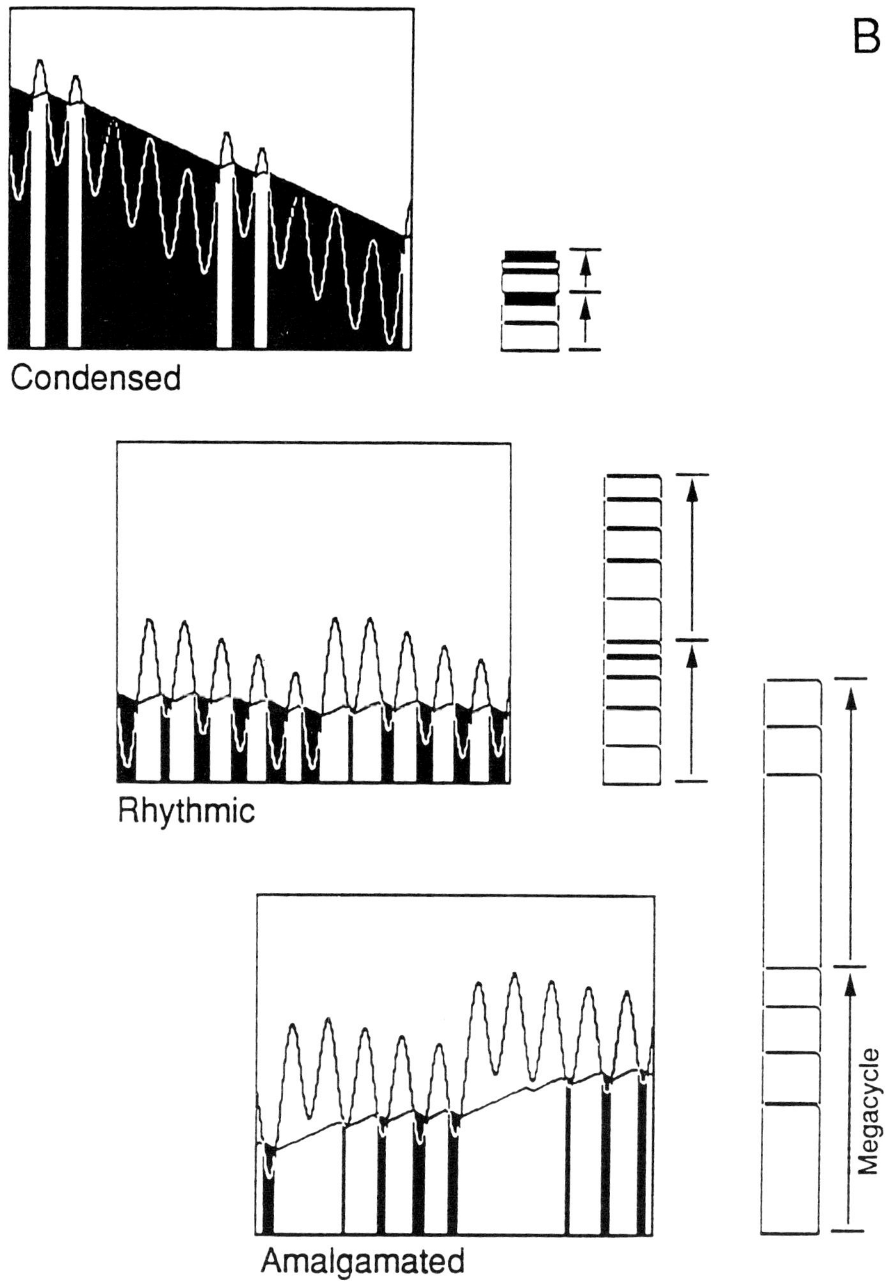

Fig. 178.-(Continued) (B) Graphical representation of high frequency, composite sea level oscillations (superimposed 100 ky asymmetric 4th order and 20 ky sinusoidal 5th order) and platform aggradation (panels on left) and the resulting megacycle packaging (columns on right). High frequency composite eustasy combined with long-term (3rd order) trends (see A) yields three classes of megacycles: (1) *condensed megacycles* are produced when one or more 20 ky oscillations are missed below the platform top (rising 3rd order sea level); (2) *rhythmic megacycles* are produced when each successive 20 ky oscillation is registered (at crest and trough of 3rd order sea level curve); (3) *amalgamated megacycles* are produced when 20 ky oscillations are missed above the platform (falling 3rd order sea level).

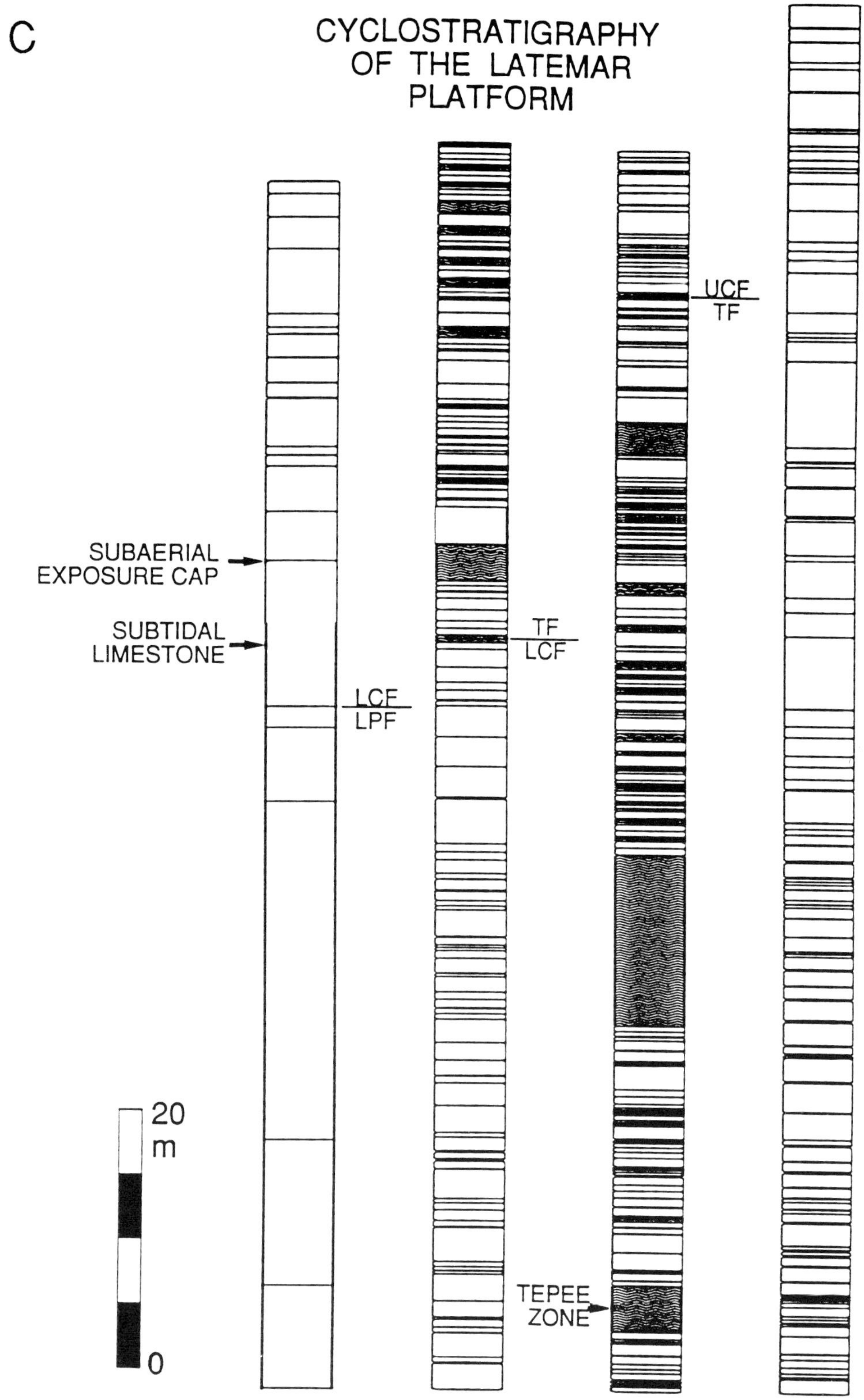

Fig. 178.-(Continued) (C) High frequency cyclostratigraphy of the Latemar platform carbonates based on several measured sections correlated by lateral tracing of key cycles. This composite section includes the uppermost part of the lower platform facies (LPF). the complete lower cyclic facies (LCF), the complete tepee facies (TF), and the lower part of the upper cyclic facies (UCF). The stacking progression from amalgamated megacycles at the base (LPF) through rhythmic megacycles (LCF) to condensed megacycles (TF) and back to rhythmic megacycles (UCF) is well displayed in the approximately 200 m thick section. Cycles and packages of cycles are disrupted by tepees are shown in zigzag pattern. The thick tepee zone in the middle of the tepee facies is shown in Figure 142. Compare with computer simulations in (A) and (B).

Shelf Margin Facies of the Lower Ordovician Grove Limestone, Central Appalachians

The Lower Ordovician Grove Limestone exposed in the Frederick Valley of western Maryland is approximately 450 m thick and can be correlated with the lower portions of the Beekmantown Group exposed some 50 km (restored) to the west in the Hagerstown Valley and 250 km west (restored) in the Nittany Arch of central Pennsylvania. Logs of these rocks measured using the subfacies and facies approach outlined here can be used to construct a cross-strike facies mosaic (Fig. 179) correlated using macrofossil and conodont zones (Nguyen and others, 1985; Hardie and Shinn, 1986). The "storm-dominated" shelf facies is the Stonehenge Limestone described above. The tidal flat facies has been described by Nguyen (1986) (see also Read and Goldhammer, 1988). The outer shelf facies consists of *Renalcis-Girvanella* bioherms, cross-stratified skeletal-peloidal grainstones, and ribbon carbonates, and resembles the "current dominated" shelf lagoon facies of the Conococheague Limestone described above. Two small wedges of Lower Ordovician shelf margin facies are found in the Grove Limestone. These shelf margin facies of the Grove Limestone are made up of two subfacies: a bioherm subfacies and a grainstone subfacies (Nguyen, 1986). Figure 99 is a two dimensional representation of a typical outcrop pattern of this facies. This shelf margin facies closely resembles the current-dominated shelf lagoon facies of the Concococheague Limestone but differs in a number of important aspects. First, bioherms up to a meter in diameter tend to be stacked together producing complex biohermal masses up to 5 m thick, riddled by pockets and channels filled with grainstones. Second, the clotted fabrics of the bioherms are made up of plate-like masses of *Girvanella* tubules tens of millimeters long and encrusting bushes of *Epiphyton* up to a few millimeters in diameter. These problematic microfossils are surrounded by sand-sized skeletal grains and peloids (not mud), isopachous bladed cements, and microspar. Thus, these bioherms have a distinctly different biota than those from the platform interiors. It is important to note that these bioherms closely resemble the *Epiphyton-Girvanella* clasts found in slope breccias in equivalent Cambro-Ordovician basinal deposits such as the Cow Head Breccia (James, 1981) and the Frederick Limestone (Reinhardt, 1974; Demicco, 1985). These Lower Ordovician *Epiphyton-Girvanella* bioherms of the Grove Limestone together with their associated ooid grainstones clearly record an ancient carbonate shelf margin rimmed by a reef tract-sand shoal.

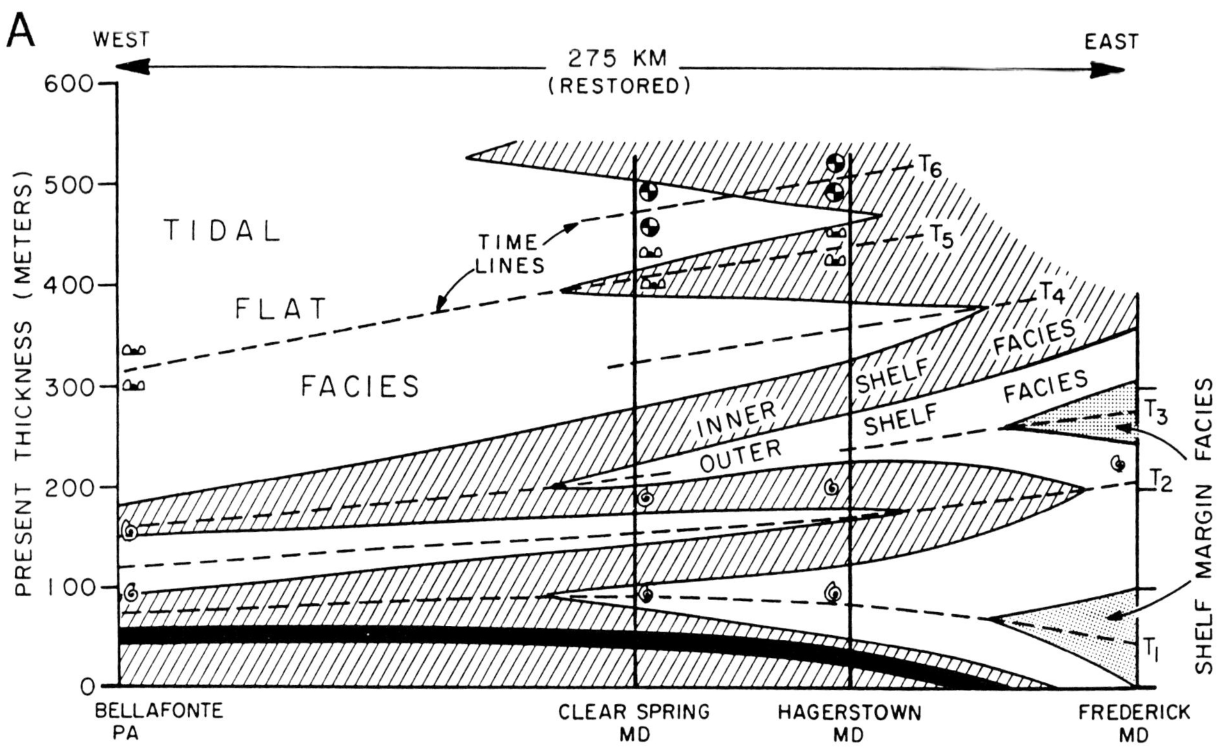

Fig. 179.-Facies stratigraphy of the Lower Ordovician platform carbonates (Beekmantown Group) of the central Appalachians. (A) Two-dimensional facies mosaic showing systematic facies zonation across strike as well as three 3rd order cycles of transgression and regression. The time lines based on facies tongues agree extremely well with time zones based on macrofossils (time line T6 is based on *Archeoscyphia*, T5 is based on *Lecanospira compacta*, and the time band between T1 and T3 is based on *Ophileta complicata*). The facies mosaic was constructed by Chau Nguyen and Robert K. Goldhammer. Shelf and margin facies are discussed elsewhere. (B) Interpretation of the Lower Ordovician facies stratigraphy in terms of 3rd order cyclostratigraphy. CB = cycle boundary (see Hardie, 1989). (C) Interpretation by Robert K. Goldhammer of Lower Ordovician Beekmantown Group facies stratigraphy in terms of sequence stratigraphy (see Hardie, 1989). SB = sequence boundary; TST = transgressive systems tract; HST = high stand systems tract; mfs = maximum flooding surface.

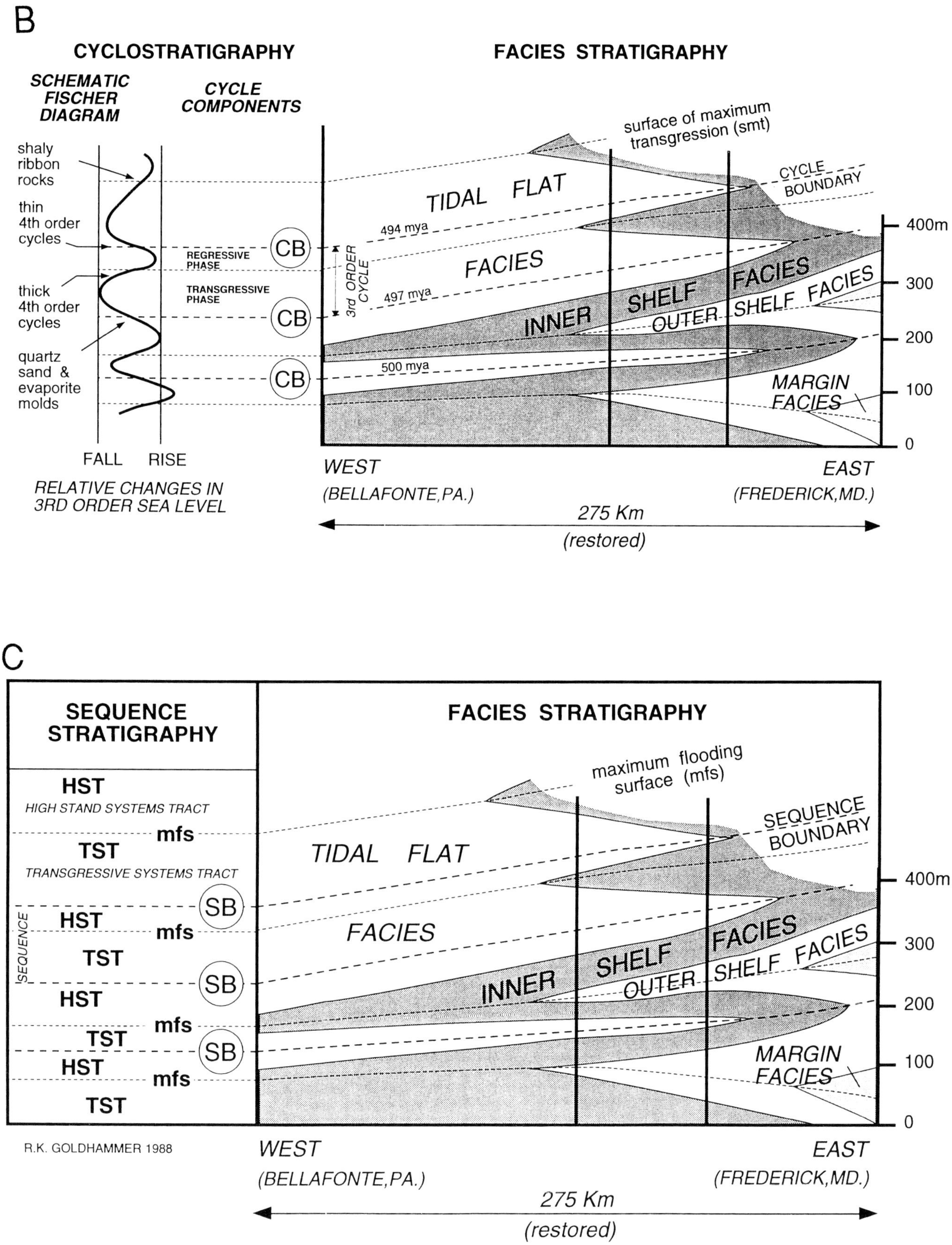
B
CYCLOSTRATIGRAPHY
SCHEMATIC FISCHER DIAGRAM
CYCLE COMPONENTS
shaly ribbon rocks
thin 4th order cycles
thick 4th order cycles
quartz sand & evaporite molds
REGRESSIVE PHASE
TRANSGRESSIVE PHASE
CB
CB
CB
FALL
RISE
RELATIVE CHANGES IN 3RD ORDER SEA LEVEL
FACIES STRATIGRAPHY
surface of maximum transgression (smt)
CYCLE BOUNDARY
TIDAL FLAT
FACIES
INNER SHELF FACIES
OUTER SHELF FACIES
MARGIN FACIES
3rd ORDER CYCLE
494 mya
497 mya
500 mya
400m
300
200
100
0
WEST
(BELLAFONTE,PA.)
EAST
(FREDERICK,MD.)
275 Km
(restored)
C
SEQUENCE STRATIGRAPHY
HST
HIGH STAND SYSTEMS TRACT
mfs
TST
TRANSGRESSIVE SYSTEMS TRACT
SB
SEQUENCE
HST
mfs
TST
SB
HST
mfs
TST
SB
HST
mfs
TST
FACIES STRATIGRAPHY
maximum flooding surface (mfs)
SEQUENCE BOUNDARY
TIDAL FLAT
FACIES
INNER SHELF FACIES
OUTER SHELF FACIES
MARGIN FACIES
400m
300
200
100
0
R.K. GOLDHAMMER 1988
WEST
(BELLAFONTE,PA.)
EAST
(FREDERICK,MD.)
275 Km
(restored)

CONCLUDING REMARKS

In searching for an apt concluding statement that would both summarize our view of the importance of sedimentary structures in unraveling the genesis and history of shallow marine carbonate deposits and, at the same time, make a plea for all students of carbonate sedimentology to become truly proficient in the recognition and analysis of sedimentary structures, we found that we had perhaps already said it best in the introduction. So, if the gentle reader would forgive us for repeating ourselves, we would like to end with a quotation from the introduction, namely:

> **"Overall, an understanding of the origin and significance of primary sedimentary structures and early diagenetic features is vital to unraveling the origin and significance of carbonate deposits in the geologic record. Without such an understanding at the individual sedimentary structure scale, we cannot hope to accurately reconstruct the large scale accumulation history of carbonate deposits or to decipher the roles of sea-level changes, sedimentation rates, subsidence rates, and tectonics in determining the facies stratigraphy, cyclostratigraphy, and sequence stratigraphy of these buildups."**

The message seems crystal clear, if we do not get the little things right we may not be able get the big things right.

ACKNOWLEDGMENTS

We would like to express our appreciation to Ed Clifton, the editor of this SEPM series, for his invitation to prepare this summary of sedimentary structures and early diagenetic features in shallow marine carbonate deposits and for his editorial encouragement and patience during the writing and review processes. Special thanks go to Ray Mitchell and Bob Loucks for their detailed and thoughtful reviews of the first draft of the manuscript. Their comments on the scientific content and style of presentation were all taken to heart and incorporated in one way or another in the revised text. Dana Ulmer-Scholle and Tim Lowenstein patiently reviewed the final draft. Dana Ulmer-Scholle also compiled the index. Anne Hull drafted the figures and David Tuttle printed many of the photographs. Karen Kuff-Demicco helped us with the design of the final draft.

We are grateful to our fellow sedimentologists who allowed us the use of their photographs, color slides, and samples for the illustrations in this atlas. To Robert Adams, Wayne Ahr, John Bridge, Godfrey Butler, Glen Dolphin, Bob Goldhammer, John Grotzinger, Paul Harris, Paul Hoffman, Miner Long, Tim Lowenstein, Ray Mitchell, Chau Nguyen, Francis Pettijohn, Fred Read, Hans Erich Reineck, Ron Spencer, and Brent Waters, our sincerest appreciation for your help. Finally, we would like to thank the Springer-Verlag Publishing Company for granting us permission to reproduce Plates 14, 28, 50B, 89, 97, and 102B that originally appeared in the *Atlas and Glossary of Primary Sedimentary Structures* by F. J. Pettijohn and P. E. Potter.

REFERENCES

Adams, J. E., and Frenzel, N. H., 1950, Capitan barrier reef, Texas and New Mexico: Journal of Geology, v. 58, p. 289-312.

Adams, R. W., 1970, Loyalhanna Limestone- cross-bedding and provenance, *in* Fisher, G. W., Pettijohn, F. J., and Reed, J. C. Jr., eds., Studies of Appalachian Geology: Central and Southern: New York, Interscience Publishers, p. 83-100.

Aigner, T., 1985, Storm Depositional Systems: New York, Lecture Notes in Earth Sciences, v. 3, Springer-Verlag, 174 p.

Aitken, J. D., 1966, Middle Cambrian to Middle Ordovician cyclic sedimentation, southern Canadian Rocky Mountains of Alberta: Bulletin of Canadian Petroleum Geology, v. 14, p. 405-442.

Aitken, J. D., 1967, Classification and environmental significance of cryptalgal limestones and dolomites with illustrations from the Cambrian and Ordovician of southwestern Alberta: Journal of Sedimentary Petrology, v. 37, p. 1163-1178.

Aitken, J. D., 1971, Control of Lower Paleozoic sedimentary facies by the Kicking Horse Rim, southern Rocky Mountains: Bulletin of Canadian Petroleum Geology, v. 19, p. 557-569.

Aitken, J. D., 1978, Revised models for depositional grand cycles, Cambrian of the southern Rocky Mountains, Canada: Bulletin of Canadian Petroleum Geology, v. 26, p. 515-542.

Alexandersson, T., 1972, Mediterranean beachrock cementation: Marine precipitation of Mg-calcite, *in* Stanley, D. J., ed., The Mediterranean Sea: A Natural Sedimentation Laboratory: Stroudsburg, Drowden, Hutchinson & Ross, Inc., p. 203-223.

Alexandersson, T., 1974, Carbonate cementation in coralline algal nodules in the Skagerrak, North Sea: Biochemical precipitation in undersaturated waters: Journal of Sedimentary Petrology, v. 44, p. 7-26.

Ali, Y. A., and West, I. M., 1983, Relationships of modern gypsum nodules in sabkhas of loess to compositions of brines and sediments in northern Egypt: Journal of Sedimentary Petrology, v. 53, p. 1151-1168.

Allen, J. R. L., 1963, The classification of cross-stratified units with notes on their origin: Sedimentology, v. 2, p. 93-114.

Allen, J. R. L., 1970, A quantitative model of climbing ripples and their cross-laminated deposits: Sedimentology, v. 14, p. 5-26.

Allen, J. R. L., 1973, A classification of climbing-ripple cross-stratification: Journal of the Geological Society of London, v. 129, p. 537-541.

Allen, J. R. L., 1974, Studies in fluviatile sedimentation: implications of pedogenic carbonate units, Lower Old Red Sandstone, Anglo-Welsh outcrop: Geological Journal, v. 9, p. 181-208.

Allen, J. R. L., 1984, Sedimentary Structures: their Character and Physical Basis (unabridged one-volume edition): New York, Developments in Sedimentology 30, Elsevier, 1256 p.

Allen, J. R. L., 1985, Principles of Physical Sedimentology: Boston, George Allen and Unwin, 272 p.

Allen, J. R. L., 1987, Reworking of muddy intertidal sediments in the Severn Estuary, southwestern U.K.-a preliminary survey: Sedimentary Geology, v. 50, p. 1-23.

Andrews, J. E., Shepard, F. P., and Hurley, R. J., 1970, Great Bahama Canyon: Geological Society of America Bulletin, v. 81, p. 1061-1078.

Anketell, J. M., Cegla, J., and Dzulynski, S., 1970, On the deformational structures in systems with reversed density gradients: Polskie Towarzystwo Geologiczne, Rocznik, v. 40, p. 3-30.

Arakel, A. V., 1980, Genesis and diagenesis of Holocene evaporitic sediments in Hutt and Leeman Lagoons, Western Australia: Journal of Sedimentary Petrology, v. 50, p. 1305-1326.

Arnott, R. W., and Southard, J. B., 1990, Exploratory flow-duct experiments on combined-flow bed configurations, and some implications for interpreting storm-event stratification: Journal of Sedimentary Petrology, v. 60, p. 211-219.

Assereto, R., and Folk, R. L., 1980, Diagenetic fabrics of aragonite, calcite, and dolomite in an ancient peritidal-spelean Bertrand-Sarfati, J., 1976, An attempt to classify Late Preenvironment: Triassic Calcare Rosso, Lombardia, Italy: Journal of Sedimentary Petrology, v. 50, p. 371-394.

ASSERETO, R. L., AND KENDALL, C. G. ST. C., 1977, Nature, origin and classification of peritidal tepee structures and related breccias: Sedimentology, v. 24, p. 153-210.

AULSTEAD, K. L., AND SPENCER, R. J., 1985, Diagenesis of the Keg River Formation, northwestern Alberta: fluid inclusion evidence: Bulletin of Canadian Petroleum Geology, v. 33, p. 167-183.

AULSTEAD, K. L., SPENCER, R. J., AND KROUSE, H. R., 1988, Fluid inclusion and isotopic evidence on dolomitization, Devonian of western Canada: Geochimica et Cosmochimica Acta, v. 52, p. 1027-1035.

AWRAMIK, S. M., 1971, Precambrian columnar stromatolite diversity: reflection of metazoan appearance: Science, v. 174, p. 825-826.

AWRAMIK, S. M., 1984, Ancient stromatolites and microbial mats, *in* Cohen, Y., Castenholz, R. W., and Halvorson, H. O., eds., Microbial Mats: Stromatolites: New York, A.R. Liss Inc., p. 1-22.

BADGLEY, P. C., 1965, Structural and tectonic principles: New York, Harper and Row, 521 p.

BALL, M. M., 1967, Carbonate sand bodies of Florida and the Bahamas: Journal of Sedimentary Petrology, v. 37, p. 556-591.

BARNES, I., 1965, Geochemistry of Birch Creek, Inyo County, California: a travertine depositing creek in an arid climate: Geochimica et Cosmochimica Acta, v. 29, p. 85-112.

BASAN, P. B., ed., 1978, Trace Fossil Concepts: Tulsa, Society of Economic Paleontologists and Mineralogists Short Course No. 5, 181 p.

BATHURST, R. G. C., 1959, The cavernous structure of some Mississippian "stromatactis" reefs in Lancashire, England: Journal of Geology, v. 67, p. 506-521.

BATHURST, R. G. C., 1975, Carbonate Sediments and their Diagenesis: New York, Elsevier, 658 p.

BATHURST, R. G. C., 1980a, Lithification of carbonate sediments: Science Progress Oxford, v. 66, p. 451-471.

BATHURST, R. G. C., 1980b, Stromatactis- origin related to submarine cemented crusts in Paleozoic mud mounds: Geology, v. 8, p. 131-134.

BATHURST, R. G. C., 1982, Genesis of stromatactis cavities between submarine crusts in Paleozoic carbonate mud buildups: Journal of the Geological Society of London, v. 139, p. 165-181.

BATHURST, R. G. C., 1987, Diagenetically enhanced bedding in argillaceous platform limestones: stratified cementation and selective compaction: Sedimentology, v. 34, p. 749-778.

BAUERMAN, H., 1885, Report on the geology of the country near the forty-ninth parallel of North Latitude west of the Rocky Mountains: Canadian Geological Survey Reports of Progress, 1882-1884, part B, p. 1-42.

BAULD, J., 1981, Occurrence of benthic microbial mats in saline lakes: Hydrobiologia, v. 81, p. 87-111.

BEALES, F. W., AND HARDY, J. W., 1980, Criteria for the recognition of diverse dolomite types with an emphasis on studies of host rocks for Mississippi Valley-type ore deposits, *in* Zenger, D. H., Dunham, J. R., and Ethington, R. L., eds., Concepts and Models of Dolomitization: Tulsa, Society of Economic Paleontologists and Mineralogists Special Publication 28, p. 197-214.

BEBOUT, D. G., AND MAIKLEM, W. R., 1973, Ancient anhydrite facies and environments, Middle Devonian Elk Point Basin, Alberta: Bulletin of Canadian Petroleum Geology, v. 21, p. 287-343.

BERTRAND-SARFATI, J., 1976, An attempt to classify Late Precambrian stromatolite microstructure, *in* Walter, M. R., ed., Stromatolites: New York, Developments in Sedimentology 20, Elsevier, p. 251-259.

BLACK, M., 1933, The algal sediments of Andros Island, Bahamas: Royal Society of London Philosophical Transactions, Series B, v. 122, p. 165-191.

BLATT, H., MIDDLETON, G., AND MURRAY, R., 1980, Origin of sedimentary rocks (2nd ed.): Englewood Cliffs, Prentice-Hall, 782 p.

BLOUNT, D. N., AND MOORE, C. H., 1969, Depositional and non-depositional carbonate breccias, Chiantla Quadrangle, Guatemala: Geological Society of America Bulletin, v. 80, p. 429-442.

BOERSMA, J. R., 1969, Internal structure of some tidal megaripples on a shoal in the Westerschelde estuary, the Netherlands. Report of a preliminary investigation: Geologie en Mijnbouw, v. 48, p. 409-414.

BOERSMA, J. R., AND TERWINDT, J. H. J., 1981, Neap-spring tide sequences of intertidal shoal deposits in a mesotidal estuary: Sedimentology, v. 28, p. 151-170.

BORNHOLD, B. D., AND PILKEY, O. H., 1971, Bioclastic turbidite sedimentation in Columbus Basin, Bahamas: Geological Society of America Bulletin, v. 82, p. 1341-1354.

BOSELLINI, A., 1984, Progradation geometries of carbonate platforms: examples from the Triassic of the Dolomites, northern Italy: Sedimentology, v. 31, p. 1-24.

BOSELLINI, A., AND HARDIE, L. A., 1973, Depositional theme of a marginal marine evaporite: Sedimentology, v. 20, p. 5-27.

BOSELLINI, A., MASETTI, D., AND SARTI, M., 1981, A Jurassic 'Tongue of the Ocean' infilled with oolitic sands: the Belluno Trough, Venetian Alps, Italy: Marine Geology, v. 44, p. 59-95.

BOSELLINI, A., AND ROSSI, D., 1974, Triassic carbonate buildups of the Dolomites, northern Italy, *in* Laporte, L. F., ed., Reefs in Space and Time: Tulsa, Society of Economic Paleontologists and Mineralogists Special Publication 13, p. 209-233.

BOTTJER, D. J., ed., 1987, New Concepts in the Use of Biogenic Sedimentary Structures for Paleoenvironmental Interpretation, Volume and Guidebook: Los Angeles, Society of Economic Paleontologists and Mineralogists, Pacific Section, 65 p.

BOTVINKINA, L.N., 1959, Morphological classification of bedding in sedimentary rocks: Akademia Nauk S.S.S.R., Izvestiya of the Academy of Sciences of the U.S.S.R., Geologic Series No. 6, p. 13-30.

BOUMA, A. H., 1968, Distribution of minor structures in Gulf of Mexico sediments: Transactions of the Gulf Coast Association of Geological Societies, v. 18, p. 26-33.

BRADLEY, W. H., 1929a, The varves and climate of the Green River epoch: United States Geological Survey Professional Paper 158-E, p. 87-110.

BRADLEY, W. H., 1929b, Algal reefs and oolites of the Green River Formation: United States Geological Survey Professional Paper 154-G, p. 203-223.

BRADLEY, W. H., 1931, Non-glacial marine varves: American Journal of Science, v. 222, p. 318-330.

BRADLEY, W. H., AND EUGSTER, H. P., 1969, Geochemistry and paleolimnology of the Trona deposits and associated authigenic minerals of the Green River Formation of Wyoming: United States Geological Survey Professional Paper 496-B, 71 p.

BRAITHWAITE, C. J. R., 1979, Crystal textures of Recent fluvial pisoliths and laminated crystalline crusts in Dyfed, South Wales: Journal of Sedimentary Petrology, v. 49, p. 181-194.

BRANTLEY, S. L., CRERAR, D. A., MOLLER, N. E., AND WARME, J. H., 1984, Geochemistry of a modern marine evaporite: Bocana de Virrila, Peru: Journal of Sedimentary Petrology, v. 54, p. 447-462.

BRAYSHAW, A. C., 1984, Characteristics and origin of cluster bedforms in coarse-grained alluvial channels, *in* Koster, E. H., and Steel, R. J., eds., Sedimentology of Gravels and Conglomerates: Calgary, Canadian Society of Petroleum Geologists Memoir 10, p. 77-85.

BRETZ, J. H., AND HORBERG, L., 1949, Caliche in southeastern New Mexico: Journal of Geology, v. 57, p. 491-511.

BRICKER, O. P., 1971, Carbonate Cements: Baltimore, The Johns Hopkins University Press, 376 p.

BRIDGE, J. S., AND DROSER, M. L., 1985, Unusual marginal-marine lithofacies from the Upper Devonian Catskill clastic wedge: Geological Society of America Special Paper 201, p. 143-161.

BRIDGE, J. S., AND GORDON, E. A., 1985, Quantitative interpretation of ancient river systems in the Oneonta Formation, Catskill Magnafacies: Geological Society of America Special Paper 201, p. 163-181.

BROECKER, W. S., AND TAKAHASHI, T., 1966, Calcium carbonate precipitation on the Bahama Banks: Journal of Geophysical Research, v. 71, p. 1575-1602.

BROMLEY, R. G., 1967, Some observations on burrows of thalassinidean Crustacea in Chalk hardgrounds: Quarterly Journal of the Geological Society of London, v. 123, p. 157-177.

BROMLEY, R. G., 1968, Burrows and borings in hardgrounds: Meddelelser Dansk Geologisk Forening (København), v. 18, p. 247-250.

BROMLEY, R. G., 1975, Trace fossils at omission surfaces, in Frey, R. W., ed., Study of Trace Fossils: New York, Springer-Verlag, p. 399-229.

BROWN, C. N., 1956, The origin of caliche on the north-eastern Llana Estacado, Texas: Journal of Geology, v. 64, p. 1-15.

BUICK, R., DUNLOP, J. S. R., AND GROVES, D. I., 1981, Stromatolite recognition in ancient rocks: an appraisal of irregularly laminated structures in an Early Archean chert-barite unit from North Pole, Western Australia: Alcheringa, v. 5, p. 161-181.

BURKY, D., KING, S. A., HORN, M. K., AND MANHEIN, F. T., 1970, Geological significance of cocoliths in fine-grained carbonate bands of post-glacial Black Sea sediments: Nature, v. 226, p. 156-158.

BURST, J. F., 1965, Subaqueously formed shrinkage cracks in clay: Journal of Sedimentary Petrology, v. 35, p. 348-353.

BUTLER, G. P., 1969, Modern evaporite deposition and geochemistry of coexisting brines, the sabkha, Trucial Coast, Arabian Gulf: Journal of Sedimentary Petrology, v. 39, p. 70-89.

BUTLER, G. P., 1970, Holocene gypsum and anhydrite of the Abu Dhabi sabkha, Trucial Coast: an alternative explanation of origin, *in* Rau, J. L., and Dellwig, L. F., eds., Proceedings of the Third Salt Symposium: Cleveland, Northern Ohio Geological Society, p. 120-152.

CALVERT, S. E., 1966, Origin of diatom-rich varved sediments from the Gulf of California: Journal of Geology, v. 74, p. 546-565.

CAMPBELL, C. V., 1967, Lamina, laminaset, bed and bedset: Sedimentology, v. 8, p. 7-26.

CARBALLO, J. D., LAND, L. S., AND MISER, D. E., 1987, Holocene dolomitization of supratidal sediments by active tidal pumping, Sugarloaf Key, Florida: Journal of Sedimentary Petrology, v. 57, p. 153-165.

CARPENTER, A. B., 1980, The chemistry of dolomite formation I: The stability of dolomite, *in* Zenger, D. H., Dunham, J. B., and Ethington, R. L., eds., Concepts and Models of Dolomitization: Tulsa, Society of Economic Paleontologists and Mineralogists Special Publication 28, p. 111-121.

CASAS, E., AND LOWENSTEIN, T. K., 1989, Diagenesis of saline pan halite: comparison of petrographic features of modern, Quaternary and Permian halites: Journal of Sedimentary Petrology, v. 59, p. 724-739.

CASTENS-SEIDELL, B., 1984, The anatomy of a modern marine siliciclastic sabkha in a rift valley setting: northwest Gulf of California tidal flats, Baja California: Unpublished Ph.D. Dissertation, The Johns Hopkins University, Baltimore, 386 p.

CASTENS-SEIDELL, B., AND HARDIE, L. A., 1983, Gypsum-anhydrite deposition in sabkhas: new observations from the Holocene tidal flats of the northwest Gulf of California (abs.): Geological Society of America Abstracts with Programs, v. 15, p. 540.

CHAFETZ, H. S., 1986, Marine peloids: a product of bacterially induced precipitation of calcite: Journal of Sedimentary Petrology, v. 56, p. 812-817.

CHAFETZ, H.S., AND BUTLER, J.C., 1980, Petrology of recent caliche pisolites, spherulites and speleothem deposits from central Texas: Sedimentology, v. 27, p. 497-518.

CHAFETZ, H. S., AND FOLK, R. L., 1984, Travertines, depositional morphology and the bacterially constructed constituents: Journal of Sedimentary Petrology, v. 54, p. 289-316.

CHOQUETTE, P. W., AND JAMES, N. P., 1990, Limestones-The burial diagenetic environment: Geoscience Canada Reprint Series 4, Geological Association of Canada, p. 75-112.

CHOQUETTE, P. W., AND PRAY, L. C., 1970, Geologic nomenclature and classification of porosity in sedimentary carbonates: American Association of Petroleum Geologists Bulletin, v. 54, p. 207-250.

CLIFTON, H. E., 1967, Solution-collapse and cavity filling in the Windsor Group, Nova Scotia, Canada: Geological Society of America Bulletin, v. 78, p. 819-832.

CLIFTON, H. E., 1976, Wave-formed structures- a conceptual model, *in* Davis, R. A., and Ethington, R. L., eds., Beach and Nearshore Sedimentation: Tulsa, Society of Economic Paleontologists and Mineralogists Special Publication 24, p. 126-148.

CLOUD, P. E., 1960, Gas as a sedimentary and diagenetic agent: American Journal of Science, v. 258, p. 35-45.

CLOUD, P. E., 1962, Environment of calcium carbonate deposition west of Andros Island, Bahamas: United States Geological Survey Professional Paper 350, p. 1-138.

CLOYD, K. C., DEMICCO, R. V., AND SPENCER, R. J., 1990, Tidal channel, levee, and crevasse-splay deposits from a Cambrian tidal channel system: a new mechanism to produce shallowing-upwards sequences: Journal of Sedimentary Petrology, v. 60, p. 73-83.

CLOYD, K. C., RAGLAND, D. A., JONES, L., AND DONOVAN, R. N., 1986, Dolomite with evaporitic connections in the Ordovician Cool Creek Formation, Southwestern Oklahoma, *in* Gilbert, M. C., and Donovan, R. N., eds., Geology of the Eastern Wichita Mountains, Southwestern Oklahoma- Fragments of an Aulacogen?: Stillwater, Oklahoma Geological Survey Guidebook 24, p. 29-34.

CODY, R. C., 1979, Lenticular gypsum: occurrences in nature and experimental determinations of effects of soluble green plant material on its formation: Journal of Sedimentary Petrology, v. 49, p. 1015-1028.

COHN, F., 1864, Uber die Entstehung des travertin in den Wasserfallen von Tivoli: Neues Jahrbuch fur Minerologie, Geologie, und Palaeontologie, v. 67, p. 580-610.

COLLINSON, J. D., 1970, Bedforms of the Tana River, Norway: Geografiska Annaler, v. 52-A, p. 31-56.

COLLINSON, J. D., AND THOMPSON, D. B., 1989, Sedimentary Structures (2nd ed.): London, Allen and Unwin Ltd., 207p.

COLTON, G. W., 1970, The Appalachian Basin- its depositional sequences and their geologic relationships, *in* Weaver, K. N., Fisher, G.W., Pettijohn, F. J., and Reed, J. C., eds., Studies in Appalachian Geology, Central and Southern: New York, Wiley Interscience, p. 5-48.

CONIGLIO, M., AND DIX, G. R., 1992, Carbonate slopes, *in* Walker, R. G., and James, N. P., Facies Models: Response to Sea Level Change: St. Johns, Newfoundland, Geological Associationnada, p. 349-373.

CONIGLIO, M., AND HARRISON, R. S., 1983, Holocene and Pleistocene caliche from Big Pine Key, Florida: Bulletin of Canadian Petroleum Geology, v. 31, p. 3-13.

COOK, H. E., HINE, A. C., AND MULLINS, H. T., 1983, Platform Margin and Deep Water Carbonates: Tulsa, Society of Economic Geologists and Mineralogists Short Course Notes No. 12, 573 p.

COOK, H. E., AND MULLINS, H. T., 1983, Basin margin environment, *in* Scholle, P. A., Bebout, D. G., and Moore, C. H., eds., Carbonate Depositional Environments: Tulsa, American Association of Petroleum Geologists Memoir 33, p. 540-617.

CRAFT, J. H., AND BRIDGE, J. S., 1987, Shallow-marine sedimentary processes in the Late Devonian Catskill Sea, New York State: Geological Society of America Bulletin, v. 98, p. 338-355.

CREVELLO, P. D., AND SCHLAGER, W., 1980, Carbonate debris sheets and turbidites, Exuma Sound, Bahamas: Journal of Sedimentary Petrology, v. 50, p. 1121-1148.

CRIMES, D. S., AND HARPER, J. C., eds., 1970, Trace fossils: Geological Journal Special Issue Number 6, 547 p.

CRIMES, D. S., AND HARPER, J. C., eds., 1977, Trace fossils 2: Geological Journal Special Issue Number 9, 351 p.

CROWELL, J. C., HOPE, R. A., KAHLE, J. E., OVENSHINE, A. T., AND SAMPS, R. H., 1966, Deep-water sedimentary structures, Pliocene Pico Formation, Santa Paula Creek, Ventura Basin, California: California Division of Mines Geologic Special Report 89, p. 1-40.

CUMINGS, E. R., 1932, Reefs or bioherms?: Geological Society of America Bulletin, v. 43, p. 331-352.

CUMINGS, E. R., AND SCHROCK, R. R., 1928, Niagaran coral reefs of Indiana and adjacent states and the stratigraphic relations: Geological Society of America Bulletin, v. 39, p. 579-620.

CURRAN, H. A., ed., 1985, Biogenic Structures: Their use in Interpreting Depositional Environments: Tulsa, Society of Economic Paleontologists and Mineralogists Special Publication 35, 347 p.

CURTIS, R., EVANS, G., KINSMAN, D. J. J., AND SHEARMAN, D. J., 1963, Association of dolomite and anhydrite in the recent sediments of the Persian Gulf: Nature, v. 197, p. 679-680.

DAVIES, D. K., 1968, Carbonate turbidites, Gulf of Mexico: Journal of Sedimentary Petrology, v. 38, p. 1100-1109.

DAVIES, D. K., 1972, Deep sea sediments and their sedimentation, Gulf of Mexico: American Association of Petroleum Geologists Bulletin, v. 56, p. 2212-2239.

DAVIES, G. R., 1970, Algal-laminated sediments, Gladstone Embayment, Shark Bay, Western Australia, *in* Logan, B. W., Davies, G. R., Read, J. F., and Cebulski, D. E., eds., Carbonate Sedimentation and Environments, Shark Bay, Western Australia: Tulsa, American Association of Petroleum Geologists Memoir 13, p. 169-205.

DEAN, W. E., DAVIES, G. R., AND ANDERSON, R. Y., 1975, Sedimentological significance of nodular and laminated anhydrite: Geology, v. 3, p. 367-372.

DEAN, W. E., AND EGGLESTON, J. R., 1975, Comparative anatomy of marine and freshwater algal reefs, Bermuda and central New York: Geological Society of America Bulletin, v. 86, p. 665-676.

DEAN, W. E., AND FOUCH, T. D., 1983, Lacustrine, *in* Scholle, P. A., Bebout, D. G., and Moore, C. H., eds., Carbonate Depositional Environments: Tulsa, American Association of Petroleum Geologists Memoir 33, p. 98-130.

DE BOER, P. L., 1979, Convolute lamination in modern sands of the estuary of the Oosterschelde, the Netherlands, formed as the result of entrapped air: Sedimentology, v. 26, p. 283-294.

DEELMAN, J. C., 1972, On mechanisms causing birdseye structures: Neues Jahrbuch fur Geologie und Palaontologie, Monatschefte, v. 1972, p. 582-595.

DEFFEYES, K. S., LUCIA, F. J., AND WEYL, P. K., 1965, Dolomitization of Recent and Plio-Pleistocene sediments by marine evaporite waters on Bonaire, Netherlands Antilles, *in* Pray, L. C., and Murray, R. C., eds., Dolomitization and Limestone Diagenesis-A Symposium: Tulsa, Society of Economic Paleontologists and Mineralogists Special Publication 13, p. 71-88.

DEGENS, E. T., AND ROSS, D. A., eds., 1969, Hot Brines and Recent Heavy Metal Deposits in the Red Sea: New York, Springer-Verlag, 600 p.

DEMICCO, R. V., 1981, Comparative sedimentology of an ancient carbonate platform: the Conococheague Limestone of the central Appalachians: Unpublished Ph.D. Dissertation, The Johns Hopkins University, Baltimore, 333 p.

DEMICCO, R. V., 1983, Wavy and lenticular-bedded carbonate ribbon rocks of the Upper Cambrian Conococheague Limestone, central Appalachians: Journal of Sedimentary Petrology, v. 53, p. 1121-1132.

DEMICCO, R. V., 1985, Platform and off-platform carbonates of the Upper Cambrian of western Maryland, U.S.A.: Sedimentology, v. 32, p. 1-22.

DEMICCO, R. V., AND HARDIE, L. A., 1981, Patterns of platform and off-platform carbonate sedimentation in the upper Cambrian of the central Appalachians and their implication for sea level history: 2nd International Symposium on the Cambrian System: United States Geological Survey Open-File Report 81-743, p. 67-70.

DEMICCO, R. V., AND MITCHELL, R. W., 1982, Facies of the Great American Bank in the central Appalachians, *in* Lyttle, P. T., ed., Central Appalachian Geology, Field Trip Guidebook: Falls Church, American Geological Institute, p. 171-266.

DICKINSON, K. A., 1968, Petrology of the Buckner Formation in adjacent parts of Texas, Louisiana, and Arkansas: Journal of Sedimentary Petrology, v. 38, p. 555-567.

DILL, R. F., KENDALL, C. G. ST. C., AND SHINN, E. A., 1989, Giant subtidal stromatolites and related sedimentary features, Lee Stocking Island, Exumas, Bahamas: Washington, D.C., IGC Field Trip T 373, American Geophysical Union, 33 p.

DILL, R. F., SHINN, E. A., JONES, A. T., KELLY, K., AND STEINEN, R. P., 1986, Giant subtidal stromatolites forming in normal salinity water: Nature, v. 324, p. 55-58.

DONOVAN, R. N., AND FOSTER, R. J., 1972, Subaqueous shrinkage cracks from the Caithness Flagstone Series (Middle Devonian) of northeast Scotland: Journal of Sedimentary Petrology, v. 42, p. 309-317.

DOTT, R. H., AND BATTEN, R. L., 1988, Evolution of the Earth (4th ed.): New York, McGraw-Hill, 643 p.

DRAVIS, J. J., 1979, Rapid and widespread generation of Recent oolitic hardgrounds on a high energy Bahamian Platform, Eleuthera Bank, Bahamas: Journal of Sedimentary Petrology, v. 49, p. 195-208.

DRAVIS, J. J., 1983, Hardened subtidal stromatolites, Bahamas: Science, v. 219, p. 385-386.

DROSER, M. L., AND BOTTJER, D. J., 1986, A semiquantitative classification of ichnofabric: Journal of Sedimentary Petrology, v. 56, p. 558-569.

DROSER, M. L., AND BOTTJER, D. J., 1989, Ichnofabric of sandstones deposited in high-energy nearshore environments: measurement and utilization: Palaios, v. 4, p. 598-604.

DUKE, W. L., 1980, Upper Cretaceous (Turonian) Cardium Formation, southern Alberta: Hamilton, Technical Memorandum 80-4, McMaster University, 59 p.

DUKE, W. L., 1985, Hummocky cross-stratification, tropical hurricanes, and intense winter storms: Sedimentology, v. 32, p. 167-194.

DUNHAM, R. J ., 1969a, Early vadose silt in Townsend mound (reef), New Mexico, *in* Friedman, G. M., ed., Depositional Environments in Carbonate Rocks: Tulsa, Society of Economic Paleontologists and Mineralogists Special Publication 14, p. 139-182.

DUNHAM, R. J., 1969b, Vadose pisolite in the Capitan reef (Permian), New Mexico and Texas, *in* Friedman, G. M., ed., Depositional Environments in Carbonate Rocks: Tulsa, Society of Economic Paleontologists and Mineralogists Special Publication 14, p. 182-190.

DUNN, J. R., 1953, The origin of the deposits of tufa in Mono Lake: Journal of Sedimentary Petrology, v. 23, p. 18-23.

DUNN, P. A., 1991, Diagenesis and cyclostratigraphy- an example from the Middle Triassic Latemar platform, Dolomite Mountains, northern Italy: Unpublished Ph.D. Dissertation, The Johns Hopkins University, Baltimore, 567 p.

DUPONT, M. ED., 1881, Sur l'origine des calcaires devoniens de la Belgique: Bulletin de L'Academie Royale des Sciences, v. 2, p. 264-280.

DZULYNSKI, S., 1965, New data on experimental production of sedimentary structures: Journal of Sedimentary Petrology, v. 35, p. 196-212.

DZULYNSKI, S., 1966, Sedimentary structures resulting from convection-like pattern of motion: Annales Societatis Geologorum Poloniae, v. 36, p. 3-21.

DZULYNSKI, S., AND WALTON, E. K., 1963, Experimental production of sole markings: Transactions of the Edinburgh Geological Society, v. 19, p. 279-305.

DZULYNSKI, S., AND WALTON, E. K., 1965, Sedimentary features of flysch and greywackes: Amsterdam, Elsevier, 300 p.

EARDLEY, A. J., 1938, Sediments of Great Salt Lake, Utah: American Association of Petroleum Geologists Bulletin, v. 22, p. 1305-1411.

EGGLESTON, J. R., AND DEAN, W. E., 1976, Freshwater stromatolitic bioherms in Green Lake, New York, in Walter, M. R., ed., Stromatolites: New York, Developments in Sedimentology 20, Elsevier, p. 479-488.

EINSELE, G., AND SEILACHER, A., eds., 1982, Cyclic and Event Stratification: New York, Springer-Verlag, 536 p.

EKDALE, A. A., BROMLEY, R. G., AND PEMBERTON, S. G., 1984, Ichnology, Trace Fossils in Sedimentology and Stratigraphy: Tulsa, Society of Economic Paleontologists and Mineralogists Short Course Notes No. 15, 317 p.

EKDALE, A. A., AND POLLARD, J., eds., 1991, Ichnofabric and ichnofacies: Palios (Special Issue), v. 6, p. 197-343.

ELLIOT, R. E., 1964, A classification of cross-stratification structures: Nature, v. 202, p. 451-452.

ELLIOT, R. E., 1965, A classification of subaqueous sedimentary structures based on rheological and kinematical parameters: Sedimentology, v. 5, p. 193-209.

ELLIOT, T., 1986, Siliciclastic shorelines, *in* Reading, H. G., ed., Sedimentary Environments and Facies (2nd ed.): Oxford, Blackwell, p. 155-188.

ENOS, P., AND MOORE, C. H., 1983, Fore-reef slope environment, *in* Scholle, P. A., Bebout, D. G., and Moore, C. H., eds., Carbonate Depositional Environments: Tulsa, American Association of Petroleum Geologists Memoir 33, p. 508-537.

ENOS, P., AND PERKINS, R. D., 1979, Evolution of Florida Bay from island stratigraphy: Geological Society of America Bulletin, v. 90, p. 59-83.

ERIKSSON, K. A., AND TRUSWELL, J. F., 1974, Tidal flat associations from a Lower Proterozoic carbonate sequence in South Africa: Sedimentology, v. 21, p. 293-309.

ESTEBAN, M., 1974, Caliche textures and "Microcodium": Societa Geologica Italiana, Bollettino, v. 92, p. 105-125.

ESTEBAN, M., AND KLAPPA, C. F., 1983, Subaerial exposure, *in* Scholle, P. A., Bebout, D. G., and Moore, C. H., eds., Carbonate Depositional Environments: Tulsa, American Association of Petroleum Geologists Memoir 33, p. 1-54.

ESTEBAN, M., AND PRAY, L. C., 1983, Pisoids and pisolite facies (Permian), Guadalupe Mountains, New Mexico and West Texas, *in* Peryt, T. M., ed., Coated Grains: New York, Springer-Verlag, p. 503-537.

EUGSTER, H. P., AND HARDIE, L. A., 1978, Saline lakes, *in* Lerman, A., ed., Lakes: Chemistry, Geology and Physics: New York, Springer-Verlag, p. 237-293.

EVANS, G., SCHMIDT, V., BUSH, P., AND NELSON, H., 1969, Stratigraphy and geologic history of the sabkha, Abu Dhabi, Persian Gulf: Sedimentology, v. 12, p. 145-159.

FAIRCHILD, I. J., 1980, Sedimentation and origin of Late Precambrian "dolomite" from Scotland: Journal of Sedimentary Petrology, v. 50, p. 423-446.

FENTON, C. L., AND FENTON, M. A., 1937, Belt Series of the north: stratigraphy, sedimentation, paleontology: Geological Society of America Bulletin, v. 48, p. 1873-1969.

FERGUSON, J., BURNE, R. V., AND CHAMBERS, L. A., 1982, Lithification of peritidal carbonates by continental brines at Fisherman Bay, south Australia, to form a megapolygon/spelean limestone association: Journal of Sedimentary Petrology, v. 52, p. 1127-1147.

FISCHER, A. G., 1964, The Lofer Cyclothems of the Alpine Triassic, *in* Merriam, D. F., ed., Symposium on Cyclic Sedimentation: Kansas State Geological Survey Bulletin 169, v. 1, p. 107-149.

FISCHER, A. G., 1965, Fossils, early life, and atmospheric history: National Academy of Sciences Proceedings, v. 53, p. 1205-1215.

FISCHER, A. G., 1975, Tidal deposits, Dachstein Limestone of the North-Alpine Triassic, *in* Ginsburg, R. N., ed., Tidal Deposits: New York, Springer-Verlag, p. 235-242.

FISCHER, A. G., AND GARRISON, R. E., 1967, Carbonate lithification on the sea floor: Journal of Geology, v. 75, p. 488-496.

FLUGEL, E., 1982, Microfacies Analysis of Limestones: New York, Springer-Verlag, 633 p.

FOLK, R. L., 1959, Practical petrographic classification of limestones: Geological Society of America Bulletin, v. 43, p. 1-38.

FOLK, R. L., AND PITTMAN, J. S., 1971, Length-slow chalcedony: a new testament for vanished evaporites: Journal of Sedimentary Petrology, v. 41, p. 1045-1058.

FORSTNER, V., MULLER, G., AND REINECK, H.-E., 1968, Sedimente und Sedimentgefuge des Rhein Deltas in Bodensee: Neues Jahrbuch für Geologie und Paläontologie, Abhandlungen, v. 109, p. 33-62.

FREY, R. W., 1975, The Study of Trace Fossils: Berlin, Springer-Verlag, 562 p.

FUCHS, T., 1895, Studien uber Fucoiden und Hieroglyphen: Denkschriften Akademie der Wissenschaften, Wien, v. 62, p. 369-448.

FUCHTBAUER, H., AND HARDIE, L. A., 1976, Experimentally determined homogeneous distribution coefficients for precipitated magnesium calcites: application to marine carbonate cements (abs.): Geological Society of America Abstracts with Programs, v. 8, p. 877.

FUCHTBAUER, H., AND HARDIE, L. A., 1980, Comparison of experimental and natural magnesium calcites (abs.): International Association of Sedimentologists, Bochum., p. 167-168.

GARRELLS, R. M., AND MACKENZIE, F. T., 1971, Evolution of Sedimentary Rocks: New York, W. W. Norton, 397 p.

GARRETT, P., 1970, Phanerozoic stromatolites: noncompetitive ecological restriction by grazing and burrowing animals: Science, v. 169, p. 171-173.

GARRETT, P., 1977, Biological communities and their sedimentary record, *in* Hardie, L. A., ed., Sedimentation on the modern carbonate tidal flats of northwest Andros Island, Bahamas: Baltimore, Studies in Geology No. 22, Johns Hopkins University Press, p. 124-158.

GEBELEIN, C. D., 1969, Distribution, morphology and accretion rate of recent subtidal algal stromatolites, Bermuda: Journal of Sedimentary Petrology, v. 39, p. 49-69.

GEBELEIN, C. D., 1974, Guidebook for Modern Bahamian Platform Environments: St. George, Bermuda, Geological Society of America Annual Meeting Fieldtrip Guidebook, 93 p.

GEBELEIN, C. D., 1976, Open marine subtidal and intertidal stromatolites (Florida, the Bahamas and Bermuda), *in* Walter, M. R., ed., Stromatolites: New York, Developments in Sedimentology 20, Elsevier, p. 381-388.

GEBELEIN, C. D., AND HOFFMAN, P., 1973, Algal origin of dolomite laminations and stromatolitic limestones: Journal of Sedimentary Petrology, v. 43, p. 603-613.

GERDES, G., AND KRUMBEIN, W. E., 1987, Biolaminated deposits: New York, Springer-Verlag, Lecture Notes in Earth Science 9, 183 p.

GILE, L. H., PETERSON, F. F., AND GROSSMAN, R. B., 1966, Morphological and genetic sequence of carbonate accumulation in desert soils: Soil Science, v. 101, p. 347-360.

GINSBURG, R. N., 1955, Recent stromatolitic sediments from south Florida (abs.): Journal of Paleontology, v. 29, p. 723-724.

GINSBURG, R. N., 1960, Ancient analogues of Recent stromatolites: Proceedings of International Geological Congress, 21st Session, Copenhagen, Section 22, p. 26-35.

GINSBURG, R. N., 1967, Stromatolites: Science, v. 157, p. 339-340.

GINSBURG, R. N., 1971a, Landward movement of carbonate mud: new model for regressive cycles in carbonate (abs.): American Association of Petroleum Geologists Bulletin, v. 55, p. 340.

GINSBURG, R. N., 1971b, Stromatolite Stratigraphy, *in* Ginsburg, R. N., Rezak, R., and Wray, J. L., eds., Geology of Calcareous Algae, Notes for a Short Course: Miami, Sedimenta I, Comparative Sedimentology Laboratory, Rosenstiel School of Marine and Atmospheric Sciences, p. 13.2-13.3.

GINSBURG, R. N., 1974, Introduction to comparative sedimentology of carbonates: American Association of Petroleum Geologists Bulletin, v. 58, p. 781-786.

GINSBURG, R. N., ed., 1975, Tidal Deposits: New York, Springer-Verlag, 428 p.

GINSBURG, R. N., HARDIE, L. A., BRICKER, O. P., GARRETT, P., AND WANLESS, H. R., 1977, Exposure index: a quantitative approach to defining position within the tidal zone, *in* Hardie, L. A., ed., Sedimentation on the Modern Carbonate Tidal Flats of Northwest Andros Island, Bahamas: Baltimore, Studies in Geology No. 22, The Johns Hopkins University Press, p. 7-11.

GINSBURG, R. N., ISHAM, L. B., BEIN, S. J., AND KUPERBURG, J., 1954, Laminated algal sediments of south Florida and their recognition in the fossil record: Unpublished Report (54-21), University of Miami Marine Laboratory, Coral Gables, 33 p.

GINSBURG, R. N., AND JAMES, N. P., 1976, Submarine botryoidal aragonite in Holocene reef limestones, Bermuda: Sedimentology, v. 20, p. 575-614.

GINSBURG, R. N., MARSZALEK, D. S., AND SCHNEIDERMANN, N., 1971a, Ultrastructure of carbonate cements in a Holocene algal reef of Bermuda: Journal of Sedimentary Petrology, v. 41, p. 472-482.

GINSBURG, R. N., AND SCHROEDER, J. H., 1973, Growth and submarine fossilization of algal cup reefs, Bermuda: Sedimentology, v. 20, p. 575-614.

GINSBURG, R. N., SCHROEDER, J. H., AND SHINN, E. A., 1971b, Recent synsedimentary cementation in subtidal Bermuda reefs, *in* Bricker, O. P., ed., Carbonate Cements: Baltimore, Studies in Geology No. 19, The Johns Hopkins University Press, p. 54-58.

GLEASON, P. J., AND SPACKMAN, W., 1974, Calcareous periphyton and water chemistry in the Everglades, *in* Gleason, P. J., ed., Environments of South Florida, Past and Present: Miami, Miami Geological Society Memoir 2, p. 146-181.

GOLDHAMMER, R. K., 1987, Platform carbonate cycles, Middle Triassic of northern Italy: the interplay of local tectonics and global eustasy: Unpublished Ph.D. Dissertation, The Johns Hopkins University, Baltimore, 468 p.

GOLDHAMMER, R. K., DUNN, P. A., AND HARDIE, L. A., 1987, High frequency glacio-eustatic sea level oscillations with Milankovitch characteristics recorded in Middle Triassic platform carbonates is northern Italy: American Journal of Science, v. 287, p. 853-892.

GOLDHAMMER, R. K., DUNN, P. A., AND HARDIE, L. A., 1990, Depositional cycles, composite sea-level changes, cycle stacking patterns, and the hierarchy of stratigraphic forcing: Examples from Alpine Triassic platform carbonates: Geological Society of America Bulletin, v. 102, p. 535-562.

GOLDHAMMER, R. K., HARDIE, L. A., AND NGUYEN, C. T., 1985, Compactional features in Cambro-Ordovician carbonates of the central Appalachians and their significance (abs.): American Association of Petroleum Geologists Bulletin, v. 69, p. 257.

GOLDHAMMER, R. K., AND HARRIS, M. T., 1989, Eustatic controls on the stratigraphy and geometry of the Latemar buildup (Middle Triassic), the Dolomites of northern Italy, *in* Crevello, P. D. and others, eds., Controls on Carbonate Platform and Basin Development: Tulsa, Society of Economic Paleontologists and Mineralogists Special Publication 44, p. 323-338.

GOLUBIC, S., 1976a, Organisms that build stromatolites, *in* Walter, M. R., Stromatolites: New York, Developments in Sedimentology 20, Elsevier, p. 113-126.

GOLUBIC, S., 1976b, Taxonomy of extant stromatolite-building cyanophytes, *in* Walter, M. R., ed., Stromatolites: New York, Developments in Sedimentology 20, Elsevier, p. 127-140.

GORNITZ, V. M., AND SCHREIBER, B. C., 1981, Displacive halite hoppers from the Dead Sea; some implications for ancient evaporite deposits: Journal of Sedimentary Petrology, v. 51, p. 787-794.

GOUDIE, A., 1973, Duricrusts in Tropical and Subtropical Landscapes: Oxford, Clarendon, 174 p.

GRAF, D. L., AND GOLDSMITH, J. R., 1956, Some hydrothermal syntheses of dolomite and protodolomite: Journal of Geology, v. 64, p. 173-186.

GROSS, M .G., GUELUER, S. M., CREAGER, J. S., AND DAWSON, W. A., 1963, Varved marine sediments in a stagnant fjord: Science, v. 141, p. 918-919.

GROTZINGER, J. P., 1985, Evolution of early Proterozoic passive-margin carbonate platform, Rocknest Formation, Wopmay Orogen, Northwest Territories, Canada: Unpublished Ph.D. Dissertation, Virginia Polytechnic Institute and State University, Blacksburg, 225 p.

GROTZINGER, J. P., 1986a, Cyclicity and paleoenvironmental dynamics, Rocknest platform, northwest Canada: Geological Society of America Bulletin, v. 97, p. 1208-1231.

GROTZINGER, J. P., 1986b, Evolution of Early Proterozoic passive-margin carbonate platform, Rocknest Formation, Wopmay Orogen, Northwest Territories, Canada: Journal of Sedimentary Petrology, v. 56, p. 831-847.

GROTZINGER, J. P., 1989, Facies and evolution of Precambrian carbonate depositional systems: emergence of the modern platform archetype, *in* Crevello, P. D. and others, eds., Controls on Carbonate Platform and Basin Development: Tulsa, Society of Economic Paleontologists and Mineralogists Special Publication 44, p. 79-106.

GROTZINGER, J. P., AND READ, J. F., 1983, Evidence for primary aragonite precipitation, Lower Proterozoic (1.9 Ga) Rocknest Dolomite, Northwest Canada: Geology, v. 11, p. 710-713.

GROVER, G. JR., AND READ, J. F., 1978, Fenestral and associated vadose diagenetic fabrics of tidal flat carbonates, Middle Ordovician New Market Limestone, southwestern Virginia: Journal of Sedimentary Petrology, v. 48, p. 453-473.

HALL, J., 1847, Natural history of New York organic remains of the Lower Division of New York System: Paleontology, v. 1, p. 1-338.

HALL, J., 1883, Cryptozoon N. G., *Cryptozoon proliferium* n. sp.: New York State Museum Annual Report 36, 1 p. and 2 plates.

HALLEY, R. B., 1976, Textural variations within Great Salt Lake algal mounds, *in* Walter, M. R., ed., Stromatolites: New York, Developments in Sedimentology 20, Elsevier, p. 435-445.

HALLEY, R. B., HARRIS, P. M., AND HINE, A. C., 1983, Bank margin environment, *in* Scholle, P. A., Bebout, D. C., and Moore, C. H., eds., Carbonate Depositional Environments: Tulsa, American Association of Petroleum Geologists Memoir 33, p. 464-506.

HANDFORD, C. R., 1981, Coastal sabkha and salt pan deposition of the lower Clear Fork Formation (Permian), Texas: Journal of Sedimentary Petrology, v. 29, p. 239-253.

HANDFORD, C. R., 1991, Marginal marine halite: sabkhas and salinas, *in* Melvin, J. L., ed., Evaporites, Petroleum and Mineral Resources: New York, Developments in Sedimentology 50, Elsevier, p. 1-66.

HANDFORD, C. F., KENDALL, A. C., PREZBINDOWSKI, D. R., DUNHAM, J. B., AND LOGAN, B. W., 1984, Salina-margin tepees, pisolites, and aragonitic cements, Lake MacLeod, Western Australia: their significance in interpreting ancient analogs: Geology, v. 12, p. 523-527.

HARDIE, L. A., ed., 1977a, Sedimentation on the Modern Carbonate Tidal Flats of Northwest Andros Island, Bahamas: Baltimore, Studies in Geology No. 22, The Johns Hopkins University Press, 202 p.

HARDIE, L. A., 1977b, Algal structures in cemented crusts and their environmental significance, *in* Hardie, L. A., ed., Sedimentation on the Modern Carbonate Tidal Flats of Northwestern Andros Island, Bahamas: Baltimore, Studies in Geology No. 22, The Johns Hopkins University Press, p. 159-177.

HARDIE, L. A., 1977c, A concluding note: sensitivity of the record, *in* Hardie, L. A., ed., Sedimentation on the Modern Carbonate Tidal Flats of Northwest Andros Island, Bahamas: Baltimore, Johns Hopkins University Studies in Geology No. 22, p. 188-189.

HARDIE, L. A., 1984, Evaporites: marine or nonmarine: American Journal of Science, v. 284, p. 193-240.

HARDIE, L. A., 1987, Perspectives: Dolomitization: a critical view of some current views: Journal of Sedimentary Petrology, v. 57, p. 166-183.

HARDIE, L.A., 1989, Cyclic platform carbonates in the Cambro-Ordovician of the central Appalachians, Cambro-Ordovician Carbonate Banks and Siliciclastic Basins of the United States Appalachians, Field Trip Guidebook T161, 28th International Geological Congress: Washington, D.C., American Geophysical Union, p. 51-88.

HARDIE, L. A., BOSELLINI, A., AND GOLDHAMMER, R. K., 1986, Repeated subaerial exposure of subtidal carbonate platforms, Triassic, northern Italy: evidence for high frequency sea level oscillations on a 10^4 year scale: Paleoceanography, v. 1, p. 447-457.

HARDIE, L. A., DUNN, P. A., AND GOLDHAMMER, R. K., 1991, Field and modelling studies of Cambrian carbonate cycles, Virginia Appalachians-Discussion: Journal of Sedimentary Petrology, v. 61, p. 636-646.

HARDIE, L. A., AND EUGSTER, H. P., 1970, The evolution of closed-basin brines: Mineralogical Society of America Special Paper 3, p. 273-290.

HARDIE, L. A., AND EUGSTER, H. P., 1971, The depositional environment of marine evaporites: a case for shallow, clastic accumulation: Sedimentology, v. 16, p. 187-220.

HARDIE, L. A., AND GARRETT, P., 1977, General environmental setting, *in* Hardie, L. A., ed., Sedimentation on the Modern Carbonate Tidal Flats of Northwest Andros Island, Bahamas: Baltimore, Studies in Geology No. 22, The Johns Hopkins University Press, p. 12-49.

HARDIE, L. A., AND GINSBURG, R. N., 1977, Layering: the origin and environmental significance of lamination and thin bedding, *in* Hardie, L. A., ed., Sedimentation on the Modern Carbonate Tidal Flats of Northwest Andros Island, Bahamas: Baltimore, Studies in Geology No. 22, The Johns Hopkins University Press, p. 50-123.

HARDIE, L. A., LOWENSTEIN, T. K., AND SPENCER, R. J., 1985, The problem of distinguishing between primary and secondary features in evaporites, *in* Schreiber, B. C., ed., Sixth International Symposium on Salt: Alexandria, Salt Institute, p. 11-93.

HARDIE, L.A., MITCHELL, R.W., AND DEMICCO, R.V., 1982, Influence of early cementation on dolomitization and compaction in Cambro-Ordovician carbonates of the central Appalachians (abs.): American Association of Petroleum Geologists Bulletin, v. 66, p. 577.

HARDIE, L. A., AND SHINN, E. A., 1986, Carbonate depositional environments modern and ancient, part 3: tidal flats: Colorado School of Mines Quarterly, v. 81, p. 1-74.

HARDIE, L. A., SMOOT, J. P., AND EUGSTER, H. P., 1978, Saline lakes and their deposits: a sedimentological approach, *in* Matter, A., and Tucker, M. E., eds., Modern and Ancient Lake Sediments: Oxford, Blackwell, International Association of Sedimentologists Special Publication 2, p. 7-41.

HARMS, J. C., 1969, Hydraulic significance of some sand ripples: Geological Society of America Bulletin, v. 80, p. 363-396.

HARMS, J. C., SOUTHARD, J. B., SPEARING, D. R., AND WALKER, R. G., 1975, Depositional Environments as Interpreted from Primary Sedimentary Structures and Stratification Sequences: Tulsa, Society of Economic Paleontologists and Mineralogists Short Course 2, 166 p.

HARMS, J. C., SOUTHARD, J. B., AND WALKER, R. G., 1982, Structures and Sequences in Clastic Rocks: Tulsa, Society of Economic Paleontologists and Mineralogists Short Course Notes 9, 249 p.

HARRIS, P. M., 1979, Facies anatomy and diagenesis of a Bahamian ooid shoal: Miami, Sedimenta VII, University of Miami, 163 p.

HARRIS, P. M., AND DODMAN, C. A., 1982, Jurassic evaporites of the U.S. Gulf Coast: the Smackover-Buckner contact, *in* Handford, C. R., Loucks, R. G., and Davies, G. R., eds., Depositional and Diagenetic Spectra of Evaporites- Core Workshop: Tulsa, Society of Economic Paleontologists and Mineralogists Core Workshop No. 3, p. 174-192.

HARRISON, R. S., 1977, Subaerial versus submarine discontinuity surfaces in a Pleistocene reef complex, Barbados, W.I.: Miami, Proceedings of the Third International Coral Reef

Symposium, Rosenstiel School of Marine and Atmospheric Science, University of Miami, p. 143-147.

HARRISON, R. S., AND STEINEN, R. P., 1978, Subaerial crusts, caliche profiles and breccia horizons: comparison of some Holocene and Mississippian exposure surfaces, Barbados and Kentucky: Geological Society of America Bulletin, v. 89, p. 385-396.

HARRISON, S. C., 1971, The sediments and sedimentary processes of the Holocene tidal flat complex, Delmarva Peninsula, Virginia: Unpublished Ph.D. Dissertation, The Johns Hopkins University, Baltimore, 201 p.

HAWLEY, N., 1981, Flume experiments on the origin of flaser bedding: Sedimentology, v. 28, p. 699-712.

HAY, R. L., PEXTON, R. E., TEAGUE, T. T., AND KYSER, T. K., 1986, Spring-related carbonate rocks, Mg clays, and associated minerals in Pliocene deposits of the Amargosa Desert, Nevada and California: Geological Society of America Bulletin, v. 97, p. 1488-1503.

HAYS, J. D., IMBRIE, J., AND SHACKLETON, N. J., 1976, Variations in the Earth's orbit: pacemaker of the ice ages: Geological Society of America Bulletin, v. 97, p. 1121-1132.

HECKEL, P. H., 1972, Possible inorganic origin for stromatactis in calcilutite mounds in the Tully Limestone, Devonian of New York: Journal of Sedimentary Petrology, v. 42, p. 7-18.

HESSE, R., AND READING, H. G., 1978, Subaqueous clastic fissure eruptions and other examples of sedimentary transportation in the lacustrine Horton Bluff Formation (Mississippian) Nova Scotia, Canada, *in* Matter A., and Tucker, M. E., eds., Modern and Ancient Lake Sediments: Oxford, Balckwell, Special Publications International Association of Sedimentologists 2, p. 239-256.

HINNOV, L. A., AND GOLDHAMMER, R. K., 1988, The identification of Melankovitch signals in Middle Triassic platform carbonate cycles using a super-resolution spectral technique (abs.): American Association of Petroleum Geologists Bulletin, v. 72, p. 197.

HOFFMAN, P., 1974, Shallow and deepwater stromatolites in Lower Proterozoic platform-to-basin facies change, Great Slave Lake, Canada: American Association of Petroleum Geologists Bulletin, v. 58, p. 856-867.

HOFFMAN, P., 1975, Shoaling-upward shale-to-dolomite cycles in the Rocknest Formation (Lower Proterozoic), Northwest Territories, Canada, *in* Ginsburg, R. N., ed., Tidal Deposits: New York, Springer-Verlag, p. 257-265.

HOFFMAN, P., 1976a, Stromatolite morphogenesis in Shark Bay, Western Australia, *in* Walter, M. R., ed., Stromatolites: New York, Developments in Sedimentology 20, Elsevier, p. 261-271.

HOFFMAN, P., 1976b, Environmental diversity of Middle Precambrian stromatolites, *in* Walter, M. R., ed., Stromatolites: New York, Developments in Sedimentology 20, Elsevier, p. 599-611.

HOFMANN, H. J., 1973, Stromatolites: characteristics and utility: Earth-Science Reviews, v. 9, p. 339-373.

HORODYSKI, R. J., 1976, Stromatolites of the Upper Siyeh Limestone (Middle Proterozoic), Belt Supergroup, Glacier National Park, Montana: Precambrian Research, v. 3, p. 517-536.

HOWARD, J. D., AND REINECK, H.-E., 1981, Depositional facies of high-energy beach-to-offshore sequence: comparison with low-energy sequence: American Association of Petroleum Geologists Bulletin, v. 65, p. 807-830.

HSU, K. J., AND JENKYNS, H. C., eds., 1974, Pelagic Sedimentation: On Land and Under the Sea: Oxford, Blackwell, International Association of Sedimentologists Special Publication 1, 447 p.

HSU, K. J., AND SIEGENTHALER, C., 1969, Preliminary experiments on hydrodynamic movement induced by evaporation and their bearing on the dolomite problem: Sedimentology, v. 12, p. 11-25.

HUANG, T. C., AND PIERCE, J. W., 1971, The carbonate minerals of deep-sea bioclastic turbidites, southern Blake basin: Journal of Sedimentary Petrology, v. 41, p. 251-260.

HUBERT, J. F., 1978, Paleosol caliche in the New Haven Arkose, Newark Group, Connecticut: Paleogeography, Paleoclimatology, and Paleoecology, v. 24, p. 151-168.

HUNTER, R. E., 1977, Basic types of stratification in small eolian dunes: Sedimentology, v. 24, p. 361-387.

IMBRIE, J., HAYS, J. D., MARTINSON, D. G., MCINTYRE, A., MIX, A. C., MORLEY, J. J., PISIAS, N. G., PRELL, W. L., AND SHACKLETON, N. J., 1984, The orbital theory of Pleistocene climate: support from a revised chronology of the marine ^{18}O record, *in* Berger, A., Imbrie, J., Hays, J., Kukla, G., and Saltzman, B., eds., Milankovitch and Climate: Boston, Reidel, p. 269-306.

IMBRIE, J., AND BUCHANAN, H., 1965, Sedimentary structures in modern carbonate sands of the Bahamas, *in* Middleton, G. V., ed., Primary Sedimentary Structures and their Hydrodynamic Interpretation: Tulsa, Society of Economic Paleontologists and Mineralogists Special Publication 12, p. 149-172.

IMLAY, R. W., 1940, Lower Cretaceous and Jurassic Formations of southern Arkansas and their oil and gas possibilities: Little Rock, Information Circular 12, Arkansas Resources and Development Commission, 64 p.

INDEN, R. F., AND MOORE, C. H., 1983, Beach environment, *in* Scholle, P. A., Bebout, D. G., and Moore, C. H., eds., Carbonate Depositional Environments: Tulsa, American Association of Petroleum Geologists Memoir 33, p. 212-265.

INGRAM, R.L., 1954, Terminology for the thickness of stratification and parting units in sedimentary rocks: Geological Society of America Bulletin, v. 65, p. 937-938.

INMAN, D. L., AND BOWEN, A. J., 1963, Flume experiments on sand transport by waves and currents: Proceedings of 8th Conference on Coastal Engineering, p. 137-150.

IRION, G., and Muller, G., 1968, Mineralogy, petrology and chemical composition of some calcareous tufa from the Schwabische Alb, Germany, *in* Muller, G., and Friedman, G. M., eds., Recent Developments in Carbonate Sedimentology in Central Europe: New York, Springer-Verlag, p. 157-171.

JACKA, A. D., 1974, Replacement of fossils by length-slow chalcedony and associated dolomitization: Journal of Sedimentary Petrology, v. 44, p. 421-427.

JACOB, A. F., 1973, Descriptive classification of cross-stratification: Geology, v. 1, p. 103-105.

JAMES, N. P., 1972, Holocene and Pleistocene calcareous crust (caliche) profiles: criteria for subaerial exposure: Journal of Sedimentary Petrology, v. 42, p. 817-836.

JAMES, N. P., 1981, Megablocks of calcified algae in the Cow Head Breccia, western Newfoundland: vestiges of a Cambro-Ordovician platform margin: Geological Society of America Bulletin, v. 92, p. 799-811.

JAMES, N. P., 1983, Reef, *in* Scholle, P. A., Bebout, D. G., and Moore, C. H., eds., Carbonate Depositional Environments: Tulsa, American Association of Petroleum Geologists Memoir 33, p. 345-440.

JAMES, N. P., 1984, Shallowing-upwards sequences in carbonates, *in* Walker, R. G., ed., Facies Models (2nd ed.): St. Johns, Newfoundland, Geoscience Canada Reprint Series 1, p. 213-228.

JAMES, N. P., and Bourque, P.-A., 1992, Reefs and Mounds, *in* Walker, R. G., and James, N. P., eds., Facies Models: Response to Sea Level Change: St. Johns, Newfoundland, Geological Association of Canada, p. 323-347.

JAMES, N. P., AND CHOQUETTE, P. W., eds., 1988, Paleokarst: New York, Springer-Verlag, 416 p.

JAMES, N. P., AND CHOQUETTE, P. W., 1990a, Limestones- Introduction: St. Johns, Newfoundland, Geoscience Canada Reprint Series 4, Geological Association of Canada, p. 9-12.

JAMES, N. P., AND CHOQUETTE, P. W., 1990b, Limestones- The sea floor diagenetic environment: St. Johns, Newfoundland, Geoscience Canada Reprint Series 4, Geological Association of Canada, p. 13-34.

JAMES, N. P., AND CHOQUETTE, P. W., 1990c, Limestones- The meteoric diagenetic environments: St. Johns, Newfoundland, Geoscience Canada Reprint Series 4, Geological Association of Canada, p. 35-73.

JAMES, N. P., AND GINSBURG, R. N., 1979, The Seaward Margin of Belize Barrier and Atoll Reefs: Oxford, Blackwell, Special Publications of the International Association of Sedimentologists 3, 191 p.

JAMES, N. P., GINSBURG, R. N., MARSZALEK, D. S., AND CHOQUETTE, P. W., 1976, Facies and fabric specificity of early subsea cements in shallow Belize (British Honduras) reefs: Journal of Sedimentary Petrology, v. 46, p. 523-544.

JAMES, N. P., AND KOBLUK, D. R., 1978, Lower Cambrian patch reefs and associated sediments: southern Labrador, Canada: Sedimentology, v. 25, p. 1-35.

JAMES, N. P., AND MACINTYRE, I. G., 1985, Carbonate depositional environments, modern and ancient. Part 1 - Reefs, zonation, depositional facies and diagenesis: Quarterly Journal of the Colorado School of Mines, v. 80, 70 p.

JINDRICH, V., 1969, Recent carbonate sedimentation by tidal channels in the lower Florida Keys: Journal of Sedimentary Petrology, v. 39, p. 531-553.

JOHNSON, J. H., 1961, Limestone-building algae and algal limestones: Boulder, Colorado School of Mines, Johnson Publishing Co., 297 p.

JONES, B., AND GOODBODY, Q. H., 1985, Oncolites from a shallow lagoon, Grand Cayman Island: Bulletin of Canadian Petroleum Geology, v. 32, p. 254-260.

JULIA, R., 1983, Travertines, *in* Scholle, P. A., Bebout, D. G., and Moore, C. H., eds., Carbonate Depositional Environments: Tulsa, American Association of Petroleum Geologists Memoir 33, p. 64-72.

JUNG, W., 1958, Zur Feinstratigrafie de Werraanhydrite (Zechstein I) im Bereich der Sangerhauser und Mansfelder Mulde: Geologie, Beihefte, v. 24, p. 1-88.

JUNGST, H., 1934, Zur geologischen Bedeutung der synarese: Geologiche Rundschau, v. 25, p. 312-325.

KALKOWSKY, E., 1908, Oolith und stromatolith im norddeutschen Buntsandstein: Zeitschrift der Deutschen Geologischen Gesellschraft, v. 60, p. 68-125.

KASTNER, M., 1980, Length-slow chalcedony: the end of the new testament: Transactions of the American Geophysical Union, v. 61, p. 399.

KEENE, J. B., 1983, Chalcedonic quartz and occurrence of quartzine (length-slow chalcedony) in pelagic sediments: Sedimentology, v. 30, p. 449-454.

KENDALL, A. C., 1984, Evaporites, *in* Walker, R. G., ed., Facies Models (2nd ed.): St. Johns, Newfoundland, Geoscience Canada Reprint Series 1, p. 259-296.

KENDALL, A. C., 1985, Radiaxial fibrous calcite: a reappraisal, *in* Schneidermann, N., and Harris, P. M., eds., Carbonate Cements: Tulsa, Society of Economic Paleontologists and Mineralogists Special Publication 36, p. 59-77.

KENDALL, A. C., AND TUCKER, M. E., 1973, Radiaxial fibrous calcite: a replacement after acicular carbonate: Sedimentology, v. 20, p. 365-389.

KENDALL, C. G. ST. C., AND SKIPWITH, P. A. D'E., 1968, Recent algal mats of a Persian Gulf lagoon: Journal of Sedimentary Petrology, v. 38, p. 1040-1058.

KENDALL, C. G. ST. C., AND SKIPWITH, P. A. D'E., 1969, Geomorphology of a Recent shallow-water carbonate province: Khor al Bazam, Trucial Coast, southwestern Persian Gulf: Bulletin of the Geological Society of America, v. 80, p. 865-892.

KENDALL, C. G. ST. C., AND WARREN, J., 1987, A review of the origin and setting of tepees and their associated fabrics: Sedimentology, v. 34, p. 1007-1027.

KENNARD, J. M., AND JAMES, N. P., 1986, Thrombolites and stromatolites: two distinct types of microbial structures: Palaios, v. 1, p. 492-503.

KINSMAN, D. J. J., 1966, Gypsum and anhydrite of Recent age, Trucial Coast, Persian Gulf, *in* Rau, J. L., ed., Proceedings of the Second Salt Symposium: Cleveland, Northern Ohio Geological Society, v. 1, p. 302-326.

KINSMAN, D. J. J., 1969, Modes of formation, sedimentary associations and diagnostic features of shallow water and supratidal evaporites: American Association of Petroleum Geologists Bulletin, v. 53, p. 830-840.

KINSMAN, D. J. J., AND PARK, R. K., 1976, Algal belt and coastal sabkha evolution, Trucial Coast, Persian Gulf, *in* Walter, M. R., ed., Stromatolites: New York, Developments in Sedimentology 20, Elsevier, p. 421-433.

KLAPPA, C. F., 1978, Biolithogenesis of Microcodium; elucidation: Sedimentology, v. 25, p. 489-522.

KLAPPA, C. F., 1979, Calcified filaments in Quaternary calcretes; organo-mineral interactions in the subacrial vadose environment: Journal of Sedimentary Petrology, v. 49, p. 955-968.

KLEIN, G. DEV., 1971, A sedimentary model for determining paleotidal range: Geological Society of America Bulletin, v. 82, p. 2585-2592.

KOERSCHNER, W. F. III., AND READ, J. F., 1989, Field and modeling studies of Cambrian carbonate cycles, Virginia Appalachians: Journal of Sedimentary Petrology, v. 59, p. 654-687.

KOHOUT, F. A., HENRY, H. R., AND BANKS, J. E., 1977, Hydrogeology related to geothermal conditions of the Florida Plateau, *in* Smith, D. L., and Griffin, G. M., eds., The Geothermal Nature of the Florida Plateau: Florida Bureau of Geology, Special Publication 21, p. 1-41.

KOMAR, P. D., 1974, Oscillatory ripple marks and the evaluation of ancient wave conditions and environments: Journal of Sedimentary Petrology, v. 44, p. 169-180.

KOMAR, P. D., 1976, Beach Processes and Sedimentation: Englewood Cliffs, Prentice-Hall, 429 p.

KOSTER, E. H., AND STEEL, R. J., eds., 1984, The Sedimentology of Gravels and Conglomerates: Calgary, Canadian Society of Petroleum Geologists Memoir 10, 441 p.

KREISA, R. D., 1981, Storm-generated sedimentary structures in subtidal marine facies with examples from the Middle and Upper Ordovician of southwestern Virginia: Journal of Sedimentary Petrology, v. 51, p. 823-848.

KREISA, R. D., 1986, Hummocky cross-stratification is trough cross-bedding (abs.): Abstracts, Society of Economic Paleontologists and Mineralogists Annual Midyear Meeting, v. III, p. 63.

KRUMBEIN, W. E., 1983, Stromatolites- the challenge of a term in space and time: Precambrian Research, v. 20, p. 493-531.

KUENEN, PH. H., 1952, Paleogeographic significance of graded bedding and associated features: Koninklijke Nederlandse Akademie van Wetenschappen, Afdeeling voor de Wisen Natuurkundige Wetenschappen, Verhandelingen, ser. B, v. 55, p. 28-36.

KUENEN, PH. H., 1953, Significant features of graded bedding: American Association of Petroleum Geologists Bulletin, v. 37, p. 1044-1066.

KUENEN, PH. H., 1965, Value of experiments in geology: Geologie en Mjinbouw, v. 44, p. 22-36.

LAPORTE, L. F., 1971, Paleozoic carbonate facies of the central Appalachian shelf: Journal of Sedimentary Petrology, v. 41, p. 724-740.

LAND, L. S., AND GOREAU, T. F., 1970, Submarine lithification of Jamaican reefs: Journal of Sedimentary Petrology, v. 40, p. 457-462.

LASEMI, Z., BOARDMAN, M. R., AND SANDBERG, P. A., 1989, Cement origin of supratidal dolomite, Andros Island, Bahamas: Journal of Sedimentary Petrology, v. 59, p. 249-257.

LECKIE, D. A., AND WALKER, R. G., 1982, Storm- and tide-dominated shorelines in Cretaceous Moosebar-Lower Gates Interval-outcrop equivalents of deep basin gas trap in western Canada: American Association of Petroleum Geologists Bulletin, v. 66, p. 138-157.

LEEDER, M. R., 1982, Sedimentology, Process and Product: Boston, George Allen and Unwin, 344 p.

LEES, A., 1964, The structure and origin of the Waulsortian (Lower Carboniferous) "reefs" of west-central Eire: Philosophical Transactions of the Royal Society of London, Series B, v. 247, p. 483-531.

LOGAN, B. W., 1974, Inventory of diagenesis in Holocene-Recent carbonate sediments, Shark Bay, Western Australia, *in* Logan, B. W., and others, eds., Evolution and Diagenesis of Quaternary Carbonate Sequences, Shark Bay, Western Australia: Tulsa, American Association of Petroleum Geologist Memoir 22, p. 195-249.

LOGAN, B. W., DAVIES, G. R., READ, J. F., AND CEBULSKI, D. E., 1970, Carbonate Sedimentation and Environments, Shark Bay, Western Australia: Tulsa, American Association of Petroleum Geologists Memoir 13, 223 p.

LOGAN, B. W., HOFFMAN, P., AND GEBELEIN, C. D., 1974a, Algal mats, crypt-algal fabrics and structures, Hamelin Pool, Western Australia, *in* Logan, B. W., and others, eds., Evolution and Diagenesis of Quaternary Carbonate Sequences, Shark Bay, Western Australia: Tulsa, American Association of Petroleum Geologists Memoir 22, p. 140-193.

LOGAN, B. W., READ, J. F., HAGAN, G. M., HOFFMAN, P., BROWN, R. G., WOODS, P. T., AND GEBELEIN, C. D., 1974b, Evolution and diagenesis of Quaternary carbonate sequences, Shark Bay, Western Australia: Tulsa, American Association of Petroleum Geologists Memoir 22, 358 p.

LOGAN, B. W., REZAK, R., AND GINSBURG, R. N., 1964, Classification and environmental significance of algal stromatolites: Journal of Geology, v. 72, p. 68-83.

LOGAN, B. W., AND SEMENIUK, V., 1976, Dynamic Metamorphism: Processes and Products in Devonian Carbonate Rocks, Canning Basin, Western Australia: Sydney, Geological Society of Australia Special Publication 16, 138 p.

LONGMAN, M. W., 1980, Carbonate diagenetic textures from nearshore diagenetic environments: American Association of Petroleum Geologists Bulletin, v. 64, p. 461-487.

LOUCKS, R. G., AND LONGMAN, M., 1982, Lower Cretaceous Ferry Lake Anhydrite, Fairway Field, East Texas: product of shallow-subtidal deposition, *in* Handford, C. R., Loucks, R. G., and Davies, G. R., eds., Depositional and Diagenetic Spectra of Evaporites: Tulsa, Society of Economic Paleontologists and Mineralogists, Core Workshop 3, p. 130-173.

LOUCKS, R. G., AND HANDFORD, C. R., 1992, Origin and recognition of fractures, breccias, and sediment fills in paleocave-reservoir networks, *in* Candelaria, M. P., and Reed, C. L., eds., Paleokarst, Karst Related Diagenesis and Reservoir Development: Examples from Ordovician-Devonian Age Strata of West Texas and the Mid-continent: Publication No. 92-33, Permian Basin Section of the Society of Economic Paleontologists and Mineralogists, p. 31-44.

LOWENSTEIN, T. K., 1987, Evaporite depositional fabrics in the deeply buried Jurassic Buckner Formation, Alabama: Journal of Sedimentary Petrology, v. 57, p. 108-116.

LOWENSTEIN, T. K., 1988, Origin of depositional cycles in a Permian "saline giant": the Salado (McNutt zone) evaporites of New Mexico and Texas: Geological Society of America Bulletin, v. 100, p. 592-608.

LOWENSTEIN, T. K., AND HARDIE, L. A., 1985, Criteria for the recognition of salt pan evaporites: Sedimentology, v. 32, p. 627-644.

LUCIA, F. J., 1968, Recent sediments and diagenesis of south Bonaire, Netherlands Antilles: Journal of Sedimentary Petrology, v. 38, p. 845-858.

LUCIA, F. J., 1972, Recognition of evaporite-carbonate shoreline sedimentation, *in* Rigby, K. J., and Hamblin, W. K., eds., Recognition of Ancient Sedimentary Environments: Tulsa, Society of Economic Paleontologists and Mineralogists Special Publication 16, p. 160-191.

LYELL, C., 1842, The Principles of Geology (6th ed.), Volume 1: Boston, Hilliard, Gray & Co., 442 p.

MACHEL, H. G., AND BURTON, E. A., 1991, Burial-diagenetic sabkha-like gypsum and anhydrite nodules: Journal of Sedimentary Petrology, v. 61, p. 394-405.

MACINTYRE, I. G., 1977, Distribution of submarine cements in a modern Caribbean fringing reef, Galeta Point, Panama: Journal of Sedimentary Petrology, v. 47, p. 503-516.

MACINTYRE, I. G., 1985, Submarine cements- the peloidal question, *in* Schneidermann, N., and Harris, P. M., eds., Carbonate Cements: Tulsa, Society of Economic Paleontologists and Mineralogists Special Publication Number 36, p. 109-116.

MALIVA, R. G., 1987, Quartz geodes: early diagenetic silicified anhydrite nodules related to dolomitization: Journal of Sedimentary Petrology, v. 57, p. 1054-1059.

MARKELLO, J. R., AND READ, J. F., 1981, Carbonate ramp to deeper shale shelf transitions of an Upper Cambrian intrashelf basin, Nolichucky Formation, southwest Virginia: Sedimentology, v. 28, p. 573-597.

MASSON, P. H., 1955, An occurrence of gypsum in southwest Texas: Journal of Sedimentary Petrology, v. 25, p. 72-77.

MATTER, A., 1967, Tidal flat deposits in the Ordovician of western Maryland: Journal of Sedimentary Petrology, v. 37, p. 601-609.

MAZZULLO, S. J., REID, A. M., AND GREGG, J. M., 1987, Dolomitization of Holocene Mg-calcite supratidal deposits, Ambergris Cay, Belize: Geological Society of America Bulletin, v. 98, p. 224-231.

MCCAVE, I. N., 1970, Deposition of fine-grained suspended sediments from tidal currents: Journal of Geophysical Research, v. 75, p. 4151-4159.

MCCAVE, I. N., 1971, Mud layers and deposition from tidal currents: discussion of a paper by G. de V. Klein, Tidal origin of a Pre-cambrian quartzite - the lower fine grained quartzite (Middle Dalradian) of Islay, Scotland: Journal of Sedimentary Petrology, v. 41, p. 1147-1148.

MCGANNON, D. E., 1975, Primary fluvial oolites: Journal of Sedimentary Petrology, v. 45, p. 719-727.

MCILREATH, I. A., 1977, Accumulation of a Middle Cambrian, deep water, basinal limestone adjacent to a vertical, submarine carbonate escarpment, southern Rocky Mountains, Canada, *in* Cook, H. E., and Enos, P., eds., Deep-Water Carbonate Environments: Tulsa, Society of Economic Paleontologists and Mineralogists Special Publication 25, p. 113-124.

MCILREATH, I. A., AND JAMES, N. P., 1984, Carbonate slopes, *in* Walker, R. G., ed., Facies Models (2nd ed.): Calgary, Geoscience Canada, Reprint Series 1, Geological Association of Canada, p. 245-257.

MCKEE, E.D., AND WEIR, G.W., 1953, Terminology of stratification and cross-stratification: Geological Society of America Bulletin, v. 64, p. 381-390.

MCKENZIE, J. A., HSU, K. J., AND SCHNEIDER, J. F., 1980, Movement of subsurface waters under the sabkha, Abu Dhabi, UAE, and its relation to evaporative dolomite genesis, *in* Zenger, D. H., Dunham, J. B., and Ethington, R. L., eds., Concepts and Models of Dolomitization: Tulsa, Society of Economic Paleontologists and Mineralogists Special Publication 28, p. 11-30

MENDELSOHN, F., 1976, Mineral deposits associated with stromatolites, *in* Walter, M. R., ed., Stromatolites: New York, Developments in Sedimentology 20, Elsevier, p. 645-662.

MEYERHOFF, A. A., AND HATTEN, C. W., 1974, Bahamas salient of North America, *in* Burk, C. A., and Drake, C. L., eds., The Geology of Continental Margins: New York, Springer-Verlag, p. 429-446.

MEYERS, W. J., AND HILL, B. E., 1983, Quantitative studies of compaction in Mississippian skeletal limestones, New Mexico: Journal of Sedimentary Petrology, v. 53, p. 231-242.

MIALL, A. D., 1978, Lithofacies types and vertical profile models in braided river deposits: a summary, *in* Miall, A. D., ed., Fluvial Sedimentology: Canadian Society of Petroleum Geologists Memoir 5, p. 597-604.

MIDDLETON, G. V., 1961, Evaporite solution breccias from the Mississippian of southwest Montana: Journal of Sedimentary Petrology, v. 31, p. 189-195.

MIDDLETON, G. V., 1973, Johannes Walther's law of correlation of facies: Geological Society of America Bulletin, v. 84, p. 979-988.

MILLER, M. F., EKDALE, A. A., AND PICARD, M. D., eds., 1984, Trace fossils and paleoenvironments: marine carbonate, marginal marine terrigenous and continental terrigenous settings: Journal of Paleontology (Special Issue), v. 58, p. 283-597.

MILLIMAN, J. D., 1966, Submarine lithification of carbonate sediments: Science, v. 153, p. 994-997.

MILLIMAN, J. D., 1974, Marine Carbonates: New York, Springer-Verlag, 375 p.

MILLIMAN, J. D., AND MULLER, J., 1973, Precipitation and lithification of magnesium calcite in the deep-sea sediments of the eastern Mediterranean Sea: Sedimentology, v. 20, p. 29-45.

MILLIMAN, J. D., ROSS, D. A., AND KU, T. U., 1969, Precipitation and lithification of deep-sea carbonates in the Red Sea: Journal of Sedimentary Petrology, v. 39, p. 724-736.

MITCHELL, R. W., 1981, The comparative sedimentology of the Bahamian-type shelf carbonates of the Middle Ordovician St. Paul Group of the central Appalachians: Unpublished Ph.D. Dissertation, The Johns Hopkins University, Baltimore, 495 p.

MITCHELL, R. W., 1985, Comparative sedimentology of shelf carbonates of the Middle Ordovician St. Paul Group, central Appalachians: Sedimentary Geology, v. 43, p. 1-41.

MONTANEZ, I. P., AND READ, J. F., 1992, Fluid-rock interaction history during stabilization of early dolomites, Upper Knox Group (Lower Ordovician), U.S. Appalachians: Journal of Sedimentary Petrology, v. 62, p. 753-778.

MONTY, C. L. V., 1967, Distribution and structure of recent stromatolitic algal mats, eastern Andros Island, Bahamas: Societe Geologique de Belgique Annales, v. 90, p. 55-100.

MONTY, C. L. V., 1972, Recent algal stromatolitic deposits, Andros Island, Bahamas. Preliminary report: Geologische Rundschau, v. 61, p. 742-783.

MONTY, C. L. V., 1976, The origin and development of cryptalgal fabrics, *in* Walter, M. R., ed., Stromatolites: New York, Developments in Sedimentology 20, Elsevier, p. 193-249.

MONTY, C. L. V., 1977, Evolving concepts on the nature and the ecological significance of stromatolites, *in* Flugel, E., ed., Fossil Algae, Recent Results and Developments: New York, Springer-Verlag, p. 15-35.

MONTY, C. L. V., AND HARDIE, L. A., 1976, The geological significance of the freshwater blue-green algal calcareous marsh, *in* Walter, M. R., ed., Stromatolites: New York, Developments in Sedimentology 20, Elsevier, p. 447-477.

MOORE, C. H., 1989, Carbonate Diagenesis and Porosity: New York, Developments in Sedimentology 46, Elsevier, 338 p.

MOORE, E. S., 1914, Mud cracks open under water: American Journal of Science, v. 38, p. 101-102.

MORROW, D. W., CUMMING, G. L., AND KOEPNIK, R. B., 1986, Manatoe Facies-a gas-bearing, megacrystaline, Devonian dolomite, Yukon and Northwest Territories, Canada: American

Association of Petroleum Geologists Bulletin, v. 70, p. 702-720.

MULLINS, H., NEUMANN, C., WILBER, R. J., AND BOARDMANN, M. R., 1980, Nodular carbonate sediment of Bahamian slopes: possible precursors to nodular limestones: Journal of Sedimentary Petrology, v. 50, p. 117-133.

MULTER, H. G., ed., 1975, Field guide to some carbonate rock environments, Florida Keys and Western Bahamas: Madison, Contribution No. 40, Fairleigh Dickinson University, 175 p.

MULTER, H. G., AND HOFFMEISTER, J. E., 1968, Subaerial laminated crusts of the Florida Keys: Journal of Sedimentary Petrology, v. 79, p. 183-192.

MULTER, H. G., AND HOFFMEISTER, J. E., 1975, Subaerial laminated crusts of the Florida Keys, in Multer, H. G., ed., Field Guide to Some Carbonate Rock Environments, Florida Keys and Western Bahamas: Madison, Contribution 40, Fairleigh Dickinson University, p. 119-127.

MURRAY, R. C., 1964, Origin and diagenesis of gypsum and anhydrite: Journal of Sedimentary Petrology, v. 34, p. 512-523.

MYROW, P. M., AND SOUTHARD, J. B., 1991, Combined-flow model for vertical stratification sequences in shallow marine storm-deposited editors: Journal of Sedimentary Petrology, v. 61, p. 202-210.

NEEV, D., AND EMERY, K. O., 1967, The Dead Sea: depositional processes and environments of evaporites: Israel Geological Survey Bulletin 41, 147 p.

NEMEC, W., AND STEEL, R. J., 1984, Alluvial and coastal conglomerates: their significant features and some comments on gravelly mass-flow deposits, *in* Koster, E. H., and Steel, R. J., eds., Sedimentology of Gravels and Conglomerates: Calgary, Canadian Society of Petroleum Geologists Memoir 10, p. 1-31.

NEUMANN, A. C., GEBELEIN, C. D., AND SCOFFIN, T. P., 1970, The composition, structure and erodibility of subtidal mats, Abaco, Bahamas: Journal of Sedimentary Petrology, v. 40, p. 274-297.

NEUMANN, A. C., KOFOED, J. W., AND KELLER, G. H., 1977, Lithoherms in the straits of Florida: Geology, v. 5, p. 4-10.

NEUMANN, A. C., AND LAND, L. S., 1975, Lime mud deposition and calcareous algae in the Bight of Abaco, Bahamas: a budget: Journal of Sedimentary Petrology, v. 45, p. 763-786.

NIO, S. D., SIEGENTHALER, C., AND YANG, C. S., 1983, Megaripple cross-bedding as a tool for the reconstruction of the paleohydraulics in a Holocene subtidal environment: Geologie en Mijnbouw, v. 62, p. 499-510.

NGUYEN, C. T., 1986, Depositional facies stratigraphy of the Lower Ordovician (Tremadocian) platform carbonates of the central Appalachians: Unpublished Ph.D. Dissertation, The Johns Hopkins University, Baltimore, 303 p.

NGUYEN, C. T., GOLDHAMMER, R. K., AND HARDIE, L. A., 1985, Depositional facies mosaics and their time lines in the Lower Ordovician carbonates of the central Appalachians (abs.): American Association of Petroleum Geologists Bulletin, v. 69, p. 292.

O'CONNER, M. P., 1972, Classification and environmental significance of the cryptalgal organosedimentary "molar-tooth" structure from the Late Precambrian Belt-Purcell Supergroup: Journal of Geology, v. 80, p. 592-610.

OTTO, G. H., 1938, The sedimentation unit and its use in field sampling: Journal of Geology, v. 46, p. 569-582.

PALMER, A. A., 1971, The Cambrian of the Appalachian and eastern New England regions, eastern United States, *in* Holland, C. H., ed., Cambrian of the New World: London, Wiley-Interscience, p. 169-217.

PANNELLA, G., MACCLINTOCK, C., AND THOMPSON, M. N., 1968, Paleontological evidence of variations in length of synodic month since Late Cambrian: Science, v. 162, p. 792-796.

PATTERSON, R. J., AND KINSMAN, D. J. J., 1977, Marine and continental groundwater sources in a Persian Gulf coastal sabkha: American Association of Petroleum Geologists, Studies in Geology, v. 4, p. 381-397.

PENTECOST, A., 1978, Blue green algae and freshwater carbonate deposits: Proceedings of the Royal Society of London, Series B, v. 200, p. 43-61.

PERKINS, R. D., 1977, Depositional framework of Pleistocene rocks in Florida, *in* Enos, P. and Perkins, R. D., eds., Quaternary Sedimentation in South Florida: Boulder, Geological Society of America Memoir 147, p. 131-198.

PERYT, T. M., ed., 1983, Coated Grains: New York, Springer-Verlag, 655 p.

PETTIJOHN, F. J., 1962, Paleocurrents and paleogeography: American Association of Petroleum Geologists Bulletin, v. 46, p. 1468-1493.

PETTIJOHN, F. J., 1975, Sedimentary Rocks (3rd ed.): New York, Harper and Row, 628 p.

PETTIJOHN, F. J., AND POTTER, P. E., 1964, Atlas and Glossary of Primary Sedimentary Structures: New York, Springer-Verlag, 370 p.

PETTIJOHN, F. J., POTTER, D. E., AND SIEVER, R., 1987, Sand and Sandstone (3rd ed.): New York, Springer-Verlag, 553 p.

PFEIL, R. W., AND READ, J. F., 1980, Cambrian carbonate platform margin facies, Shady Dolomite, southwestern Virginia, U.S.A.: Journal of Sedimentary Petrology, v. 50, p. 91-116.

PHILLIPS, O. M., 1991, Flow and Reactions in Permeable Rocks: New York, Cambridge University Press, 285 p.

PICARD, M. D., 1966, Oriented, linear-shrinkage cracks in Green River Formation (Eocene) Raven Ridge area, Unita Basin, Utah: Journal of Sedimentary Petrology, v. 36, p. 1050-1057.

PICARD, M. D., 1969, Oriented, linear-shrinkage cracks in Alcova Limestone Member (Triassic), southern Wyoming: Contributions to Geology of the University of Wyoming, v. 7, p. 1-7.

PICARD, M. D., AND HIGH, L. R., 1972, Criteria for recognizing lacustrine rocks, *in* Rigby, J., and Hamblin, W. K., eds., Recognition of Ancient Sedimentary Environments: Tulsa, Society of Economic Paleontologists and Mineralogists Special Publication 16, p. 108-145.

PITTMAN, J. S., AND FOLK, R. L., 1971, Length-slow chalcedony after sulphate evaporite minerals in sedimentary rocks: Nature, v. 230, p. 64-65.

PLAYFORD, P. E., 1984, Platform-margin and marginal-slope relationships in Devonian reef complexes of the Canning Basin, *in* Purcell, P. G., ed., The Canning Basin, Western Australia: Perth, Geological Society of Australia, p. 189-214.

PLAYFORD, P. E., AND COCKBAIN, A. E., 1969, Algal stromatolites: deepwater forms in the Devonian of Western Australia: Science, v. 165, p. 1008-1010.

PLAYFORD, P. E., AND COCKBAIN, A. E., 1976, Modern algal stromatolites at Hamelin Pool, a hypersaline barred basin in Shark Bay, Western Australia, *in* Walter, M. R., ed., Stromatolites: New York, Developments in Sedimentology 20, Elsevier, p. 389-411.

PLUMMER, P. S., AND GOSTIN, V. A., 1981, Shrinkage cracks: desiccation or synaeresis: Journal of Sedimentary Petrology, v. 51, p. 1147-1156.

POTTER, P. E., AND PETTIJOHN, F. J., 1963, Paleocurrents and Basin Analysis: New York, Springer-Verlag, 296 p.

PRATT, B. R., 1983, Epiphyton and Renalcis- Diagenetic microfossils from calcification of coccoid blue-green algae: Journal of Sedimentary Petrology, v. 54, p. 948-971.

PRATT, B. R., AND JAMES, N. P., 1982, Crypt-algal-metazoan bioherms of early Ordovician age in the St. George Group, western Newfoundland: Sedimentology, v. 29, p. 499-542.

PRATT, B. R., AND JAMES, N. P., 1986, The St. George Group (Lower Ordovician) of western Newfoundland: tidal flat island model for carbonate sedimentation in shallow epeiric seas: Sedimentology, v. 33, p. 313-343.

PRAY, L. C., 1960, Compaction in calcilutites (abs.): Geological Society of America Bulletin, v. 71, p. 1946.

PRAY, L.C., AND MURRAY, R.C., eds., 1965, Dolomitization and limestone diagenesis- a Symposium: Tulsa, Society of Economic Paleontologists and Mineralogists Special Publication 13, 180 p.

PREISS, W. V., 1976, Basic field and laboratory methods for the study of stromatolites, *in* Walter, M. R., Stromatolites: New York, Developments in Sedimentology 20, Elsevier, p. 5-13.

PRESLEY, M. W., AND MCGILLIS, K. A., 1982, Coastal Evaporite and Tidal Flat Sediments of the Upper Clear Fork and Glorieta Formations, Texas Panhandle: Austin, Reports of Investigations, Bureau of Economic Geology, University of Texas, v. 115, 50 p.

PURDY, E. G., 1963a, Recent calcium carbonate facies of the Great Bahama Bank, 1. Petrography and reaction groups: Journal of Geology, v. 71, p. 334-355.

PURDY, E. G., 1963b, Recent calcium carbonate facies of the Great Bahama Bank, 2. Sedimentary facies: Journal of Geology, v. 71, p. 472-497.

PURDY, E.G., 1968, Carbonate diagenesis: an environmental survey: Geologica Romana, v. 7, p. 724-736.

PURSER, B. H., 1969, Syn-sedimentary marine lithification of Middle Jurassic limestones in the Paris Basin: Sedimentology, v. 12, p. 205-230.

PURSER, B. H., 1972, Recent aragonitic encrustations and pseudo-stromatolites of the Trucial Coast, Persian Gulf: Societe Geologique de Belgique, Annales, v. 95, p. 403-406.

PURSER, B. H., ed., 1973, The Persian Gulf: Springer-Verlag, New York, 471 p.

PURSER, B.H., AND EVANS, G., 1973, Regional sedimentation along the Trucial Coast, Persian Gulf, in Purser, B. H., ed., The Persian Gulf: New York, Springer-Verlag, p. 211-232.

PURSER, B. H., AND LOREAU, J.-P., 1973, Aragonitic, supratidal encrustations on the Trucial Coast, Persian Gulf, *in* Purser, B. H., ed., The Persian Gulf: New York, Springer-Verlag, p. 343-376.

RAAF, J. F. M, DE, AND BOERSMA, J. R., 1971, Tidal deposits and their sedimentary structures: Geologie en Mijnbouw, v. 50, p. 479-505.

RAAF, J. F .J. DE, BOERSMA, J. R., AND VAN GELDER, A., 1977, Wave-generated structures and sequences from a shallow marine succession, Lower Carboniferous, County Cork, Ireland: Sedimentology, v. 24, p. 451-483.

RADBRUCH, D. H., 1957, Hypothesis regarding the origin of thinolite tufa at Pyramid Lake, Nevada: Geological Society of America Bulletin, v. 68, p. 1683-1688.

READ, J. F., 1973, Carbonate cycles, Pillara Formation (Devonian), Canning Basin, Western Australia: Bulletin of Canadian Petroleum Geology, v. 21, p. 38-51.

READ, J. F., 1974, Calcrete deposits and Quaternary sediments, Edel Province, Shark Bay, Western Australia, *in* Logan, B. W., Read, J. F., Hagan, G., M., Hoffman, P., Brown, R. G., Woods, P. J., and Gebelein, C. D., Evolution and Diagenesis of Quaternary Carbonate Sequences, Shark Bay, Western Australia: Tulsa, American Association of Petroleum Geologist Memoir 22, p. 250-282.

READ, J. F., 1976, Calcretes and their distinction from stromatolites, *in* Walter, M. R., ed., Stromatolites: New York, Developments in Sedimentology 20, Elsevier, p. 55-71.

READ, J. F., 1980, Carbonate ramp-to-basin transitions and foreland basin evolution, Middle Ordovician, Virginia Appalachians: American Association of Petroleum Geologists Bulletin, v. 64, p. 1575-1612.

READ, J. F., AND GROVER, G. A., 1977, Scalloped and planar erosion surfaces Middle Ordovician limestones, Virginia; analogues of Holocene exposed karst or tidal rock platforms: Journal of Sedimentary Petrology, v. 47, p. 956-972.

READ, J. J., AND GOLDHAMMER, R. K., 1988, Use of Fischer plots to define third-order sea-level curves in peritidal carbonates, Ordovician, Appalachians: Geology, v. 16, p. 895-899.

REEVES, C. C., 1977, Caliche: Origin, Classification, Morphology and Uses: Lubbock, Estacado Books, 233 p.

REID, R. P., AND BROWNE, K. M., 1991, Intertidal stromatolites in a fringing Holocene reef complex, Bahamas: Geology, v. 19, p. 15-18.

REINECK, H.-E., 1960, Uber die entehung von linsen und flaser schichten: Geologische Rundschau, v. 49, p. 149-161.

REINECK, H.-E., 1972, Tidal flats, *in* Rigby, J. K., and Hamblin, W. K., eds., Recognition of Ancient Sedimentary Environments: Tulsa, Society of Economic Paleontologists and Mineralogists Special Publication 17, p. 146-159.

REINECK, H.-E., AND SINGH, I. B., 1980, Depositional Sedimentary Environments (2nd ed.): New York, Springer-Verlag, 549 p.

REINECK, H.-E., AND WUNDERLICH, F., 1967a, Zeitmessungen an Gezeitenschichten: Natur und Museum, v. 97, p. 193-197.

REINECK, H.-E., AND WUNDERLICH, F., 1967b, A new method to measure rate of deposition of single lamina on tidal flats and shelf bottoms (abs.):, 7th International Sedimentological Congress, p. 333.

REINECK, H.-E., AND WUNDERLICH, F., 1968, Classification and origin of flaser and lenticular bedding: Sedimentology, v. 11, p. 99-104.

REINECK, H.-E., AND WUNDERLICH, F., 1969, Die Entstehung von Schichten und Schichtbanken im Watt: Senckenbergiana Maritima, v. 50, p. 85-106.

REINHARDT, J., 1974, Stratigraphy, sedimentology and Cambro-Ordovician paleogeography of the Frederick Valley, Maryland: Baltimore, Maryland Geological Survey Report of Investigations 23, 74 p.

REINHARDT, J., AND HARDIE, L. A., 1976, Selected Examples of Carbonate Sedimentation, Lower Paleozoic of Maryland: Baltimore, Maryland Geological Survey Guidebook, Number 5, 53 p.

REINHARZ, E., NILSEN, K. J., BOESCH, D. F., BERTELSEN, R., AND O'CONNELL, A. E., 1982, A radiographic examination of physical and biogenic sedimentary structures in the Chesapeake Bay: Baltimore, Report of Investigations No. 36, Maryland Geological Survey, 58 p.

RICH, J. L., 1951, Three critical environments of deposition, and criteria for recognition of rocks deposited in each of them: Geological Society of America Bulletin, v. 62, p. 1-20.

RIDING, R., 1979, Origin and diagenesis of lacustrine algal bioherms at the margin of the Ries Crater, Upper Miocene, southern Germany: Sedimentology, v. 26, p. 645-680.

RIDING, R., ed., 1990, Calcareous Algae and Stromatolites: New York, Springer-Verlag, 530 p.

RILEY, C. M., AND BYRNE, J. V., 1961, Genesis of primary structures in anhydrite: Journal of Sedimentary Petrology, v. 31, p. 553-559.

RISACHER, F., AND EUGSTER, H. P., 1979, Holocene pisoliths and encrustations associated with spring-fed surface, Pastos Grandes, Bolivia: Sedimentology, v. 26, p. 253-270.

ROBBIN, D. M., AND STIPP, J. J., 1979, Depositional rate of laminated soilstone crusts, Florida Keys: Journal of Sedimentary Petrology, v. 49, p. 175-180.

ROBBINS, L. L., AND BLACKWELDER, P. L., 1992, Biochemical and ultrastructural evidence for the origin of whitings: A biologically induced calcium carbonate precipitation mechanism: Geology, v. 20, p. 464-468.

ROEHL, P. O., AND CHOQUETTE, P. W., eds., 1985, Carbonate Petroleum Reservoirs: New York, Springer-Verlag, 622 p.

ROOT, S. I., 1964, Cyclicity of the Conococheague Formation: Pennsylvania Academy of Science Proceedings, v. 38, p. 157-160.

ROOT, S. I., 1968, Geology and mineral resources of southeastern Franklin County, Pennsylvania: Pennsylvania Geological Survey 4th Series Atlas 119 Edition, 118 p.

RUBIN, D. M., AND FRIEDMAN, G. M., 1977, Intermittently emergent shelf carbonates: an example from the Cambro-Ordovician of eastern New York State: Sedimentary Geology, v. 19, p. 81-106.

RUNNEGAR, B., POJETA, J., TAYLOR, M. E., AND COLLINS, D., 1979, New species of the Cambrian and Ordovician chitons *Matthevia* and *Chelodes* from Wisconsin and Queensland: evidence for the early history of polyplacophoran mollusks: Journal of Paleontology, v. 53, p. 1374-1394.

RUSNAK, G. A., AND NESTERHOFF, W. D., 1964, Modern turbidites: terrigenous abyssal plain versus bioclastic basin, *in* Miller, R. L., ed., Papers in Marine Geology, Shepard Commemorative Volume: New York, Macmillan, p. 488-507.

SANDERSON, D. J., AND DONOVAN, N. R., 1974, The vertical packing of shells and stones on some recent beaches: Journal of Sedimentary Petrology, v. 44, p. 680-688.

SANDO, W.J., 1957, Beekmantown Group (Lower Ordovician) of Maryland: Boulder, Geological Society of America Memoir 68, 161 p.

SCHLAGER, W., 1981, The paradox of drowned reefs and carbonate platforms: Bulletin of the Geological Society of America, v. 92, p. 197-211.

SCHLAGER, W., AND JAMES, N. P., 1978, Low-magnesium calcite limestones forming at the deep-sea floor, Tongue of the Ocean, Bahamas: Sedimentology, v. 25, p. 675-702.

SCHOLL, D. W., 1960, Pleistocene algal pinnacles at Searles Lake, California: Journal of Sedimentary Petrology, v, 30, p. 414-431.

SCHOLL, D. W., AND TAFT, W. H., 1964, Algae, contributors to the formation of calcareous tufa, Mono Lake, California: Journal of Sedimentary Petrology, v. 34, p. 309-319.

SCHOLLE, P. A., BEBOUT, D. G., AND MOORE, C. H., eds., 1983, Carbonate Depositional Environments: Tulsa, American Association of Petroleum Geologists Memoir 33, 708 p.

SCHOLLE, P. A., AND HALLEY, R. B., 1985, Burial diagenesis: out of sight, out of mind!, *in* Schneidermann, N., and Harris, P. M., eds., Carbonate Cements: Tulsa, Society of Economic Paleontologists and Mineralogists Special Publication 36, p. 309-334.

SCHOLLE, P. A., AND KINSMAN, D. J., 1974, Aragonitic and high-Mg calcite caliche from the Persian Gulf - a modern analog for the Permian of Texas and New Mexico: Journal of Sedimentary Petrology, v. 44, p. 904-916.

SCHREIBER, B. C., ROTH, M. S., AND HELMAN, M. L., 1982, Recognition of primary facies characteristics of evaporites and the differentiation of the forms from diagenetic overprints, *in* Handford, C. R., Loucks, R. G., and Davies, G. R., eds., Depositional and Diagenetic Spectra of Evaporites- A Core Workshop: Tulsa, Society of Economic Paleontologists and Mineralogists, Core Workshop No. 3, p. 1-32.

SCHROCK, R. R., 1948, Sequence in Layered Rocks: New York, McGraw-Hill, 507 p.

SCHWARZ, H.-U., 1975, Sedimentary Structures and Facies Analysis of Shallow Marine Carbonates: Stuttgart, Contributions to Sedimentology 3, E. Schweizerbart'sche Verlagsbuchhandlung, 100 p.

SCOFFIN, T. P., 1987, An Introduction to Carbonate Sediments and Rocks: Glasgow, Blackie, 274 p.

SELLEY, R. C., 1988, Applied Sedimentology: New York, Academic Press, 446 p.

SEMIKHATOV, M. A., GEBELEIN, C. D., CLOUD, P., AWRAMIK, S. M., AND BENMORE, W. C., 1979, Stromatolite morphogenesis - progress and problems: Canadian Journal of Earth Science, v. 16, p. 992-1015.

SEPKOSKI, J. J., 1982, Flat-pebble conglomerates, storm deposits, and the Cambrian bottom fauna, *in* Einsele, G., and Seilacher A., eds., Cyclic and Event Stratification: New York, Springer-Verlag, p. 371-385.

SHEARMAN, D. J., 1963, Recent anhydrite, gypsum, dolomite and halite from the coastal flats of the Arabian shore of the Persian Gulf: Geological Society of London, Proceedings, v. 1607, p. 63-65.

SHEARMAN, D. J., 1966, Origin of marine evaporites by diagenesis: Transactions of the American Institute of Mining, Metallurgical, and Petroleum Engineers (B), v. 75, p. 208-215.

SHEARMAN, D. J., 1970, Recent halite rock, Baja, California, Mexico: Institute of Mining and Metallurgy, Transactions, v. B79, p. 155-162.

SHEARMAN, D. J., MCGUGAN, A., STEIN, C., AND SMITH, A. J., 1989, Ikaite, $CaCO_3 \cdot 6H_2O$, precursor of the thinolites in the Quaternary tufas and tufa mounds of the Lahontan and Mono Lake Basins, western United States: Geological Society of America Bulletin, v. 101, p. 913-917.

SHEARMAN, D. J., AND SMITH, A. J., 1985, Ikaite, the parent mineral of jarrowite-type pseudomorphs: Proceedings of the Geological Association, v. 96, p. 305-314.

SHINN, E. A., 1968a, Practical significance of birdseye structures in carbonate rocks: Journal of Sedimentary Petrology, v. 38, p. 215-223.

SHINN, E. A., 1968b, Burrowing in recent lime sediments of Florida and the Bahamas: Journal of Paleontology, v. 42, p. 879-894.

SHINN, E. A., 1968c, Selective dolomitization of recent sedimentary structures: Journal of Sedimentary Petrology, v. 38, p. 612-616.

SHINN, E. A., 1969, Submarine lithification of Holocene carbonate sediments in the Persian Gulf: Sedimentology, v. 12, p. 109-144.

SHINN, E. A., 1983a, Tidal flat environment, *in* Scholle, P. A., Bebout, D. G., AND Moore, C. H., eds., Carbonate Depositional Environments: Tulsa, American Association of Petroleum Geologists Memoir 33, p. 173-210.

SHINN, E.A., 1983b, Birdseyes, fenestrae, shrinkage pores, and Loferites: a reevaluation: Journal of Sedimentary Petrology, v. 53, p. 619-628.

SHINN, E. A., 1986, Modern carbonate tidal flats: their diagnostic features: Quarterly Journal of the Colorado School of Mines, v. 81, p. 7-35.

SHINN, E. A., GINSBURG, R. N., AND LLOYD, R. M., 1965, Recent supratidal dolomite from Andros Island, Bahamas, in Pray, L. C., and Murray, R. C., Dolomitization and Limestone Diagenesis, a Symposium: Tulsa, Society of Economic Paleontologists and Mineralogists Special Publication 13, p. 112-123.

SHINN, E. A., HALLEY, R. B., HUDSON, J. H., AND LIDZ, B. H., 1977, Limestone compaction: an enigma: Geology, v. 5, p. 21-24.

SHINN, E. A., AND LIDZ, B. H., 1988, Blackened limestone pebbles: fire at subaerial unconformities, *in* James, N. P., and Choquette, P. W., eds., Paleokarst: New York, Springer-Verlag, p. 117-131.

SHINN, E. A., LLOYD, R. M., AND GINSBURG, R. N., 1969, Anatomy of a modern carbonate tidal flat, Andros Island, Bahamas: Journal of Sedimentary Petrology, v. 39, p. 1202-1228.

SHINN, E. A., AND ROBBIN, D. M., 1983, Mechanical and chemical compaction in fine-grained shallow-water limestones: Journal of Sedimentary Petrology, v. 53, p. 596-618.

SHINN, E. A., STEINEN, R. P., LIDZ, B. H., AND SWART, P. K., 1989, Perspectives: Whitings, a sedimentologic dilemma: Journal of Sedimentary Petrology, v. 59, p. 147-161.

SIMMS, M. A., 1984, Dolomitization by groundwater-flow systems in carbonate platforms: Transactions of the Gulf Coast Association of Geological Societies, v. 34, p. 411-420.

SIMMS, M. A., AND HARDIE, L. A., 1983, Reflux of seawater as a dolomitizing mass transfer process on carbonate banks (abs.): Geological Society of America Abstracts with Programs, v. 15, p. 688.

SIMONSON, B. M., 1988, Roll-up structures: evidence of microbial mats in 2.5 Ga basinal carbonates of Western Australia (abs.): Society of Economic Paleontologists and Mineralogists Annual Midyear Meeting Abstracts, Columbus, Ohio, v. V, p. 50.

SLACK, K. V., 1967, Physical and chemical description of Birch Creek, a travertine depositiong stream, Inyo County, California: United States Geological Survey Professional Paper 549-A, 19 p.

SMART, P. L., PALMER, R. J., WHITAKER, F., AND WRIGHT, V. P., 1988, Neptunian dikes and fissure fills: an overview and account of some modern examples, in James, N. P., and Choquette, P. W., eds., Paleokarst: New York, Springer-Verlag, p. 149-163.

SMITH, C. L., 1940, The Great Bahama Bank. 1. General hydrographic and chemical factors. 2. Calcium carbonate precipitation: Journal of Marine Research, v. 3, p. 1-31, p. 147-189.

SMITH, D. B., 1971, Possible displacive halite in the Permian Upper Evaporite Group of northeast Yorkshire: Sedimentology, v. 17, p. 221-232.

SMOOT, J. P., 1977, Sedimentology of a saline closed basin: the Wilkens Peak Member of the Green River Formation (Eocene), Wyoming: Unpublished Ph.D. Dissertation, The Johns Hopkins University, Baltimore, 296 p.

SMOOT, J. P., 1978, Origin of the carbonate sediments in the Wilkens Peak Member of the lacustrine Green River Formation (Eocene), Wyoming, U.S.A., *in* Matter, A., and Tucker, M. E., eds., Modern and Ancient Saline Lakes: Oxford, Blackwell, International Association of Sedimentologists Special Publication 2, p. 109-127.

SMOOT, J. P., 1983, Depositional subenvironments in an arid closed basin: the Wilkens Peak Member of the Green River Formation (Eocene), Wyoming, U.S.A.: Sedimentology, v. 30, p. 801-827

SMOOT, J. P., AND CASTENS-SEIDELL, B., 1982, Sedimentary fabrics produced in playa sediments by efflorescent salt crusts, an explanation for "adhesion ripples" (abs.): International Congress of Sedimentology, Hamilton, p. 10.

SMOOT, J. P., AND KATZ, S. B., 1982, Comparison of modern playa mudflat fabrics to cycles in the Triassic Lockatong Formation of New Jersey (abs.): Geological Society of America Abstracts with Program, v. 14, p. 83.

SMOOT, J. P., AND LOWENSTEIN, T. K, 1991, Depositional environments of nonmarine evaporites, *in* Melvin, J. L. ed., Evaporites: Petroleum and Mineral Resources: New York, Developments in Sedimentology 50, Elsevier, p. 189-347.

SORBY, H. C., 1908, On the application of quantitative methods to the study of the structure and history of rocks: Quarterly Journal of the Geological Society of London, v. 64, p. 171-233.

SOUTHARD, J. B., LAMBIE, J. M., FEDERICO, D. C., PILE, H. T., AND WEIDMAN, C. R., 1990, Experiments on bed configurations in fine sands under bidirectional purely oscillatory flow, and the origin of hummocky cross-stratification: Journal of Sedimentary Petrology, v. 60, p. 1-17.

SPENCER, R. J., AND LOWENSTEIN, T. K., 1990, Evaporites, *in* McIlreath, I. A., and Morrow, D. W., eds., Diagenesis: St. Johns, Newfoundland, Geoscience Canada Reprint Series 4, Geological Association of Canada, p. 141-163.

STEINEN, R. P., GRAY, N. H., AND MOONEY, J., 1987, A Mesozoic carbonate hot-spring deposit in the Hartford Basin of Connecticut: Journal of Sedimentary Petrology, v. 57, p. 319-326.

STOCKMAN, K. W., GINSBURG, R. N., AND SHINN, E. A., 1967, The production of lime mud by algae in south Florida: Journal of Sedimentary Petrology, v. 37, p. 633-648.

STRAATEN, L. M. J. U. VAN, 1953, Rhythmic pattern on Dutch North Sea beaches: Geologie en Mijnbouw, v. 15, p. 31-43.

STRAATEN, L. M. J. U. VAN, 1954a, Composition and structure of recent marine sediments in the Netherlands: Leidse Geologische Mededelingen, v. 19, p. 1-110.

STRAATEN, L. M. J. U. VAN, 1954b, Sedimentology of recent tidal flat deposits and the Psammites du Con droz (Devonian): Geologie en Mjinbouw, v. 16e, p. 25-47.

STRAATEN, L. M. J. U. VAN, 1954c, Radiocarbon datings and changes of sea level at Velzen (Netherlands): Geologie en Mjinbouw, v. 16e, p. 247-253.

SURDAM, R. C., AND WRAY, J. L., 1976, Lacustrine stromatolites, Eocene Green River Formation, Wyoming, *in* Walter, M. R., ed., Stromatolites: New York, Developments in Sedimentology 20, Elsevier, p. 535-542.

SWINEFORD, A., LEONARD, A. B., AND FRYE, J. C., 1958, Petrology of the Pliocene pisolitic limestone of the Great Plains: Geological Survey of Kansas Bulletin, v. 130, p. 97-116.

TAFT, W. H., ARRINGTON, F., HAINOVITZ, A., MACDONALD, C., AND WOOLHEATER, C., 1968, Lithification of modern carbonate sediments at Yellow Bank, Bahamas: Bulletin of Marine Science, v. 18, p. 762-828.

TAYLOR, J. C. M., AND ILLING, L. V., 1969, Holocene intertidal calcium carbonate cementation, Qatar, Persian Gulf: Sedimentology, v. 12, p. 69-107.

TEBBUTT, G. E., CONLEY, C. D., AND BOYD, D. W., 1965, Lithogenesis of a distinctive carbonate rock fabric: University of Wyoming Contributions to Geology, v. 4, p. 1-13.

TEDESCO, L.P., AND WANLESS, H .R., 1991, Generation of sedimentary fabrics and facies by repetitive excavation and storm infilling of burrow networks, Holocene of south Florida and Caicos Platform, B.W.I.: Palaios, v. 12, p. 326-343.

TEISSEYRE, A. K., 1977, Pebble clusters as a directional structure in fluvial gravels: modern and ancient: Geologica Sudetica, v. 12, p. 79-89.

TERWINDT, J. H. J., 1988, Palaeo-tidal reconstructions of inshore tidal depositional environments, *in* de Boer, P. L., van Gelder, A., and Nio, S. D., eds., Tide-Influenced Sedimentary Environments and Facies: Boston, D. Reidel Publishing Co., p. 233-263.

THOMPSON, R. W., 1968, Tidal Flat Sedimentation on the Colorado River Delta, Northwestern Gulf of California: Boulder, Geological Society of America Memoir 107, 133 p.

THRAILKILL, J., 1976, Speleothems, *in* Walter, M. R., ed., Stromatolites: New York, Developments in Sedimentology 20, Elsevier, p. 73-86.

TOOMEY, D. F., AND BABCOCK, J. A., 1983, Precambrian and Paleozoic algal carbonates, west Texas-southern New Mexico: Professional Contributions, Colorado School of Mines, No. 11, 345 p.

TOOMEY, D. F., MITCHELL, R. W., AND LOWENSTEIN, T. K., 1988, Algal biscuits" from the Lower Permian Herington/Krider Limestones of southern Kansas-northern Oklahoma: paleocology (*sic*) and paleodepositional setting: Palaios, v. 3, p. 285-297.

TOUREK, T. J., 1970, The depositional environments and sediment accumulation models for the Upper Silurian Wills Creek Shale and Tonoloway Limestone, central Appalachians: Unpublished Ph.D. Dissertation, The Johns Hopkins University, Baltimore, 282 p.

TSIEN, H. H., 1985, Origin of stromatactis - a replacement of colonial microbial accretions, *in* Toomey, D. F., and Nitecki, M. H., eds., Paleoalgology: Contemporary Research and Applications: Berlin, Springer-Verlag, p. 274-289.

TUCKER, M. E., AND WRIGHT, V. P., 1990, Carbonate Sedimentology: Oxford, Blackwell, 482 p.

VAN HOUTEN, F. B., 1964, Cyclic lacustrine sedimentation, Upper Triassic Lockatong Formation, central New Jersey and adjacent Pennsylvania, *in* Merriam, D. F., ed., Symposium on Cyclic Sedimentation: Kansas Geological Survey Bulletin 169, p. 495-531.

VAN TUYL, F. M., 1916, The origin of dolomite: Annual Reports of the Iowa Geological Survey, v. 25, p. 251-422.

VOGELTANZ, V. R., 1969, Fischfunde aus der Salburger Obertrias: Der Aufschluss, v. 4, p. 96-99.

WALCOTT, C. D., 1894, Paleozoic intraformational conglomerates: Geological Society of America Bulletin, v. 5, p. 191-198.

WALKER, R. G., 1984, General introduction: facies, facies sequences and facies models, *in* Walker, R. G., ed., Facies Models (2nd ed.): Calgary, Geoscience Canada Reprint Series 1, Geological Association of Canada, p. 1-9.

WALKER, R. G., AND JAMES, N. P., eds., 1992, Facies Models: Response to Sea Level Change: Calgary, Geological Association of Canada, 409 p.

WALKER, R. G., AND PLINT, A. G., 1992, Wave- and storm-dominated shallow marine systems, *in* Walker, R. G. and James, N. P., eds., Facies Model: Response to Sea Level Change: St. Johns, Newfoundland, Geological Association of Canada, p. 219-238.

WALLACE, M. W., 1987, The role of internal erosion and sedimentation in the formation of stromatactis mudstones and associated lithologies: Journal of Sedimentary Petrology, v. 57, p. 695-700.

WALLS, R. A., HARRIS, W. B., AND NUNAN, W. E., 1975, Calcareous crust (caliche) profiles and early subaerial exposure of Carboniferous carbonates, northeastern Kentucky: Sedimentology, v. 22, p. 417-440.

WALTER, M. R., 1972, Stromatolites and the biostratigraphy of the Australian Precambrian and Cambrian: Palaeontological Association of London, Special Papers in Palaeontology, No. 11, 190 p.

WALTER, M. R., ed., 1976a, Stromatolites: New York, Developments in Sedimentology 20, Elsevier, 790 p.

WALTER, M. R., 1976b, Introduction, in Walter, M. R., ed., Stromatolites: New York, Developments in Sedimentology 20, Elsevier, p. 1-3.

WALTER, M. R., 1976c, Geyserites ofYellowstone National Park: an example of abiogenic "stromatolites", in Walter, M. R., ed., Stromatolites: New York, Developments in Sedimentology 20, Elsevier, p. 87-112.

WALTER, M. R., 1976d, Hot-spring sediments in Yellowstone National Park, in Walter, M. R., ed., Stromatolites: New York, Developments in Sedimentology 20, Elsevier, p. 489-498.

WALTER, M. R., 1983, Archean stromatolites: evidence of the earth's earliest biosphere, *in* Schopf, J. W., ed., Earth's Earliest Biosphere: Princeton, Princeton University Press, p. 187-213.

WALTON, E. K., 1956, Limitations of graded bedding and alternate criteria of upward sequence in the rocks of the Southern Upland: Transactions of the Edinburgh Geological Society, v. 16, p. 267-271.

WANLESS, H. R., 1979, Limestone response to stress: pressure solution and dolomitization: Journal of Sedimentary Petrology, v. 49, p. 437-462.

WANLESS, H. R., 1982, Limestone response to stress-pressure solution and dolomitization- reply: Journal of Sedimentary Petrology, v. 52, p. 328-332.

WANLESS, H. R., TYRRELL, K. M., TEDESCO, L. P., AND DRAVIS, J. J., 1988, Tidal-flat sedimentation from Hurricane Kate, Caicos Platform, British West Indies: Journal of Sedimentary Petrology, v. 58, p. 724-738.

WARREN, J. K., 1982a, The hydrological significance of Holocene tepees, stromatolites, and boxwork limestones in coastal salinas in south Australia: Journal of Sedimentary Petrology, v. 52, p. 1171-1201.

WARREN, J. K., 1982b, The hydrologic setting, occurrence and significance of gypsum in late Quaternary salt lakes in South Australia: Sedimentology, v. 29, p. 609-637.

WARREN, J. K., 1991, Sulfate dominated sea-marginal and platform evaporite settings: sabkhas and salinas, mudflats and salterns, *in* Melvin, J. L., ed., Evaporites, Petroleum and Mineral Resources: New York, Elsevier, p. 69-187.

WARREN, J. K., AND KENDALL, C. G. ST. C., 1985, Comparison of marine (subaerial) and salina (subaqueous) evaporites: ancient and modern: American Association of Petroleum Geologists Bulletin, v. 69, p. 1013-1023.

WATERS, B. B., 1986, Sedimentology and paleogeography of the Upper Cambrian Waterfowl Formation, southern Canadian Rockies: Unpublished M.Sc. Thesis, The University of Calgary, Calgary, 216 p.

WATERS, B. B., SPENCER, R. J., AND DEMICCO, R. V., 1989, Three-dimensional architecture of shallowing-upward carbonate cycles: Middle and Upper Cambrian Waterfowl Formation, Canmore, Canada: Bulletin of Canadian Petroleum Geology, v. 37, p. 198-209.

WATTS, N. L., 1977, Pseudo-anticlines and other structures in some calcretes of Botswana and South Africa: Earth Surface Processes, v. 2, p. 63-74.

WATTS, N. L., 1978, Displacive calcite; evidence from recent and ancient calcretes: Geology, v. 6, p. 699-703.

WATTS, N. L., 1980, Quaternary pedogenic calcretes from the Kalahari (southern Africa); mineralogy, genesis and diagenesis: Sedimentology, v. 27, p. 661-686.

WEED, W. H., 1889, Formation of travertine and siliceous sinter by the vegetation of hot springs: United States Geological Survey, 9th Annual Report, 1887-1888, p. 613-676.

WEIMER, R. J., HOWARD, J. D., AND LINDSAY, D. R., 1982, Tidal flats and associated tidal channels, *in* Scholle, P. A., and Spearing, D., eds., Sandstone Depositional Environments: Tulsa, American Association of Petroleum Geologists Memoir 31, p. 191-245.

WELLER, J. M., 1959, Compaction of sediments: American Association of Petroleum Geologists Bulletin, v. 43, p. 273-319.

WEST, I. M., ALI, Y. A., AND HILMY, M. E., 1979, Primary gypsum nodules in a modern sabkha on the Mediterranean coast of Egypt: Geology, v. 7, p. 354-358.

WHITE, W. A., 1961, Colloid phenomena in sedimentation of argillaceous rocks: Journal of Sedimentary Petrology, v. 31, p. 560-570.

WILKINSON, B. H., 1979, Biomineralization, paleoceanography, and the evolution of calcareous marine organisms: Geology, v. 7, p. 524-527.

WILSON, E. N., HARDIE, L. A., AND PHILLIPS, O. M., 1990, Dolomitization front geometry, fluid flow patterns, and the origin of massive dolomite: the Triassic Latemar Buildup, northern Italy: American Journal of Science, v. 290, p. 741-796.

WILSON, J. L., 1975, Carbonate Facies in Geologic History: New York, Springer-Verlag, 469 p.

WILSON, J. L., AND JORDAN, C., 1983, Middle shelf environment, *in* Scholle, P. A., Bebout, D. G., and Moore, C. H., eds., Carbonate Depositional Environments: Tulsa, American Association of Petroleum Geologists Memoir 33, p. 298-343.

WINLAND, H .D., AND MATTHEWS, R. K., 1974, Origin and significance of grapestone, Bahama Islands: Journal of Sedimentary Petrology, v. 44, p. 921-927.

WRIGHT, V. P., 1982, The recognition and interpretation of paleokarsts: two examples from the Lower Carboniferous of South Wales: Journal of Sedimentary Petrology, v. 52, p. 83-94.

WRIGHT, V. P., 1989, A micromorphological classification of fossil and Recent calcic and petrocalcic microstructures: New York, Proceedings of the International Working Meeting on Soil Micromorphology, San Antonio, 1988, Elsevier, 315 p.

YOUNG, G. M., AND LONG, D. G. F., 1977, Carbonate sedimentation in a Late Precambrian shelf sea, Victoria Island, Canadian Arctic Archipelago: Journal of Sedimentary Petrology, v. 47, p. 943-955.

YOUNG, R. B., 1935, A comparison of certain stromatolitic rocks in the limestones of the Dolomite Series of South Africa with marine algal sediments in the Bahamas: Transactions Geological Society of South Africa, v. 37, p. 153-162.

F

G

H

N

O